GEOLOGY AND OFFSHORE RESOURCES OF PACIFIC ISLAND ARCS— NEW IRELAND AND MANUS REGION, PAPUA NEW GUINEA

CIRCUM-PACIFIC COUNCIL FOR ENERGY AND MINERAL RESOURCES EARTH SCIENCE SERIES, VOLUME 9

GEOLOGY AND OFFSHORE RESOURCES OF PACIFIC ISLAND ARCS— NEW IRELAND AND MANUS REGION, PAPUA NEW GUINEA

Circum-Pacific Council for Energy and Mineral Resources

Earth Science Series, Volume 9

GEOLOGY AND OFFSHORE RESOURCES OF PACIFIC ISLAND ARCS— NEW IRELAND AND MANUS REGION, PAPUA NEW GUINEA

Edited By

Michael S. Marlow, Shawn V. Dadisman, and Neville F. Exon

Published by the Circum-Pacific Council
for Energy and Mineral Resources
Houston, Texas, U.S.A.
1988

Copyright © 1988 by
The Circum-Pacific Council for Energy and Mineral Resources
All Rights Reserved
Published August, 1988

ISBN: 0-933687-10-9

Cover Design: Ben Servino, U.S. Geological Survey

Cover Illustration: Tectonic diagram of New Ireland and Manus region, Papua New Guinea, after W. D. Stewart and M. J. Sandy (this volume) with inserts of seismicity in the Bismarck Sea region (McCue, this volume), the "bright spot" along seismic-reflection line 401 (Exon and Marlow, petroleum potential, this volume), and a photography of field work on New Ireland (M. Marlow). Layout work by Phyllis Swenson.

Circum-Pacific Council for Energy and Mineral Resources Earth Science Series
F. L. Wong and H. G. Greene, Editors

1. Tectonostratigraphic Terranes of the Circum-Pacific Region
    edited by D. G. Howell

2. Geology and Offshore Resources of Pacific Island Arcs—Tonga Region
    compiled and edited by D. W. Scholl and T. L. Vallier

3. Investigations of the northern Melanesian Borderland
    edited by T. M. Brocher

4. Geology and Offshore Resources of Pacific Island Arcs—Central and Western Solomon Islands
    edited by J. G. Vedder, K. S. Pound, and S. Q. Boundy

5A. The Antarctic Continental Margin: Geology and Geophysics of Offshore Wilkes Land
    edited by S. L. Eittreim and M. A. Hampton

5B. The Antarctic Continental Margin: Geology and Geophysics of the Western Ross Sea
    edited by A. K. Cooper and F. J. Davey

6. Geology and Resource Potential of the Continental Margin of Western North American and Adjacent Ocean Basins—Beaufort Sea to Baja California
    edited by D. W. Scholl, A. Grantz, and J. G. Vedder

7. Marine Geology, Geophysics, and Geochemistry of the Woodlark Basin—Solomon Islands
    edited by B. Taylor and N. F. Exon

8. Geology and Offshore Resources of Pacific Island Arcs—Vanuatu Region
    edited by H. G. Green and F. L. Wong

9. Geology and Offshore Resources of Pacific Island Arcs—New Ireland and Manus Region, Papua New Guinea
    edited by M. S. Marlow, S. V. Dadisman, and N. F. Exon

Circum-Pacific Council publications are available from:
The AAPG Bookstore • P.O. Box 979 • Tulsa, Oklahoma 74101-0979 • USA
and its international distributors.

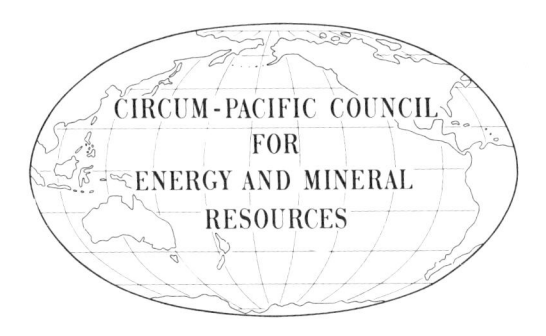

FOREWORD

The Earth Science Series of the Circum-Pacific Council for Energy and Mineral Resources (CPCEMR) is designed to convey the results of geologic research in and around the Pacific Basin. Topics of interest include framework geology, petroleum geology, hard minerals, geothermal energy, environmental geology, volcanology, oceanography, tectonics, geophysics, geochemistry, and applications of renewable energy. The CPCEMR, supports and publishes results of scientific research that will advance the knowledge of energy and mineral resources potential in the Circum-Pacific region. The Earth Science Series is specifically designed to publish papers that include new data and new maps, report on CPCEMR-sponsored symposia and workshops, and describe the results of onshore and marine geological and geophysical explorations.

This volume reports the results of one of fourteen internationally sponsored surveys to investigate the energy and mineral resources in the Southwest Pacific. The 1982, 1984 and 1986 surveys were fostered by the Australia-New Zealand-United States Tripartite Agreement. Geophysical and geological data were collected aboard the U.S. Geological Survey's (USGS) R/V *S.P. Lee* and the Hawaii Institute of Geophysics' (HIG) R/V *Kana Keoki* and R/V *Moana Wave* in the waters of Fiji, Kiribati, Papua New Guinea, Solomon Islands, Tonga, Western Samoa, and Vanuatu.

Funding for ship time was made available through the U.S. Agency for International Development, the USGS, the U.S. Office of Naval Research (for HIG's 1982 work), the Australian Development Assistance Bureau, and the New Zealand Ministry of Foreign Affairs. Coordination of the program was provided by the U.S. Department of State and the United Nations-sponsored Committee for Coordination of Joint Prospecting for Mineral Resources in South Pacific Offshore Areas (CCOP/SOPAC). Over 150 scientists and technicians participated in the cruises and represented the South Pacific island nations and funding countries mentioned above, as well as CCOP/SOPAC, New Caledonia, the United Kingdom and France.

Michel T. Halbouty

Chairman and President

PREFACE

The scientific results presented in this volume are part of a major collaboration of the U.S. Geological Survey (USGS) with government geological agencies of Australia, New Zealand, and nations in the South Pacific to investigate the geology and to evaluate potential oil, gas, and mineral resources of the Circum-Pacific region.

This collaboration began in 1982 with SOPAC Tripartite I which investigated the potential for offshore oil and gas resources in Solomon Islands, Tonga, and Vanuatu and was responsible for spearheading renewed resource-oriented marine geological and geophysical exploration in the Pacific. The joint surveys were continued in 1984 as SOPAC Tripartite II, which is part of a larger USGS program (Operation Deep Sweep) designed, in part, to investigate the Exclusive Economic Zones of United States territories in the Pacific and to study other selected parts of the Pacific with an international team composed primarily of scientists from the Tripartite countries.

The studies reported in this volume are beneficial to the Australia, New Zealand, and United States geoscientific research partners and the member countries of the United Nations-sponsored Committee for Co-ordination of Joint Prospecting for Mineral Resources in South Pacific Offshore Areas (CCOP/SOPAC). The benefits to the participating groups include obtaining a comprehensive update and evaluation of the Pacific Basin energy and mineral resources; assisting developing nations of the Pacific in assessing their potential resources; providing scientists of the participating countries an opportunity to study analogous geologic provinces and then develop models, hypotheses and theories that will be useful in unravelling complex geologic problems in their home waters; and bringing together a team of international scientists that work cooperatively in sharing information and ideas that are germane to the scientific understanding of the Circum-Pacific geology. The benefits of this work to the general geoscientific community are also great as new information about remote areas of the Pacific is being added to our knowledge and understanding of geology and geologic processes worldwide. Results of other investigations that are part of Operation Deep Sweep are, or will be, reported in other volumes of the Earth Science Series of the Circum-Pacific Council for Energy and Mineral Resources.

Dallas Peck

Director, U.S. Geological Survey

The research recorded here is a major contribution by the countries in the CCOP/SOPAC region and by the wider international community to the understanding of the natural resource potential and the geological origin and significance of the seabed of national Exclusive Economic Zones (EEZ) and the oceanic areas of the South Pacific. At the same time, it is an outstanding example of international partnership and co-operation, with mutual benefits to all concerned.

CCOP/SOPAC warmly welcomes the success of this collaborative enterprise, and, on behalf of the member governments and its Technical Secretariat, I wish to place on record our sincere appreciation for the very generous contribution by the Tripartite Partners—the governments of Australia, New Zealand, and the United States—and by the Circum-Pacific Council for Energy and Mineral Resources.

Jioji Kotobalavu

Director
CCOP/SOPAC Technical Secretariat

TABLE OF CONTENTS

PART 1 - INTRODUCTION

Tripartite Study of the New Ireland - Manus Region, Papua New Guinea:
An Introduction 1
    N.F. Exon, M.S. Marlow

PART 2 - TOPICAL STUDIES

Geology of New Ireland and Djaul Islands, Northeastern Papua New Guinea 13
    W.D. Stewart, M.J. Sandy

Stratigraphy of Manus Island, Western New Ireland Basin, Papua New Guinea 31
    G. Francis

Earthquakes and Crustal Stress in the North Bismarck Sea 41
    K.F. McCue

Sedimentology and Morphology of the Insular Slope—New Ireland to Manus Islands,
Papua New Guinea 47
    P.R. Carlson, D.M. Rearic, P.J. Quinterno

Late Tertiary and Quaternary Foraminifera and Paleobathymetry of Dredge and
Core Samples from the New Ireland Basin (Cruise L7-84-SP) 65
    D.J. Belford

Neogene Foraminifera as Time-Space Indicators in New Ireland, Papua New Guinea 91
    D.W. Haig, P.J. Coleman

Volcanism in the New Ireland Basin and Manus Island Region: Notes on the
Geochemistry and Petrology of some Dredged Volcanic Rocks from a Rifted-Arc Region 113
    R.W. Johnson, M.R. Perfit, B.W. Chappell, A.L. Jaques, R.D. Shuster, W.I. Ridley

Geochemistry of Bathyal Ferromanganese Deposits from the New Ireland Region
in the Southwest Pacific Ocean 131
    B.R. Bolton, N.F. Exon

Offshore Structure and Stratigraphy of New Ireland Basin in Northern Papua
New Guinea 137
    M.S. Marlow, N.F. Exon, H.F. Ryan, S.V. Dadisman

Hydrocarbon Gas in Bottom Sediment from Offshore the Northern Islands of
Papua New Guinea 157
    K.A. Kvenvolden

Petroleum Source Rock Study, Miocene Rocks of New Ireland, Papua New Guinea 161
    M. Glikson

The Petroleum Potential of the New Ireland Basin, Papua New Guinea 185
    N.F. Exon, M.S. Marlow

Multichannel Seismic-Reflection Data Collected at the Intersection of the Mussau
and Manus Trenches, Papua New Guinea 203
    H.F. Ryan, M.S. Marlow

Geophysical Study of a Magma Chamber Near Mussau Island, Papua New Guinea 211
    S.V. Dadisman, M.S. Marlow

Widespread Lava Flows and Sediment Deformation in a Forearc Setting North of
Manus Island, Northern Papua New Guinea 221
    M.S. Marlow, N.F. Exon, D.L. Tiffin

PART 3 - SUMMARY

Geology and Offshore Resource Potential of the New Ireland - Manus Region:
A Synthesis 241
    N.F. Exon, M.S. Marlow

PART 4 - APPENDIX

Appendix 1. Marine Topography of the Papua New Guinea Region 265
    T.E. Chase, B.A. Seekins, J.D. Young, S.V. Dadisman

Appendix 2. Geophysical Data Near Manus and New Ireland Islands, Papua New Guinea 273
    J.R. Childs, M.S. Marlow

PART 1 – INTRODUCTION

Marlow, M.S., Dadisman, S.V., and Exon, N.F., editors, 1988, Geology and offshore resources of Pacific island arcs—New Ireland and Manus region, Papua New Guinea, Circum-Pacific Council for Energy and Mineral Resources Earth Science Series, v. 9: Houston, Texas, Circum-Pacific Council for Energy and Mineral Resources.

TRIPARTITE STUDY OF THE NEW IRELAND-MANUS REGION, PAPUA NEW GUINEA: AN INTRODUCTION

N.F. Exon
Bureau of Mineral Resources, Geology and Geophysics, Canberra, Australia 2601

M.S. Marlow
U.S. Geological Survey, Menlo Park, California 94025, USA

INTRODUCTION

In 1984, the U.S. Geological Survey Research Vessel *S.P. Lee* carried out a three-week geoscience cruise in Papua New Guinea, over the New Ireland Basin north of New Ireland and the Manus forearc north of Manus, an area 900 km long and 200 km wide. About 4300 km of underway geophysical data were recorded, including 2000 km of multichannel seismic-reflection profiles. In addition, 21 sampling stations were occupied. The results from the cruise have been combined with those from earlier cruises, and with known outcrop geology, to enable us to present a complete regional study in this volume.

BACKGROUND

The R/V *S.P. Lee* carried out a petroleum-oriented research cruise (L7-84-SP) in the New Ireland - Manus region in 1984, as part of a Tripartite Marine Geoscience Program funded by Australia, New Zealand, and the United States of America. This program, which was designed to assist Pacific Island countries, was co-ordinated by CCOP/SOPAC (the Committee for Co-ordination of Joint Prospecting for Mineral Resources in South Pacific Offshore Areas) - an intergovernmental organization based in Suva, Fiji. The cruise started from Kieta on Bougainville on June 28 and ended in Rabaul on New Britain on July 17. Cruise tracks and sampling stations are shown in Figure 1.

Scientific participants on the cruise were:

- M.S. Marlow, U.S. Geological Survey (USGS) - co-chief scientist
- N.F. Exon, Bureau of Mineral Resources (BMR), Australia - co-chief scientist
- D.J. Belford, BMR - micropaleontologist
- D.L. Tiffin, CCOP/SOPAC, Fiji - geophysicist
- P.R. Carlson, USGS - sedimentologist
- J.R. Childs, USGS - geophysicist
- R.C. Culotta, USGS - geologist
- S.V. Dadisman, USGS - geologist
- K.L. Kinoshita, USGS - navigation analyst
- B.L. Krause, USGS - navigation analyst
- K.A. Kvenvolden, USGS - geochemist
- C.A. Madison, USGS - computer assistant
- H.F. Ryan, USGS - geologist
- E. Frankel[1], University of Queensland - sedimentologist

The scientific crew was ably assisted by marine technicians L.D. Kooker, D.J. Hogg, H.D. Williams, and J.F. Stampfer of the USGS as well as the officers and crew of the R/V *S.P. Lee* under the command of Captain A. McClenaghan.

Immediately after the cruise, Exon, Marlow, Belford, Carlson, Frankel and Kvenvolden flew to Kavieng on western New Ireland to join a four-day field excursion led by W.D. Stewart and G. Francis of the Geological Survey of Papua New Guinea

---
[1] Now at New South Wales Institute of Technology

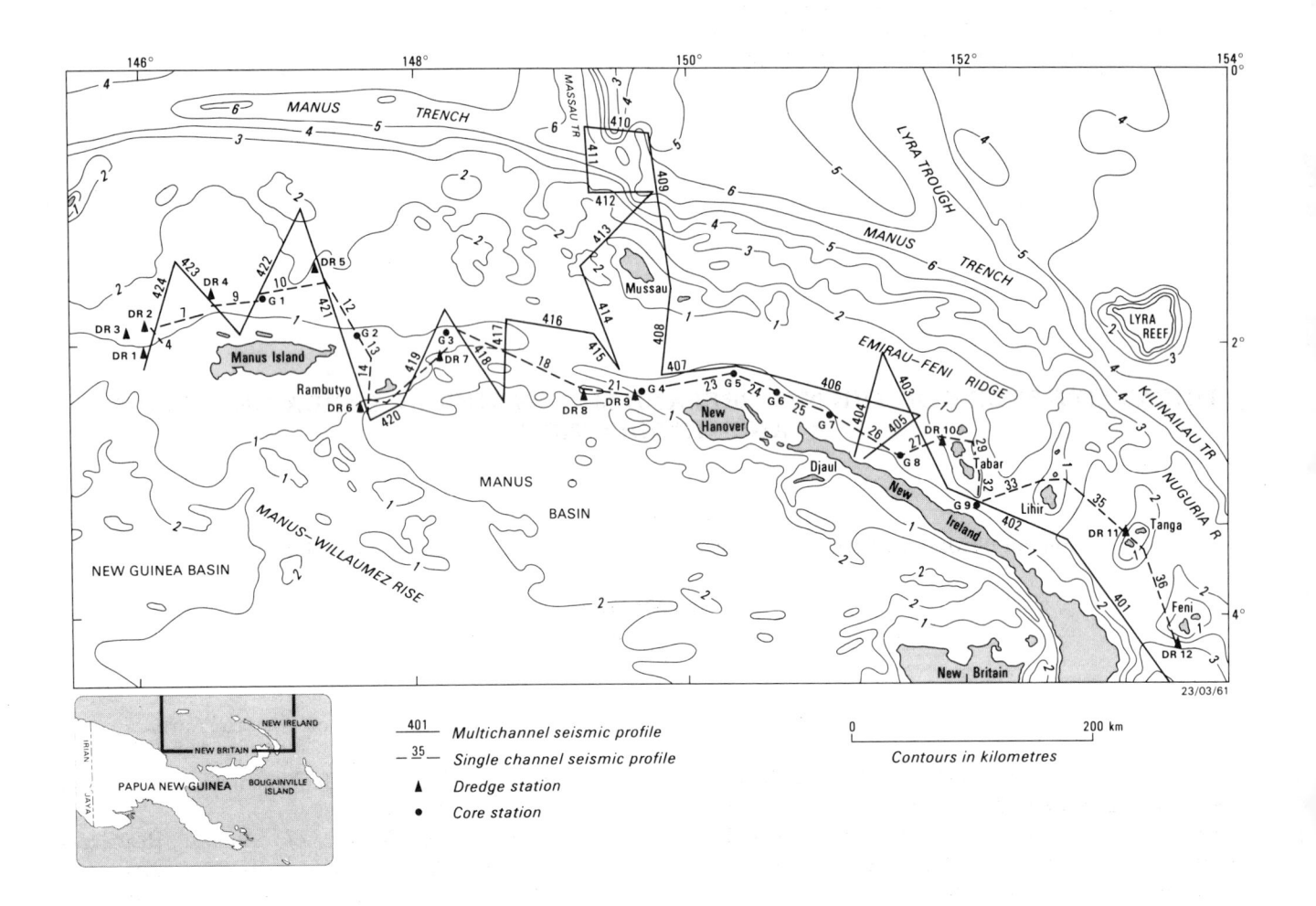

Figure 1. Map relating the geoscience cruise of the R/V *S.P. Lee* to the physiography of the New Ireland-Manus region of Papua New Guinea, and showing the location of seismic-reflection profiles and sampling stations. The New Ireland Basin proper extends from Mussau to beyond Feni Island, and lies between the Manus and Kilinailau Trenches and the Manus Basin. DR = *S.P. Lee* dredge station, G = gravity core station.

(GSPNG). On the excursion they examined the onshore equivalents of sequences surveyed offshore in the New Ireland Basin. These included the Miocene Lossuk River beds and Lelet Limestone, the late Miocene to early Pliocene Punam Limestone, and the Pliocene Rataman Formation (Figure 2). The excursion ended in Namatanai in central New Ireland, and led to a close working partnership between the shipboard scientists and the GSPNG geologists, who were involved in a detailed stratigraphic study of parts of New Ireland and a revision of the geology of Manus Island (Stewart and Sandy, 1986; Francis, this volume).

The land-based studies, carried out by the GSPNG, have been combined with the offshore results from this and earlier cruises to produce a complete basin study. The study volume consists of 17 scientific papers, in which considerable emphasis is placed on the petroleum potential of the completely untested New Ireland Basin.

Other Investigations

Onshore geological mapping and related studies have been extensive. On Manus Island, such studies include those of Thompson (1952) and company geologists, culminating in the regional mapping of Jaques (1980), which has been revised by Francis (this volume). The stratigraphy on New Ireland and Djaul Island was largely established by French (1966), Hohnen (1978) and Brown (1982), and revised in 1984-86 by Stewart and Sandy (1986, and this volume). The stratigraphy of New Hanover was outlined by Brown (1982). The geology of the Tabar-to-Feni groups of islands north and east of New Ire-

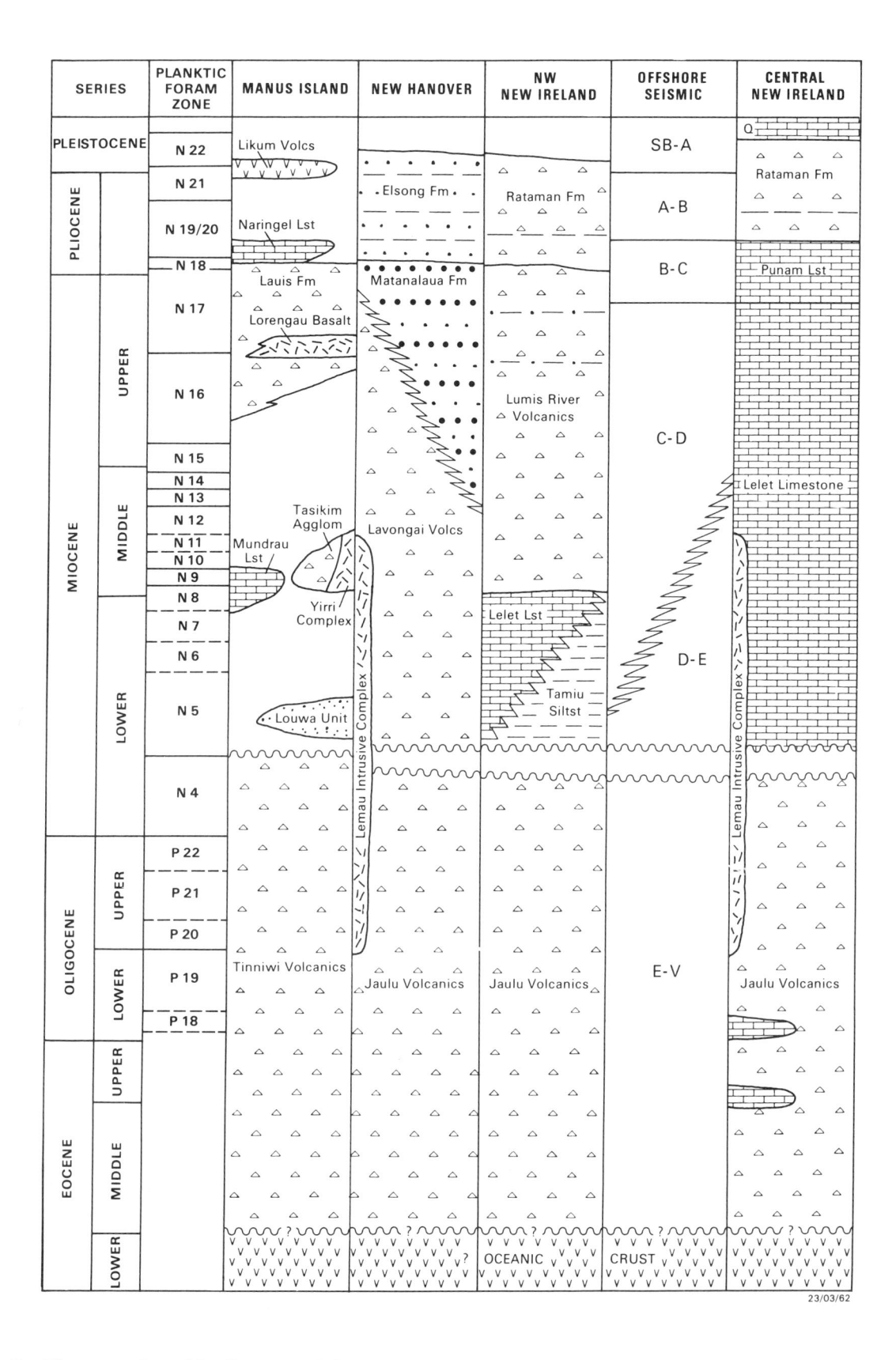

Figure 2. Time-stratigraphic diagram of the New Ireland-Manus region, relating land stratigraphy to paleontological zones, and offshore seismic stratigraphy. Diagram is based on papers by Stewart and Sandy, and Francis, in this volume.

land was described in detail by Wallace et al. (1983).

The tectonic background of the New Ireland - Manus region has been discussed by a number of authors, including Connelly (1976), Johnson (1979), Johnson, Mutter, and Arculus (1979), Taylor (1979), Wallace et al. (1983), Exon and Tiffin (1984), Falvey and Pritchard (1984), Kroenke (1984), and Exon et al. (1986). Offshore surveys have been conducted by Lamont-Doherty Geological Observatory (*Vema* cruises 24 and 43 and *Robert D. Conrad* cruise 10), the Hawaii Institute of Geophysics (*Mahi* cruise 73/04), BMR (Connelly, 1976), CCOP/SOPAC (Eade 1979, Exon 1981, Tiffin, 1981), the French IFP-CEPM-ORSTOM group (de Broin, Aubertin and Ravenne, 1977), and Gulf Research and Development Company (1973) as reported by Exon and Tiffin (1984) and Exon et al. (1986).

Syntheses of onshore and offshore data were made by Wallace et al. (1983), Exon and Tiffin (1984), and Exon et al. (1986). The GSPNG has recently produced reports on the stratigraphy and structure of New Ireland and Djaul Island (Stewart and Sandy, 1986), the hydrocarbon potential of the New Ireland Basin (Stewart, Francis, and Pederson, 1986), and on the reservoir and source-rock potential of New Ireland outcrops (Sandy, 1986).

Regional Setting

Exon and Tiffin (1984) described the "New Ireland Basin" as an arcuate feature, extending 900 km west-northwestward from the Feni Islands in the east to Manus Island in the west (Figure 1), and bounded by the 6000-m deep Manus-Kilinailau Trench in the north, and the 2500-m deep Manus Basin in the south. We here use the general term "New Ireland-Manus region" for Exon and Tiffin's (1984) "New Ireland Basin", restricting the New Ireland Basin to the structural basin extending east from Mussau Island. The western part of the region, near Manus, is markedly deformed, and different in character from the New Ireland Basin. The New Ireland Basin is 600 km long and as much as 200 km wide and includes New Ireland, New Hanover, and Mussau Islands (Exon et al., 1986). The part of the Manus region that we studied is 300 km long and 200 km wide and extends from west of Mussau to northwest of Manus. Geomorphically, the Manus region is a forearc, and hence, we use the term "Manus forearc" for that region between the West Melanesian volcanic arc and the western Manus Trench.

Overall, the New Ireland Basin slopes gently down to the northeast, but axial water depths increase from about 1500 m off New Hanover to 3000 m in the southeast (Figure 1). The basin is truncated in the southwest by a series of major southeast-trending onshore and offshore faults having an overall vertical displacement of more than 2000 m (e.g. Connelly, 1976); these faults are Pliocene to Holocene in age and are related to the formation of the Manus Basin, a spreading marginal basin (Taylor, 1979). Several of the faults clearly have a transcurrent motion. Others appear to be purely normal faults related to subsidence as the crust of the Manus Basin cooled and sank.

The New Ireland Basin was shown by Exon and Tiffin (1984) to be a simple downwarp with as much as 5 km of Oligocene and younger sedimentary fill. The entire New Ireland-Manus region was a forearc basin to the Manus-Kilinailau Trench in Eocene and Oligocene times. Two outer-arc highs to the north, the Emirau-Feni Ridge, and the Nuguria Ridge farther offshore from Lihir, Tanga and Feni, limit much of the depositional basin (Figure 1). Evidence from Mussau indicates recent uplift of the northwesterly part of the Emirau-Feni Ridge. Eocene to early Miocene volcanic rocks form the basement of the basin. Further periods of volcanism occurred in the Miocene, Pliocene, and Pleistocene. The pre-middle Miocene, Pliocene, and younger volcanics of the Tabar-to-Feni groups of islands (Wallace et al., 1983) have cut through the southeastern part of the basin, destroying its simple structure.

Paleomagnetic data, with other information, suggest that New Britain, Manus, New Hanover, and New Ireland Islands were all part of a northwest-facing volcanic arc in the Eocene-early Oligocene (Falvey and Pritchard, 1984). By the late Oligocene, the arc had rotated to face northeast, perhaps as the result of formation of the Solomon Sea back-arc basin. Volcanism related to this arc ceased in the early Miocene. There followed a long period of quiescence, during which limestone was widely deposited on the arc. After a short period in the early Pliocene of southwest-facing arc volcanism, probably involving the New Britain Trench, the back-arc Manus Basin started to form in the mid Pliocene. This led to the relative movement of New Britain southeastwards past New Ireland. A hot light mantle anomaly, associated with the opening of the Manus Basin, may have caused the uplift of the southwestern side of the New Ireland Basin by as much as 2000 m (Johnson, Mutter, and Arculus, 1979).

Few signs of compressional features exist in the New Ireland Basin, although southward-dipping

INTRODUCTION

thrusts exist beneath the old Manus forearc north of Manus Island (Marlow, Exon and Tiffin, this volume). Exon and Tiffin (1984) noted that fault systems affecting the basin itself fall into four categories: those along the southwestern side of New Ireland, normal faults parallel to and within the basin, normal faults perpendicular to the basin axis, and normal faults associated with the volcanic Tabar-to-Feni islands. Although seismic-reflection evidence indicates that some of the faults are long-lived growth faults, many have formed recently. The spectacular fault system on the southwestern flank of the New Ireland Basin separates the oceanic crust of the Manus Basin from the island-arc crust of the New Ireland-New Hanover region. The fault system extends northwestward from southernmost New Ireland (Weitin and Sapom Faults), along the southwest side of the island, and into the gap between New Hanover and Manus Island. The fault zone represents left-lateral transform movement, accompanied by vertical adjustments down to the Manus Basin. A parallel set of faults separates the Willaumez-Manus Rise south of Manus Island from the Manus Basin; these faults cut Manus Island at Manai Bay.

Many normal faults parallel the basinal axis and such faults are commonly visible on New Ireland and New Hanover. Offshore in the New Ireland Basin, these faults are common only near land, where displacements of more than 100 m occur in the Pleistocene section. Major north-to-northwest trending faults occur between New Hanover and Manus Island, especially west of Mussau where fault scarps are as much as 2000 m high.

Some fault systems are normal to the basin's axis, such as the Andalom, Ramat, and Matakan Faults on New Ireland (Hohnen, 1978). Gulf seismic-reflection profiles parallel to and near the New Ireland coast indicate that minor faults extend offshore. The north-south elongation of the volcanic Tabar and Lihir Islands suggests that the islands have formed over such faults (Exon et al., 1986). Large-scale uplift associated with the emplacement of these volcanoes has caused normal faulting and some folding, displacements being as much as 2000 m.

R/V S.P. LEE SHIPBOARD OPERATIONS.

During the first 11 days of the cruise, geophysical profiles were run from east to west, including multichannel seismic-reflection profiles. During the next 7 days geological sampling and single-channel seismic-reflection profiling were carried out from west to east (Figure 1). The multichannel seismic-reflection lines were placed so that they would complement the pre-existing CCOP/SOPAC, BMR and Gulf seismic profiles used by Exon and Tiffin (1984).

Navigational control for the survey consisted of satellite positioning, supplemented by radar fixes from nearby islands, doppler sonar, and dead-reckoning. The satellite system used satellites in polar orbit, which gave reliable fixes for our ship on an average of every four hours.

Details of shipboard geophysical systems are given by Scholl and Vallier (1985). Seismic-reflection data collected on the cruise included 24-channel and single-channel profiles. About 2000 km of 24-channel reflection data were collected, by using a five-air-gun array totalling 1265 cubic inches (21 liters) as a sound source, and a seismic cable 2400 m long. These data were recorded on magnetic tape at a 2-ms sample rate and an average record length of 10 s of sub-bottom time. One channel was displayed on a graphic recorder as an aid to selecting sampling sites. About 3200 km of single-channel seismic-reflection data were collected with the multichannel data and, later, between sampling sites. The results were displayed on a graphic recorder. Two crossings of the Manus Trench yielded reflections from oceanic crust beneath the inner trench wall. Tie lines to existing multichannel seismic-reflection surveys were completed over the New Ireland Basin, allowing us to make structure and isopach maps of the basin and detailed seismic stratigraphic studies of the basin fill.

Sonobuoys were used to collect seismic-refraction data at selected sites along the multifold reflection profiles. The refraction data were recorded on magnetic tape, played back in filtered form on a graphic recorder, and analyzed aboard the ship. Although the system operated well during the first part of the cruise, moisture in the cable to the receiving antenna (which could not be fixed at sea) reduced the quality of results toward the end of the cruise.

Magnetic and gravity data were collected during underway geophysical operations. A total of 4080 km of gravity and 3880 km of magnetic data were recorded on magnetic tape and strip-chart recorders. Magnetic data were sampled at 4-s intervals and gravity data at 20-s intervals. Large anomalies were observed in both the gravity and magnetic data on crossings of the island arcs in the New Ireland and Manus Island regions.

Graphic recordings of 3.5 kHz and 12 kHz bathymetric data were collected along 4300 km of trackline during the cruise, in water depths ranging to 6400 meters. These data were digitized on board onto magnetic tape. In several flat-lying areas of the New Ireland Basin, reflectors were detected to a few tens of meters sub-bottom on the 3.5 kHz records.

The latter part of the cruise was devoted to the dredging of outcrop sequences revealed on seismic-reflection records, and to the coring of surface sediment. Dredging was attempted 16 times at 12 stations, and rocks were recovered in 11 hauls (Table 1). Gravity cores were taken by using a 3-m core-barrel and an 800 pound weight stand. Cores were attempted 12 times at 9 stations and 6 cores were

Table 1. Dredge stations on R/V *S. P. Lee* New Ireland-Manus region cruise.

Number	Latitude (S)	Longitude (E)	Water depth (m)	Recovery	Comments
DR 1	2°03.87′	146°04.41′	600-800	None	Lost dredge after strong bites
DR 1A	2°03.73′	146°04.93′	550-600	None	Empty after strong bites
DR 1B	2°03.25′	146°04.64′	600-750	None	Lost dredge after strong bites
DR 2	1°52.39′	146°04.26′	1100-1400	1 cobble	Mn-coated ocean island basalt
DR 3	1°53.92′	145°56.89′	1100-1250	3 kg	Highly-altered Mn-coated volcanics; foraminiferal ooze
DR 4	1°37.44′	146°34.29′	1300-2000	None	Foraminiferal ooze smear
DR 5	1°20.94′	147°18.26′	2000-2300	40 kg	White chalk, late Miocene (N18); foraminiferal ooze, Holocene
DR 6	2°28.70′	147°40.10′	1200-1750	None	Lost dredge after strong bites
DR 6A	2°27.68′	147°41.16′	1000-1500	2 kg	Calcareous mudstone, marl, late Miocene (N18); mud, late Pliocene (N21); andesite; yellow brown foraminiferal ooze, Holocene
DR 7	2°05.63′	148°12.25′	1000-1250	10 kg	Buff foraminifera-rich chalk, late Pliocene (N21); volcanic sandstone
DR 7A	2°02.91′	148°12.61′	800-900	40 kg	Foraminifera-rich marl (Pliocene, N19) interbedded with ashy siltstone; volcanic marl; calcareous mudstone, late Pliocene (N21); pumice
DR 8	2°23.46′	149°16.34′	1400-1500	20 kg	Foraminiferal calcarenite, Pliocene (N19); calcareous volcanic siltstone; pumice
DR 9	2°24.69′	149°37.14′	950-1020	15 kg	Buff chalk, late Pliocene (N21); yellowish calcilutite; pumice
DR 10	2°40.64′	151°48.67′	900-1350	50 kg	Calcareous mudstone, marl, calcilutite, chalk, ashstone; late Miocene (N17), Pliocene (N19, N21)
DR 11	3°20.71′	153°09.01′	900-950	100 kg	Massive trachyte, trachybasalt, basalt and agglomerate
DR 12	4°10.36′	153°33.70′	2000-2250	250 kg	Trachyte, trachyandesite, trachybasalt, agglomerate, reef rock, coral

recovered; they varied in length from 85 to 277 cm (Table 2). All cores were sampled for hydrocarbon gases. The results are given by Kvenvolden elsewhere in this volume.

ACKNOWLEDGMENTS

Non-shipboard scientists who have contributed to this volume include field geologists M. Sandy, G. Francis and W.D. Stewart from the GSPNG; igneous petrologists and geochemists R.W. Johnson and A.L. Jaques of BMR, M. Perfit and R.D. Shuster of Florida State University, B.W. Chappell of the Australian National University, and W.I. Ridley of the USGS, Denver; organic geochemists Miryam Glikson of the Australian National University, Canberra, and D. Rigby of the Australian Commonwealth Scientific and Industrial Research Organisation (CSIRO), Sydney; manganese specialist B.R. Bolton of Adelaide University; micropaleontologists D.W. Haig and P.J. Coleman of the University of Western Australia; and seismologist K. McCue of BMR. We are grateful to them all.

The extensive stratigraphic work of the GSPNG staff, in conjunction with paleontological studies by D.W. Haig and P.J. Coleman (this volume), and D. Belford (this volume), has led to a better understanding of island geology (summarized in Figure 2), and hence to a better interpretation of the offshore, largely geophysical, data. The structure contour and isopach maps of the New Ireland Basin

Table 2. Core stations on R/V *S. P. Lee* New Ireland-Manus region cruise.

Number	Latitude (S)	Longitude (E)	Water depth (m)	Recovery (cm)	Comments
G 1	1°39.60′	146°51.12′	1514	227	Pale brown to greenish gray foraminiferal ooze, Holocene
G 2	1°53.38′	147°35.39′	833	None	
G 2A	1°53.90′	147°34.96′	842	None	Soft sediment washed out at surface
G 3	1°50.69′	148°16.17′	1086	85	Pale yellowish brown to greenish gray foraminiferal ooze, Holocene
G 4	2°20.94′	149°41.58′	607	Very minor	Minor lag deposit of pumice, solitary corals, foraminifera, molluscs
G 4A	2°21.14′	149°40.97′	617	Very minor	Minor lag deposit of pumice, solitary corals, foraminifera, molluscs
G 5	2°10.82′	150°18.82′	749	Minor	Olive gray foraminifera-pteropod ooze, Holocene
G 5A	2°10.62′	150°18.79′	785	None	
G 6	2°18.79′	150°39.16′	847	166	Brownish green to olive gray foraminifera-pteropod ooze, Holocene
G 7	2°29.21′	151°03.17′	1036	277	Olive gray foraminiferal ooze to calcareous mud, Holocene
G 8	2°48.39′	151°31.28′	1236	151	Light brown to olive gray foraminiferal ooze to calcareous mud, Holocene
G 9	3°08.46′	152°05.66′	1320	100	Light brown to greenish gray foraminiferal ooze to calcareous mud, Holocene

illustrated by Marlow et al. (this volume) were prepared jointly with Robertson Research (Australia) Pty. Ltd., and Flower Doery Buchan Pty. Ltd., who used them as part of a report on petroleum potential to the Papua New Guinea Government. In essence, we interpreted all Gulf and R/V *S.P. Lee* seismic profiles, and they prepared the maps from this interpretation.

This paper was reviewed by J.B. Colwell (BMR) and T.K. Bruns (USGS), and edited by C. Campbell (USGS). The figures were drawn by A. Murray and the text was typed by J. Brushett (both BMR). We thank them all. Exon publishes with the permission of the Director, Bureau of Mineral Resources, Canberra.

Finally, we thank the sponsors of the entire Tripartite Marine Geoscience Program:

Australian Development Assistance Bureau, Department of Foreign Affairs, Australia
Bureau of Mineral Resources, Geology and Geophysics, Australia
Department of Geology, Mines and Rural Water Supplies, Vanuatu
Department of Scientific and Industrial Research, New Zealand
Geological Survey, Papua New Guinea
Ministry of Lands, Energy and Natural Resources, Solomon Islands
Ministry of Lands, Survey and Natural Resources, Kingdom of Tonga
New Zealand Ministry of Foreign Affairs, External Aid Division
United Nations Economic and Social Commission for Asia and the Pacific, Committee for Co-ordination of Joint Prospecting for Mineral Resources in South Pacific Offshore Areas
U.S. Agency for International Development (Office of Energy), Department of State
U.S. Geological Survey, Department of the Interior

REFERENCES

Brown, C.M., 1982, Kavieng, Papua New Guinea - 1:250,000 Geological Series: Geological Survey of Papua New Guinea, Explanatory Notes SA/56-9, 26 p.

Connelly, J.B., 1976, Tectonic development of the Bismarck Sea based on gravity and magnetic modelling: Geophysical Journal of the Royal Astronomical Society, v. 46, p. 23-40.

de Broin, C.E., Aubertin, F., and Ravenne, C., 1977, Structure and history of the Solomon-New Ireland region, *in* International symposium on geodynamics in South-West Pacific, Noumea, 1976: Paris, Edition Technip, p. 37-50.

Eade, J.V., 1979, Papua New Guinea offshore survey, Cruise SI-79(1): CCOP/SOPAC Cruise Report 24 (unpublished), 10 p.

Exon, N.F., 1981, Papua New Guinea geophysical survey: Hydrocarbons, Cruise PN-81(1): CCOP/SOPAC Cruise Report 52 (unpublished), 5 p.

Exon, N.F., and Tiffin, D.L., 1984, Geology and petroleum prospects of offshore New Ireland Basin in northern Papua New Guinea, *in* Watson, S.T. ed., Transactions of the Third Circum-Pacific Energy and Mineral Resources Conference, Honolulu, Hawaii, 1982, p. 623-630.

Exon, N.F., Stewart, W.D., Sandy, M.J., and Tiffin, D.L., 1986, Geology and offshore petroleum prospects of the eastern New Ireland Basin, northeastern Papua New Guinea: Bureau of Mineral Resources Journal of Australian Geology and Geophysics, v. 10, p. 39-51.

Falvey, D.A., and Pritchard, T., 1984, Preliminary paleomagnetic results from northern Papua New Guinea: evidence for large microplate rotations, *in* Watson, S.T., ed., Transactions of the Third Circum-Pacific Energy and Mineral Resources Conference, Honolulu, Hawaii, 1982, p. 593-599.

French, D.J., 1966, The geology of southern New Ireland: Bureau of Mineral Resources, Australia, Record 1966/179, 30 p.

Gulf Research and Development Company, 1973, Regional marine geophysical reconnaissance of Papua New Guinea, 1973. (Unpublished profiles available from BMR Division of Marine Geosciences and Petroleum Geology, Canberra).

Hohnen, P.D., 1978, Geology of New Ireland, Papua New Guinea: Bureau of Mineral Resources, Australia, Bulletin 194, 39 p.

Jaques, A.L., 1980, Admiralty Islands, Papua New Guinea - 1:250,000 Geological Series: Geological Survey of Papua New Guinea Explanatory Notes SA/55-10 and SA/55-11, 25 p.

Johnson, R.W., 1979, Geotectonics and volcanism in Papua New Guinea: a review of the late Cenozoic: Bureau of Mineral Resources Journal of Australian Geology and Geophysics, v. 4, p. 181-207.

Johnson, R.W., Mutter, J.C., and Arculus, R.J., 1979, Origin of the Willaumez-Manus Rise, Papua New Guinea: Earth and Planetary Science Letters, v. 44, p. 247-260.

Kroenke, L.W., 1984, Cenozoic tectonic development of the southwest Pacific: UN ESCAP, CCOP/SOPAC, Technical Bulletin 6, 122 p.

Sandy, M.J., 1986, Reservoir and source rock potential of New Ireland outcrop samples: Geological Survey of Papua New Guinea Report 86/19, 21 p.

Scholl, D.W., and Vallier, T.L., comp. and eds., 1985, Geology and offshore resources of Pacific island-arcs - Tonga region: Circum-Pacific Council for Energy and Mineral Resources Earth Science Series 2, 488 p.

Stewart, W.D., Francis, G., and Pederson, S.L., 1986, Hydrocarbon potential of the Bougainville and southeastern New Ireland Basins, Papua New Guinea: Geological Survey of Papua New Guinea Report 86/11, 13 p.

Stewart, W.D., and Sandy, M.J., 1986, Cenozoic stratigraphy and structure of New Ireland and Djaul Island, New Ireland Province, Papua New Guinea: Geological Survey of Papua New Guinea Report 86/12, 96 p.

Taylor, B., 1979, Bismarck Sea: evolution of a back-arc basin: Geology, v. 7, p. 171-174.

Thompson, J.E., 1952, Report on the geology of Manus Island, Territory of Papua and New Guinea, with reference to the occurrence of bauxite: Bureau of Mineral Resources, Australia, Record 1952/82 (unpublished), 11 p.

Tiffin, D.L., 1981, Papua New Guinea offshore hydrocarbon and phosphate survey, Cruise PN-81(2): UN ESCAP, CCOP/SOPAC Cruise Report 53 (unpublished), 17 p.

Wallace, D.A., Chappell, B.W., Arculus, R.J., Johnson, R.W.,

Perfit, M.R., Crick, I.H., Taylor, G.A.M., and Taylor, S.R., 1983, Cainozoic volcanism in the Tabar, Lihir, Tanga and Feni islands, Papua New Guinea: geology, whole-rock analyses, and rock-forming mineral composition: Bureau of Mineral Resources, Australia, Report 243; BMR Microform MF 197, 27 p.

PART 2 – TOPICAL STUDIES

Marlow, M.S., Dadisman, S.V., and Exon, N.F., editors, 1988, Geology and offshore resources of Pacific island arcs—New Ireland and Manus region, Papua New Guinea, Circum-Pacific Council for Energy and Mineral Resources Earth Science Series, v. 9: Houston, Texas, Circum-Pacific Council for Energy and Mineral Resources.

GEOLOGY OF NEW IRELAND AND DJAUL ISLANDS, NORTHEASTERN PAPUA NEW GUINEA

W.D. Stewart,[1] M.J. Sandy

Geological Survey of Papua New Guinea, Port Moresby, Papua New Guinea

ABSTRACT

New Ireland is dissected by northeasterly and northwesterly-trending normal faults, which have fragmented the island into a series of northeasterly-tilted fault blocks. An exception to this regional pattern is the Weitin Fault Zone, which appears to represent the onshore extension of a ridge-trench transform fault system.

Basement in New Ireland is formed by the Eocene(?) to Oligocene Jaulu Volcanics, and the Oligocene to middle Miocene Lemau Intrusive Complex. The basement rocks are unconformably overlain by the Miocene Lelet Limestone and the Lelet's partial lateral equivalents, the Tamiu siltstone and Lossuk River beds. In northwestern New Ireland, the lower Lelet Limestone and Lossuk River beds are succeeded conformably by the middle to late Miocene Lumis River volcanics. To the south, where the Lelet Limestone has a longer age range, the formation is overlain by the late Miocene to early Pliocene Punam Limestone, and the Pliocene to earliest Pleistocene Rataman Formation. The youngest rock units on New Ireland are Pleistocene limestones (Qc) and their lateral equivalent, the Maton Conglomerate.

The Jaulu Volcanics and Lemau Intrusive Complex are the products of island arc volcanism associated with subduction along the Manus-Kilinailau Trench. Uplift and erosion of the arc was followed by widespread carbonate platform sedimentation during the early Miocene (Lelet Limestone). At the same time, fine-grained clastic sediment accumulated in adjacent, deep water areas (Tamiu siltstone and Lossuk River beds). Local volcanism, accompanied by marked subsidence, ended carbonate sedimentation in northwestern New Ireland during the middle Miocene. Bathyal pyroclastic and epiclastic sediment accumulated in that area until the late Miocene or early Pliocene (Lumis River volcanics). To the south, carbonate platform sedimentation probably continued unabated until the late Miocene, when rapid regional subsidence drowned the platform. This event appears to have coincided with reversal of the New Ireland portion of the original island arc system. Bathyal pelagic deposition subsequently became established over a wide area (Punam Limestone). Regional volcanism commenced in the early Pliocene, probably in response to subduction along the newly-formed New Britain Trench. During the remainder of the Pliocene, bathyal sedimentation continued as pyroclastic and epiclastic sediment was deposited on submarine fans (Rataman Formation). Rapid regional uplift of the southwestern margin of the New Ireland Basin followed during the early Pleistocene. Coarse-grained terrigenous sediment accumulated against active fault scarps (Maton Conglomerate), while neritic carbonates were deposited nearby (Qc).

---

[1] Present address: Department of Geology, University of Ottawa, Ottawa, Ontario, Canada K1N 6N5

INTRODUCTION

New Ireland is one of the major islands in northeastern Papua New Guinea (Figure 1). The sedimentary rocks exposed on this island, together with those cropping out on New Hanover, Manus Island and adjacent smaller islands, form part of the Tertiary New Ireland Basin. As no exploratory wells have been drilled in the area, these onshore exposures provide important information about the composition of the sedimentary sequence in the offshore part of the basin.

In 1983, the Geological Survey of Papua New Guinea (GSPNG) commenced detailed mapping and stratigraphic studies in New Ireland and Djaul Island. The project was supported throughout by extensive micropaleontological work by D.W. Haig and P.J. Coleman (foraminifera; this volume), and H.K. Hekel (calcareous nannofossils and palynology). Preliminary stratigraphic results based on data available up to July 1984 were reported by Exon et al. (1986). Field studies have since been extended to cover the remainder of the island, and as a result some revisions have been made to earlier geological ideas.

This paper summarizes the stratigraphy and structural geology of New Ireland and Djaul Island. We concentrate mainly on the stratigraphy of the sedimentary cover rocks, and thus provides a guide for the interpretation of seismic-reflection profiles in the adjacent offshore area. Informal nomenclature is used for some units, because of the lack of complete sequences in outcrop or boreholes, and hence of suitable type sections. Units in this category are the Lossuk River beds, Tamiu siltstone and Lumis River volcanics, and subdivisions of the Lelet Limestone (Bagatere, Kimidan and Matakan units), Lumis River volcanics (Kaut, Lumis River and Kapsu units), and Rataman Formation (Ramat, Gogo, Uluputur and Djaul units). A more complete account of the geology can be found in the final GSPNG report and accompanying 1:100,000 scale geological map of the islands (Stewart and Sandy, 1986).

PREVIOUS WORK

The earliest geological observations in New Ireland were made during the period of German administration by Dr. Karl Sapper (Sapper, 1910), who traversed much of the island on foot. Systematic geological mapping by the Bureau of Mineral Resources began in southern New Ireland in 1964 (French, 1966), and coverage was extended to the remainder of the island five years later (Hohnen, 1978). New Hanover was subsequently mapped in 1973 (Brown, 1982). The resultant 1:250,000 scale map sheets and accompanying explanatory notes served, until recently, as the standard reference works on the geology of the New Ireland area.

Geological investigations by the private sector have provided a wealth of data from all parts of New Ireland. Exploration for base metal mineralization was undertaken by C.R.A. Exploration Pty. Ltd. (Conzinc Riotinto Australia Exploration Pty. Ltd., 1965), Swiss Aluminum Mining of Australia Pty. Ltd. (various reports, 1969-1972) and Esso Papua New Guinea Inc. (Mitchell and Weiss, 1982). Stratigraphic studies were also carried out by the Continental Oil Company and New Guinea Cities Service Inc. in an early effort to define the petroleum potential of the area (Ripper and Grund, 1969).

REGIONAL GEOLOGICAL SETTING AND TECTONIC EVOLUTION

New Ireland, New Hanover and the Admiralty Islands represent the uplifted southwestern margin of the New Ireland Basin. The submerged part of this feature trends for about 900 km from Feni Island in the east to Manus island in the west (Figure 2). The platform is bounded on the south by young oceanic crust of the Manus Basin, and on the north by the Emirau-Feni Ridge. The Manus-Kilinailau Trench lies to the north of the ridge, and marks the southern boundary of oceanic crust of the Caroline and Pacific plates.

New Ireland is situated in a zone of complex interaction between the Pacific and Indo-Australian plates. The boundary between these two major features is occupied by a microplate complex, consisting of the North and South Bismarck microplates and the Solomon Sea Microplate (Figure 2). In the eastern part of the Manus Basin, the boundary between the North and South Bismarck microplates is marked by an asymmetric spreading center bounded by two transform fault systems (Taylor, 1979). The easternmost of these fault systems is known to trend along the southwestern margin of New Ireland, but the fault's exact configuration is poorly defined. The distribution of earthquake epicenters in the area (see Figure 15 *in* Johnson, 1979) suggests that more than one transform fault segment may be present. This system is represented in the southeast by the submarine faults bounding St. Georges Channel, and major parallel fault systems in northeastern New Britain and southern New Ireland

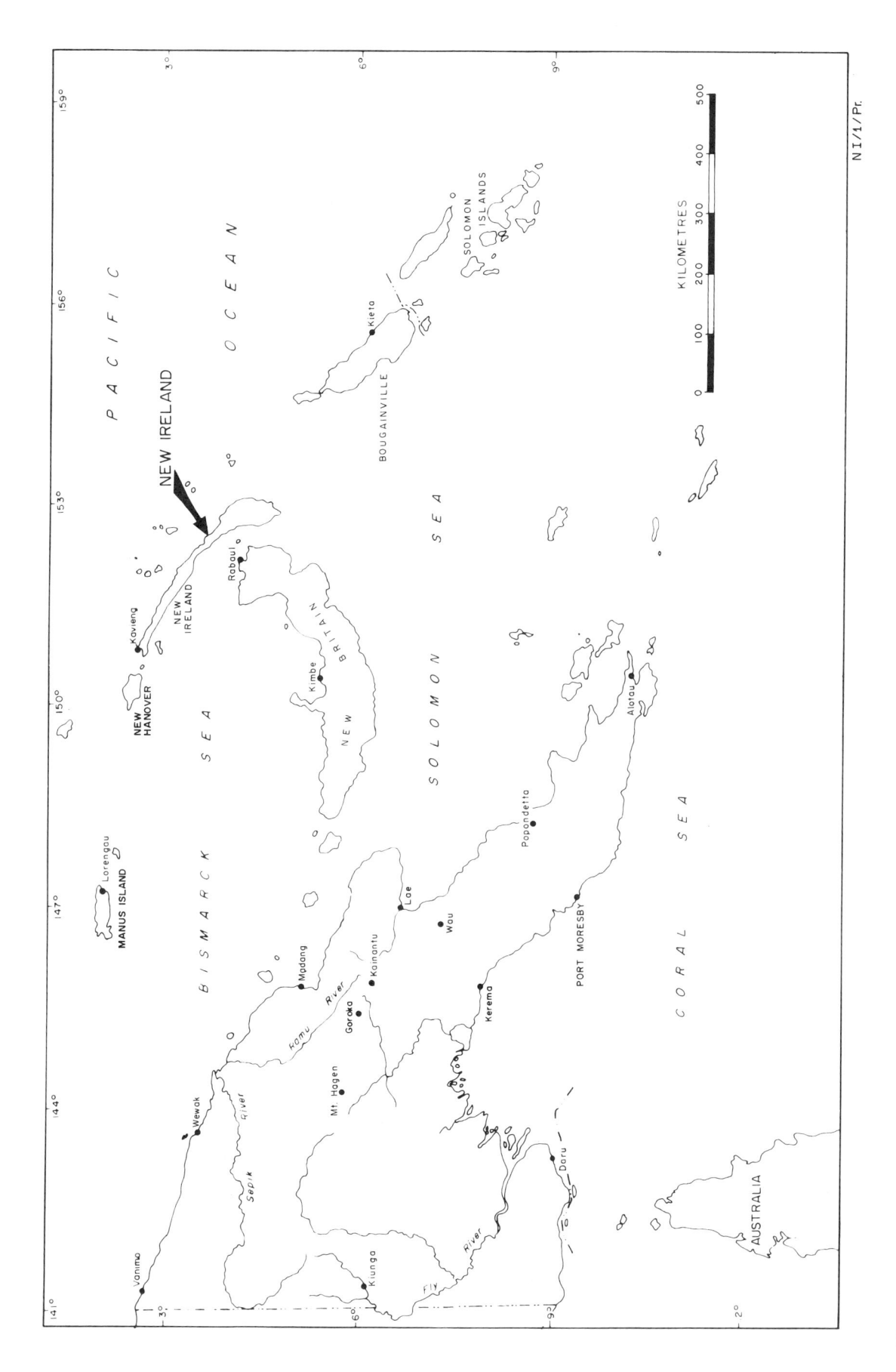

Figure 1. Location of New Ireland in Papua New Guinea.

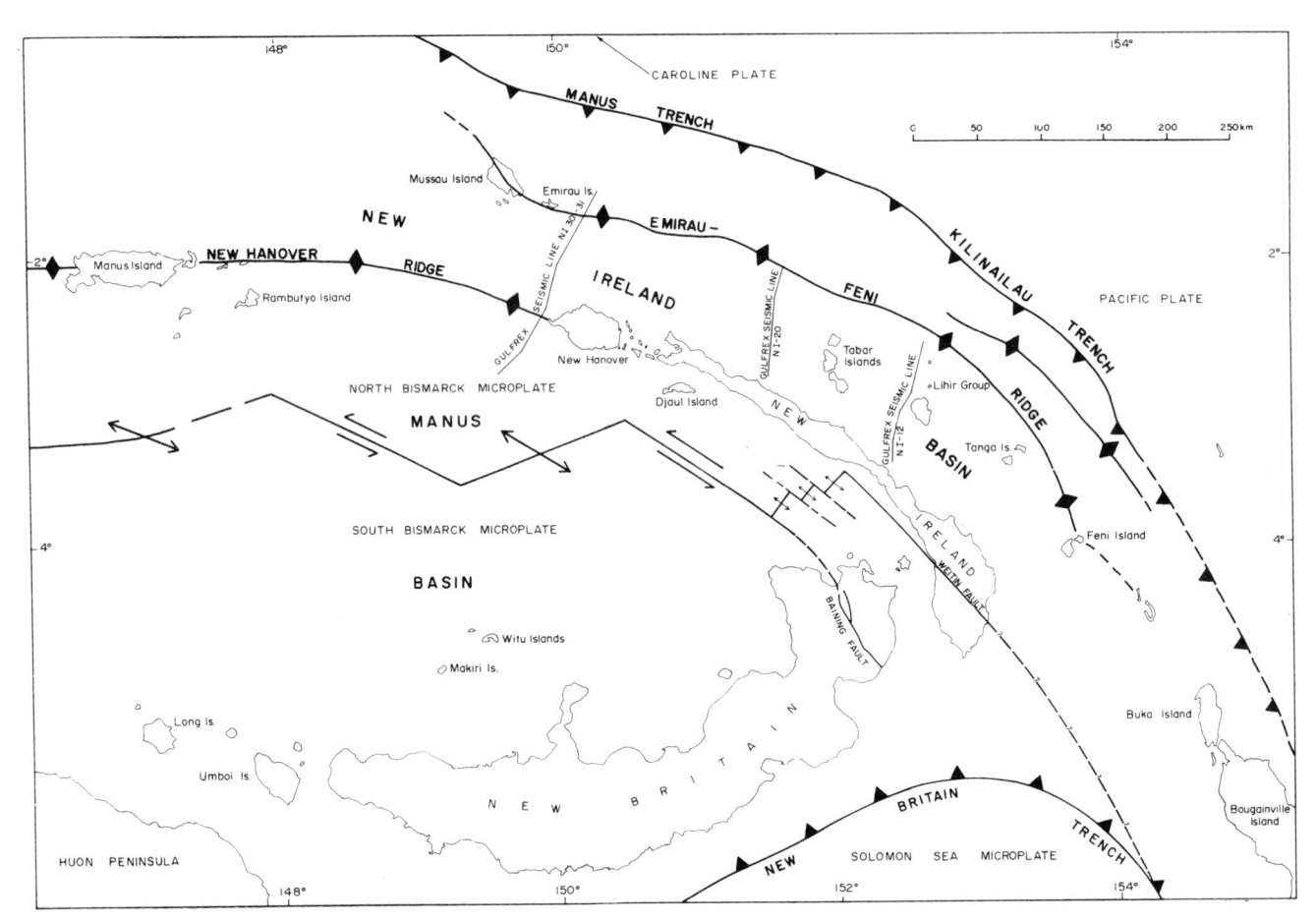

Figure 2. Regional tectonic elements (after Jonson, 1979 and Exon et al., 1986, incorporating unpublished data from the 1985 cruise of the R/V *Moana Wave*).

(Figure 2).

During the Cenozoic, the New Ireland region was affected by large microplate rotations. Paleomagnetic reconstructions by Falvey and Pritchard (1984) suggest that New Ireland originally formed part of a northwest-facing island arc system during the Eocene, with subduction of oceanic crust occurring along the Manus-Kilinailau Trench. Back-arc spreading rotated the island arc system to face northeastward by the late Oligocene. The subduction zone eventually failed because of collision with the anomalously thick oceanic crust of the Ontong Java Plateau, and the island arc system was transferred from the Indo-Australian Plate to the Pacific Plate. North and northeasterly-directed subduction subsequently commenced along the New Britain Trench on the opposite side of the old island arc.

Falvey and Pritchard's (1984) data suggest that the reversal was not synchronous throughout the arc. Reversal of the Manus-New Britain-Huon portion of the arc apparently took place during the late Oligocene, followed by reversal of the New Ireland-North Solomons portion of the arc during the late Miocene. Later, during the mid-Pliocene, back-arc spreading commenced in the Manus Basin. The subsequent northwestward displacement of New Ireland relative to New Britain, along a transform fault system, coincided with rapid regional uplift of the southwestern margin of the New Ireland Basin.

STRUCTURAL GEOLOGY

New Ireland is dissected by two major sets of normal faults (Figure 3). The first of these trends westerly to northwesterly, and appears to control the island's coastlines. The second fault set consists of northerly to northeasterly-trending faults, which are oriented nearly perpendicular to the trend of New Ireland. The intersection of these two fault sets has resulted in the fragmentation of New Ireland into a series of discrete, variably-sized fault blocks.

On a regional scale, the major fault blocks in New Ireland are tilted to the northeast. This is clearly expressed by the asymmetric relief profile of

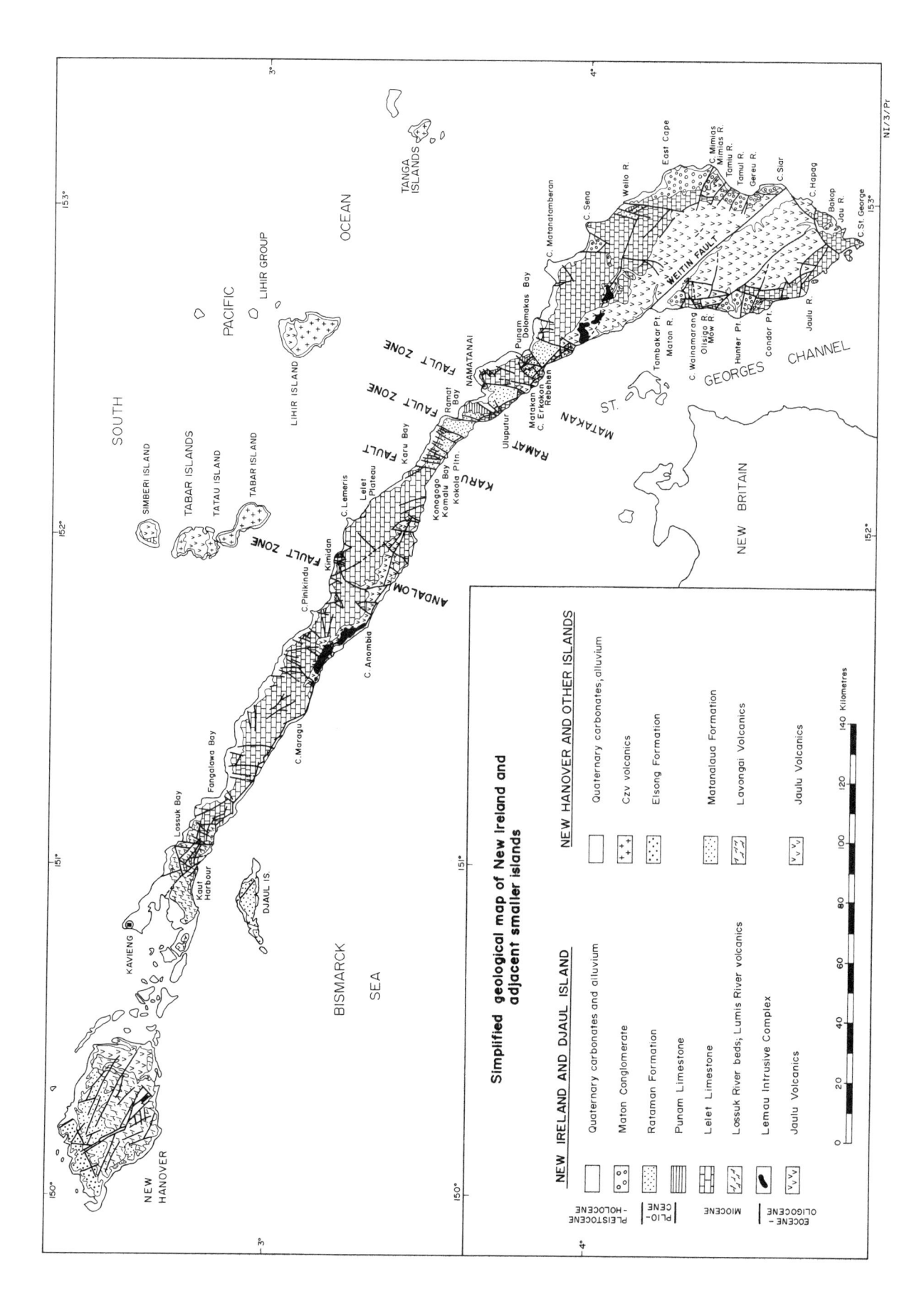

Figure 3. Simplified geological map of New Ireland and adjacent smaller islands.

the island. Along the southwestern coast of the island, however, individual fault blocks are locally tilted in the opposite direction.

The dominant structural feature of southern New Ireland is the Weitin Fault Zone, which obliquely traverses the island for 62 km on a trend of 320 degrees. The fault zone varies from about 0.5 to 3 km in width, and is bounded by high fault scarps. Short fault segments, some of which affect Quaternary cover, are visible in parts of the fault zone. These features, together with the remarkably straight surface trace of the fault zone, are consistent with the characteristics of a strike-slip fault. This interpretation is also supported by regional evidence. According to Taylor (1979), the Weitin Fault Zone appears to represent the southern, onshore extension of his seismic segment 4, which he inferred to be a ridge-trench transform fault system.

The sense of motion along the Weitin Fault Zone has not been established. Johnson (1979) noted that the general trend of seismic segment 4, together with four earthquake focal mechanism solutions from the northwestern end of the segment, are consistent with left-lateral strike-slip movement. This sense of movement was also suggested for the Weitin Fault by French (1966) and Hohnen (1978), although their attempts to relate fault movement to stream course displacement were inconclusive. However, motion along the fault zone could have been right-lateral, which is also geometrically possible. The geology of the Cape Wainamarang area on the southwestern side of the Weitin Fault bears distinct similarities to that of the East Cape-Cape Siar area on the opposite side of the fault to the southeast (Figure 3). If a match is postulated between these two areas, then the match would imply offset by right-lateral wrench movement. To the northwest, the offshore extension of the Weitin Fault Zone is probably intersected by the northerly-trending Karu, Ramat and Matakan faults. These features may correspond to one of the sets of conjugate fractures normally formed in response to wrench movement. If this inference is correct, then their orientation is consistent with that expected for right-lateral wrench movement along the Weitin Fault Zone.

STRATIGRAPHY

Jaulu Volcanics

Definition, Distribution and Thickness

The basement rocks of New Ireland are the Jaulu Volcanics (Hohnen, 1978; Brown, 1982; Figure 4). A type section was designated for this unit along the Jaulu River in southwestern New Ireland by Hohnen (1978), who estimated a maximum exposed thickness of about 2000 m. The total thickness of the Jaulu Volcanics is unknown.

The Jaulu Volcanics are widely exposed in southern New Ireland, and along the southwest coast of the island between Cape Maragu and the Lelet Plateau (Figure 3). They also crop out in eastern New Hanover, and equivalent rocks are present on the Tabar Islands and Mussau Island.

Lithology

The main rock types in the Jaulu Volcanics are massive andesitic agglomerate, lapilli tuff, crystal lithic tuff, porphyritic pyroxene andesite lava, and ignimbrite. Rare tuffaceous limestone, calcareous tuff, volcanolithic conglomerate, and volcanolithic sandstone are also interbedded in the sequence.

French (1966) and Hohnen (1978) reported large limestone lenses in the Jaulu Volcanics. These are now known to be in-faulted Lelet Limestone. All fossils recovered so far from the formation have come from thin limestone or clastic interbeds, or from small allochthonous limestone blocks.

Stratigraphic Relationships and Age

The lower contact of the Jaulu Volcanics is not exposed. They are presumed to overlie oceanic crust of Eocene age.

The Jaulu Volcanics are unconformably overlain by the Lelet Limestone in most areas. In northwestern New Ireland, they are probably overlain locally by the Lossuk River beds, although the contact is not exposed. In southeastern New Ireland, photogeological interpretation suggests that the Jaulu Volcanics are locally unconformably overlain by the Tamiu siltstone (Figure 4). Offshore, the formation is inferred to correlate in part with the E-V sequence recognized by Exon and Tiffin (1984) in seismic-reflection profiles.

Combined isotopic and paleontological evidence from New Ireland and Manus Island suggests an Eocene to Oligocene age range for the sequence. However, the Jaulu Volcanics may range upward in age into the earliest Miocene (Figure 4).

On New Ireland, no direct isotopic or fossil evidence is available to confirm exposures of Eocene strata. A stock belonging to the Lemau Intrusive

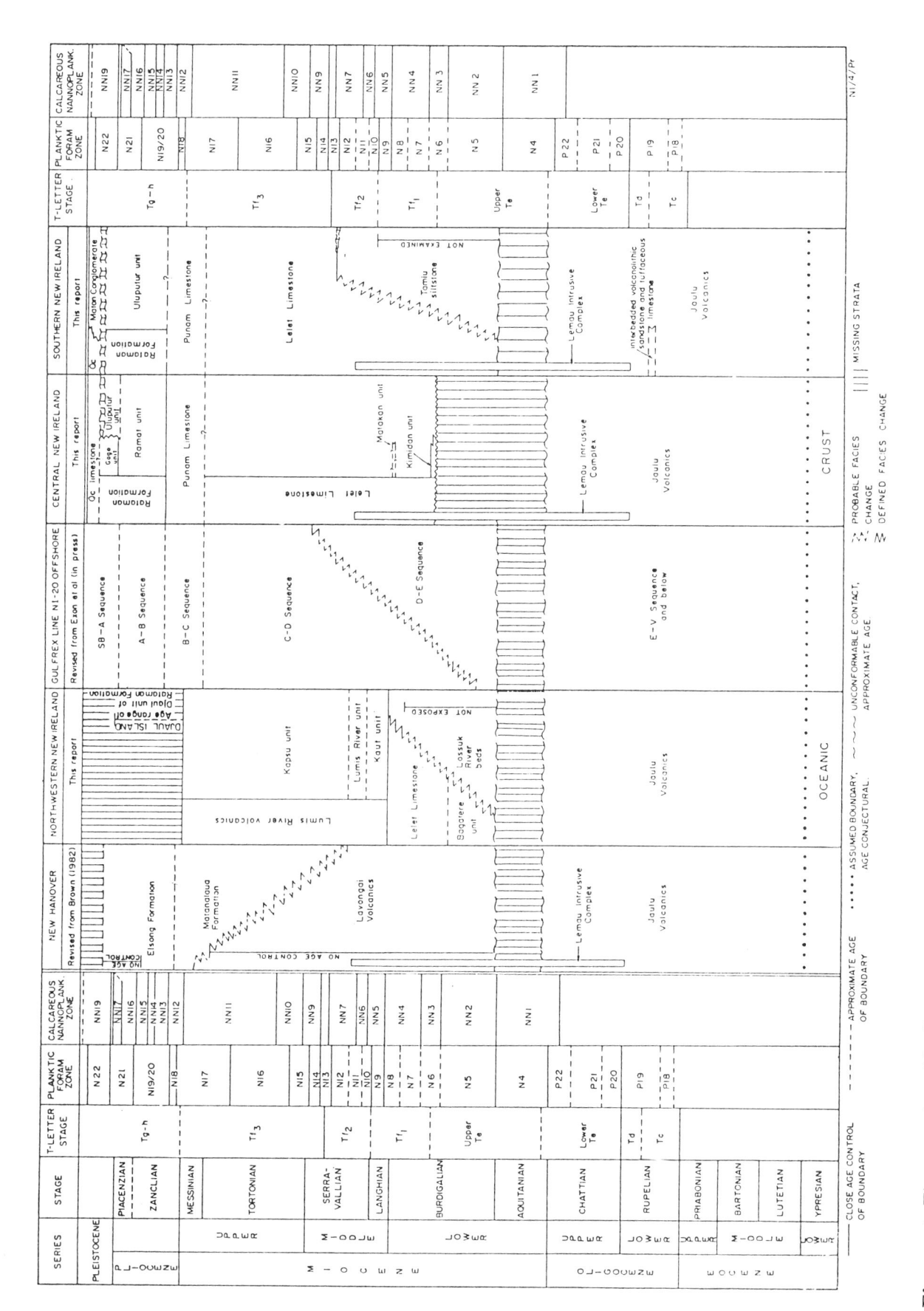

Figure 4. Time-stratigraphic diagram for the New Ireland area.

Complex north of Cape Anomia has yielded a K-Ar age of 37.4±1.7 Ma (Mitchell and Weiss, 1982). Hence, the intruded Jaulu Volcanics must be either Eocene or earliest Oligocene in age at that locality. In the absence of more definitive evidence, an Eocene lower age limit has been tentatively inferred through correlation with equivalent basement rocks on Manus Island, the Tinniwi Volcanics. The oldest K-Ar date reported by Jaques (1980) for that unit is 49.0±5.0 Ma (original date corrected using revised constants of Steiger and Jager, 1977 and the tables of Dalrymple, 1979).

Strata of Oligocene age are known to be present in the Jaulu Volcanics because of both fossil and isotopic evidence. Early Oligocene (Tc-Td) foraminiferal assemblages have been recovered from three separate localities (McGowran *in* Mitchell and Weiss, 1982; Binnekamp *in* Hohnen, 1978; Haig *in* Stewart and Sandy, 1986). In addition, a single K-Ar determination on an andesite was reported by Hohnen (1978) to have given a mid-Oligocene age of 31.5±1.0 Ma.

The upper age limit of the Jaulu Volcanics is uncertain. Early Miocene (upper Te) foraminiferal assemblages have been recovered from the basal Lelet Limestone. This indicates only that the uppermost Jaulu Volcanics are no younger than earliest Miocene in age.

Lemau Intrusive Complex

The Lemau Intrusive Complex was defined by Hohnen (1978) to encompass intrusive rocks formerly included in the "Older Volcanics" by Sapper (1910), the Jaulu Volcanics by French (1966), and the "Lemau Igneous Belt" by Ripper and Grund (1969). The most extensive exposures of this unit occur along the southwest coast of New Ireland (Figure 3).

The Lemau Intrusive Complex comprises stocks and dikes of gabbro, diorite, quartz diorite, and leucocratic rocks of variable compositions. Weak porphyry copper mineralization has been found at two localities.

Float samples of intrusive rocks from southern New Ireland were reported by Hohnen (1978) to have yielded K-Ar ages ranging from 32.6±1.0 Ma to 14.2±0.5 Ma (early or middle Oligocene to middle Miocene). An additional K-Ar date of 37.4±1.7 Ma (earliest Oligocene) was obtained by Mitchell and Weiss (1982) for a stock exposed north of Cape Anomia.

Cumulative evidence from New Ireland and Manus Island suggests that the island arc was affected by an initial intrusive phase during the early to middle Oligocene (about 38-32 Ma), followed by a second intrusive phase during the early to middle Miocene (about 18-14 Ma). The initial phase was magmatically related to the Jaulu Volcanics, whereas the second phase was probably magmatically related to the Kaut and Lumis River units of the Lumis River volcanics in northwestern New Ireland (Figure 4). The intrusive rocks exposed along the southwest coast of New Ireland were probably all emplaced during the initial phase, as they are unconformably overlain by the Lelet Limestone. No intrusions belonging to the younger phase have been positively identified. However, French (1966, p. 3) reported that the limestones exposed near King Bay (north of Cape Wainamarang) showed "evidence of intrusion". These strata are now known to belong to the Lelet Limestone. If French's observation is correct, then unmapped intrusions belonging to the younger intrusive phase probably occur in the area. No other examples of an intrusive relationship with the sedimentary cover rocks have been reported in New Ireland.

Lelet Limestone

Definition

Hohnen (1978) used the term Lelet Limestone for a thick sequence of neritic carbonate rocks exposed in the northern half of New Ireland. He designated a type section on the northern margin of the Lelet Plateau (Figure 3). The same author proposed that a similar, geographically separated succession exposed in southern New Ireland be named the Surker Limestone because of the exposure's apparently shorter age range and different geomorphological expression. Recent work by the GSPNG failed, however, to discern any significant difference in either the lithology or the age range of the two units. As a result, the term Surker Limestone is considered to be redundant, and the name Lelet Limestone has been applied to the Miocene neritic carbonate succession throughout New Ireland.

Stewart and Sandy (1986) have included three laterally discontinuous clastic units in the Lelet Limestone (Figure 4). Two of these, the Bagatere and Kimidan units, occur at the base of the Lelet Limestone in northwestern and central New Ireland respectively. The stratigraphic position of the third, the Matakan unit, is uncertain. This unit is exposed only in the Matakan area on the southwest coast.

Distribution and Thickness

The Lelet Limestone is widely exposed between Lossuk Bay and Cape Mimias (Figure 3). The formation is also exposed in a series of fault blocks along the west coast between the Weitin Fault Zone and Cape St. George.

Thickness estimates for the Lelet Limestone remain conjectural because of extensive fault repetition and difficulties in obtaining precise biostratigraphic control. Hohnen's (1978) original estimate, of up to 1400 m in the Lelet Plateau area, may be excessive in light of depth-to-basement calculations undertaken by Robertson Research (Australia), using aeromagnetic data acquired by Esso (Mitchell and Weiss, 1982). Preliminary results suggest that residual thicknesses may range from about 250 to 750 m in northwestern New Ireland, about 250 to 500 m in the area immediately southeast of Lelet Plateau, and up to a maximum of about 1000 m near Cape Matanatamberan (E. Bowen, pers. comm).

The inferred offshore equivalent of the Lelet Limestone (C-D sequence) appears to range in thickness from about 1000 m to more than 2000 m in parts of the basin (Exon et al., 1986). Partial equivalents of this formation are also exposed on Simberi Island (Wallace et al., 1983).

Lithology

The Lelet Limestone is composed mainly of white, cream and light brown, massive to thick bedded limestone. Foraminiferal and algal-foraminiferal biomicrite and microsparite are the dominant rock types, and tend to occur in thick, monotonous sequences. Benthic foraminifera are common, and often occur together with variable proportions of fragmented coralline algae or *Halimeda* green algal plates.

Associated allochems, when present, include bivalves, gastropods, echinoid plates and spines, occasional whole echinoids, sponge spicules, bryozoa, phylloid algal fragments, *Dascycladacean* algal fragments, pellets, and scattered *Acropora* branching coral fragments. In parts of the lower Lelet Limestone, packed algal-foraminiferal biomicrite and microsparite are common. These contain abundant fragmented and encrusting coralline algae associated with foraminiferal and other skeletal debris.

Massive colonial corals are rare in the onshore sequence, and appear to occur only locally in the sequence's lower part. No framework reefs were identified during recent GSPNG field studies on the island.

The Bagatere unit is composed mainly of tuff and tuffaceous volcanolithic sandstone with interbeds of limestone and minor agglomerate. The unit has a maximum exposed thickness of about 400 m. The sandstone commonly contains pelecypod shell debris, carbonaceous fragments, silicified wood, and minor fragmented *Acropora(?)* corals. The limestone interbeds are composed mainly of slightly tuffaceous, red algal-foraminiferal microsparite. They contain locally abundant whole and fragmented bivalves, *Acropora*(?) fragments, and occasional massive colonial corals.

Only incomplete sections of the Kimidan unit have been observed, but the total thickness of the unit probably does not exceed a few tens of meters. In the Andalom Fault Zone, the Kimidan unit consists of interbedded calcareous, muddy sandstone, and calcareous, carbonaceous sandy mudstone with pelecypod and gastropod fragments.

The exposed part of the Matakan unit is about 30 m thick and consists mainly of friable, highly calcareous, carbonaceous silty mudstone with abundant bivalves, benthic foraminifera, gastropods and *Acropora*(?) finger coral fragments. Large allochthonous blocks of sub-bituminous coal have been incorporated in the mudstone.

Stratigraphic Relationships and Age

The Lelet Limestone rests unconformably on the Jaulu Volcanics or Lemau Intrusive Complex in most areas. In southeastern New Ireland, the Lelet locally overlies unconformably the Tamiu siltstone (Figure 4).

The age of the lower contact of the Lelet Limestone probably varies, because of onlap against rugged basement topography (Figure 4). Foraminiferal evidence suggests that the basal Lelet Limestone (Kimidan unit) near Lelet Plateau is probably late Burdigalian in age (upper upper Te). A much thicker sequence of upper Te sediment, incorporating both the Bagatere unit and the lower part of the limestone succession, is present in northwestern New Ireland. Thus, basement in the Lelet Plateau area may have remained temporarily emergent while neritic clastic and carbonate sedimentation was taking place elsewhere.

Paleontological evidence indicates that the lower contact of the Lelet Limestone is also diachronous in a basinward direction (Figure 4). In southwestern New Ireland, the basal Lelet Limestone is early Miocene (upper Te) in age. Along the Tamiu

River in southeastern New Ireland, the Lelet Limestone overlies the Tamiu siltstone with local unconformity. The latter unit ranges upward in age into the late middle Miocene, and is therefore a lateral equivalent of the lower Lelet Limestone to the west.

In northwestern New Ireland, only the lower Lelet was deposited. Here, the Lelet is conformably and gradationally overlain by the Kaut unit of the Lumis River volcanics. Combined foraminiferal and nannofossil evidence indicates an earliest middle Miocene age for the contact (Figure 4).

Elsewhere in New Ireland, the upper contact of the Lelet Limestone is poorly understood because of ambiguous and apparently conflicting evidence. At the only locality where this contact has been observed (Mow River), the Lelet Limestone is apparently conformably and gradationally overlain by the Punam Limestone. However, tenuous foraminiferal evidence appears to suggest a marked age discrepancy between the uppermost Lelet Limestone (possibly mid-middle Miocene, based on a questionable identification of *Globorotalia foshi foshi* in thin section) and the basal Punam Limestone (latest Miocene or earliest Pliocene, N17-N18). Hence, a subtle, undetected paraconformity may intervene between the two formations at that locality. The paraconformity could be either local or regional in extent. Equally tenuous evidence from another locality suggests a longer age range for the Lelet Limestone. A fault-bounded exposure of the Punam Limestone near Cape St. George contains carbonate turbidites consisting partly of algal material. The sequence has been dated as N17 (probably upper N17). If the algal material was derived from a time-equivalent carbonate platform, then such derivation may constitute indirect evidence that neritic carbonate deposition persisted at least into the latest Miocene.

Depositional Environment

The predominance of biomicrite in the Lelet Limestone suggests that deposition occurred either in a wide, protected lagoon behind some form of barrier, or on an open shelf experiencing only low to moderate wave action. The virtual absence of frame-building massive colonial corals, and the presence in some facies of abundant fragmented and encrusting coralline algae or *Halimeda* plates, indicate that red algal-foraminiferal banks and *Halimeda* banks may have been the most common organic buildups. In such an environment, processes such as baffling, encrusting, binding, and organic sediment production would have been more important than frame-building. Associated branching *Acropora* coral colonies probably formed discrete patch reefs or grew on top of the banks. Production and accumulation of carbonate sediment apparently took place more quickly than could be accommodated by subsidence, resulting in progradation of the carbonate platform over the platform's deep-water clastic equivalents.

Fossil evidence indicates that the Bagatere unit accumulated in a neritic environment. Intermittent volcanism introduced pyroclastic sediment, which was partially reworked during quiescent periods. As the volcanism waned, carbonate sedimentation gradually became established.

Both the Kimidan and Matakan units appear to represent sedimentation in local, restricted lagoons, at times under reducing conditions. Analysis of the microflora in the Matakan unit indicates substantial contributions by mangroves (Hekel *in* Stewart and Sandy, 1986), suggesting the proximity of mangrove swamps. Considerable accumulation of organic material in these areas led to the formation of coal beds, transported blocks of which were subsequently incorporated in nearby lagoonal sediment.

Lossuk River Beds

Definition

The term Lossuk River beds was first used by Ripper and Grund (1969) and later adopted by Hohnen (1978) for a sequence of siltstone, limestone, labile sandstone and conglomerate exposed along the upper reaches of the Lossuk River in northwestern New Ireland. Recent re-examination of the section by the GSPNG suggested that these strata are composed mainly of Quaternary alluvium and large allochthonous blocks of limestone shed from uplifted fault blocks to the south. However, exposures of fine clastic sediment of earliest middle Miocene and older age are present in nearby creeks, and these were assigned to the Lossuk River beds by Stewart and Sandy (1986). This definition of the Lossuk River beds differs substantially from that used by Brown (1982), who included rocks now assigned to the Kapsu unit of the Lumis River volcanics and the Djaul unit of the Rataman Formation.

Distribution and Thickness

The Lossuk River beds are exposed only in a small area near Lossuk Bay in northwestern New Ireland. This unit has been correlated with the D-E seismic sequence in the adjacent offshore area by Exon et al. (1986).

The total thickness of the Lossuk River beds is unknown. The exposed portion of the sequence is estimated to be about 150 m thick, but this figure may be unreliable because of extensive faulting in the area. Seismic-reflection profiles indicate that the D-E sequence thickens basinward, from a feather edge against basement to more than 1000 m in distal parts (Exon et al., 1986).

Lithology

The Lossuk River beds consist of interbedded calcareous, carbonaceous mudstone, siltstone, tuffaceous sandy siltstone, and minor fine-grained, tuffaceous sandstone. Volcanolithic pebbly sandstone and tuffaceous limestone are rarely present. The sedimentary units contain abundant planktonic foraminifera together with derived gastropod and pelecypod shell fragments, echinoid spines, coral fragments, benthic foraminifera and carbonaceous material.

Stratigraphic Relationships, Age and Depositional Environment

The exposed part of the Lossuk River beds is latest early Miocene to earliest middle Miocene (N8) in age. The unit is sharply but conformably overlain by the Kaut unit of the Lumis River volcanics. Foraminiferal and calcareous nannofossil evidence indicate that this contact is time-equivalent to the contact between the lower Lelet Limestone and Kaut unit near Kaut Harbor. Thus, the Lossuk River beds are a lateral equivalent of at least part of the lower Lelet Limestone (Figure 4).

The Lossuk River beds are also partly equivalent to the Tamiu siltstone in southeastern New Ireland. Although these two units were deposited in similar environments, they are differentiated because of lithological dissimilarities and their geographical separation.

The Lossuk River beds accumulated at outer neritic to upper bathyal depths (50-500 m) near the carbonate platform margin. The shallow water detritus found in this sediment was probably swept off the platform by storm-generated currents.

Tamiu Siltstone

Definition

The name Tamiu siltstone was proposed by Stewart and Sandy (1986) for a newly recognized succession of early(?) to middle Miocene fine clastic sediment underlying the Lelet Limestone in southeastern New Ireland. The exposed part of this sequence has been divided into two distinctive units.

Distribution and Thickness

The Tamiu siltstone was first recognized in the Tamiu River south of Cape Mimias (Figure 3). Similar rocks cropping out in the upper Mimias and Gereu rivers have also been assigned to this unit. The Tamiu siltstone, like the Lossuk River beds, is probably correlative with the D-E seismic sequence recognized by Exon and Tiffin (1984) offshore.

The total thickness of the Tamiu siltstone was not established directly in the field because of extensive faulting and incomplete exposure. Photogeological interpretation suggests that the siltstone is at least 300 m thick near the southeastern end of the Weitin Fault Zone.

Lithology

The lower unit of the Tamiu siltstone is composed dominantly of massive, calcareous siltstone. Interbedded lithologies include calcareous, volcanolithic sandstone, carbonaceous, pelecypodal, pebbly or sandy mudstone, and local, rhythmically interbedded, tuffaceous biomicrite and algal-pelecypodal mudstone. Fossils recovered from this unit include abundant planktonic foraminifera, benthic foraminifera, pelecypod fragments, echinoid spines, algal plates, and minor carbonaceous material.

The upper unit is composed of interbedded calcareous, tuffaceous siltstone, sandy siltstone, argillaceous sandstone, sandy mudstone, pelecypodal siltstone, and laminated tuffaceous biomicrite. Disturbed lenses and paralithons of sub-bituminous coal are associated with these sedimentary units in a major fault zone along the Mimias River. The fossil assemblage includes pelecypod shell fragments, benthic foraminifera, echinoid fragments, minor coral fragments, and rare planktonic foraminifera.

Stratigraphic Relationships, Age and Depositional Environment

Photogeological interpretation suggests that the

Tamiu siltstone unconformably overlies the Jaulu Volcanics. Along the Tamiu River, the upper unit of the Tamiu siltstone is overlain with low angular unconformity by the Lelet Limestone.

Middle Miocene (N10 and N12) foraminiferal assemblages have been recovered from the lower unit. Since the Tamiu siltstone appears to overlie the Jaulu Volcanics to the south, the total age range of this unit is probably early to middle Miocene. The upper unit is no younger than late middle Miocene, because the basal Lelet Limestone overlying the upper unit contains a lower Tf foraminiferal assemblage. Thus, the time gap represented by the unconformity is short (Figure 4).

Foraminiferal assemblages from the lower unit indicate deposition at depths ranging from upper bathyal (150-500 m) to outer neritic (50-150 m). Similar evidence suggests an inner neritic environment (water depths of less than about 50 m) for the upper unit. The Tamiu siltstone is therefore a shallowing upward sequence that accumulated near the carbonate platform margin. At least locally, the overall regression was aided by syn-sedimentary upfaulting and tilting before progradation of the main carbonate platform.

Lumis River Volcanics

Definition

The name Lumis River volcanics was proposed by Stewart and Sandy (1986) for a newly recognized middle Miocene and younger succession of pyroclastic and epiclastic rocks overlying the lower Lelet Limestone and Lossuk River beds in northwestern New Ireland. The sequence has been informally subdivided into the Kaut, Lumis River and Kapsu units in ascending stratigraphic order (Figure 4). Rocks now assigned to the Lumis River volcanics were referred to as the "unnamed succession" by Exon et al. (1986), and include strata mapped originally as part of the Jaulu Volcanics and Lossuk River beds by Hohnen (1978) and Brown (1982).

Distribution and Thickness

The Lumis River volcanics crop out only in northwestern New Ireland (Figure 3). This sequence has not been specifically recognized in offshore seismic-reflection profiles.

The sequence is estimated to have a total thickness of at least 800 m. This includes thicknesses of about 100 m for the Kaut unit, 400 m for the Lumis River unit, and a minimum of about 300 m for the Kapsu unit.

Lithology

The Kaut unit is composed mainly of andesitic crystal-lithic tuff, sparsely fossiliferous volcanolithic sandstone, fossiliferous dolomitic limestone, and minor agglomerate. The limestones occur only in the lower part of the unit, and are composed of tuffaceous, algal-foraminiferal biomicrite and biosparite. They contain abundant fragmented and encrusting coralline algae and benthic foraminifera in association with echinoid spines, gastropods, bivalves, and minor planktonic foraminifera. The upper part of the sequence is dominated by pyroclastic rocks, which coarsen upward and eventually grade into Lumis River unit agglomerates.

The Lumis River unit is composed predominantly of massive andesitic agglomerate and lesser volcanic breccia. Rare interbeds of crystal-lithic tuff and tuffaceous, fossiliferous siltstone also occur. West of Kaut Harbor, rhyolitic ignimbrite is interbedded with the agglomerate near a group of presumed former volcanic vents. The upper contact of the Lumis River unit is sharp but apparently conformable.

The Kapsu unit contains interbedded andesitic crystal-lithic tuff, fossiliferous, calcareous, tuffaceous siltstone and sandstone, non-calcareous mudstone, minor lapilli tuff, and rare agglomerate. The lower part of the unit is unfossiliferous, but abundant planktonic foraminifera occur higher in the sequence.

Stratigraphic Relationships and Age

The Lumis River volcanics are lateral equivalents of the upper Lelet Limestone and lower Punam Limestone (Figure 4). They conformably overlie both the Lossuk River beds and the lower Lelet Limestone, and are unconformably overlain by Quaternary alluvium.

The age range of the Lumis River volcanics is earliest middle Miocene to latest Miocene or earliest Pliocene (about N9 to upper N17; Figure 4). Combined calcareous nannofossil and foraminiferal evidence indicates an earliest middle Miocene age (about N9 to N10) for the Kaut unit. An early middle Miocene (N11) foraminiferal assemblage has been recovered from a thin siltstone interbed about 20 m below the top of the Lumis River unit. The total age range of the Lumis unit probably corresponds roughly to the upper part of zone N10 to about zone N12. All productive samples from the upper Kapsu unit have yielded upper N17 foraminiferal assemblages. As the lower contact of the Kapsu unit appears to be conformable, the unit's age range is probably mid-middle Miocene to latest Miocene or earliest Pliocene (about N12 to upper N17).

Depositional Environment

The Lumis River volcanics accumulated in environments ranging from the outer shelf to the continental slope. Foraminiferal assemblages suggest deposition at depths ranging from middle to outer neritic (50-150 m) for the lower Kaut unit, outer neritic to upper bathyal (50-500 m) for the upper Kaut unit and Lumis River unit, and middle to lower bathyal (greater than 500 m) for the upper Kapsu unit. These observations suggest that marked subsidence occurred in the northwestern New Ireland area from the early middle Miocene onwards.

Punam Limestone

Definition

Sapper (1910) used the term "Punam layers" for a sequence of foraminiferal-rich "chalk-like limestones" exposed near Punam Village in central New Ireland (Figure 3). French (1966) revised the name to Punam Limestone, and Hohnen (1978) formally designated a type section at Punam.

The Punam Limestone was assigned a Pliocene and younger age by Hohnen (1978), and was thought by him to unconformably overlie the Rataman Formation. However, we now know that the Punam Limestone conformably underlies the Rataman Formation.

Distribution and Thickness

The Punam Limestone is exposed in fault-bounded blocks along the northeast coast of New Ireland between Cape Pinikindu and Cape Sena (Figure 3). The most extensive exposures occur near Namatanai and Punam. Small, isolated fault blocks are also present in the Cape Mimias area and near Cape St. George. The extensively exposed limestone sequence between Condor Point and Cape St. George, which was originally mapped as Punam Limestone by French (1966) and Hohnen (1978), is now known to be almost exclusively Lelet Limestone (Figure 3).

No complete sections of the Punam Limestone have been located, so no reliable estimates of the limestone's total thickness can be made. The best exposures suggest a minimum thickness of about 200 m, but this figure may be inaccurate because of pervasive mesoscale faulting. The inferred offshore equivalent of the Punam Limestone (B-C sequence) ranges in thickness from about 200 m to 500 m in parts of the southeastern New Ireland Basin (Exon et al., 1986).

Lithology

The Punam Limestone is typically composed of massive, friable, packed foraminiferal biomicrite with a variable, though minor, tuffaceous component. Exposures are commonly finely laminated, and parts of the sequence display ripple cross-lamination and low angle cross-bedding. Convolute laminae and syntaphral faulting are rarely present.

Stratigraphic Relationships, Age and Depositional Environment

The Punam Limestone overlies the Lelet Limestone, and sharply but conformably underlies the Rataman Formation. The probable age range of the sequence is late Miocene to early Pliocene (Figure 4). The oldest foraminiferal assemblages recovered from the Punam Limestone have been referred to zone N17 (probably upper N17). Assemblages belonging to the lower part of the N19/20 planktonic zone have been recovered immediately below the upper contact of the formation.

The Punam Limestone accumulated as a foraminiferal ooze in a middle to lower bathyal environment (water depths more than about 500 m). Occasional influence by bottom currents is indicated by sedimentary structures in parts of the sequence. Pelagic deposition was accompanied by limited airfall input from distant pyroclastic volcanism.

Rataman Formation

Definition

Sapper (1910) initially used the term "Rataman

layers" for a sequence cropping out in an area named the Rataman District by the former German administration. This sequence was renamed the Rataman Formation by Hohnen (1978), who designated a type section near Kokola Plantation west of Ramat Bay. Strata originally assigned by Sapper (1910) and French (1966) to the Tamul beds in southeastern New Ireland were also included in this formation.

Rocks equivalent to those exposed in Hohnen's (1978) type section and nearby reference sections have been recognized throughout central and southern New Ireland, and as far north as Djaul Island. Lateral facies variations are readily apparent outside the type area, so the Rataman Formation has been subdivided into the Ramat, Gogo, Uluputur and Djaul units.

The Ramat unit corresponds to rocks included in Hohnen's (1978) type section, together with those found in reference sections near Komalu Bay, at Rebehen near Cape Erkokon, and in an unnamed river that passes under the Gogo Bridge near Ramat Bay (Figure 3). The contact between this unit and the overlying Gogo unit is visible at the last locality.

The Uluputur unit is named after the Uluputur beds, which were defined originally by Hohnen (1978) from a type section along a tributary to the Sae River north of Uluputur Village. Similar rocks, though of a longer age range, are present in reference sections along the Weilo, Mow and Olisigo rivers of southern New Ireland.

The Djaul unit corresponds to rocks mapped originally by Hohnen (1978) and Brown (1982) as the Lossuk River beds on Djaul Island. Extensively faulted exposures of this unit are present along several creeks in the hilly parts of the island.

Distribution and Thickness

Exposures of the Rataman Formation occur in three discrete areas. The Ramat and Gogo units are exposed in the narrow central portion of New Ireland between Karu Bay and Dolomakas Bay (Figure 3). The Uluputur unit is exposed mainly in a series of fault blocks along the east coast of the island between Cape Sena and Cape Siar, and along the west coast near Cape Wainamarang. The Djaul unit is exposed only on Djaul Island.

Estimation of stratigraphic thicknesses is complicated by a lack of complete exposures and extensive mesoscale fault repetition. The Ramat unit appears to have a minimum thickness of between 400 and 600 m. The minimum thicknesses of the Uluputur and Djaul units are probably similar. The exposed part of the Gogo unit probably does not exceed about 20-30 m in thickness. The inferred offshore equivalent of the Rataman Formation (A-B sequence) is estimated to range in thickness from 500 to 1000 m (Exon et al., 1986).

Lithology

The Ramat unit is composed largely of rhythmically interbedded, resistant, andesitic to dacitic crystal-lithic tuff or tuffaceous sandstone, and recessive, tuffaceous siltstone and mudstone. The last rock type becomes progressively more calcareous upward, and eventually grades into friable, packed planktonic foraminiferal biomicrite. Massive tuff, lapilli tuff and rare volcanic breccia occur near the base of the unit.

The coarser interbeds exhibit a range of sedimentary features, including normal grading, horizontal lamination, small-scale cross-lamination, convolute laminae, dish structures and fluid escape structures. The intervening mudstones are massive and structureless, and contain sparse foraminiferal assemblages.

The overlying Gogo unit is composed of massive, friable, packed foraminiferal biomicrite. Planktonic and deep water benthic foraminifera are abundant in the lower part of the sequence. Isolated exposures, representing the upper part of the unit, contain transported shallow water detritus including algal, coral and molluscan fragments.

The Uluputur unit is a lithologically variable succession containing both fine and coarse-grained epiclastic sediment. Much of the sequence is composed of interbedded calcareous, friable, silty sandstone, lithic sandstone, siltstone, argillaceous biomicrite, muddy or sandy volcanolithic pebble conglomerate, volcanolithic pebbly mudstone, and minor mudstone. Parts of the sequence have a less diverse composition, and consist mainly of monotonously interbedded siltstone, silty sandstone and sandstone. Sedimentary features include horizontal and convolute lamination, small-scale cross-lamination, and syntaphral folding and faulting. Carbonaceous laminae and wood fragments are locally abundant. The fossil assemblage is dominated by planktonic and deep water benthic foraminifera, but transported macrofossil fragments are commonly found either in thin lenses or dispersed throughout some beds.

The upper part of the Uluputur unit characteristically contains massive bedded, poorly sorted, matrix-supported conglomerate in units up to 10 m thick. The conglomerate is interbedded with sandstone units containing moderately abundant planktonic foraminifera. Clasts range from about 5 cm to more than 2 m, and are often incorporated in a sandy, shell-rich matrix. In central New Ireland, the underlying Ramat unit was the main contributor of the clasts. In southern New Ireland, the clasts were derived mainly from the Jaulu Volcanics, with occasional contributions from the Punam and Lelet limestones.

The Djaul unit contains andesitic lava flows in addition to pyroclastic and epiclastic sediment. Stratigraphic relationships are obscured by structural complications, but the lower part of the succession appears to be dominated by interbedded vesicular andesite lava, crystal tuff, and crystal-lithic tuff. Interbeds of tuffaceous foraminiferal siltstone and sandstone appear higher in the sequence. In some exposures, the lavas are absent, and the sediment occurs either alone or interbedded with thick beds of lapilli tuff.

Stratigraphic Relationships and Age

The Rataman Formation ranges from early Pliocene to early Pleistocene in age (lower N19/20 to lower N22; Figure 4). In central and southern New Ireland, the Rataman Formation sharply but conformably overlies the Punam Limestone, and is overlain by both the Maton Conglomerate and neritic carbonate rocks (Qc).

In central New Ireland, the Ramat unit is conformably and gradationally overlain by the Gogo unit (Figure 4). The age of the contact is latest Pliocene or earliest Pleistocene (upper N21). Locally, however, the Gogo unit is absent because of facies change, and is replaced by coarse-grained rocks of the Uluputur unit.

The upper contact of the Gogo unit has not been located. However, the isolated exposures containing derived shallow water detritus may be transition beds between this unit and neritic Qc limestones.

The Uluputur unit of southern New Ireland is a lateral equivalent of both the Ramat and Gogo units. The Uluputur unit is unconformably overlain by the Maton Conglomerate and Qc limestone.

The exposed portion of the Djaul unit is of late Pliocene (N21) age and the lower contact is not exposed. The unit is unconformably overlain by Qc limestone.

Depositional Environment

The Rataman Formation was deposited in a bathyal regime during a major Pliocene volcanic episode. The timing of this event is expressed clearly by the Ramat-Gogo sequence, which suggests an abrupt beginning to pyroclastic volcanism in the early Pliocene, followed by a gradual waning of activity until the latest Pliocene or earliest Pleistocene.

Foraminiferal evidence suggests that the Ramat and Gogo units were deposited in a middle to lower bathyal environment (water depths more than about 500 m). The lithological character and sedimentary structures of the Ramat unit are consistent with deposition mainly by turbidity currents and modified grain flows on the lower part of a submarine fan. The Gogo unit marks a return to solely pelagic deposition before regional uplift.

The Uluputur unit is a regressive sequence deposited mainly in a submarine fan environment. The sequence lacks pyroclastic rocks, and was deposited away from major volcanic centers. Foraminiferal assemblages indicate that most of the unit was laid down in a middle to lower bathyal environment (water depths more than about 500 m). The character of these sedimentary bodies suggests that they represent middle fan channel and interchannel deposits. Only the uppermost part of the sequence accumulated in upper bathyal to outer neritic environments. The coarse-grained conglomerates in this part of the unit are probably channel-fill deposits from the uppermost part of a submarine fan.

Foraminiferal assemblages from the Djaul unit indicate deposition in a middle to lower bathyal regime. The submarine lava flows in the sequence reflect proximity to eruptive centers.

Maton Conglomerate

Definition

The name Maton Conglomerate was used by French (1966) for a widely exposed succession of conglomerate and pebbly sandstone in southern New Ireland. The type section is located along the Maton River (Figure 3).

Distribution and Thickness

The Maton Conglomerate is widely exposed along the west coast of New Ireland between Tambakar Point and Condor Point, and along the east coast between East Cape and Cape Siar (Figure 3). The formation is also exposed in a major graben immediately west of Cape Sena. A thickness of 200-300 m was estimated by French (1966).

Lithology

The Maton Conglomerate is composed of coarse clastic sediment together with minor claystone and coal. Because of extensive faulting and a lack of age control, no stratigraphic sequence has been determined.

The formation is composed dominantly of weakly lithified, sandy, volcanolithic pebble, cobble and boulder conglomerate. The conglomerate contains lenses and beds of coarse-grained volcanolithic sandstone and pebbly sandstone. Thick units of boulder conglomerate and boulder-pebble breccia with a friable sandy clay matrix are interbedded in parts of the sequence. Most of the clasts in these rocks were derived from the Jaulu Volcanics, with minor contributions from the Lelet and Punam limestones.

Exposures along the Tamul River contain fining-upward sequences involving pebble or cobble conglomerate, trough cross-bedded, medium to coarse-grained sandstone, and clayey sandstone. The sandstones contain carbonaceous streaks, wood chunks, and locally abundant coal fragments. These sedimentary units are associated with units of sandy, carbonaceous claystone, which are interbedded with sub-bituminous coal seams ranging from 3 cm to 2 m in thickness.

Stratigraphic Relationships and Age

The Maton Conglomerate unconformably overlies the Uluputur unit, the Lelet Limestone and the Jaulu Volcanics. The upper contact of the conglomerate is not exposed. The formation is a lateral equivalent of neritic Qc limestones.

No fossils have been recovered from this formation. A mid-Pleistocene and younger age is inferred for the Maton Conglomerate on the basis of its stratigraphic position.

Depositional Environment

The Maton Conglomerate was deposited on and adjacent to alluvial fans developed against fault scarps formed during latest Pliocene to Pleistocene regional uplift. The major lithofacies in the formation probably correspond to upper fan channel, subaerial debris flow, and braided stream deposits. The associated claystones and coals may have accumulated in inactive slough channels or in overbank areas in distal parts of the fans.

SEDIMENTATION HISTORY

Southward subduction along the Manus-Kilinailau Trench resulted in island arc volcanism during the Eocene or earliest Oligocene. Throughout the remainder of the Oligocene, coarse-grained pyroclastic rocks and lavas were laid down in proximal areas (Jaulu Volcanics). Brief quiescent periods during the early Oligocene allowed local deposition of neritic clastic sediment and tuffaceous limestone. About the same time, intrusive rocks were emplaced during the initial intrusive phase of the Lemau Intrusive Complex.

By the latest Oligocene or earliest Miocene, regional volcanism effectively had ceased. The arc was faulted, uplifted and eroded, exposing rocks of the Lemau Intrusive Complex. Carbonate sedimentation commenced in shallow water areas over the rugged volcanic substrate (basal Lelet Limestone). Intermittent local volcanism persisted near northwestern New Ireland, and initially impeded carbonate deposition (Bagatere unit). At the same time, predominantly fine-grained epiclastic sediment accumulated seaward of the incipient shelf edge (Lossuk River beds and lower Tamiu siltstone).

By the late early Miocene, remnant positive areas in central New Ireland had become the sites of restricted lagoonal clastic sedimentation (Kimidan unit). As subsidence continued, carbonate sediment encroached over the area. Meanwhile, bathyal clastic deposition persisted in parts of southeastern New Ireland until the late middle Miocene. Under the influence of at least local upfaulting near the shelf edge, these sediments eventually built upward to neritic depths (upper Tamiu siltstone), and were buried by prograding carbonates of the Lelet Limestone.

In northwestern New Ireland, deposition of the Lelet Limestone and Lossuk River beds ended in the earliest middle Miocene because of a local resurgence of volcanism. This event was probably genetically related to the younger intrusive phase of the Lemau Intrusive Complex. Initial activity resulted in the deposition mainly of tuffs and related sediment (Kaut unit), and culminated in the accumulation of a thick agglomerate sequence (Lumis River unit). Ignimbrite flows were laid down adjacent to eruptive centers west of Kaut Harbor. Volcanism waned from the late middle Miocene onward, but combined epiclastic and pyroclastic deposition continued at least until the latest Miocene (Kapsu unit). This volcanic episode coincided with marked local subsidence, which ultimately depressed the northwestern New Ireland area to middle or lower bathyal depths.

In central and southern New Ireland, carbonate platform sedimentation continued unhindered by the volcanism to the northwest, and probably persisted into the late Miocene in most areas. About this time, reversal of the New Ireland portion of the original island-arc system took place. This event appears to have coincided with rapid regional subsidence, which eventually drowned the carbonate platform. Bathyal pelagic carbonate sedimentation (Punam Limestone) ensued, accompanied by only minimal input from distant pyroclastic volcanism.

Volcanism recommenced on a regional scale during the early Pliocene, probably in response to subduction along the New Britain Trench. At that time New Ireland still formed part of the newly-formed, southwest-facing island arc system. This major volcanic episode is recorded by bathyal sediment deposition of the Rataman Formation, which accumulated on submarine fan systems adjacent to relative highs. The Djaul Island region was situated near the eruptive centers, and received both lava flows and coarse pyroclastic sediment (Djaul unit). The Pliocene paleogeographic position of the island relative to New Ireland is not known, as Djaul Island may have been transported a considerable distance during subsequent transform faulting. The central New Ireland area received substantial quantities of tuff, which were redistributed by turbidity currents and modified grain flows to the distal parts of submarine fans (Ramat unit). As volcanic activity waned during the latest Pliocene, pelagic carbonate sedimentation returned to this area. Meanwhile, epiclastic sediment accumulated in middle to upper submarine fan environments in distal areas to the south. By the latest Pliocene or earliest Pleistocene, proximal parts of the fan were receiving large quantities of coarse-grained detritus shed from older stratigraphic units. These sedimentary beds may have been derived initially from newly-formed submarine scarps, marking the inception of the major tectonic movements responsible for the emergence of present-day New Ireland.

The remainder of the Pleistocene was marked by rapid regional uplift of the southwestern margin of the New Ireland Basin. This event coincided with northwestward displacement of New Ireland relative to New Britain along a transform fault system. By about mid-Pleistocene time, the northern and central parts of New Ireland had been uplifted to neritic depths, allowing deposition of fringing reefs and associated shelf sediment over wide areas (Qc limestone). In southern New Ireland, vertical movements were apparently more pronounced. The sedimentary cover rocks were rapidly stripped away, exposing the Jaulu Volcanics. Thick accumulations of clastic detritus were deposited on alluvial fans adjacent to active fault scarps (Maton Conglomerate). These tectonic movements had little effect on areas further northeast, where dominantly pelagic sedimentation has continued until the present day.

ACKNOWLEDGMENTS

We thank the referees, H.A. Davies and C.M. Brown from the Bureau of Mineral Resources, for their comments on the paper, which has been considerably strengthened as a result of their input.

REFERENCES

Brown, C.M., 1982, Kavieng, Papua New Guinea, geological series - explanatory notes: Geological Survey of Papua New Guinea, Sheet SA/56-9, scale 1:250,000.

C.R.A. Exploration Party Ltd., 1965, Final report on prospecting activities in Special Prospecting Authorities 15-29 inclusive and 39 - New Guinea Islands: C.R.A. Exploration Party Ltd. Report (unpublished), Geological Survey of Papua New Guinea Open File Report, 2 p.

Dalrymple, G.B., 1979, Critical tables for conversion of K-Ar ages from old to new constants: Geology, v. 7, n. 11, p. 558-560.

Exon, N.F., and Tiffin, D.L., 1984, Geology and petroleum prospects of offshore New Ireland Basin in northern Papua New Guinea, in Watson, S.T., ed., Transactions of the Third Circum-Pacific Energy and Mineral Resources Conference, p. 623-630.

Exon, N.F., Stewart, W.D., Sandy, M.J., and Tiffin, D.L., 1986, Geology of the eastern New Ireland Basin, northeastern Papua New Guinea: Bureau of Mineral Resources Journal of Australian Geology and Geophysics, v. 10, n. 1, p 39-51.

Falvey, D.A., and Pritchard, T., 1984, Preliminary paleomagnetic results from northern Papua New Guinea: evidence for large microplate rotations, in Watson, S.T., ed., Transactions of the Third Circum-Pacific Energy and Mineral Resources Conference, p. 593-599.

French, D.J., 1966, The geology of southern New Ireland: Bureau of Mineral Resources Australia, Record 1966/179, 30 p.

Hohnen, P.D., 1978, The geology of New Ireland: Bureau of Mineral Resources Australia, Bulletin 176, 39 p.

Jaques, A.L., 1980, Admiralty Islands, Papua New Guinea, geological series - explanatory notes: Geological Survey of Papua New Guinea, Sheets SA/55-10 and SA/55-11, scale 1:250,000.

Johnson, R.W., 1979, Geotectonics and volcanism in Papua New Guinea: a review of the late Cainozoic: Bureau of Mineral Resources Journal of Australian Geology and Geophysics, v. 4,

p. 181-207.

Mitchell, P.A., and Weiss, T.V., 1982, Prospecting Authority 485, New Ireland, Papua New Guinea. Final report: Esso Papua New Guinea Inc. report (unpublished), Geological Survey of Papua New Guinea Open File Report, 31 p.

Ripper, D., and Grund, R., 1969, The geology of Permit 48 (Territory of Papua and New Guinea): Continental Oil Company Australia Ltd. Report (unpublished), Geological Survey of Papua New Guinea Open File Report, 23 p.

Sapper, K., 1910, Beitrag zur Landeskunde von Neu-Mecklenburg und seinen Nachbarinseln: Mitteilungen deutsche Schutzgebiete Ergebnisse, v. 3, p. 1-130.

Steiger, R.H., and Jager, B., 1977, Subcommission on geochronology: convention on the use of decay constants in geo and cosmochronology: Earth and Planetary Science Letters, v. 36, p. 359-362.

Stewart, W.D., and Sandy, M.J., 1986, Cenozoic stratigraphy and structure of New Ireland and Djaul Island, New Ireland Province, Papua New Guinea (with micropaleontological appendices by D.W. Haig, P.J. Coleman and H.K. Hekel): Geological Survey of Papua New Guinea Report 86/12, 96 p.

Taylor, B., 1979, Bismarck Sea: evolution of a back-arc basin: Geology, v. 7, n. 4, p. 171-174.

Wallace, D.A., Chappell, B.W., Arculus, R.J., Johnson, R.W., Perfit, M.R., Crick, I.H., Taylor, G.A.M., and Taylor, S.R., 1983, Cainozoic volcanism in the Tabar, Lihir, Tanga and Feni islands, Papua New Guinea: geology, whole-rock analyses, and rock-forming mineral composition: Bureau of Mineral Resources Report 243, BMR Microform MF 197.

Marlow, M.S., Dadisman, S.V., and Exon, N.F., editors, 1988, Geology and offshore resources of Pacific island arcs—New Ireland and Manus region, Papua New Guinea, Circum-Pacific Council for Energy and Mineral Resources Earth Science Series, v. 9: Houston, Texas, Circum-Pacific Council for Energy and Mineral Resources.

STRATIGRAPHY OF MANUS ISLAND, WESTERN NEW IRELAND BASIN, PAPUA NEW GUINEA

Geoffrey Francis
Geological Survey of Papua New Guinea, Port Moresby, Papua New Guinea

ABSTRACT

Although recent geological studies of Manus Island have confirmed the general stratigraphic succession established by earlier 1:250,000 mapping, they have led to some significant changes in our understanding of the island's stratigraphy. The island is underlain by middle Eocene to earliest Miocene island-arc basement: the Tinniwi Volcanics. The newly-named "Louwa unit", of early Miocene age, is limited in surface extent and consists of bathyal calcareous siltstone and biomicrite. The andesitic Tasikim Agglomerate is late(?) early Miocene and middle Miocene in age and is widely distributed in western, northern and southeastern Manus. The Mundrau Limestone, of late early Miocene to earliest middle Miocene age, onlaps the Tinniwi Volcanics in the central part of the island. It consists largely of algal-foraminiferal biomicrite and biomicrudite, and formed a carbonate platform. The calcalkaline subvolcanic Yirri Intrusive Complex is an early to middle Miocene batholith in central Manus, intruding Tinniwi Volcanics, Louwa(?) unit and Mundrau Limestone.

The late Miocene Lauis Formation consists mainly of bathyal siltstone, sandstone, conglomerate and tuff, and is more extensive in the east than the west. It includes the "Rambutyo Beds" of earlier mappers. The "Lorengau basalt" appears to be interbedded with the Lauis Formation on eastern Manus. The early Pliocene Naringel Limestone occurs as raised reefs or carbonate banks, largely on Los Negros Island in the east. The Likum Volcanics form a Pliocene-Pleistocene caldera on southwestern Manus Island.

INTRODUCTION

Manus Island is located at the western end of the Outer Melanesian Island Arc and is the largest island in the Admiralty Islands group (Figure 1). Its geological basement consists of island arc andesitic and minor basaltic volcanic rocks (Tinniwi Volcanics) which are as old as Eocene and are thought to overlie early Paleogene oceanic crust. The basement is overlain by Neogene strata here considered to be part of the New Ireland Basin, a mainly offshore fore-arc basin lying between the Outer Melanesian Arc and the Manus (West-Melanesian) Trench. Early reconnaissance geological traverses of Manus were carried out by Thompson (1952). Subsequently, more detailed studies of copper prospects associated with the Yirri Intrusive Complex were conducted by Exoil, International Mining Corporation (IMC) and Australian-Anglo-American in 1969-1975 (Exoil, 1973; Kirwin, 1974; Australian-Anglo-American, 1975).

Regional mapping of the island by the Geological Survey of Papua New Guinea in 1973-74 resulted in the production of 1:100,000 scale (Jaques, 1975) and 1:250,000 scale (Jaques, 1976; 1980) geological maps. Detailed mapping was confined to areas of central Manus within or adjacent to the Yirri Intrusive Complex, whereas areas of stratified rocks

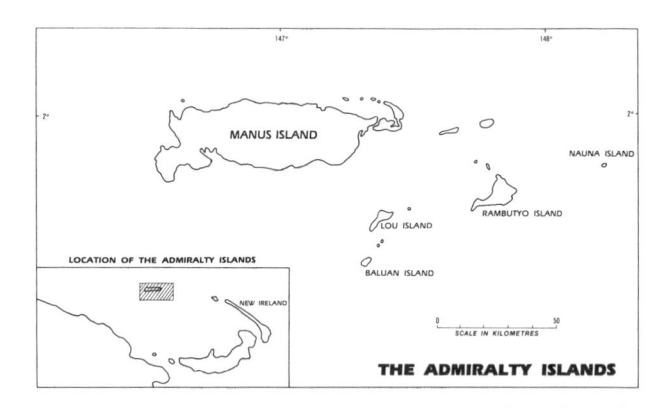

Figure 1. Location of the Admiralty Islands.

in western, northern and eastern Manus were mapped only at reconnaissance level (Jaques, 1975). This reconnaissance mapping concentrated mainly on establishing the gross distribution of rock units rather than on determining stratigraphic boundary relationships. Such relationships were often inferred from macroscale rock distribution patterns, rather than from actual observations of contacts or dating of samples from horizons close to lithostratigraphic boundaries.

More detailed geological and geomorphological studies of limestone areas in central and eastern Manus were carried out by Francis in 1975, but only preliminary results have been published (Francis, 1975, 1977). Recent work includes further exploration of porphyry copper-gold deposits (Rosengren, 1982), paleomagnetic studies (Nion, 1980; Falvey and Pritchard, 1984) and engineering geological studies of southwestern Manus (King et al., 1984). In 1983 the Geological Survey commenced work on the stratigraphy and petroleum potential of the New Ireland Basin, with fieldwork on Manus being undertaken by G. Francis, B. Pawih and M.J. Sandy. This study has confirmed the gross rock distribution patterns and the general nature of the stratigraphic succession established by Jaques (1980). However, new observations of stratigraphic relationships and micropaleontological age determinations indicate that some interpretations of stratigraphic relationships proposed by earlier workers such as Jaques (1980) need revision. Because micropaleontological analyses of samples by D.W. Haig and H.K. Hekel have not been completed, this paper is a report of work in progress rather than a finished product.

The generalized geology of Manus Island, modified after Jaques (1980), is shown in Figure 2, and time-stratigraphic relationships between rock units are presented in Figure 3, which also shows microfossil zones. K-Ar ages referred to in the text and used for the construction of Figure 3 have been converted to the constants of Steiger and Jager (1977), using the critical tables of Dalrymple (1979). Correlations between foraminiferal ages and isotopic ages are based partly on recent work by Chaproniere (1981), Adams (1984) and Haig (1985), and thus differ in some respects from those accepted by Jaques (1980). Grid references mentioned in the text refer to the 1:100,000 Lorengau (8393) and Malai (8293) topographic sheets published by the National Mapping Bureau of Papua New Guinea and the Royal Australian Survey Corps.

DESCRIPTIONS OF STRATIGRAPHIC UNITS

Tinniwi Volcanics

Definition, Distribution and Thickness

The Tinniwi Volcanics form the basement that underlies the Neogene sedimentary sequence on Manus Island, and are extensively distributed throughout central Manus (Figure 2). Their thickness is difficult to estimate because of pervasive faulting, but is likely to be at least 700 m.

Lithology

The Tinniwi Volcanics are composed of hornblende and pyroxene andesite, basaltic andesite and basalt flows, lava breccias and pyroclastics, with some interbeds of argillaceous tuffaceous lithic sandstone, tuffaceous siltstone and foraminiferal(?) biomicrite (Kirwin, 1974; Jaques, 1980). Petrographic studies of the volcanics by Jaques (1980) indicate that they are subalkaline rocks having affinities with both island-arc tholeiites and low-K calcalkaline lineages. The pyroclastics include agglomerate, lapilli tuff and medium-grained lithic-crystal tuff. Both lavas and pyroclastics are strongly indurated, and are often altered, containing veins of secondary epidote and calcite.

Stratigraphic Relationships and Age

The Tinniwi Volcanics probably overlie early Paleogene oceanic crust and are unconformably overlain by early or middle Miocene Mundrau Limestone. Webb (1974) obtained middle Eocene K-Ar ages of 49.0 ± 5.0 Ma and 45.8 ± 5.0 Ma for samples collected by Jaques (1980) along the Leiwa River. Two

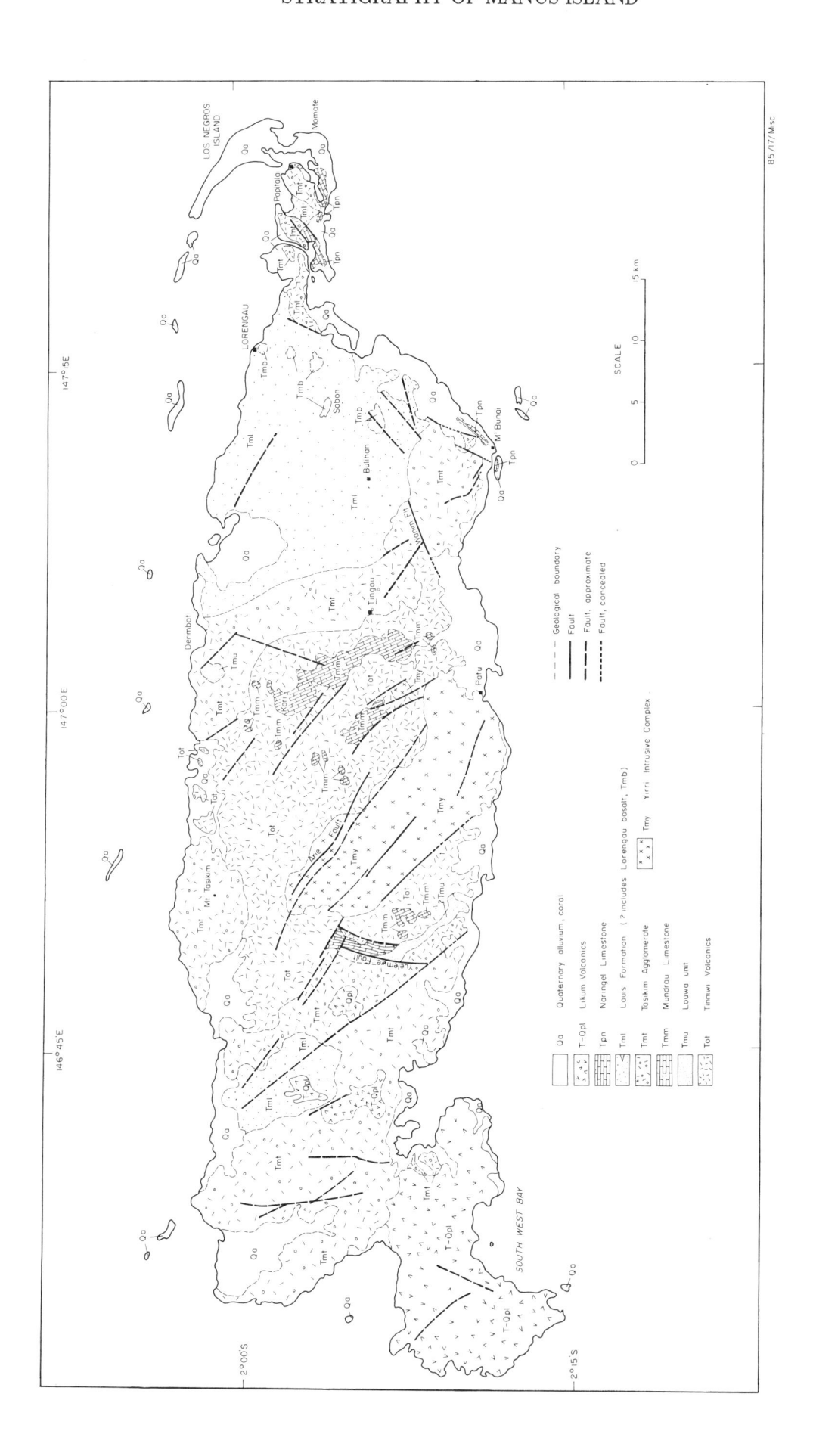

Figure 2. Generalized geology of Manus Island.

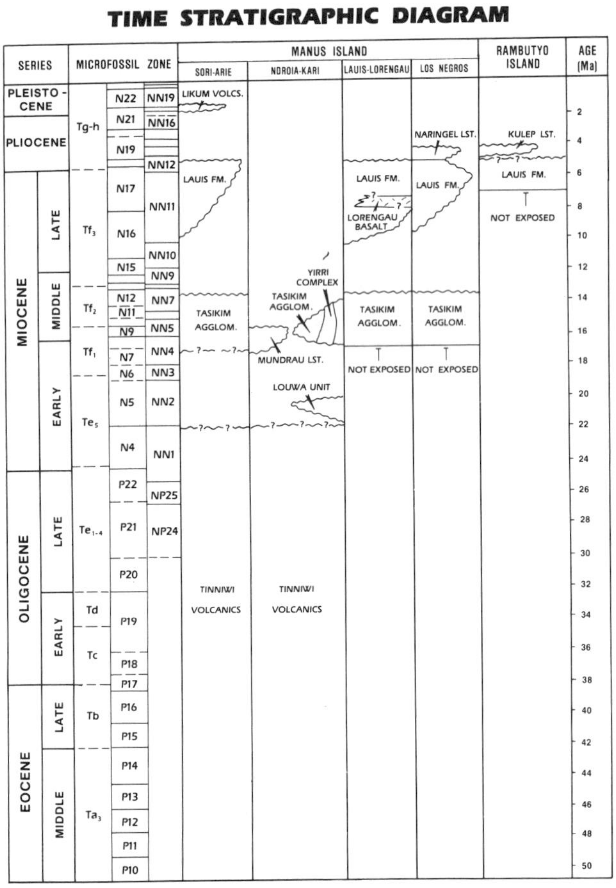

Figure 3. Time stratigraphic diagram, Manus and Rambutyo Islands.

additional K-Ar ages of 23.8 ± 0.5 Ma and 20.7 ± 0.8 Ma were determined for samples from structurally disturbed areas adjacent to the Tinniwi and Yuelemwe faults, which Jaques (1980) assigned to the Tinniwi Volcanics. It seems likely from relative ages that the first of these samples came from the Tinniwi Volcanics, but the latter sample could be basal Tasikim Agglomerate. The problem of the stratigraphic relationships between the Tinniwi Volcanics and the Tasikim Agglomerate is discussed in more detail below.

In the headwaters of the Wamatja River, andesitic pyroclastics occur in association with belts of partly recrystallized grey algal-foraminiferal biomicrite which were considered by Kirwin (1974) and Jaques (1980) to be interbeds within the Tinniwi Volcanics. However, Dr. G.P. Robinson identified *Orbulina* sp. in a sample collected by the writer from this locality in 1975, suggesting that the biomicrite is no older than middle Miocene (N9). Either the pyroclastics here are Tasikim Agglomerate or the biomicrite is infaulted Mundrau Limestone.

The Tinniwi Volcanics are intruded by batholiths and stocks of the early to middle Miocene Yirri Intrusive Complex which in some cases have extensively metamorphosed the volcanic rocks. Webb (1973) determined a K-Ar age of 30.8 ± 0.4 Ma for a hornblende diorite intruding the Tinniwi Volcanics. It seems likely that this Oligocene intrusion is magmatically related to the Tinniwi Volcanics rather than to the younger Yirri Complex. Collectively the evidence suggests that the Tinniwi Volcanics range from middle Eocene to earliest Miocene in age.

Louwa Unit (new informal unit)

Definition, Distribution and Thickness

This new stratigraphic unit has a type area in the valley of the Louwa River, northern Manus (G.R. 030796, Lorengau). It was previously included in the Mundrau Limestone by Jaques (1980). The thickness of the Louwa unit is difficult to determine because of poor outcrop and faulting but is likely to be at least 80 m. The unit has been definitely recorded only in areas adjacent to the type locality. However, sediment mapped by Jaques (1980) as "Tingau Conglomerate" along the Yuelemwe River could be Louwa unit. The Louwa unit could be more extensive in the subsurface, but no drillholes have penetrated the base of the overlying(?) Tasikim Agglomerate.

Lithology

The Louwa unit consists of bathyal grey tuffaceous calcareous siltstone and sandy siltstone, grey to pinkish-grey argillaceous algal-foraminiferal biomicrite and mottled white/grey algal-foraminiferal biomicrite. Occasional diamictites are present. The argillaceous limestones are carbonate turbidites with mixed assemblages of inner neritic and bathyal foraminiferids, and occur as interbeds in a fine clastic sequence. The mottled biomicrites contain an exclusively inner neritic fauna and are probably olistoliths. However, some could be local carbonate buildups associated with rugged paleobathymetric highs.

Stratigraphic Relationships and Age

Micropaleontological determinations by D.J. Belford and G.C.H. Chaproniere (Jaques, 1980), by

D.W. Haig (written comm., 1985) and H.K. Hekel (1985) indicate that the Louwa unit is of early Miocene (Te5 and N5) age. Stratigraphic relationships with the adjacent Tasikim Agglomerate cannot be determined with certainty because of poor outcrop and faulting, but their relative ages suggest that the Louwa unit probably underlies the Tasikim Agglomerate and overlies the Tinniwi Volcanics. The Louwa unit is similar in facies to but slightly older than the N8 Lossuk River beds (Stewart and Sandy, this volume) of northwestern New Ireland.

Mundrau Limestone

Definition, Distribution and Thickness

The type locality of the Mundrau Limestone is here designated as Poworei, a major stream sink of the Worei River in south central Manus (G.R. 050662, Lorengau). The limestone occurs mainly in a belt extending northwest from Poworei to Kari, but is also found in the headwaters of the Tinniwi River and along the Yuelemwe River. It has a maximum residual thickness of 220 m but the thickness is extremely variable because of onlap of the Mundrau Limestone onto rugged basement.

Lithology

The Mundrau Limestone is composed of massive to crudely bedded inner neritic white to grey algal-foraminiferal biomicrite, biomicrudite and fossiliferous micrite, with rare bioturbated dismicrite. In places the limestone has been extensively recrystallized to form stylobedding and stylobreccias. Calcareous volcanolithic pebble-granule conglomerate and pebbly sandstone occur at the base of this unit near Kari. The limestone is dense and compact with porosities generally 1-3% but occasionally as much as 7%.

Stratigraphic Relationships and Age

The Mundrau Limestone unconformably overlies the Tinniwi Volcanics, showing pronounced onlap on rugged basement highs possibly formed by penecontemporaneous faulting. On Tawi the volcanics rise 150 m above the top of the adjacent Mundrau Limestone. The onlap margin of the limestone northwest of this locality dips at about 40° to the north, whereas the limestone dips at about 15° north.

Foraminiferal data on the Mundrau Limestone (Kicinski and Belford, 1956; Chaproniere, 1973; D.J. Belford, oral comm., 1984), reassessed in the light of recent work on benthic foraminiferal age ranges by Adams (1984), suggest a late/early Miocene or earliest middle Miocene (Tf1) age for the formation. Sample 73290540 from the lower part of the limestone at its type locality contains a *Miogypsina thecidaeformis* - *Marginopora vertebralis* assemblage which is latest early Miocene/earliest middle Miocene (upper Tf1), whereas sample 73290501 from Mataworei contains a *Miogypsina thecidaeformis* - *Austrotrillina howchini* assemblage which is mid or upper Tf1. In addition, the range of overlap for the less specific or more tentative ages is also within Tf1. Samples R5952 and R5954 containing *Miogypsina kotoi* are Tf1 or lower Tf2, and tentative identifications of *Miogypsina thecidaeformis* in 73290075 and 73290535 suggest that these samples are probably no younger than Tf1. The Mundrau Limestone is in part a lateral equivalent of the Lelet Limestone (Stewart and Sandy, this volume) of New Ireland.

Tasikim Agglomerate

Definition, Distribution and Thickness

The type locality for the Tasikim Agglomerate is around Mt. Tasikim in northern Manus (Jaques, 1980). The unit is widely distributed in western, northern and eastern Manus, with a maximum residual thickness of about 400 m. The Tasikim Agglomerate here includes most of the "Tingau Conglomerate" and the older phase of the "Lorengau Basalt" in the sense of Jaques (1980).

Lithology

This unit consists of andesitic and basaltic agglomerate, lapilli tuff and tuff, with occasional interbeds of hornblende andesite lava, volcanolithic conglomerate, tuffaceous sandstone and siltstone.

Stratigraphic Relationships and Age

K-Ar hornblende ages of 15.3 ± 0.7 Ma and 13.9 ± 0.7 Ma have been determined by A.W. Webb for samples from the Tasikim Agglomerate as defined by Jaques (1980). Jaques (1980) considered that the Tasikim Agglomerate overlies the late Miocene to earliest Pliocene Lauis Formation, and that the middle Miocene K-Ar ages were anomalous. He apparently based this conclusion on the occurrence of the Tasikim Agglomerate as topographic highs rising above adjacent lowlands developed on the Lauis Formation. However, it is unlikely that K-Ar analyses on hornblende would yield ages that are anomalously old, and the topographic relationships can equally well be explained in terms of a younger Lauis Formation onlapping highs in the Tasikim Agglomerate. Such relationships occur on Los Negros Island and to the east of Lorengau, where latest Miocene (N17) Lauis Formation forms topographic lows and unconformably onlaps ridge-forming middle Miocene volcanic rocks (see below). For these reasons the Tasikim Agglomerate is here considered to pre-date the Lauis Formation and to be largely of middle Miocene age. It is probably the extrusive equivalent of one or more phases within the early to middle Miocene (18-13 Ma) Yirri Intrusive Complex of central Manus. There is overlap in the geochemical compositions of the Tasikim Agglomerate and the Yirri Intrusive Complex, but the latter complex spans a wider compositional range (A.L. Jaques, written comm. 1986).

Unfortunately there are no age determinations for the lower part of the Tasikim Agglomerate, and no unfaulted, well-exposed contact with the older Tinniwi Volcanics or Louwa unit has yet been found. Thus the age range of the Tasikim Agglomerate is uncertain, though the ages for the magmatically (?)-related Yirri Complex suggest that the Tasikim Agglomerate might range back into the latest early Miocene. It is likely that the Tasikim Agglomerate is, at least in part, a lateral equivalent of the Mundrau Limestone, and that the tuffaceous andesitic detritus in the basal Mundrau Limestone at Kari was derived from penecontemporaneous Tasikim volcanism. In southwestern Manus the Tasikim Agglomerate is unconformably overlain by Pliocene-Pleistocene Likum Volcanics.

Jaques (1980) mapped an extensive area in northeastern Manus as "Tingau Conglomerate", on the basis of three reconnaissance traverses: in the headwaters of the Tingau River, along the Drangot River, and near Derimbat. At its type locality on the Tingau River, the unit consists of volcanolithic conglomerate and coarse tuffaceous sandstone, with occasional interbeds of calcareous tuffaceous siltstone. He considered that the unit was in part of middle Miocene age. Additional traverses in this region since 1974 have revealed that much of it consists of coarse pyroclastics with subordinate interbedded volcaniclastics, these sequences being more similar in character to the type Tasikim Agglomerate than to the type "Tingau Conglomerate". There does not appear to be any significant difference in topography or photo-pattern between the pyroclastics and the volcaniclastics. In this paper most of the "Tingau Conglomerate" is included within the Tasikim Agglomerate, as currently available data do not allow it to be consistently or reliably mapped as a distinct unit. Another belt of "Tingau Conglomerate" was mapped by Jaques along the Yuelemwe River. However, photographs and field descriptions of these rocks by Dr. G.P. Robinson suggest that they are similar to the Louwa unit and on this basis they are here tentatively mapped as Louwa unit (Figure 2). However, more precise micropaleontological dating is required to confirm this correlation.

The Tasikim Agglomerate is a lateral and facies equivalent of the Kaut and Lumis River units (Stewart and Sandy, this volume) of northeastern New Ireland.

Yirri Intrusive Complex

The Yirri Intrusive Complex forms a major northwesterly trending batholith in central Manus and a few smaller stocks in areas farther to the north and east. It is a calcalkaline subvolcanic complex composed mainly of microdiorite and micromonzonite with some quartz diorite and monzonite, monzodiorite, tonalite and microtrondhjemite (Kirwin, 1974; Jaques and Webb, 1975). The Dremsel alunitic phase of this complex is composed of brecciated quartz micromonzonite and quartz-alunite-zeolite breccia that intrude magmatically-related andesitic pyroclastics and lavas (Jaques, 1980).

Stratigraphic Relationships and Age

K-Ar ages ranging from 18.1 ± 0.7 to 10.4 ± 0.8 Ma have been determined for rocks from this complex by Jaques and Webb (1975). However, these workers consider that some of the ages are anomalously young because of hydrothermal alteration, and on this basis it seems likely that the age range of the Yirri Complex is 18-13 Ma (early to middle Miocene). The Yirri Complex intrudes the Tinniwi Volcanics and in places intrudes overlying carbonates of the Mundrau Limestone or Louwa unit. It probably includes the intrusive equivalents of the Tasikim Agglomerate.

The Yirri Intrusive Complex is a correlative of the younger (18-14 Ma) phase of the Lemau Intrusive Complex (Stewart and Sandy, this volume) of New Ireland. Hypabyssal intrusives in the Lavongai Volcanics of New Hanover (Brown, 1982) are similar in lithology and hydrothermal alteration to those in the Yirri Complex and might be related to the same phase of magmatic activity.

Lauis Formation

Definition, Distribution and Thickness

The type section of the Lauis Formation designated by Jaques (1980) extends along the Lauis River to the north of Bulihan. Jaques (1980) estimated the thickness of this formation to be up to 4500 m, but recent work with more precise micropaleontological age control suggests that its maximum residual thickness onshore is about 500 m. The Lauis Formation has an extensive distribution on eastern Manus, and occurs in more restricted areas of northwestern and southwestern Manus. Preliminary results from the 1984 R/V *S.P. Lee* cruise (Belford, this volume) suggest that the Lauis Formation has an extensive offshore distribution to the north and southeast of Manus, as samples assigned to zone N18 with lithologies typical of the Lauis Formation were dredged from sites DR 5 and DR 6A. The Lauis Formation is now considered to include the "Rambutyo Beds", a unit of the same facies and age recognized by Jaques (1980) on Rambutyo Island. If these beds were to be retained as a distinct unit it would be necessary to postulate an arbitrary offshore boundary that really reflected a lateral change in nomenclature.

Lithology

This formation is composed of well-bedded grey calcareous tuffaceous siltstone and sandstone, volcanolithic pebble conglomerate and lithic-crystal tuff, with occasional foraminiferal biomicrites, basalt flows(?) and dikes (Jaques, 1980). Locally it contains carbonaceous material and inner neritic molluscs, but these have been transported into a bathyal environment (D.W. Haig, written comm., 1985; D.J. Belford, oral comm., 1985). Graded bedding, laminites, load casts and other turbidite features are present.

Stratigraphic Relationships and Age

Planktonic foraminiferal and calcareous nannofossil dating indicate that the Lauis Formation is of late Miocene to earliest Pliocene (N16-N18; NN11-NN12) age. Jaques (1980) considered that the Lauis Formation was in part a lateral equivalent of the older volcanics in the "Lorengau Basalt" and was locally overlain by younger basalts of this unit. The type area for the older phase of the "Lorengau Basalt" in the sense of Jaques (1980) was in a fault zone along the Wanim River in southeastern Manus. A K-Ar age of 13.9 ± 0.8 Ma was determined for a sample collected by Jaques (1980) about 4 km east of the type area, and a K-Ar age of 13.5 ± 0.2 Ma has recently been determined by Webb (1985) for a sample from Papitalai near the eastern end of Manus. However, new exposures created by the construction of the Tingau-M'bunai road in 1975-1976, suggest that it is composed mainly of agglomerate and lapilli tuff similar to that of the Tasikim Agglomerate, and the K-Ar ages determined for it do not differ significantly from those determined from the Tasikim Agglomerate. For these reasons the older phase of the "Lorengau basalt" is here included in the Tasikim Agglomerate.

At Sabon, Warembu and Lorengau, fine clastics of the Lauis Formation are overlain by plagioclase-phyric basalts which Jaques (1980) included within the "Lorengau Basalt". The type locality designated for this younger phase of the basalt lies on the left bank of the Lorengau River adjacent to its mouth. K-Ar plagioclase ages of 8.8 ± 0.6 Ma and 8.1 ± 0.5 Ma have been determined for samples of the younger basalt collected by Jaques (1980). In its type locality the younger basalt overlies the Lauis Formation with an apparently conformable contact, the sequence dipping offshore at about 12° in a northeasterly direction. However, in other localities such as the Lauis River southwest of Sabon, the Lauis Formation occurs topographically well above adjacent basalt, and although the contacts are poorly exposed there is no evidence of faulting. On this basis Braybrooke and Pieters (1969) concluded that the basalt disconformably overlies the Lauis Formation, locally infilling erosional depressions.

An alternative explanation for these topographic relationships is that the basalt is interbedded with the Lauis Formation. The following evidence supports this interpretation :

(i) At the mouth of the Lorengau River the top of the basalt is concealed beneath late Quaternary littoral deposits. However, Lauis Formation crops out on an intertidal platform to the east of the river mouth. If these beds continued along trend beneath the Quaternary cover then they would overlie the basalt.
(ii) The youngest microfossil ages determined for the Lauis Formation are earliest Pliocene (N18 and NN12), which is significantly younger than the 8.2 Ma and 7.6 Ma younger uncertainty limits on the K-Ar ages. If the basalt were really a younger unit disconformable on the Lauis Formation, these anomalous ages would have to be caused by excess radiogenic argon in the plagioclase.

For these reasons it is here considered that the younger basalts are probably interbedded with the Lauis Formation, but poor outcrop makes it unlikely that this could be proven conclusively without drilling or magnetic surveys. Hence in this paper "Lorengau basalt" is used as an informal name for the late Miocene plagioclase-phyric basalts associated with the Lauis Formation. Jaques (1980) also believed that these basalts were in part a lateral equivalent of the Lauis Formation.

The Matanalaua Formation of New Hanover (Brown, 1982) and the Kapsu unit of northwestern New Ireland (Stewart and Sandy, this volume) are facies equivalents and at least in part time equivalents of the Lauis Formation.

Naringel Limestone

Definition, Distribution and Thickness

The Naringel Limestone occurs as a series of raised reefs or carbonate banks on the eastern extremity of Manus, on Los Negros Island, and at M'bunai in southeastern Manus. Its type locality is Lolak on Los Negros Island (G.R. 417706, Lorengau), where an incomplete thickness of 30 m is exposed in Lolak quarry and in Nge-Pelimat Cave. The maximum residual thickness of this limestone is 80 m.

Lithology

The Naringel Limestone consists mainly of massive to crudely bedded white to pinkish-grey algal-foraminiferal biomicrite, algal-foraminiferal-coralline biomicrudite and transitional biomicrudite/biosparrudite, with occasional coralgal biolithite and red-brown calcareous, argillaceous biogenic dolomite or recrystallized limestone. Intergranular porosities of rock chips generally range from 5 to 13% but there is considerable mesoscale mouldic and vugular porosity. The scarcity of *in situ* colonial organisms suggests that either there has been extensive destruction of the original framework by boring organisms, or the limestone bodies were largely banks protected by seagrass rather than wave-resistant coralgal frameworks.

Stratigraphic Relationships and Age

The Naringel Limestone unconformably overlies the Tasikim Agglomerate and overlies the Lauis Formation. The latter contact was considered to be unconformable by Jaques (1980) from macroscale rock distribution patterns, but more recent work has revealed that some areas mapped as Naringel Limestone by Jaques (1980) are actually Tasikim Agglomerate. At Papitalai the contact is a conformable and gradational one, with several meters of calcareous ferruginous mudstone and ferruginous argillaceous biomicrite occurring between the Lauis Formation and the Naringel Limestone. A preliminary foraminiferal determination by D.W. Haig (written comm., 1985) indicates that the passage beds are earliest Pliocene (N18 or lower N19).

Deposition of the Naringel Limestone was initiated as marginal reefs or banks on the flanks of northeasterly trending coastal highs of Tasikim Agglomerate. However, a minor transgression during Naringel deposition led to the development of patch and barrier structures up to one kilometer seaward from the volcanic highs. These structures subsequently became emergent, and now form low ridges separated from the volcanic hinterland by late Quaternary lagoons and alluvial plains. Because of the dominant inner neritic facies, no age-diagnostic foraminiferids have been found in the Naringel Limestone. From the age determination for the passage beds, and the well-preserved reef or bank morphology suggesting deposition during a single transgressive phase, the limestone is here considered to be of early Pliocene age. This is somewhat older than the Pliocene-Pleistocene or early Pleistocene age postulated by Jaques (1976) and Francis (1977).

The Kulep Limestone of Rambutyo and Nauna Islands is lithologically very similar to the Naringel and is probably a lateral equivalent. However inner neritic carbonates of this type are likely to be confined to the flanks of fault-controlled highs, and bathyal equivalents of the Naringel Limestone are probably present in more basinal offshore areas. The dredging of a later Pliocene (N21) tuffaceous mud from site DR 6A by the R/V *S.P. Lee* (Belford, this volume) suggests that the Lauis Formation is here overlain by a veneer of Rataman Formation.

Likum Volcanics (revised name)

Definition, Distribution and Thickness

The type area for this formation is Southwest Bay, which is probably a caldera (Jaques, 1980). Jaques (1980) named the formation "Likum Basalt", but recent work by King et al. (1984) has shown that it also contains intermediate flows and abundant pyroclastics of variable composition. For these reasons the formation is here renamed "Likum Volcanics". The thickness of the volcanics in their type area is about 200 m, and thinner cover occurs at Lepatuan to the northeast.

Lithology

The Likum Volcanics consist of vesicular

olivine basalt and mafic andesite, dacite, vitrophyre and pitchstone with vitric-crystal tuff and lapilli tuff. The basalts are mildly alkaline and transitional varieties (Jaques, 1980, 1981).

Stratigraphic Relationships and Age

The Likum Volcanics unconformably overlie the Lauis Formation and Tasikim Agglomerate. A K-Ar age of 1.79 ± 0.3 Ma has been determined for a flow from these volcanics (Jaques, 1980) and the uncertainty on this age straddles the Pliocene/Pleistocene boundary as defined by Aguirre and Pasini (1985).

SEDIMENTATION HISTORY

The Manus segment of the Outer Melanesian Arc developed between middle Eocene and earliest Miocene times, and was probably related to a southerly dipping subduction zone along the Manus (West Melanesian) Trench. Volcanism waned during the early Miocene and deposition of tuffaceous fine clastics and carbonate turbidites prevailed (Louwa unit). Although this unit was deposited largely in a bathyal environment, transported inner neritic detritus provides evidence for carbonate buildups on local bathymetric highs. Towards the end of the early Miocene, or early in the middle Miocene, carbonate platforms of Mundrau Limestone developed. The platforms were located mainly along the southwestern flank of a northwesterly trending fault-controlled basement high of Tinniwi Volcanics.

There was renewed intensive volcanism at the end of the early Miocene or during the middle Miocene, with extrusion of the Tasikim Agglomerate and intrusion of the subvolcanic Yirri Complex. This intrusive activity was largely controlled by northwesterly trending fault systems (Jaques, 1980). The Tasikim volcanism probably ended deposition of the Mundrau Limestone, and built up volcanic piles about 400 m thick near major eruptive centers such as Mt. Tasikim.

In late middle Miocene times further faulting occurred, followed by the deposition of bathyal tuffaceous clastics of the late Miocene to earliest Pliocene Lauis Formation unconformably overlying the Tasikim Agglomerate. Depocenters of northeasterly or northwesterly trend were separated by highs of Tasikim Agglomerate. During the early Pliocene there was a pronounced regression, with buildups of Naringel and Kulep limestones being deposited on the flanks of fault-controlled highs, and inner neritic to terrestrial Pliocene-Pleistocene Likum Volcanics being extruded from a major eruptive center at South West Bay. This regression immediately preceded the opening of the Manus Basin, which started at 3.5 Ma (Taylor, 1979). In Pliocene and Pleistocene times there was faulting along northwesterly trends in central and western Manus, with mesoscale conjugate shears suggesting that these movements were largely strike-slip. During this period these faults probably functioned as landward projections of transform faults involved in the opening of the Manus Basin. However evidence for Miocene faulting along similar trends indicates that this process possibly involved re-activation of older faults, and that individual faults could have complex histories with different senses of movement occurring at different times.

ACKNOWLEDGMENTS

The reviews of this paper by A.L. Jaques (Bureau of Mineral Resources, Australia) and J. Vedder (U.S. Geological Survey) are gratefully acknowledged.

REFERENCES

Adams, C.G., 1984, Neogene larger foraminifera, evolutionary and geological events in the context of Indo-Pacific datum planes, *in* Tsuchi, R., ed., Final Report on International Geological Correlation Project (IGCP) 114, Japan, p. 47-67.

Aguirre, E., and Pasini, G., 1985, The Pliocene-Pleistocene Boundary: Episodes, v. 8, no. 2, p. 116-120.

Australian-Anglo-American, 1975, Manus Island - Review of Exploration to 14th May, 1975: Quarterly Report, (unpublished).

Braybrooke, J.C. and Pieters, P.E., 1969, Geological reconnaissance for a possible hydro-electric scheme on the Lauis River, Manus Island T.P.N.G.: Australia, Bureau of Mineral Resources Record 1969/6, (unpublished).

Brown, C.M., 1982, Kavieng, Papua New Guinea - 1:250,000 Geological Series : Geological Survey of Papua New Guinea, Explanatory Notes SA/56-9, 26 p.

Chaproniere, G.C.H., 1973, Micropaleontology Report, Manus Island samples: Report to Geological Survey Papua New Guinea, 3 p. (unpublished).

Chaproniere, G.C.H., 1981, Australasian mid-Tertiary larger foraminiferal associations and their bearing on the East Indian Letter Classification: Bureau of Mineral Resources Journal of Australian Geology and Geophysics, v. 6, p. 145-151.

Dalrymple, G.B., 1979, Critical tables for the conversion of K-Ar ages from old to new constants: Geology, v. 7, no. 11, p. 558-560.

Exoil, N.L., 1973, PA 83 (NG) Manus Island, Final Report, (unpublished).

Falvey, D.A., and Pritchard, T., 1984, Preliminary paleomagnetic results from northern Papua New Guinea: evidence for large microplate rotations; *in* Watson, S.T., ed., Transactions of the Third Circum-Pacific Energy and Mineral Resources Conference, p. 593-599.

Francis, G., 1975, Karst on Manus Island: Niugini Caver 3, p. 67-91.

Francis, G., 1977, Caves in the Quaternary raised reefs of eastern Manus, Papua New Guinea: Australian Speleological Federation, 11th Biennial Conference, Proceedings, p. 146-157.

Haig, D.W., 1985, *Lepidocyclina* associated with Early Miocene planktic foraminiferids from the Fairfax Formation, Papua New Guinea: South Australian Department of Mines and Energy, Special Publications, v. 5, p. 117-131.

Hekel, H.K., 1985, Examination of sediment samples from the Papuan Basin, New Ireland and Manus Island for nannofossils and palynology: Report to Geological Survey Papua New Guinea, 19 p. (unpublished).

Jaques, A.L., 1975, 1:100,000 Geological Map, Manus Island: Geological Survey Papua New Guinea, Preliminary Edition (unpublished).

Jaques, A.L., 1976, Explanatory notes on the Admiralty Islands Geological Map: Geological Survey Papua New Guinea Report 76/15, 31 p.

Jaques, A.L., 1980, Admiralty Islands, Papua New Guinea - 1:250,000 Geological Series: Geological Survey Papua New Guinea, Explanatory Notes SA/55-10 & SA55-11, 25 p.

Jaques, A.L., 1981, Quaternary volcanism on Manus and M'Buke Islands: Geological Survey Papua New Guinea Memoir 10, p. 213-220.

Jaques, A.L., and Webb, A.W., 1975, Geochronology of "porphyry copper" intrusives from Manus Island, Papua New Guinea: Geological Survey of Papua New Guinea Report 75/5, 19 p.

Kicinski, F.M., and Belford, D.J., 1956, Note on the Tertiary succession and foraminifera of Manus Island: Australia, Bureau of Mineral Resources Report 25, p. 71-75.

King, J., Ghartey, E., Buleka, J., and Anderson, G., 1984, Road materials investigation in south west Manus: Geological Survey Papua New Guinea Report 84/7, 17 p.

Kirwin, D.J., 1974, Preliminary report on the geology of the central Manus Is. porphyry copper deposits: Australian-Anglo-American Report, 38 p. (unpublished).

Nion, S.S., 1980, Bismarck Archipelago Palaeomagnetic Project Phase One Summary 0 1979: Geological Survey Papua New Guinea Report 80/3, 3 p.

Rosengren, P., 1982, Geology, *in* O'Grady, I.R., Kelly, M.R. and Rosengren, P., Final Report P.A. 436 (NG) Manus Island: Conzinc Riotinto Australia (CRA) Exploration Report, (unpublished).

Steiger, R.H., and Jager, E., 1977, Subcommission on geochronology; Convention on the use of decay constants in geo- and cosmochronology: Earth and Planetary Science Letters, v. 36, p. 359-362.

Taylor, B., 1979, Bismark Sea: evolution of a back-arc basin: Geology, v.7, no. 4, p. 171-174.

Thompson, J.E., 1952, Report on the geology of Manus Island, Territory of Papua and New Guinea, with reference to the occurrence of bauxite: Australia, Bureau of Mineral Resources Record 1952/82, (unpublished).

Webb, A.W., 1973, AMDEL (Australian Mineral Development Laboratories) Geochronological Report 2837/73, 2 p. (unpublished).

Webb, A.W., 1974, AMDEL (Australian Mineral Development Laboratories) Geochronological Report 1788/74, 3 p. (unpublished).

Webb, A.W., 1985, K-Ar dating of Tasikim Agglomerate: AMDEL (Australian Mineral Development Laboratories) Report G 6360/85, (unpublished).

Marlow, M.S., Dadisman, S.V., and Exon, N.F., editors, 1988, Geology and offshore resources of Pacific island arcs—New Ireland and Manus region, Papua New Guinea, Circum-Pacific Council for Energy and Mineral Resources Earth Science Series, v. 9: Houston, Texas, Circum-Pacific Council for Energy and Mineral Resources.

EARTHQUAKES AND CRUSTAL STRESS IN THE NORTH BISMARCK SEA

K.F. McCue
Australian Seismological Centre, Bureau of Mineral Resources, Canberra, A.C.T. 2601, Australia

ABSTRACT

Three small plates, the North and South Bismarck Sea plates and the Solomon Sea plate, separate the converging Pacific and Australian plates in the Papua New-Guinea - Solomon Islands region. Most of the shallow seismic energy released in the region occurs along the Solomon/South Bismarck Sea plate boundary south of New Britain. The least seismically active boundary is that between the Pacific/Caroline and North Bismarck Sea plates. Neither the eastern nor the western end of the North Bismarck Sea plate is clearly defined, but both may be unique trench-trench-trench-transform quadruple junctions.

Focal mechanisms of three earthquakes along the northern boundary of the North Bismarck plate have a significant north-northeasterly thrust component. By comparison, earthquakes along the Solomon/South Bismarck Sea and Solomon/Pacific plate boundaries, and those within the Caroline plate, have similar thrust-type mechanism along a northeasterly direction.

Seismological evidence does not show that the Mussau Trench along the Caroline/Pacific boundary is an active subduction zone.

INTRODUCTION

Earthquake activity in the Papua New Guinea (PNG) region is intense and has been studied by many authors including Sieberg (1932), Gutenberg and Richter (1954), Brookes (1965), Denham (1971), Johnson and Molnar (1972), Curtis (1973), Everingham (1974), Taylor (1979), Weissel et al. (1982), Ripper and McCue (1983), Letz (1983), and others. Epicenters in PNG are preferentially distributed along plate boundaries that are characterized by widely varying levels of activity and different tectonic styles. For example, the seismic zones along the south coast of New Britain and the west coast of Bougainville are very active with Benioff zones extending to depths of five hundred kilometers or more (Ripper and McCue, 1982). Large shallow earthquakes are typified by thrust-type mechanisms (Johnson and Molnar, 1972). Seismic zones through the central Bismarck and southern Solomon Seas are much less active; the earthquakes are all shallow and characteristically of strike-slip or near-normal fault type respectively (Ripper, 1975a, b; 1982).

This paper focuses on the earthquake zone across the north Bismarck Sea, which is diffuse and less active than any other continuous seismic zone in Papua New Guinea. Johnson (1979) gives a useful summary of the plate tectonic setting of the region, which can now be supplemented by more recent epicenter information and three new fault-plate solutions.

PLATE TECTONIC SETTING

A detailed map of the shallow seismicity in the Papua New Guinea region outlines three small plates within the zone of interaction between the Australian

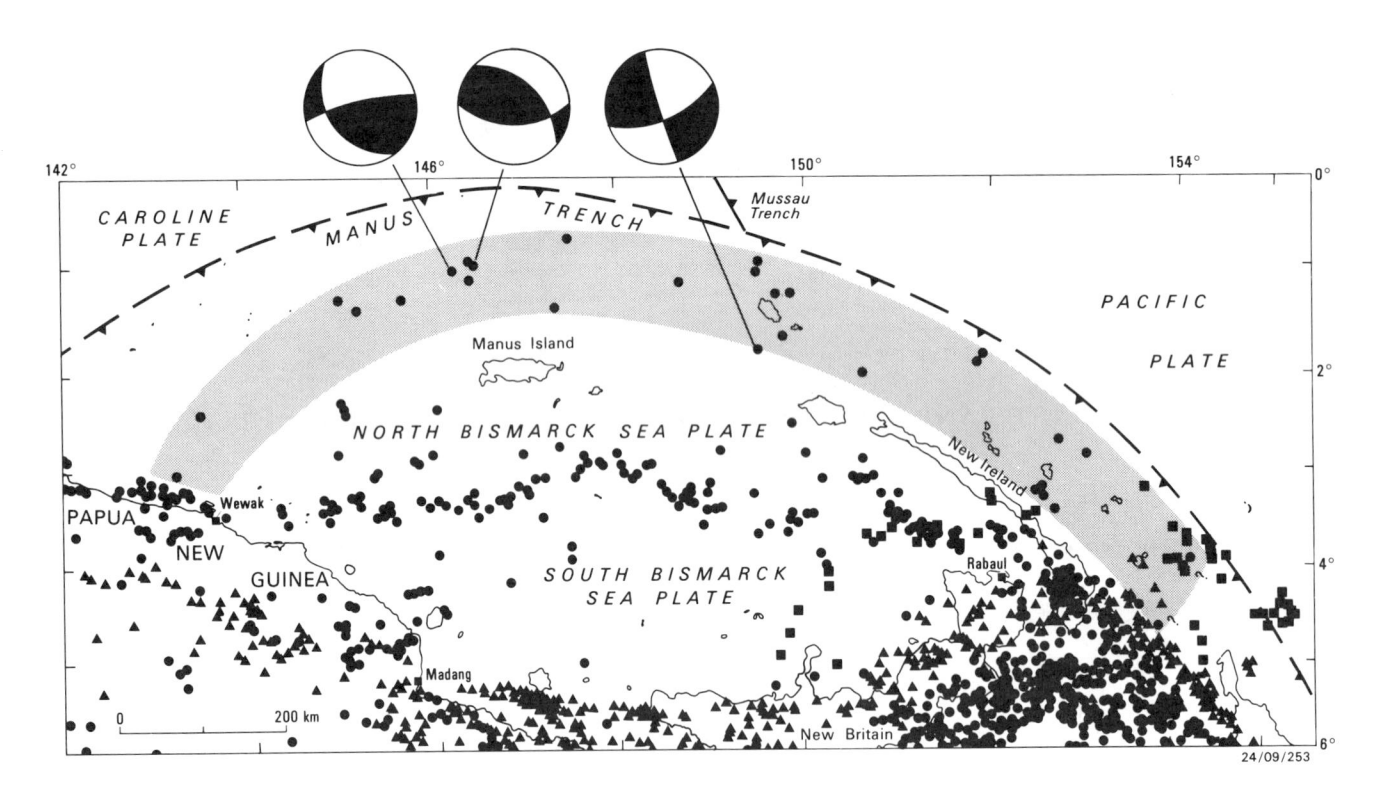

Figure 1. Seismicity in the Bismarck Sea region (0° - 6°S, 142° - 155°E) using 10 or more recording stations. Time span is 1964 to 1985 (from the Geophysical Observatory, Port Moresby). Circles denote focal depths less than 70 km. Triangles denote focal depths 70 to 300 km. Squares denote focal depths greater than 300 km.

and Pacific plates (Ripper and McCue, 1982; Figure 1). The best defined are the Solomon Sea and South Bismarck Sea plates. The style of faulting varies around the boundaries of these two plates; from normal along the spreading center across the Solomon Sea, to intermediate angle thrusting under New Britain, and to sinistral strike-slip along the Bismarck Sea en echelon transforms (Ripper, 1975a, b).

Another plate which has a bearing on the tectonics of the region is the Caroline Sea plate (Weissel and Anderson, 1978; Hegarty, Weissel, and Hayes, 1986). The eastern boundary of the Caroline plate is not well defined but is presumed to be associated with the meridional-trending Mussau Trench, which intersects the Manus or West Melanesian Trench at about 1°S, 149.5°E (Hegarty, Weissel, and Hayes, 1986). No earthquakes have occurred along the Mussau Trench in the last two decades, although infrequent large shallow earthquakes have occurred 150 km or so to the west, one at 5.6°N, 146.9°E on August 20, 1968 and another at 1.0°N, 147.6°E on August 30, 1976. Fault mechanisms of both events (Weissel and Anderson, 1978; McCue, 1981; Hegarty, Weissel, and Hayes, 1986) show a strong right lateral strike-slip component of faulting on the nodal plate paralleling the trench, and in each case the principal stress direction was near horizontal, striking north-northeast to south-southwest at a very acute angle to the trench. The August 30, 1976 earthquake mechanism is reproduced in Figure 2a from McCue (1981). These solutions pose severe problems for the Hegarty, Weissel, and Hayes (1986) model of easterly subduction of the Caroline plate at the Mussau Trench.

Indeed the lack of seismic activity along the Mussau Trench indicates either that the Caroline plate is no longer active, or that activity is so low that existing seismographs are not close enough, or have not been in operation long enough, to detect any earthquakes.

EARTHQUAKE EPICENTERS AND FOCAL MECHANISMS

A diffuse belt of approximately thirty shallow earthquakes extends over a 90 km wide, 1500 km long, arcuate zone from the northern tip of Bougainville to the north coast of Papua New Guinea at about 3.2°S, 143.5°E as shown by the shaded region

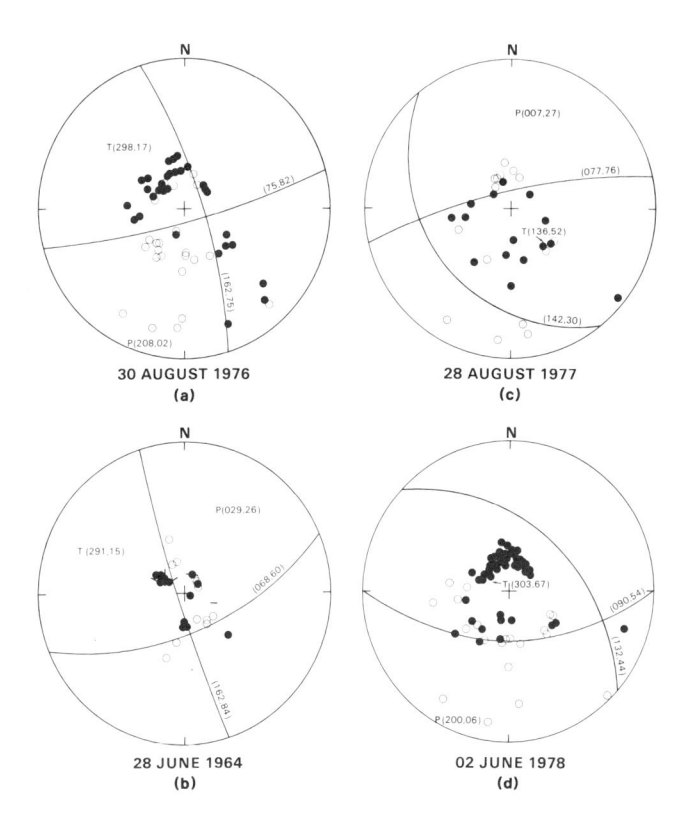

Figure 2. (a). Fault plane solution for the Caroline Islands earthquake of August 30, 1976. P, T (azimuth, dip) are principal stress axes. Closed circles are short period compressions. Open circles are short period dilatations. (b). Fault plane solution for the North Bismarck Sea earthquake of June 28, 1964 (from Ripper, 1971). Symbols are the same as above. (c). Fault plane solution for the North Bismarck Sea earthquake of August 28, 1977. Symbols are the same as above. (d). Fault plane solution for the North Bismarck Sea earthquake of June 2, 1978. Symbols are the same as above.

of Figure 1. The southern limit of the seismicity is poorly defined. The absence of shallow foci off the east coast of Bougainville suggests that the southern plate junction is between New Ireland and Bougainville. The seismic belt appears to be continuous except for a 200 km long gap near the western end. A similar, inexplicable, gap occurs in the center of the Solomon Sea lineament, at about 9.5°S, 152.5°E at the southern margin of the Solomon Sea plate (Ripper and McCue, 1982). These gaps may, or may not, be the most likely sites of future activity and are too short to cast doubt on the nature of either zone as a plate boundary.

Denham (1971) was the first to consider the shaded zone of Figure 1 to be the northern boundary of a North Bismarck Sea plate. The relative motion of the North Bismarck Sea plate with the Pacific or Caroline plates is small as evidenced by the low seismicity. Krause (1973) assumed zero relative velocity across the seismic zone, in his analysis of the triple junction interactions between crustal plates of the Bismarck and Solomon Seas.

The nature of the northern boundary and how or whether the earthquakes are related to the Manus Trench is unclear. All foci are crustal and the three fault plane solutions have a large thrust component. All of the epicenters are on the concave side (south) of the Manus Trench and diverge further from the Trench in the west than the east. Seismic reflection profiles show strata broken by thrust faults in the Manus forearc (Marlow et al. this volume), and deformed sediment in the Manus Trench east of its intersection with the Mussau Trench but not west of this point (Ryan and Marlow, this volume). Subduction may have recently recommenced but supporting evidence is weak.

The plate junctions at either end of the West Melanesian arc are not clearly defined by the seismicity, but appear to be complex junctions of four plates; the North Bismarck, South Bismarck, Solomon and Pacific plates in the east and the North Bismarck, South Bismarck, Caroline (or Pacific) and Australian plates in the west (Figure 3).

The West Melanesian arc earthquakes plotted in Figure 1 are, with one exception, all smaller than magnitude 6.0 but larger than magnitude 5.0 which is about the detection threshold. The seismicity lies beyond the aperture of the Papua New Guinea seismograph network and no regional seismographs exist north or east of the zone. This results in poor epicenter locations that may be subject to systematic errors. This could account for some of the spatial scatter in epicenters.

Only three earthquakes were large enough and sufficiently widely recorded that their focal mechanisms could be studied. Ripper (1971) analyzed the mechanism of the June 28, 1964 earthquake between Manus Island and New Ireland, to which the Berkeley Seismograph station (USA) assigned a magnitude of 5.7 to 6. Ripper's (1971) solution is strongly strike-slip with the principal stress oriented at N29°E (Figure 2b).

The two more recent events occurred on August 28, 1977 and June 02, 1978, close together in the zone approximately 100 km north of Manus Island. These were both felt on the island. Hypocentral details for the three earthquakes and a large historical earthquake are listed in Table 1 from Ripper (1971), Everingham (1977) and McCue (1979; 1982).

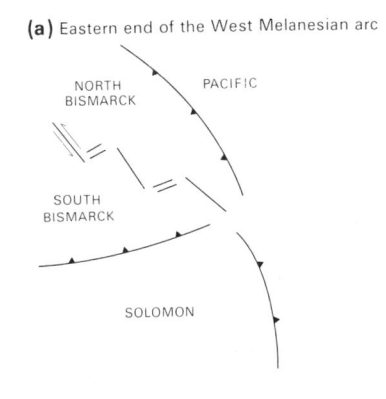

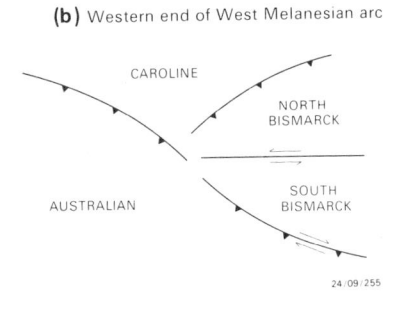

Figure 3. Possible quadruple junctions of the (a) eastern end of the West Melanesian arc, and (b) western end of the West Melanesian arc.

Fault plane solutions of the four events discussed in this paper are shown in Figure 2 and relevant parameters are summarized in Table 2. The mechanisms of the 1977 and 1978 earthquakes have nodal planes that dip at intermediate angles to the horizontal but are not tightly constrained in strike direction because of the limited azimuthal station coverage. Both solutions are predominantly thrusts with the principal stress horizontal and striking in a north-northeast to south-southwest direction. The 1978 earthquake was also analyzed (by McCaffrey and McCue, unpublished), using an iterative procedure, to compare theoretical and observed radiation patterns from both long period vertical and horizontal seismograms, in the distance range 30 to 90 degrees (Nábelek, 1984; McCaffrey and Nábelek, 1987).

The resultant mechanism and stress direction were almost identical to those determined here from the P-wave radiation pattern. The waveform analysis also computes focal depth, independently of travel times, from the interference between the P and pP (surface reflection of P near the epicenter) waves, and confirmed further that these earthquakes are very shallow, no deeper than 5 km. Only the 1978 solution has a nodal plane striking parallel to the nearby trench. Movement on this east-west plane is principally down-dip to the south, with a small component of left-lateral strike-slip.

A small tsunami was reported on the northwest coast of Manus Island following the 1977 earthquake, which is further evidence for the dip-slip nature of that earthquake. Tectonic displacement of the sea floor should have a substantial dip-slip component to generate tsunamis (Wiegel, 1970). The 1977 tsunami could have been caused by an earthquake-generated submarine landslide or an underwater volcanic eruption associated with the earthquake. The latter two scenarios are unlikely. The earthquake was considerably smaller than those that generate submarine landslides in the Gogol River sediments off Madang (Everingham, 1977), and an underwater eruption

Table 1. Hypocentral details of largest earthquakes.

Date U.T.	Time	Latitude °S	Longitude °E	Depth km	Magnitude mb	MS	N	Comments
23 Dec 1930	213536	1.4	144.3	S	5.5		-	Tsunamigenic
28 Jun 1964	125135	1.7	149.6	S	5.8			
28 Aug 1977	201005	1.08	145.23	S	5.2	5.5	108	Felt MM4 W. Manus
02 Jun 1978	113006	1.05	146.45	10	5.9	5.3	164	Felt MM2 Momote

S - implies shallow depth (<20 km) but poorly constrained.
mb - is the body wave magnitude
MS - is the surface wave magnitude
MM - is the Modified Mercalli Intensity
N - is the number of recording seismographs.

Table 2. Details of fault plane solutions.

Year	Plane 1 dip az	Plane 2 dip az	P axis dip az	T axis dip az	B axis dip az
1964	60 068	84 162	26 029	15 291	60 173
1976	82 075	75 162	02 208	17 298	77 111
1977	76 077	30 142	27 007	52 136	25 263
1978	54 090	44 132	06 200	67 303	25 106

would have produced floating mats of pumice and ash in the Bismarck Sea. Such mats would have been readily visible from the air and reported when washed ashore.

On March 23, 1930 an earthquake in the western Bismarck Sea at 1.4°S, 144.3°E, generated a tsunami which devastated some islands in the Ninigo group, west of Manus Island, and a section of the Papua New Guinea north coast (Everingham, 1977). This is the only earthquake in the zone which has exceeded magnitude 6.0. The tsunami is taken to be evidence that shallow thrusting was associated with this earthquake.

DISCUSSION

A definitive tectonic model of the north Bismarck Sea region cannot be deduced from only thirty epicenters and three fault plane solutions. The epicenters, though broadly spread, do form a nearly continuous zone along a 1500 km arc parallel to the Manus Trench. The frequency of locatable earthquakes (above magnitude 5.0) is so low that details of the junctions of this zone with the New Guinea north coast zone in the west and the Bismarck arc zone in the east are obscure.

The only consistent feature of the fault-plane solutions for three earthquakes in the center of the zone is the principal stress direction, which strikes approximately NNE-SSW. Two of the earthquakes had principally thrust type mechanisms, while one appears to have had a predominantly strike-slip mechanism with only a small component of thrust.

The lack of subcrustal foci (Ripper and McCue, 1982) indicates that no Benioff zone exists south of the Manus Trench, and if subduction is recurring there, then it must be of very recent origin.

REFERENCES

Brookes, J.A., 1965, Earthquake activity and seismic risk in Papua and New Guinea: Bureau of Mineral Resources, Australia, Report 74, 30 p.

Curtis, J.W., 1973, Plate Tectonics of the Papua New Guinea - Solomon Islands Region: Journal of the Geological Society of Australia, v. 20, p. 1-19.

Denham, D., 1971, Seismicity and tectonics of New Guinea and the Solomon Islands in Recent crustal movements: Royal Society of New Zealand, Bulletin, no. 9, p. 31-38.

Everingham, I.B., 1974, Large earthquakes in the New Guinea-Solomon Islands region, 1873 - 1972: Tectonophysics, v. 23, p. 323-338.

Everingham, I.B., 1977, Preliminary catalogue of tsunamis for the New Guinea/Solomon Islands region, 1768 - 1972: Bureau of Mineral Resources, Australia, Report 180, 78p.

Gutenberg, B., and Richter, C.F., 1954, Seismicity of the Earth and associated phenomena: Princeton, Princeton University Press, 310 p.

Hegarty, K., Weissel, J.K., and Hayes, D.E., 1986, Convergence at the Caroline-Pacific plate boundary: collision and subduction, in Hayes, D.E., ed., Tectonics of southeast Asian seas and islands: American Geophysical Union, Monograph 27, p. 326-348.

Johnson, T. and Molnar, P., 1972, Focal mechanism and plate tectonics of the southwest Pacific: Journal of Geophysical Research, v. 77, p. 5000-5032.

Johnson, R.W., 1979, Geotectonics and volcanism in Papua New Guinea: a review of the late Cainozoic: Bureau of Mineral Resources Journal of Australian Geology and Geophysics, v. 4, p. 181-207.

Krause, D.C., 1973, Crustal plates of the Bismarck and Solomon Seas, in Fraser, R., compiler, Oceanography of the South Pacific: Wellington, New Zealand National Commission for UNESCO, p. 271-280.

Letz, H., 1983, Seismotectonic map of Irian Jaya (West New Guinea), Indonesia: Berlin, Freie University.

McCue, K.F., 1979, Seismicity of the Papua New Guinea and neighbouring regions for 1977: Geological Survey of Papua New Guinea, Report 1979/32, 35 p.

McCue, K.F., 1981, Seismicity of the Papua New Guinea and neighbouring regions for 1976: Geological Survey of Papua New Guinea, Report 1981/10, 34 p.

McCue, K.F., 1982, Seismicity of the Papua New Guinea and neighbouring regions for 1978: Geological Survey of Papua New Guinea, Report 1982/4, 32 p.

McCaffrey, R., and Nábelek, J., 1987, Earthquakes, gravity, and the origin of the Bali Basin: an example of a nascent continental fold-and-thrust belt: Journal of Geophysical Research, v. 92, no. B1, p. 441-460.

Nábelek, J., 1984, Determination of earthquake source parameters from inversion of body waves: Cambridge, Massachusetts Institute of Technology, Ph.D. thesis, unpublished.

Ripper, I.D., 1971, Earthquake focal mechanism in the east New Guinea region: Bureau of Mineral Resources, Australia, Record 1971/27, 29 p.

Ripper, I.D., 1975(a), Earthquake focal mechanisms in the New Guinea-Solomon Islands region, 1963 - 1968: Bureau of Mineral Resources, Australia, Report 178, 120 p.

Ripper, I.D., 1975(b), Seismicity and earthquake focal mechanisms in the New Guinea-Solomon Islands region: Bulletin of the Australian Society of Exploration Geophysicists, v. 6, p. 80-81.

Ripper, I.D., 1982, Seismicity of the Indo-Australia/Solomon Sea plate boundary in the Southeast Papua region: Tectonophysics, v. 87, p. 355- 369.

Ripper, I.D., and McCue, K.F., 1982, Seismicity of the New Guinea region, 1964-1980, computer plots: Geological Survey of Papua New Guinea, Report 1982/10, 36 p.

Ripper, I.D. and McCue, K.F., 1983, The seismic zone of the Papuan Fold Belt: Bureau of Mineral Resources Journal of Australian Geology and Geophysics, v. 8, p. 147-156.

Sieberg, A., 1932, Erdbeben geographie: Handbuch der Geophysik, v. IV: Berlin, Verlag, p. 687-1005.

Taylor, B., 1979, Bismarck Sea: Evolution of a back-arc basin: Geology, v. 7, p. 171-174.

Weissel, J.K. and Anderson, R.M., 1978, Is there a Caroline plate?: Earth and Planetary Science Letters, v. 41, p. 143-158.

Weissel, J.K., Taylor, B., and Karner, G.D., 1982, The opening of the Woodlark Basin, subduction of the Woodlark spreading system, and the evolution of northern Melanesia since mid-Pliocene time: Tectonophysics, v. 87, p. 253-277.

Wiegel, R.L., 1970, Tsunamis, in Wiegel, R.L., ed., Earthquake engineering: Englewood Cliffs, Prentice-Hall, p. 253-306.

Marlow, M.S., Dadisman, S.V., and Exon, N.F., editors, 1988, Geology and offshore resources of Pacific island arcs—New Ireland and Manus region, Papua New Guinea, Circum-Pacific Council for Energy and Mineral Resources Earth Science Series, v. 9: Houston, Texas, Circum-Pacific Council for Energy and Mineral Resources.

SEDIMENTOLOGY AND MORPHOLOGY OF THE INSULAR SLOPE — NEW IRELAND TO MANUS ISLANDS, PAPUA NEW GUINEA

P.R. Carlson, D.M. Rearic, P.J. Quinterno
U.S. Geological Survey, Menlo Park, California 94025, USA

ABSTRACT

Surficial sediment of the insular slopes north of the Cenozoic island arc of Manus, New Hanover, and New Ireland varies from silty sand to clayey silt. The sand fraction is dominated by planktonic foraminifers and the mud by calcareous nannoplankton. Volcanic glass is common in the sand fraction and increases from west to east. However, in spite of the close proximity of the islands, volcaniclastics and moderate to low (0.83 - 0.08 %) organic carbon are secondary constituents. Much of the island-derived detritus apparently bypassed the slope, reaching the intraslope and trench basins through the numerous gullies that are incised into the Neogene-Quaternary sediment. Slides and slumps seen on some of the seismic records provide additional evidence of downslope sediment transport. A thin mantle of Holocene sediment covers the slope, except on the steep scarps where rocks as old as late Miocene crop out.

INTRODUCTION

On the 1984 cruise of the R/V *S.P. Lee* (L7-84-SP) to the New Ireland Basin, a series of short gravity cores were collected along the insular slopes north of the islands of Manus, New Hanover, and New Ireland (Figure 1). The cores were collected to determine the hydrocarbon content of the surficial sediment (see Kvenvolden, this volume) and to study the Quaternary environments of deposition on the slopes that separate the island platform from the abyssal floor of the New Ireland basin. The Quaternary depositional environments may provide clues to help determine earlier depositional environments. The cored sediment and associated seismic-reflection profiles provide evidence regarding active sea-floor processes.

The purposes of this report are to describe and discuss the slope sediment obtained in the cores and to discuss the sea-floor environment at and between coring stations using high-resolution seismic-reflection records and available bathymetric data.

REGIONAL SETTING

The New Ireland Basin is an elongate sedimentary basin (900 x 160 km) (Exon and Tiffin, 1984) that is part of the north Bismarck lithospheric plate, one of several small plates that underlies a collage of small oceanic basins and island areas found between the Caroline and Pacific plates to the north and the Indo-Australian plate to the south (Hamilton, 1979). The arcuate northwest-southeast trending New Ireland Basin, bounded on the north by the Manus Trench and on the south by the Bismarck Sea, includes the Cenozoic islands of Manus, New Hanover, and New Ireland (Figure 1). According to Exon and Tiffin (1984), this area was part of an island arc from at least Eocene until the opening of the Manus Basin in Pliocene time.

The islands of the Bismarck archipelago consist of early Tertiary andesitic tuffs and agglomerates unconformably overlain by intercalated units of shallow-water limestone and volcaniclastics, except on New Hanover where the Pliocene units are

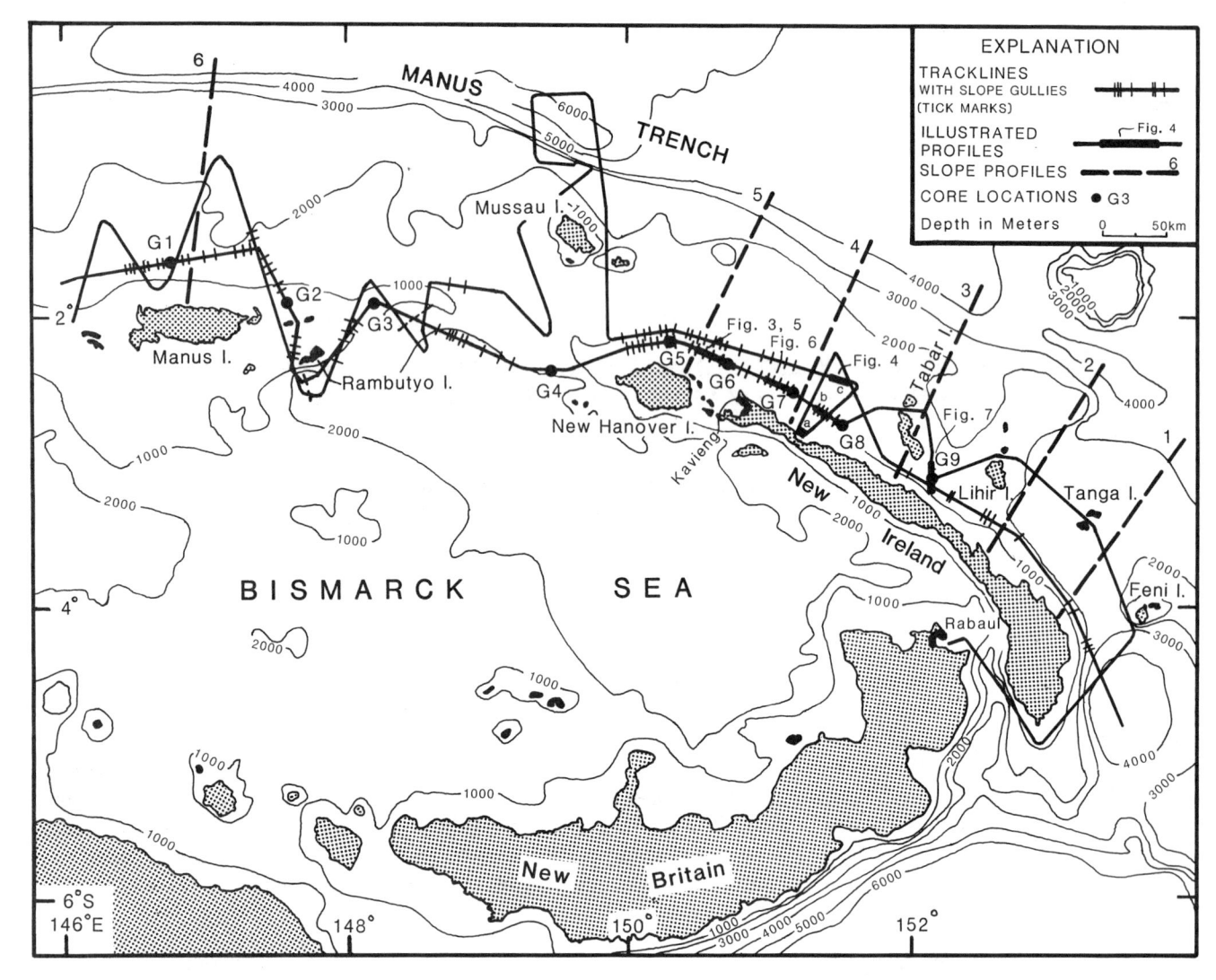

Figure 1. Location map of New Ireland to Manus Island insular slope area showing tracklines and locations of cores, profiles, and slope gullies.

primarily volcanogenic sediment (Kroenke, 1985; Exon and Tiffin, 1984). Pleistocene strata include limestone deposited as fringing reefs, thin alluvial deposits, and some basalt on Manus Island.

DATA COLLECTION

On the R/V *S.P. Lee* cruise 3200 km of single-channel seismic-reflection air-gun (20,730 cm^3) data and 4300 km of 3.5 kHz and 12 kHz bathymetric data were collected (Figure 1). Nine coring stations were occupied; gravity cores (365 kg weight and 3 m barrel) were collected at six stations, and only trace amounts of bottom sediment were retained in the core catcher at two other stations (Table 1).

The 8-cm diameter gravity cores, with plastic liners, were cut into 1 m lengths and immediately subsampled for hydrocarbons (Kvenvolden, this volume). A subsample was also taken from the core catcher to determine the maximum age of the core (Belford, this volume). The capped and taped cores were kept refrigerated until onshore analyses were begun. (See appendix for explanation of laboratory methods.)

RESULTS

Morphology of the Insular Margin

The insular shelves on the north side of the Tertiary island arc that includes the islands of

Table 1. Gravity cores for cruise L7-84-SP.

Core	Location	Latitude (S)	Longitude (E)	Water Depth (m)	Core Length (cm)	Core Descriptions
G1	N. Manus	1°39.60′	146°51.12′	1514	237	Yellowish-brown to olive-gray foraminiferal ooze.
G2	E. Manus	1°53.38′	147°35.39′	833	None	Soft sediment; washed out at sea surface.
G2a	E. Manus	1°53.38′	147°34.96′	842	None	Soft sediment; washed out at sea surface.
G3	NE Rambutyo	1°50.69′	148°16.17′	1086	85	Yellowish brown to olive-gray foraminiferal ooze.
G4	W. New Hanover	2°20.94′	149°41.58′	607	None	Minor lag deposit of pumice, solitary corals, foraminiferal pteropod ooze.
G4A	W. New Hanover	2°21.14′	149°40.97′	617	Minor	Minor lag deposit of pumice, solitary corals, foraminifers, molluscs.
G5	N. New Hanover	2°10.82′	150°18.82′	749	Minor	Olive-gray foraminiferal pteropod ooze.
G5a	N. New Hanover	2°10.62′	150°18.79′	785	None	Olive-gray foraminiferal pteropod ooze.
G6	NE New Hanover	2°18.79′	150°39.16′	847	166	Yellowish-brown to olive-gray foraminiferal-pteropod ooze.
G7	NE Kavieng	2°29.21′	151°03.17′	1036	277	Yellowish-brown to olive-gray foraminiferal ooze to calcareous mud.
G8	NE New Ireland	3°48.39′	151°31.28′	1236	151	Yellowish-brown to olive-gray foraminiferal ooze to calcareous mud.
G9	NE New Ireland	3°08.46′	152°05.66′	1320	100	Yellowish-brown to olive-gray foraminiferal ooze to calcareous mud.

Manus, New Hanover, and New Ireland are narrow (2-10 km) and contain numerous patch and fringing reefs. The slope on the north and northeast side of the islands begins less than 10 km off the islands and extends about 160-200 km to the bottom of the arcuate Manus Trench (Figure 1), which exceeds a depth of 6000 m north of New Hanover, but shoals east and west to about 4000 m off Manus and New Ireland. This broad arc-trench region is interrupted by another arc of Quaternary volcanoes positioned midway between the older island arc and the trench. Between the Tertiary and Quaternary island arcs, deep broad troughs or depressions form intraslope basins as deep as 2800 m off northeastern New Ireland. North of Manus Island these intraslope depressions shoal toward the west to about 2000 m.

The average gradient of the entire slope (Tertiary island arc to trench axis) is about 2°, whereas the gradient of the insular slope (Tertiary island arc to the floor of the intraslope basin) ranges from 1.4° off New Hanover (Figure 2, profile 5) to 5.5° off eastern New Ireland near Feni Island (Figure 2, profile 1). Much steeper gradients occur locally attaining a maximum of 10° off the eastern end of New Ireland.

The slopes off the large islands (Manus, New Hanover, and New Ireland) are incised by many small gullies (Figure 3). These gullies or small canyons extend downslope into the intraslope basins and some appear to lead into the Manus Trench. Although the spacing of seismic-reflection and bathymetric lines is not adequate to trace these gullies, some down-slope continuity is suggested by their numbers and spacing (Figure 1). Off the western third of New Ireland, three geophysical lines roughly parallel the shoreline (Figure 1) and tracings of the 3.5 kHz records show gullies on the middle and lower slope and a shallow valley or channel in the intraslope basin (Figure 4).

A total of 53 gullies (Manus-19, New Hanover-18, and New Ireland-16) were found along the midslope trackline (Figure 1); a few cores and dredges were collected along the trackline. These gullies

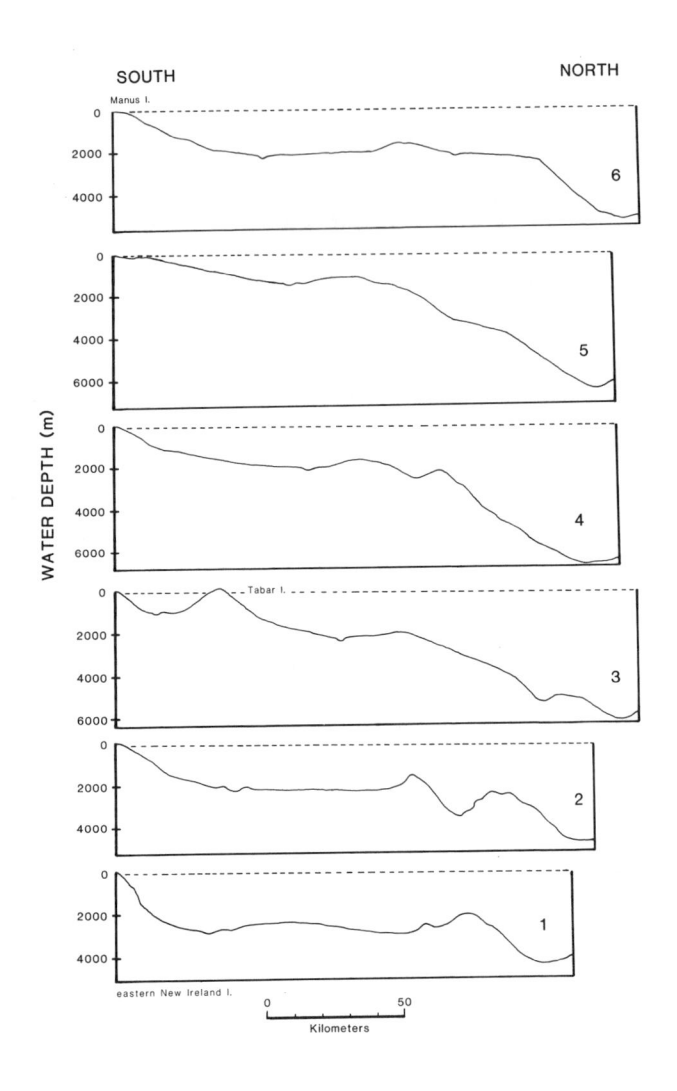

Figure 2. Insular slope profiles from New Ireland to Manus Island. Locations of profiles shown on Figure 1.

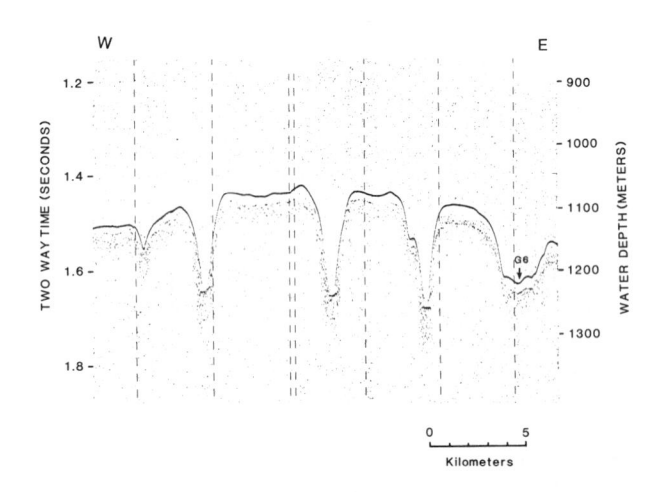

Figure 3. 3.5 kHz profile across slope gullies off New Hanover Island. See Figure 1 for location. Vertical exaggeration is about 33:1.

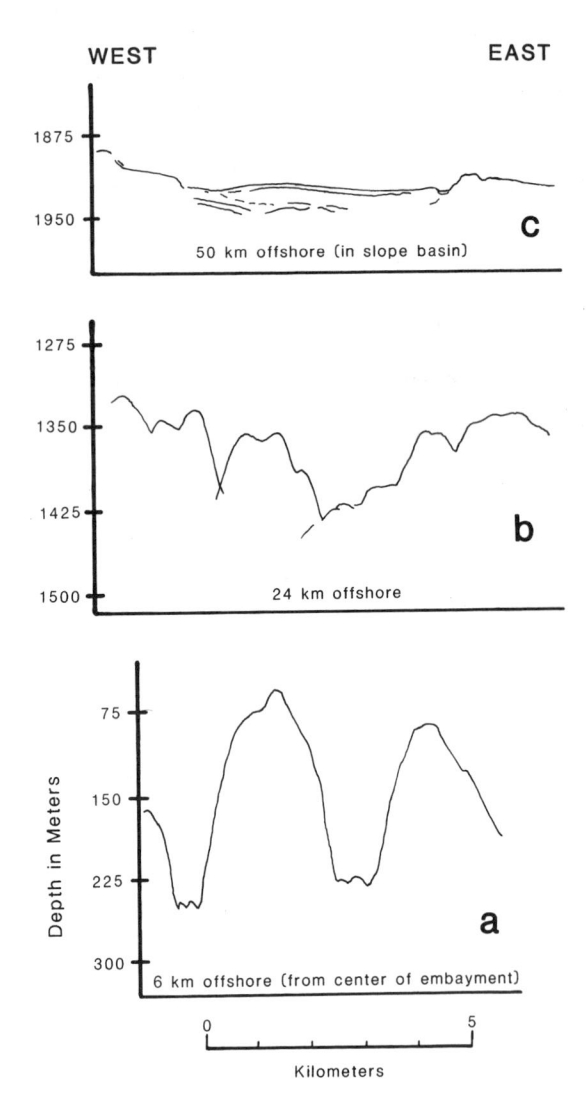

Figure 4. Tracings of 3.5 kHz profiles off New Ireland (Figure 1) showing morphologic changes down slope (a-c).

range in width from 0.3 km to 3 km (average 1.35 km) with relief from 10 m to 330 m (average 100 m). The gullies vary in cross section on the 3.5 kHz profiles from V-shaped and sediment free to flat floored and partly sediment filled (Figure 3). Seismic-reflection profiles show layering in the sediment fill of the gullies, as well as buried gullies and associated cut-and-fill features (Figure 5).

Description of Core-sites

Six gravity cores were collected along the insular slope from Manus to New Ireland (Figure 1). The cores varied in length from 85 cm to 277 cm (Table 1).

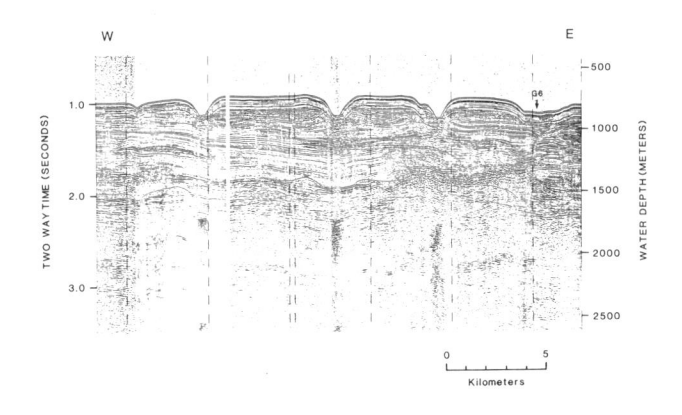

Figure 5. Seismic-reflection profiles across slope gullies off New Hanover Island (compare with Figure 3). See Figure 1 for location. Vertical exaggeration is 6.1:1. Note location of gravity core 6.

Core G1 was acquired about 30 km north of Manus Island from a small sediment-filled depression or graben in water depth of 1514 m. The thickness of sedimentary material in the 2 km-wide graben is about 1.5 sec (> 1200 m).

Core G3 was obtained 65 km northeast of Rambutyo, the largest island in the gap between Manus and New Hanover Islands, in water 1086 m deep.

Core G6 was collected 40 km northwest of Kavieng, New Ireland, at a depth of 847 m, apparently from the floor of one of the numerous gullies that incise the insular slope. This 1.6 km wide gully with walls 120 m high, contains about 200 m of sediment (Figure 5). Beneath the partially filled gully, offset about 1 km to the west, we found an older, filled gully that is related to a buried erosion surface overlain by about 300 m of sedimentary material. Clearly gullying is not just a modern phenomenon.

Core G7 was recovered about 30 km northeast of Kavieng, New Ireland, in water 1036 m deep on an insular slope dissected by numerous small gullies (Figure 6). Sediment at the bottom of the core appears to be older than in any of the other cores considered in this report; therefore, we suspect this core was collected from the wall of a gully where a thin mantling layer of Holocene foraminiferal ooze has accumulated (Figure 6).

Core G8 was collected 30 km off New Ireland and 85 km west of the Tabar Islands in water 1236 m deep. This location is on a highly dissected part of the New Ireland slope, which, according to profiles interpreted by Exon and Tiffin (1984), is underlain by a 2-3-second-thick (1,500 to 2,400 m) sedimentary section of Oligocene through Holocene age.

Core G9 was taken 15 km off New Ireland and 18 km SSE of Tabar Island in 1320 m of water in a topographic saddle between the two islands near the base of the New Ireland slope. The seismic-reflection profile that crosses the core site (Figures 1 and 7) shows an accumulation of about 120 m of indistinctly bedded and disturbed sediment overlying a well-bedded older unit that is about 180 m thick. The well-bedded unit in turn overlies a lenticular acoustic unit about 180 m thick containing numerous hyperbolics and diffractions (Figure 7). The irregular surface and distorted, disrupted internal reflectors of the upper 75-80 m of this sedimentary section suggest that it may be a submarine landslide or debris flow unit that slumped, slid, or flowed into its present position from either of the two adjacent insular slopes (New Ireland or Tabar).

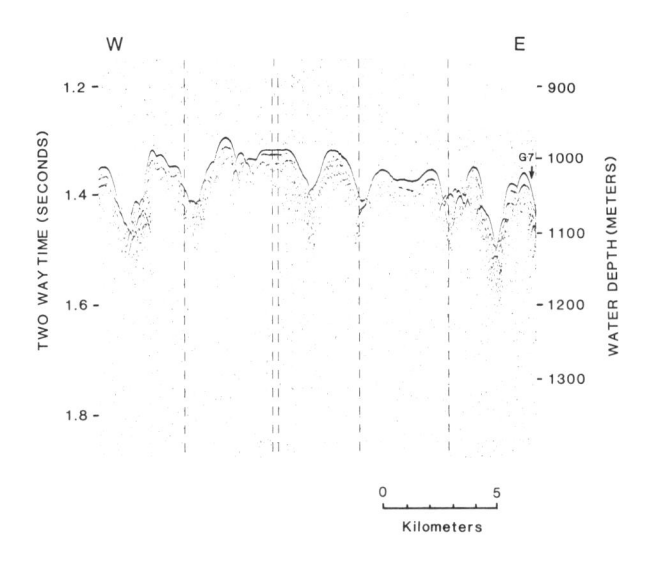

Figure 6. 3.5 kHz profile showing slope gullies off New Ireland and location of core G7. See Figure 1 for locations. Vertical exaggeration is about 33:1.

Sediment Characteristics

The six gravity cores are similar in appearances. All are capped by a 12-20-cm-thick oxidized layer of yellow-brown (10YR4/2 to 5Y6/4) calcareous ooze underlain by an olive-gray (5Y5/1 to 5Y4/2) calcareous ooze. Exon, Colwell, and Bolton (1984) also found an oxidized layer, about 20 cm thick, in insular slope cores collected off the Solomon Islands. However, in water depths of 1500 to 3500 m, the oxidized layer comprised only the upper few centimeters. Deeper than 3500 m, they found that the oxidized layer was again thicker, which suggests an oxygen minimum zone exists between 1500-3500 m. We have no cores below 1514 m, so we cannot identify such an oxygen minimum zone. The surficial oxidized layer in our cores commonly contains intermixed bands, layers, or burrows that vary from light to dark yellowish brown. The boundary between the oxidized and the olive-gray ooze is diffuse, irregular, and often disrupted by burrows. The olive-gray ooze varies from light to dark; the hue of the ooze varies from pale olive (10Y6/2) to grayish olive-green (5GY3/2). Macroscopic and X-ray inspection of the cores shows numerous burrows, occasional shell fragments and pumice fragments, and many pteropods and foraminifers.

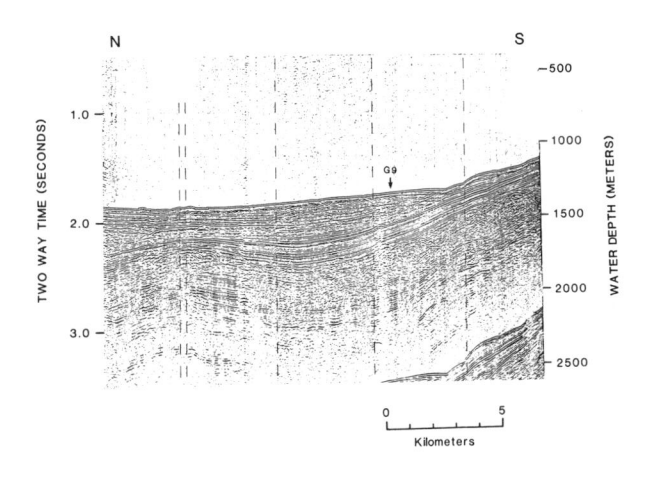

Figure 7. Seismic-reflection profile off New Ireland between Tabar and Lihir Islands (Figure 1). Note location of core G9. Vertical exaggeration is 5.6:1.

Size Analysis

Forty-seven grain-size analyses (Table 2) show that the six cores are uniform from top to bottom, except for cores G6 and G7 where several samples contain > 50 % sand (Figure 8). A plot of sizes on a sand-silt-clay ternary diagram (Figure 9) shows a weak trend toward finer samples at each end of the sampled area and coarser samples off northwestern New Ireland. The dominant sediment types are sandy silt (15 subsamples, 37 %) and silty sand (14 subsamples, 34 %). The percentage of clay is low in all cores and ranges from 10.3 to 36.0 % and averages 25.5 %, whereas sand averages 20.3 % (13.8 to 68.1 %) and silt averages 54.2 % (21.6 to 63.3 %).

Table 2. Grain size analyses.

Core	Depth (cm)	Sand %	Silt %	Clay %	Classification
G1	5-8	37.8	30.4	31.8	clayey sand
G1	15-16	36.1	33.7	30.2	silty sand
G1	30-32	37.9	40.3	21.8	sandy silt
G1	48-50	42.5	26.8	30.7	clayey sand
G1	50-52	42.4	28.4	29.2	clayey sand
G1	78-80	40.9	32.7	26.4	silty sand
G1	87-89	36.4	37.4	26.2	sandy silt
G1	102-104	47.7	28.0	24.3	silty sand
G1	110-113	40.6	30.8	28.6	silty sand
G1	150-152	37.6	28.4	34.0	clayey sand
G1	202-204	40.1	27.6	32.3	clayey sand
G1	AVERAGE	40.0	31.3	28.7	silty sand
G3	4-7	33.3	36.1	30.6	sandy silt
G3	28-30	38.1	33.5	28.4	silty sand
G3	47-50	36.9	41.0	22.1	sandy silt
G3	70-72	45.2	28.6	26.2	silty sand
G3	AVERAGE	38.4	34.8	26.8	silty sand
G6	8-10	28.9	50.0	21.1	sandy silt
G6	28-30	29.7	49.2	21.1	sandy silt
G6	50-52	40.2	39.3	20.5	silty sand
G6	87-90	31.3	48.8	19.9	sandy silt
G6	101-103	25.6	53.5	20.9	sandy silt
G6	108-110	28.9	53.2	17.9	sandy silt
G6	147-150	56.3	31.8	11.9	silty sand
G6	AVERAGE	34.4	46.5	19.0	sandy silt
G7	5-8	31.1	45.3	23.6	sandy silt
G7	23-27	31.2	48.2	20.6	sandy silt
G7	48-52	45.7	38.8	15.5	silty sand
G7	78-80	45.2	35.3	19.5	silty sand
G7	104-108	63.6	23.4	13.0	silty sand
G7	148-152	68.1	21.6	10.3	silty sand
G7	177-179	58.2	24.3	17.5	silty sand
G7	204-206	59.2	27.2	13.6	silty sand
G7	250-252	33.9	50.1	16.0	sandy silt
G7	AVERAGE	48.5	34.9	16.6	silty sand
G8	5-8	19.0	49.9	31.1	clayey silt
G8	28-30	16.6	47.4	36.0	clayey silt
G8	48-52	20.7	53.2	26.1	clayey silt
G8	78-80	13.8	59.9	26.3	clayey silt
G8	101-103	25.9	50.5	23.6	sandy silt
G8	130-132	34.4	44.9	20.7	sandy silt
G8	AVERAGE	21.7	51.0	27.3	clayey silt
G9	5-8	21.2	50.5	28.3	clayey silt
G9	28-30	19.9	47.1	33.0	clayey silt
G9	48-52	20.1	63.3	16.6	sandy silt
G9	78-82	19.8	56.0	24.2	clayey silt
G9	AVERAGE	20.3	54.2	25.5	clayey silt

Sediment Composition

A smear slide representative of the entire sample, obtained from specific intervals in each core, was viewed under a petrographic microscope, the sand or coarse fraction (> 63 μ) was studied under a low-power dissecting microscope (50 x), the clays were analyzed with an X-ray diffractometer, and the carbonate and organic carbon contents were also measured. (See appendix for procedural details).

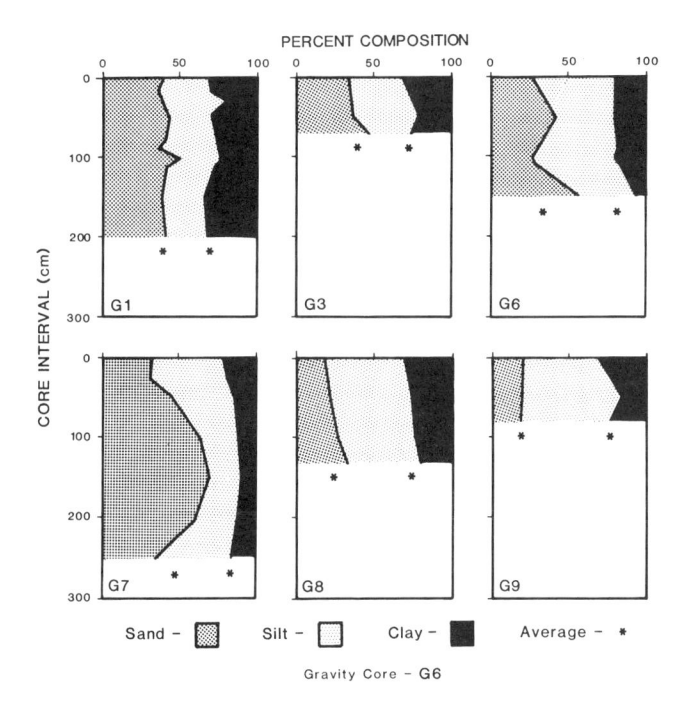

Figure 8. Plots of sand-silt-clay percents versus depth for the six gravity cores. See Figure 1 for locations.

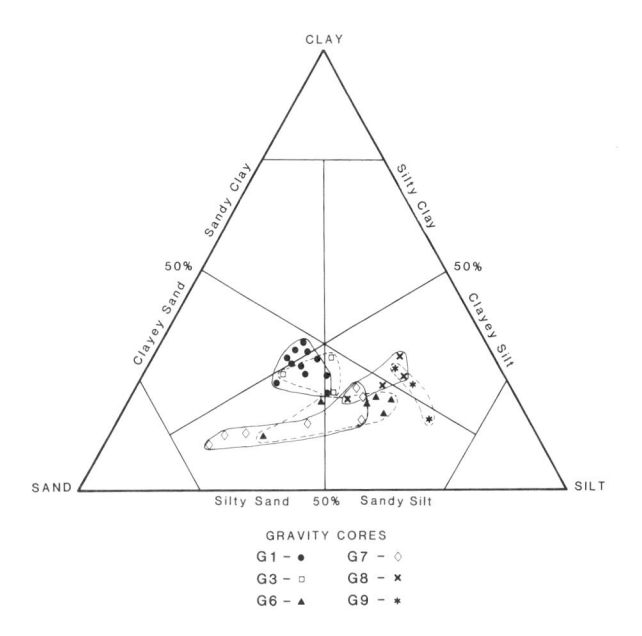

Figure 9. Ternary diagram showing textural nomenclature for the core subsamples.

Smear Slides

Estimates of composition of the total subsample based on smear slides are listed in Table 3. Nannofossils are the most abundant biogenic constituent (25-75 %) in all of the cores, followed by planktonic foraminifers (5-25 %) and spicules (1-25 %). The spicules are primarily sponge, but some tunicate spicules are present. Other biogenic material present in rare (< 5 %) to trace (< 1 %) amounts includes radiolarians, diatoms, silicoflagellates, and plant fragments. Detrital mineral grains are present in small amounts, about 1-5 %, becoming somewhat more plentiful in the east. Volcanic glass content also increases to the east. Several types of glass are present and include clear to white bubble-filled shards; light brown shards; and clear glass shards that contain green lath-shaped crystallite inclusions. The green inclusions, probably pyroxene, are most common in the more easterly cores (G7, 8 and 9).

Coarse Fraction

Counts of the sand size components of each subsample are listed in Table 4. Immediately apparent is the dominance of biologic components, especially planktonic foraminifers that make up 50-90 % of each coarse fraction (Figure 10). Pteropods are also important contributors, making up 5-20 % of the sand-size material except in the tops of the cores, where the percentages of pteropods decrease to < 5 %, and often to < 1 % in the oxidized zone.

Volcanic glass is the most plentiful detrital constituent in most of the subsamples, reaching values of 30 to 40 % of the sand fraction in some cores (Table 4). No discrete glass or ash layers were found, but in two cores (3 and 6), pieces of pumice of pebble size were found. Figure 10 illustrates a pronounced increase of volcanic glass in surface sediment from the west to the east, an increase also apparent in the average percentage of glass in the cores (Table 4).

The higher incidence of volcanic glass in the eastern cores correlates with their proximity to the islands of Lihir and Tabar (Figure 1), both of which contain late Cenozoic volcanoes (Johnson, Mackenzie, and Smith, 1978). Late Cenozoic volcanoes near the western core sites are potential sources of the glass and pumice found in cores 1 and 3. Johnson and Smith (1974) report that Tuluman volcano, 35 km southeast of Manus Island, erupted between 1953 and 1957. However, the Tertiary island arc including

Table 3. Smear slide composition estimates.

Core	Depth (cm)	Nano*	PFm	Spic	Rads	Diat	Det	VGl
G1	5-7	A-D**	C	R	T	T	T	R
G1	15-17	A	C-A	R-C			R	T
G1	27-29	A	C	R-C			R	
G1	33-34	A	C	R-C				T
G1	48-49	A	C	C	T		R	T
G1	83-84	A	C	R-C			T	T
G1	87-89	A	C	R-C			R	T
G1	101-102	A	C	R-C			R	T
G1	102-103	A	C	R	T	T	R-C	R
G1	130-131	A	A	R			R	
G1	153-154	A	C	R			R	R-C
G1	165-166	A	C	T	R	T	R	T
G1	177-178	A	C	R-C			T	R
G1	201-202	A	C	R-C			R	
G1	225-226	A	C	R-C			R	R
G3	4-5	A	C	R-C			R	R
G3	9-10	A	C	R-C			R	R
G3	18-19	A	C	C				R
G3	23-24	A	C	R-C			R	T-R
G3	29-30	A	A	R-C				T
G3	37-38	A	C	C	T		R	T-R
G3	52-53	A	C-A	C			R	T-R
G3	71-72	C	A	R-C	T		R	T
G6	6-7	A	C		R-C		R	R
G6	12-13	A	C	R-C			R	R-C
G6	15-16	A	C	R-C			R	C
G6	32-33	A	C	R				R-C
G6	35-36	A	C	R-C			R	R
G6	40-41	A-D	C	R			T	
G6	101-102	A	C					T
G6	109-110	A	R-C				T	T-R
G6	123-124	A	C	R-C			R	T
G6	126-127	A	C	R	T		R-C	T
G6	150-151	A	C-A	R-C			R	T
G7	6-7	A	C	R			R	T
G7	25-26	A	C				R-C	R
G7	50-51	A	C	R-C			R-C	R
G7	60-61	A	C	C			R	T
G7	73-74	A	C	R			R-C	R-C
G7	76-77	A	C	R			T	C
G7	90-91	A	A				T	T
G7	108-109	A	C	R			R-C	A
G7	140-141	A	C				R	R
G7	180-181	A	C	C			R	R
G7	203-204	A	C	R			R	
G7	225-226	A	A				R	
G7	228-229	A	C	C			R-C	T
G7	235-236	A	C	R			R-C	R
G7	240-241	A	A	R			R-C	T
G7	258-259	A	C	R-C			C	T
G8	6-7	A	C		R-C	R	R	R
G8	25-26	A	C	R	T		R-C	R
G8	75-76	A	C	R	R			R
G8	82-83	A	C	R	T			C
G8	105-106	A	C	R			R	T
G9	6-7	A	C		T	R	R-C	R
G9	11-12	A	C				R-C	C-A
G9	17-19	A	C	T-R	T-R	T	R	C
G9	50-51	A	C	R			R	R
G9	53-54	A	C	R			R-C	R-C
G9	75-76	A	C	R			C	T-R
G9	81-83	A	C	R			R-C	R-C

* Nano = nannoplankton, PFm = planktonic foraminifers, Spic = spicules, Rads = radiolarians, Diat = diatoms, Det = detrital grains, VGl = volcanic glass

** D = Dominant (>75%) A = Abundant (25-75%) C = Common (5-25%) R = Rare (1-5%) T = Trace (<1%)

Manus Island, forms a ridge that acts as a barrier to sediment transport between the volcanoes to the

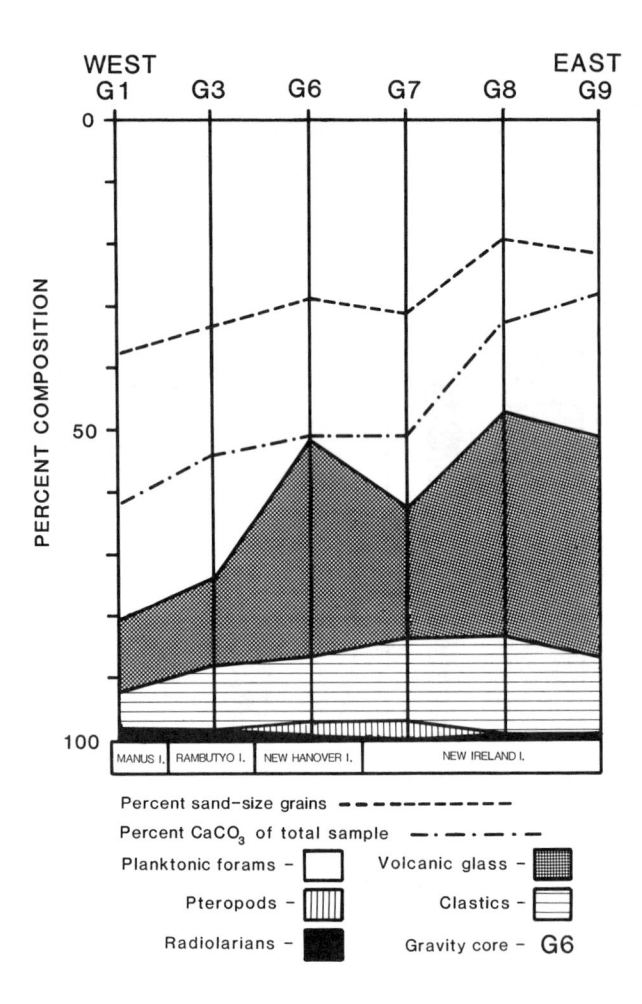

Figure 10. Composition of sand fraction of surface sediment from each core. Note also percentages of sand-size grains and of calcium carbonate.

south and the cores to the north. This could result in significantly lower glass content in cores 1 and 3.

Clay Mineralogy

The clay-size fraction ($< 2\ \mu$) ranged from 10 to 36 % of the total sample (Table 2). X-ray diffractograms of the sediment from the top, bottom, and 50 cm interval of the cores showed at least 50 % of the clay-sized fraction was composed of smectite (Table 5). The smectite represents a mixed layer smectite-illite phase with about 70 to 75 % expandable layers. Vermiculite and other mixed layer clays (possibly vermiculite-illite) were also recorded in the percentage calculations.

The amount of mixed layer clays increases eastward (Figure 11). The westernmost cores (G1 and G3), collected north of Manus and Rambutyo

Table 4. Identification of sand fraction (percent composition).

Core	Depth (cm)	Lts*	Hvy	Mica	VGl	Pter	PFm	BFm	Rads	Det	Biol
G1	5-8	5.2	0.2	0.0	12.1	0.2	80.8	0.0	1.5	17.5	82.5
G1	15-16	3.1	0.6	0.3	10.4	0.3	84.4	0.0	0.9	14.4	85.6
G1	30-32	3.2	0.9	0.0	3.2	0.0	92.1	0.0	0.6	7.3	92.7
G1	48-50	1.7	0.0	0.0	2.0	4.7	91.1	0.5	0.0	3.7	96.3
G1	50-52	5.7	1.3	0.6	6.0	5.7	80.4	0.0	0.3	13.6	86.5
G1	78-80	5.8	0.0	0.3	5.5	0.6	87.8	0.0	0.0	11.6	88.4
G1	87-89	11.6	0.6	0.3	5.7	0.3	80.0	0.0	1.5	18.2	81.8
G1	102-104	24.1	0.3	0.0	8.4	0.3	66.0	0.3	0.6	32.8	67.2
G1	110-113	14.5	0.0	0.0	4.8	0.0	79.0	0.0	1.7	19.3	80.7
G1	150-152	10.8	0.0	0.0	5.9	0.0	83.0	0.0	0.3	16.7	83.3
G1	202-204	10.1	0.3	0.0	4.0	0.0	85.6	0.0	0.0	14.4	85.6
G1	AVERAGE	8.7	0.4	0.1	6.2	1.1	82.7	0.1	0.7	15.4	84.6
G3	4-7	10.5	0.0	0.0	13.7	0.0	74.3	0.0	1.5	24.2	75.8
G3	28-30	3.0	0.3	0.7	5.6	5.0	85.4	0.0	0.0	9.6	90.4
G3	47-50	2.4	0.3	0.0	1.2	17.3	78.8	0.0	0.0	3.9	96.1
G3	70-72	1.1	0.0	0.0	2.0	10.0	86.9	0.0	0.0	3.1	96.9
G3	AVERAGE	4.3	0.2	0.2	5.6	8.1	81.4	0.0	0.4	10.2	89.8
G6	8-10	8.6	0.0	0.3	33.4	3.4	53.4	0.0	0.9	42.3	57.7
G6	28-30	8.2	0.0	0.0	27.4	5.7	58.1	0.0	0.6	35.6	64.4
G6	50-52	2.9	0.3	0.5	9.2	18.3	68.5	0.3	0.0	12.9	87.1
G6	87-90	3.8	0.3	0.0	19.4	4.3	72.2	0.0	0.0	23.5	76.5
G6	101-103	7.1	0.0	0.0	25.9	7.1	59.6	0.0	0.3	33.0	67.0
G6	108-110	10.8	0.0	0.6	20.9	7.0	60.7	0.0	0.0	32.3	67.7
G6	147-150	8.2	0.0	0.0	7.3	6.3	78.2	0.0	0.0	15.5	84.5
G6	AVERAGE	7.1	0.1	0.2	20.5	7.4	64.4	0.0	0.3	27.9	72.1
G7	5-8	10.5	0.9	0.0	20.7	4.4	63.3	0.0	0.2	32.1	67.9
G7	23-27	5.2	0.0	0.0	31.5	5.8	57.5	0.0	0.0	36.7	63.3
G7	48-52	5.8	0.8	0.3	34.4	13.3	45.4	0.0	0.0	41.3	58.7
G7	78-80	9.9	0.0	0.0	21.1	7.6	61.4	0.0	0.0	31.0	69.0
G7	104-108	4.7	0.0	0.0	8.0	8.6	78.7	0.0	0.0	12.7	87.3
G7	148-152	8.1	0.0	0.0	10.2	4.8	76.9	0.0	0.0	18.3	81.7
G7	177-179	7.0	0.0	0.0	8.1	1.2	83.7	0.0	0.0	15.1	84.9
G7	204-206	8.0	1.0	0.0	11.1	1.3	78.6	0.0	0.0	19.1	80.9
G7	250-252	20.3	0.3	0.5	17.0	7.3	54.1	0.0	0.5	38.1	61.9
G7	AVERAGE	8.8	0.3	0.1	18.0	6.0	66.6	0.0	0.1	27.2	72.8
G8	5-8	14.7	0.0	0.0	35.9	0.0	48.2	0.0	1.2	50.6	49.4
G8	28-30	8.5	0.0	0.0	14.2	0.3	77.0	0.0	0.0	22.7	77.3
G8	48-52	13.1	3.1	0.0	10.6	15.4	57.5	0.0	0.3	27.0	73.0
G8	78-80	16.6	0.6	0.0	22.5	3.7	53.5	0.0	3.1	39.7	60.3
G8	101-103	6.6	0.0	0.9	4.6	5.8	82.1	0.0	0.0	12.1	87.9
G8	130-132	10.6	0.6	0.3	40.1	2.2	46.2	0.0	0.0	51.6	48.4
G8	AVERAGE	11.7	0.7	0.2	21.3	4.6	60.8	0.0	0.8	34.0	66.0
G9	5-8	8.8	3.1	0.0	34.8	0.3	52.4	0.0	0.6	46.7	53.3
G9	28-30	12.4	0.0	0.3	17.8	0.0	69.5	0.0	0.0	30.5	69.5
G9	48-52	11.6	1.2	0.6	27.4	9.5	49.4	0.0	0.3	40.8	59.2
G9	78-82	23.4	0.0	0.3	13.9	5.3	57.1	0.0	0.0	37.6	62.4
G9	AVERAGE	14.1	1.1	0.3	23.5	3.8	57.1	0.0	0.2	38.9	61.1

* Lts = light minerals (quartz and feldspar) and rock fragments,
Hvy = heavy minerals, VGl = volcanic glass, Pter = pteropods,
PFm = planktonic foraminifers, BFm = benthic foraminifers, Rads = radiolarians,
Det = detrital grains (total), Biol = biological grains (total)

Islands, contained more than twice as much illite (> 10 %) than did the other four cores. Kaolinite plus chlorite, which ranged from 23 to 40 %, is more irregularly distributed (Figure 11). Other minerals identified on the diffractograms were amphibole, present only at the bottom of the cores, and zeolites, present in small amounts in most samples.

Table 5. Clay mineralogy (percent composition).

Core	Depth (cm)	S/I*	Exp	I	K+C	(K;C)
G1	10-11	50	75	16	34	(15;19)
G1	50-52	55	71	12	33	(16;17)
G1	224-226	56	70	19	25	(11;14)
G3	8-9	61	78	10	29	(13;16)
G3	50-52	62	75	11	27	(14;13)
G3	72-73	61	75	13	26	(11;15)
G6	5-7	58	78	2	40	(Indet)
G6	50-52	63	77	1	36	(17;19)
G6	149-150	63	75	5	32	(15;17)
G7	5-7	73	77	3	24	(12;13)
G7	50-52	62	74	2	36	(17;19)
G7	258-259	60	77	0	40	(Indet)
G8	8-9	60	83	2	38	(16;22)
G8	50-52	61	77	2	37	(19;18)
G8	131-132	71	84	4	25	(Indet)
G9	0-2	63	85	4	33	(Indet)
G9	50-51	74	75	3	23	(Indet)

* S/I = Mixed Layer Smectite-Illite
Exp = Expandable layers in S/I
I = Illite
(K;C) = % Kaolinite and % Chlorite from K+C
Indet = Indeterminate

Carbon Content

The total carbon values obtained from subsamples taken at < 1 m spacings in each of the six gravity cores range from 3.78 % to 9.75 % and average 7.23 % (Table 6). The organic carbon component is moderate, ranging from 0.08 % to 0.83 % and averaging 0.47 %. The inorganic or carbonate carbon is the overwhelming contributor to the high carbon values and ranges from 3.29 % to 9.37 % and averages 6.77 %. If we express the inorganic carbon as $CaCO_3$, the cores show a dramatic decrease from west to east. This trend in relation to the planktonic foraminifer content can readily be seen by comparing the percent of $CaCO_3$ in surface samples to the composition of the coarse fraction (Figure 10).

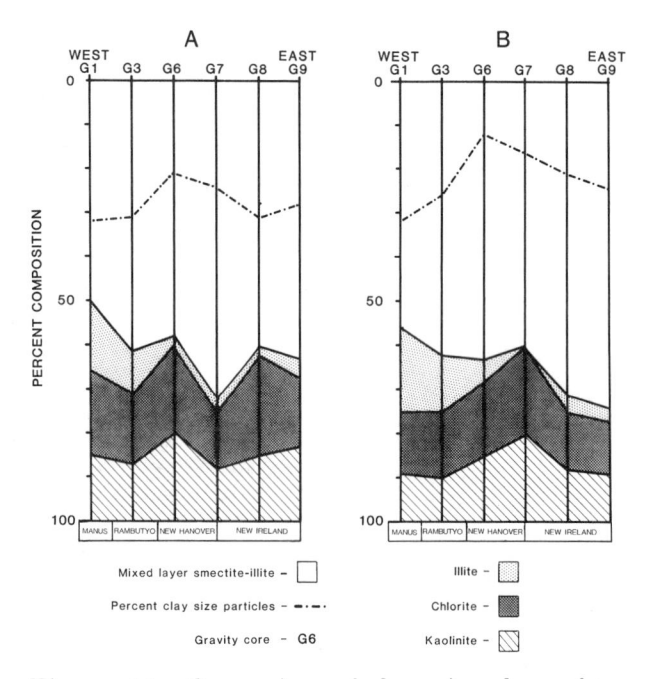

Figure 11. Comparison of clay mineralogy of tops (A) and bottoms (B) of cores. See Table 5 for sample intervals.

Geotechnical Properties

The geotechnical properties measured on the split gravity cores include water content, undisturbed shear strength, and remolded shear strength (Table 7). The water content, expressed as a percentage of the dry weight of the sediment, ranges from 73.6 to 130.5 % (Figure 12a). These values are comparable to water content for continental-slope sediment from the eastern United States, which ranged from 37 to 140 % and averaged 85 % (Keller, Lambert, and Bennett, 1979). Lee's (1978) measurements of the water content of $CaCO_3$-rich (35-90 %) cores from the eastern equatorial Pacific ranged from 80 to > 200 %. The peak undrained shear strength of our cores increases with depth in the core (Figure 12b), as would be expected; however, core G7 shows several abrupt changes with depth. The strengths measured range from a low of 4.5 kPa in core 9 to 34.9 kPa in core 7 (average 12.3 kPa). These shear strength values are somewhat greater than those measured by Lee (1978) on eastern equatorial Pacific cores (4.1-24.5 kPa; ave. 8.9 kPa) and for continental slope sediment off the eastern United States, which ranged from 0.9 to 24.1 kPa and yielded a mean undrained shear strength of 8.3 kPa (Keller, Lambert, and Bennett, 1979).

Calculations of sediment sensitivity (St) [peak undisturbed van shear strength (Su)/remolded vane

Table 6. Carbon / carbonate content.

Core	Depth (cm)	Total C %	Organic C %	Inorganic C %	CaCO$_3$ %
G1	11-12	7.74	0.27	7.47	62.18
G1	21-23	9.16	0.52	8.65	72.01
G1	102-105	7.98	0.70	7.29	60.68
G1	202-204	8.76	0.83	7.94	66.10
G1	Average	8.41	0.58	7.84	65.24
G3	5-7	6.88	0.42	6.47	53.85
G3	21-26	8.46	0.22	8.24	68.60
G3	71-73	9.42	0.46	8.96	74.64
G3	Average	8.25	0.37	7.89	65.70
G6	6-8	6.68	0.60	6.08	50.65
G6	20-22	7.32	0.55	6.78	56.44
G6	88-90	8.34	0.52	7.83	65.18
G6	102-103	6.71	0.41	6.31	52.52
G6	149-150	9.75	0.38	9.37	78.05
G6	Average	7.76	0.49	7.27	60.57
G7	5-10	6.47	0.39	6.08	50.60
G7	23-27	6.49	0.60	5.89	49.02
G7	102-106	9.28	0.08	9.20	76.64
G7	185-189	8.49	0.43	8.06	67.10
G7	205-206	8.47	0.38	8.09	67.39
G7	256-257	5.77	0.36	5.41	45.07
G7	Average	7.50	0.37	7.12	59.30
G8	6-9	4.60	0.68	3.92	32.65
G8	87-90	5.43	0.45	4.98	41.44
G8	100-105	6.44	0.54	5.90	49.15
G8	128-132	5.51	0.45	5.06	42.15
G8	Average	5.50	0.53	4.97	41.35
G9	5-10	3.78	0.50	3.29	27.36
G9	78-82	5.68	0.48	5.20	43.27
G9	Average	4.73	0.49	4.25	35.32
	Total Average	7.23	0.47	6.77	56.36

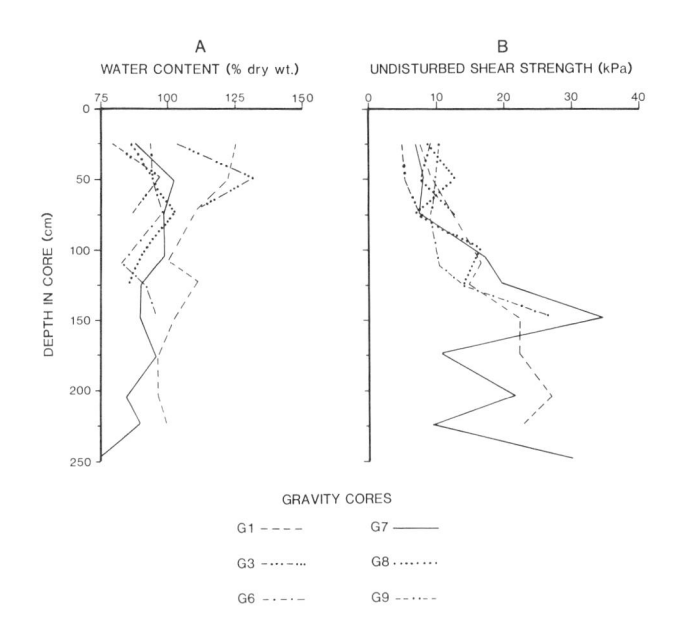

Figure 12. A) Variations of water content with depth in core. B) Undisturbed shear strength (miniature vane) versus depth in core.

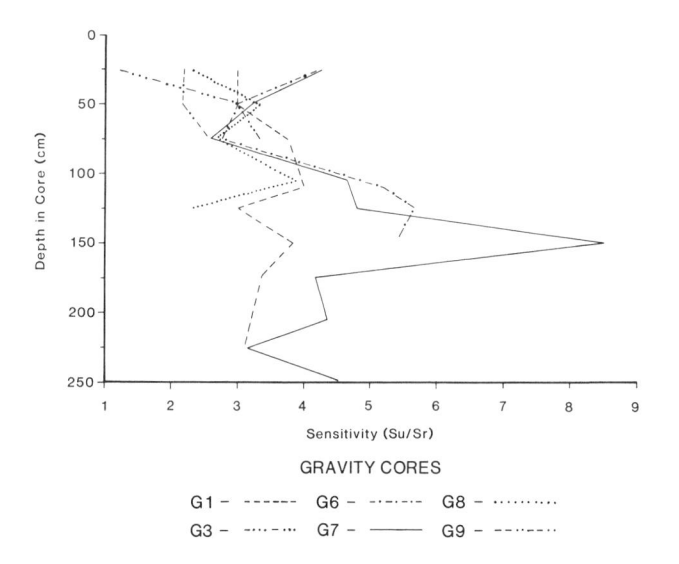

Figure 13. Changes in sediment sensitivity with depth in cores.

shear strength (Sr)] categorize most of the sediment as "medium sensitive" (2 < St < 4) (Mitchell, 1976, p. 208); however, most of the sediment from the two centrally located cores (G6, G7; Figure 13) is "very sensitive" (4 < St < 8) and thus subject to considerable strength loss as a result of disturbance or shock (such as might be generated by an earthquake).

To compare the results from the six New Ireland Basin cores of different lengths, we averaged the results of the top 75 cm (Table 8). From east to west the water content increases, as do peak undrained shear strength, CaCO$_3$, and planktonic foraminifer percentages, but the percentage of silt in the cores decreases. The centrally located cores (G6 and G7) show slightly greater sensitivity than the other cores, especially deeper in the cores (Figure 13).

Calcareous Nannoplankton

Table 9 shows the distribution of calcareous nannofossils and other microfossils that are present in smear slides taken from the tops and bottoms of the cores. Quaternary calcareous nannoplankton, dominated by *Gephyrocapsa oceanica*, are abundant in all cores. This species first appears at the base of

Table 7. Physical properties.

Core #	Depth cm	Water Cont. %	Bulk Density g/cc	Porosity %	Undist. Su* kPa	Remold Sr** kPa	Sensitivity St=Su/Sr	$CaCO_3$ %	Organic Carbon %
G1	11	----	----	----	----	----	----	62.18	0.27
G1	25	124.4	1.41	76.4	7.38	2.46	3.00	72.01	0.52
G1	50	120.8	1.42	75.9	9.24	3.28	3.00	----	----
G1	75	108.9	1.45	73.9	12.30	3.28	3.75	----	----
G1	110	99.7	1.48	72.2	16.40	4.10	4.00	60.68	0.70
G1	125	110.9	1.45	74.3	14.76	4.92	3.00	----	----
G1	150	102.5	1.47	72.7	22.14	5.74	3.86	----	----
G1	175	95.1	1.50	71.3	22.14	6.56	3.38	----	----
G1	205	96.3	1.49	71.5	27.06	----	----	66.10	0.83
G1	225	99.7	1.48	72.2	22.96	7.38	3.11	----	----
G1	Ave.	106.5	1.46	73.4	17.15	4.72	3.39	65.24	0.58
G3	05	----	----	----	----	----	----	53.85	0.42
G3	25	101.5	1.48	72.5	9.02	2.46	1.22	68.60	0.22
G3	50	130.5	1.40	77.3	7.38	2.46	3.00	----	----
G3	75	109.6	1.45	74.1	12.30	3.69	3.33	74.64	0.46
G3	Ave.	113.9	1.44	74.6	9.57	2.87	2.52	65.70	0.37
G6	06	----	----	----	----	----	----	50.65	0.60
G6	25	92.6	1.51	70.7	10.25	2.46	4.17	56.44	0.55
G6	50	93.4	1.50	70.9	9.84	3.28	3.00	----	----
G6	75	97.5	1.49	71.7	9.02	3.28	2.75	----	----
G6	88	----	----	----	----	----	----	65.18	0.52
G6	110	82.3	1.55	68.2	10.66	2.05	5.20	52.52	0.41
G6	125	91.4	1.51	70.4	13.94	2.46	5.67	----	----
G6	145	94.7	1.50	71.2	26.65	4.92	5.42	60.57	0.49
G6	Ave.	92.0	1.51	70.5	13.39	3.08	4.37	57.07	0.51
G7	05	----	----	----	----	----	----	50.60	0.39
G7	25	87.4	1.53	69.5	6.97	1.64	4.25	49.02	0.60
G7	50	102.2	1.47	72.7	7.79	2.46	3.17	----	----
G7	75	97.4	1.49	71.7	7.38	2.87	2.57	----	----
G7	106	97.7	1.49	71.8	17.22	3.69	4.67	76.64	0.08
G7	125	89.7	1.52	70.0	19.68	4.10	4.80	----	----
G7	150	87.9	1.53	69.6	34.85	4.10	8.50	----	----
G7	175	95.3	1.50	71.3	10.25	2.46	4.17	----	----
G7	185	----	----	----	----	----	----	67.10	0.43
G7	205	83.7	1.54	68.5	21.32	4.92	4.33	67.39	0.38
G7	225	89.2	1.52	69.9	9.02	2.87	3.14	----	----
G7	250	73.6	1.59	65.7	29.93	6.56	4.56	45.07	0.36
G7	Ave.	90.4	1.52	70.1	16.44	3.57	4.42	59.30	0.37
G8	06	----	----	----	----	----	----	32.65	0.68
G8	25	85.8	1.53	69.1	8.61	3.69	2.33	----	----
G8	50	94.3	1.50	71.1	12.30	3.69	3.33	----	----
G8	75	100.2	1.48	72.3	6.56	2.46	2.67	----	----
G8	87	----	----	----	----	----	----	41.44	0.45
G8	105	89.1	1.52	69.9	15.99	4.10	3.90	49.15	0.54
G8	125	84.3	1.54	68.7	14.35	6.15	2.33	42.15	0.45
G8	Ave.	90.7	1.51	70.2	11.56	4.02	2.91	41.35	0.53
G9	05	----	----	----	----	----	----	27.36	0.50
G9	25	78.3	1.57	67.1	4.51	2.05	2.20	----	----
G9	50	96.9	1.49	71.6	5.33	2.46	2.17	----	----
G9	75	85.3	1.54	69.0	7.38	2.87	2.57	43.27	0.48
G9	Ave.	86.8	1.53	69.2	5.74	2.46	2.31	35.32	0.49

* Su = Undisturbed shear strength   ** Sr = Remolded shear strength

the Quaternary *G. caribbeanica* Zone (CN14) and ranges to the present (Boudreaux and Hay, 1967, Okada and Bukry, 1980); therefore, all cores examined are assigned an age of CN14 or younger. Core G7 (258 cm) differs from the other cores in that *G. sp. cf. G. caribbeanica* is more abundant than *G. oceanica* (Table 9). This could indicate that the bottom of this core, where *G. oceanica* is sparse relative to *G. sp. cf. G. caribbeanica*, is slightly older than the other cores, because *G. caribbeanica* occurs earlier in the Quaternary than does *G. oceanica*. An alternative explanation for this change in dominance may be that water temperature was lower; *G. caribbeanica* is more tolerant of cooler water, being found in subpolar as well as tropical water (McIntyre, Be, and Roche, 1970).

Table 8. Average of physical properties from upper 75 cm of cores.

Core #	Water Content %	Undist. Strength kPa	Sensitivity St=Su/Sr	CaCO$_3$ %	Plank. Forams %	Detri. %	Sand %	Silt %	Clay %
G1	118.0	9.64	3.25	67.1	85.8	11.3	39.3	31.9	28.7
G3	113.9	9.57	2.52	65.7	79.5	12.6	36.1	36.9	27.0
G6	94.5	9.70	3.31	53.5	60.0	30.3	32.9	46.2	20.9
G7	95.7	7.38	3.33	49.8	55.4	36.7	36.0	44.1	19.9
G8	93.4	9.16	2.78	32.7	60.9	33.4	18.8	50.2	31.1
G9	86.8	5.74	2.31	35.3	57.1	39.3	20.4	53.6	26.0

Cores G1, G3, G6, and the top of G7 contain specimens that resemble *Emiliania huxleyi*, but are small, and therefore difficult to identify with a light microscope. Cores that contain *E.* sp. aff. *E. huxleyi* can only be questionably assigned to the late Quaternary nannofossil zone CN15. The lack of *E.* sp. aff. *E. huxleyi* in the lower part of G7 (258 cm) is further evidence that the lower part of this core may be older than the other cores.

Discoaster, a pre-Quaternary genus, is present in four cores; however, sparseness and poor preservation indicate that the specimens are reworked.

Planktonic Foraminifers

Although Belford (this volume) assigns all 6 cores to Zone N 23 (late? Pleistocene-Holocene), some evidence exists that the bottom of G7 may be slightly older than the other cores. As shown on Belford's Table 5, G7 is the only core to contain pink specimens of *Globigerinoides ruber*. Thompson and others (1979) report that pink forms of this species disappeared from the Pacific and Indian Oceans 120,000 years ago; if this is the case, then the bottom of core G7 would be older than 120,000 years. However, Chaproniere (1985) reported a few specimens of the pink form just north of Fiji in sediment younger than 120,000 years.

Other Microfossils

Other microfossils present in the cores include: benthic foraminifers, diatoms, radiolarians, sponge spicules, ascidian spicules, and silicoflagellates. The ascidian spicules may indicate downslope transport of sediment, as they are known to be common from intertidal depths and less common in deeper water down to 500 m (Hekel, 1973).

DISCUSSION

The surficial sediment found on the northern insular slope from Manus to New Ireland is dominated by silt and fine sand, and varies from sandy silt to silty sand to clayey silt. The dominant constituent of the sand fraction is planktonic foraminifers, but pteropods and volcanic glass are also important components of most of the cores. Glass or ash is more plentiful in the more easterly cores located off New Ireland. The mud fraction (< 63 μ) of all the cores is dominated by calcareous nannoplankton. The overall dominance of planktonic foraminifers and nannoplankton is reflected by the high CaCO$_3$ content (average 56 %).

Thus, despite the proximity of the insular slope samples to the islands (15-40 km), the sediment is dominated by planktonic constituents with secondary input of volcaniclastics from the islands. Because the nearby tropical islands are lushly vegetated, one might expect large amounts of organic material in the slope sediment. However, the organic carbon values measured in our core samples range from moderate to low (0.83 to 0.08 %). These low carbon values are consistent with the normal oceanic background amounts of methane (260-1200 nL/L) and other hydrocarbons measured in our gravity cores (Kvenvolden, this volume).

A possible explanation for these low carbon values is that much of the island-derived sedimentary material (clastic and organic) is carried downslope through gullies into the intraslope and trench basins, bypassing the slope. This explanation is supported by studies in an adjacent area where Colwell and Tiffin (1986) report the common occurrence of turbidity-current deposits in the deeper parts of the Central Solomons Trough. Other explanations for the low organic carbon could be the dominance of chemical weathering on land, oxidation at the sea floor, or ingestion by bottom fauna. None of the sediment

Table 9. Calcareous nannoplankton distribution.

Species	Core (cm)	G1 10	G1 224	G3 05	G3 72	G6 05	G6 149	G7 05	G7 258	G8 05	G8 131	G9 02	G9 81
Ceratolithus cristatus		F*	R	R		R	R	F		R	R	R	R
C. telesmus		R	R	F	R	R	R	R	R	R	R	R	R
Coccolithus pelagicus										R			
Cyclococcolithina leptopora		F	C	C	C	C	R	C	R	R	R	F	
Discoaster cf. *D. asymmetricus*												R	
D. brouweri									R				
D. cf. *D. variabilis*							R						
D. sp.											R		
Discolithina cf.*D. millepuncta*					R							F	
D. spp.		R	F		R				R	R	R		R
Emiliania aff. *E. annula*							R	R		R			
E. aff. *E. huxleyi*		R	C	R	R	T	R	R					
E. aff. *E. ovata*							R	R	R		R		R
Gephyrocapsa cf. *G. caribbeanica*		R	R	R	R	R	R	R	C	F	R		R
G. oceanica		A	A	A	A	C	C	A	F	A	A	A	A
G. cf. *G. omega*		R	R	R								R	
Helicosphaera spp.		F	F	R	F	F		F		R	R	F	R
Rhabdosphaera			R	R		R	R			R			
Scyphosphaera			R		R								
Sphenolithus spp.		R	R			R		R	R	R			
Thoracosphaera		R	R		F	R		F	R	R		F	R
Umbilicosphaera cf. *U. mirabilis*			R										
foraminifers		T	T	T	C	C	F	R	R	R	T	T	T
diatoms							R		R		R	R	
radiolarians		T	T	T	R	R	R	R	R	R	R	R	R
sponge spicules		R	T	T	T	R	R	R	R	R	R	R	R
ascidian spicules				T		R	R	R	R			T	R
silicoflagellates							T						

* A = Abundant   C = Common   F = Few   R = Rare   T = Trace

cores have any apparent laminae, and the mottles and burrows present on newly split core surfaces suggest considerable bioturbation by bottom fauna.

The sea-floor morphology, characterized by numerous gullies on the slopes off Manus and New Ireland, indicates widespread downslope transport and erosion. The high-resolution seismic-reflection records show that older gullies are also present, often slightly offset from modern gullies and buried by up to 300 m of sediment. Some of the present-day gullies show evidence of partial filling, whereas others appear to be devoid of sediment. Some of the seismic-reflection records also show evidence of submarine sliding and slumping of the slope sediment. Hence, sediment has undergone various types of downslope transport; the locations of mass movement and sites of deposition have varied with time.

Although we could not determine the Pleistocene/Holocene boundary in any of the cores, G7 appears to be the oldest of the six gravity cores. This is substantiated by both planktonic foraminiferal and calcareous nannofossil evidence. We conclude that late Pleistocene and Holocene sediment mantles much of the insular slope; exceptions are steep scarps from which were dredged rocks as old as late Miocene (Belford, this volume; Frankel and Exon, this volume).

REFERENCES

Biscaye, P.E., 1965, Mineralogy and sedimentation of recent deep-sea clay in the Atlantic Ocean and adjacent sea and oceans: Geological Society American Bulletin, v. 76, p. 803-832.

Boudreaux, J.E. and Hay, W.W., 1967, Zonation of the latest Pliocene-Recent interval: Gulf Coast Association of Geological Societies Transactions, v. 17, p. 443-445.

Carver, R.E., 1971, Procedures in sedimentary petrography; New York, John Wiley and Sons, Inc., 653 p.

Chaproniere, G.C.H., 1985, Late Tertiary and Quaternary foraminiferal biostratigraphy and paleobathymetry of cores and dredge samples from cruise KK820316 Leg 2, in Brocher, T.M., (ed.), Geological Investigations of the Northern Melanesian Borderland: Circum-Pacific Council for Energy and Mineral Resources Earth Science Series, v. 3, p. 103-122.

Colwell, J.B., and Tiffin, D.L., 1986, Recent depositional patterns in the Central Solomons Trough of the Solomon Islands, in Vedder, J.G., Pound, K.S., and Boundy, S.Q., (eds.), Geology and offshore resources of Pacific island arcs--central and western Solomon Islands, Circum-Pacific Council for Energy and Mineral Resources Earth Science Series, v4, p. 243-254.

Exon, N.F., Colwell, J.B., and Bolton, B.R. 1984, Sedimentology of Quaternary cores from the Solomon Islands offshore region, in Exon, N., and Taylor, B., (eds.), Seafloor spreading, ridge subduction, volcanism and sedimentation in the offshore Woodlark-Solomons region and Tripartite cruise report for *Kana Keoki* cruise 82-03-16 Leg 4: CCOP/SOPAC Technical Report n. 34, p. 209-243.

Exon, N.F., and Tiffin, D.L., 1984, Geology and petroleum prospects of offshore New Ireland Basin in northern Papua New Guinea: Transactions, third Circum-Pacific Energy and Mineral Resources Conference, p. 623-630.

Hamilton, Warren, 1979, Tectonics of the Indonesian Region: U.S. Geological Survey Professional Paper 1078, 345 p.

Hein, J.R., Scholl, D.W., and Gutmacher, C.E., 1976, Neogene clay minerals of the far northwest Pacific and southern Bering Sea: Sedimentation and diagenesis, in Bailey, S.W., (ed.); Proceedings of the 1975 International Clay Conference: Willmette, Illinois, Applied Publications Limited, p. 71-80.

Hekel, H., 1973, Late Oligocene to Recent nannoplankton from the Capricorn Basin (Great Barrier Reef Area); Geological Survey of Queensland, Publication n. 359, Palaeontological Papers, n. 33, 24 p.

Johnson, R.W., Mackenzie, D.E., and Smith, I.E.M., 1978, Volcanic rock associations at convergent plate boundaries; Reappraisal of the concept using case histories from Papua New Guinea: Geological Society of America Bulletin, v. 89, p. 96-106.

Johnson, R.W. and Smith, I.E., 1974, Volcanoes and rocks of St. Andrew Strait, Papua New Guinea; Journal of Geological Society of Australia, v. 21, p. 333-351.

Keller, G.H., Lambert, D.N., and Bennett, R.H., 1979, Geotechnical properties of continental slope deposits - Cape Hatteras to Hydrographer Canyon, in Doyle, L.J., and Pilky, O.H., Jr., (eds.) Geology of Continental Slopes: Society of Economic Paleontologists and Mineralogists, Special Publication no. 27, p. 131-152.

Kroenke, L., 1984, Papua New Guinea: a montage of island arcs, in Kroenke, L., (ed.), Cenozoic Tectonic Development of the Southwest Pacific: CCOP/SOPAC Technical Bulletin, n. 6, p. 29-46.

Lee, H.J., 1978, Physical properties of biogenous sediments from the eastern equatorial Pacific Ocean: U.S. Navy Civil Engineering Laboratories Technical Memorandum, TM n. M-42-78-4, 44 p.

McIntyre, Andres, Be, Allen, and Roche, Michel, 1970, Modern Pacific coccolithophorida: a paleontological thermometer: New York Academy of Science Transactions, ser. 2, v. 32, p. 720-731.

Mitchell, J.K., 1976, Fundamentals of soil behavior: New York, J. Wiley and Sons, Inc., 422 p.

Okada, Hisatake, and Bukry, David, 1980, Supplementary modification and introduction of code numbers to low latitude coccolith biostratigraphic zonation (Bukry, 1973; 1975): Marine Micropaleontology, v. 5, p. 321-325.

Thompson, P.R., Be, A.W.H., Duplessy, J.C., and Shackleton, M.J., 1979, Disappearance of pink-pigmented *Globigerinoides ruber* at 120,000 yr BP in the Indian and Pacific Oceans: Nature, v. 280 (5723), p. 554-558.

APPENDIX: LABORATORY METHODS

Six gravity cores, ranging in length from 85 to 277 cm, were analyzed using standard techniques and equipment. The refrigerated cores were split longitudinally, described, and photographed. Subsample intervals were determined in part by the visual characteristics of each core.

Grain-Size Distribution

Sediment grain-size distributions were determined by methods similar to those described in Carver (1971). Each sample was treated with hydrogen peroxide to oxidize organic carbon and disperse the sediment. Excess peroxide remaining after oxidation was gently boiled off. The sample was then washed with distilled water and centrifuged to remove solubles.

Sand ($>$ 0.063 mm) and combined silt (0.063-0.044 mm) and clay ($<$ 0.004 mm) size classes were separated by wet sieving. No gravel size material was present. The sand fraction was dried and weighed.

The silt-clay fraction was placed in 1000 ml graduated cylinders and treated with 10 ml of a 0.1 % solution of sodium hexametaphosphate (calgon) to prevent flocculation. The remainder of the cylinder was filled to 1,000 ml with distilled water. The cylinder was then agitated with a stirring rod for two minutes to homogenize the fluid-sediment column. Twenty seconds after stirring, a 20 ml pipetted aliquot was withdrawn from a depth of 20 cm, dried, and weighed. To calculate the total weight of silt and clay, the dry weight was multiplied by a factor of 50 (20 ml = 1/50th of 1,000 ml) and the weight of the calgon subtracted to give a corrected value for the silt-clay fraction.

The clay fraction was determined by withdrawing a 20 ml pipetted aliquot from a depth of 5 cm at 58 minutes after stirring. This sample was dried and weighed and the corrected weight was calculated. The silt fraction weight was determined by subtracting the weight of the clay from the weight of the silt-clay fraction.

Sand-Fraction Composition

Visual estimates of the components of the sand fraction were determined from grain mounts of the samples. Between 300 and 400 grains were identified for each sample, and the relative percent of components computed. The eight components delineated as best representing the composition of the coarse fraction were light minerals plus rock fragments, heavy minerals, mica, volcanic glass, pteropods, planktonic foraminifers, benthic foraminifers, and radiolarians. Visual estimates were performed with a binocular microscope at 50x magnification.

Whole-Sample Identification

To assess the complete composition of each core, smear slides were made from subsamples taken at varying intervals. Compositional identification was made with a petrographic microscope using varying magnifications.

Clay-Mineralogy Methods

X-ray diffraction was used to identify clay minerals. The techniques used are described in Hein, Scholl, and Gutmacher (1976) and begin with the treatment of the subsamples with Morgan's Solution (sodium acetate + glacial acetic acid) to remove $CaCO_3$ and 30 % hydrogen peroxide to remove organic matter. The samples were washed with more Morgan's Solution and centrifuged to separate the clays ($< 2 \mu$) from the coarser fraction. The clay fraction was saturated with $MgCl_2$ and X-rayed. The samples were then treated with ethylene glycol and X-rayed again. Another split of the clay fraction was saturated with $BaCl_2$ and X-rayed to determine if vermiculite was present. Peak areas were determined for the clay minerals and relative percentages were calculated and summed to equal 100. Biscaye's (1965) weighting factors, two, four, and one were used for the kaolinite plus chlorite, illite, and smectite peak areas, respectively.

Carbon/Carbonate Content

Carbon analyses were performed using a Coulometrics CO_2 Coulometer model 5010[1] which provides an absolute determination of CO_2 when the gas is introduced into the Coulometer cell and displays the result in micrograms of carbon. Samples were dried, ground into a fine powder, and stored in a desiccator to ensure dryness before weighing. Values for total carbon and inorganic carbon ($CaCO_3$) were determined by averaging the values of at least two analyses per sample.

Total carbon was calculated using a Coulometrics Total Carbon Apparatus, model 5020, which burns the sample in oxygen at 1,000 degrees Celsius to produce CO_2 gas. This gas is passed through the CO_2 Coulometer and the micrograms of total carbon determined. The micrograms of total carbon are converted to a percentage of the sample weight.

Inorganic carbon was determined with a Coulometrics Carbonate Carbon Apparatus, model 5030, which acidifies the sample and passes the evolved CO_2 to the CO_2 Coulometer where the micrograms of carbonate are determined before being converted to a percentage of the sample weight. The percentage of organic carbon is found by subtracting the percentage of the carbonate from the percentage of the total carbon. The resulting organic-carbon content is reported as a percentage of the sample by weight.

Physical Properties

Physical index properties have been determined

---

[1] Any use of trade names is for purposes of identification only and does not imply endorsement by the U.S. Geological Survey, the New Zealand Geological Survey, or the Bureau of Mineral Resources, Australia.

for the six cores at 25 cm intervals. These properties include miniature vane shear strength, sensitivity, water content, porosity, and bulk density.

Miniature vane shear strength was determined using a 12.7-mm diameter by 12.7-mm-high vane driven by a motorized vane shear device. Peak undisturbed vane shear strength (Su) and remolded vane shear strength (Sr) were measured with the axis of vane rotation perpendicular to the core axis. Strength values in kiloPascals (kPa) were read directly from a calibrated strip chart. Sensitivity (St=Su/Sr) was derived from the shear-strength values.

The water content was determined by weighing the sample before and after drying, correcting for an assumed salinity at 34°/oo, and calculating the percentage of water by weight of the whole sample. The physical properties of bulk density and porosity were calculated using a computer program developed by F. Wong and B. Edwards of the U.S. Geological Survey. Assumptions included 100 % saturation and grain density of 2.7 gm/cc.

Marlow, M.S., Dadisman, S.V., and Exon, N.F., editors, 1988, Geology and offshore resources of Pacific island arcs—New Ireland and Manus region, Papua New Guinea, Circum-Pacific Council for Energy and Mineral Resources Earth Science Series, v. 9: Houston, Texas, Circum-Pacific Council for Energy and Mineral Resources.

LATE TERTIARY AND QUATERNARY FORAMINIFERA AND PALEOBATHYMETRY OF DREDGE AND CORE SAMPLES FROM THE NEW IRELAND BASIN (CRUISE L7-84-SP)

D.J. Belford

Bureau of Mineral Resources, Geology and Geophysics, Canberra, A.C.T., 2601, Australia

ABSTRACT

Dredge and core samples from the New Ireland Basin, and from outcrop samples from New Ireland, have been examined for planktonic and benthonic foraminifera. The age of the sea-floor samples ranges from late Miocene (Zone N17) to Holocene (Zone N23).

Several species of *Globorotalia (Globorotalia)* show a tendency to develop flexuose final chambers. Specimens very similar to *Globigerina rubescens*, here referred to as *G.* sp. cf. *G. rubescens*, occur at levels below the base of Zone N20. Pink *G. rubescens* is also recorded in samples of Zone N19/20 and Zone N21 ages, but these occurrences may be caused by Recent contamination of the dredge material. Weakly fistulose forms of *Globigerinoides quadrilobatus* occur in Recent sediments. Strongly involute specimens of *Pulleniatina* occur in beds down to Zone N18 in age; the coiling direction of species of *Pulleniatina* has also been investigated. Specimens transitional from *Globorotalia (Truncorotalia) crassaformis* to the *G.(T.) tosaensis* group also have been recorded.

The faunas do not indicate any major movement of the sea-floor since the deposition of the sediments.

INTRODUCTION

Dredge samples and cores taken on the 1984 cruise (L7-84-SP) of the R.V. *S.P. Lee* in the New Ireland Basin have been examined for their foraminiferal fauna. The cruise tracks and the sample locations are shown on Figure 1, and details of the dredge and core samples are given on Tables 1 and 2. Additional dredge samples from CCOP/SOPAC Cruise PNG 79-1 and outcrop samples from New Ireland, Papua New Guinea have been examined for comparative purposes.

The foraminiferal zonation used in this paper is essentially that of Blow (1969). Some modifications have been proposed to certain of Blow's (1969) zones. For example, Bronnimann and Resig (1971) suggested that N19 and N20 be combined into a new zonal unit N19/20, because they found *Globorotalia (Turborotalia) pseudopima* Blow, the species used for the definition of Zone N20, almost down to the base of Zone N18.

Another suggested modification relevant to the present study is the division by Srinivasan and Kennett (1981) of Zone N17 into Zone N17A and Zone N17B, based on the first evolutionary appearance of *Pulleniatina primalis* at the base of Zone N17B. In the same paper these authors show a Zone N19/20 at the top of the early Pliocene of DSDP equatorial site 289, above a Zone N19. The last appearance datum of *Globorotalia margaritae* appears to be the criterion for the recognition of the base of this Zone N19/20. Essentially the same scale, with some zonal criteria slightly changed, was published by Kennett and Srinivasan (1983); the last appearance datum of *G. margaritae* as the base of Zone N19/20 was omitted, and no other criterion was given. One possible

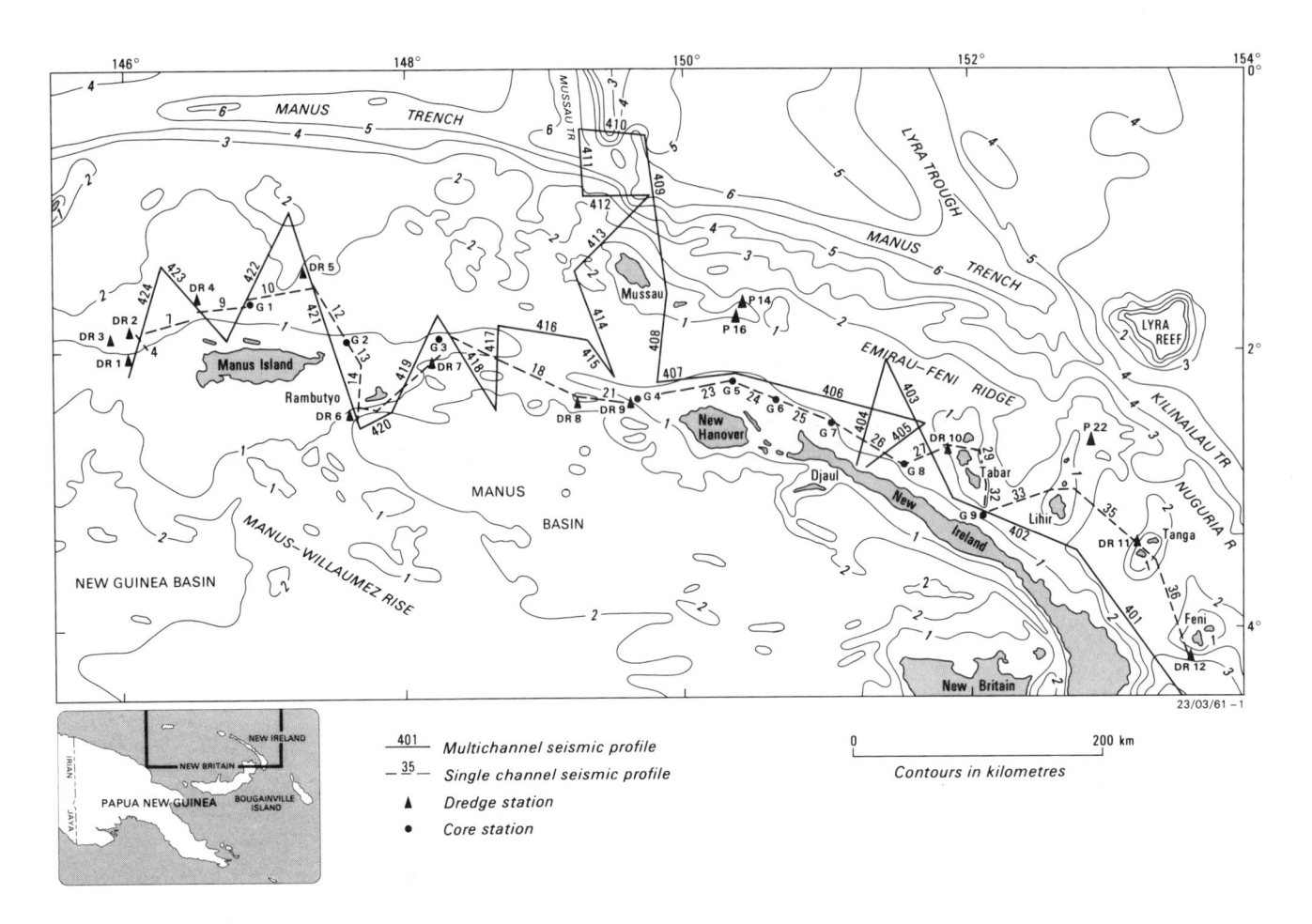

Figure 1. New Ireland-Manus region locality map showing R.V. *S.P. Lee* cruise tracks, dredge stations (DR), and gravity core stations (G). Also shows dredge stations (P) from CCOP/SOPAC cruise PN79-1. Sample PN79-1, station 24 is off map.

criterion, from their text-figure 16, is the evolutionary appearance of the *Globorotalia (Truncorotalia) crassaformis* group from *G.(T.) crassula*.

In this paper Zone N19/20 is used in the sense of Bronnimann and Resig (1971). All figured specimens are deposited in the Commonwealth Paleontological Collection, Canberra, Australia, under numbers CPC. 25344 to 25408.

DREDGE SAMPLES

DR5, 84640020.

Six samples of soft white chalk were examined; all are given a late Miocene (Zone N18) age. The fauna recorded includes *Globorotalia (G.) tumida tumida, G.(G.) tumida flexuosa, G.(G.) multicamerata, G.(G.) merotumida, Sphaeroidinellopsis seminulina, Pulleniatina primalis* and small specimens here referred to *Globigerina* sp. cf. *G. rubescens*. The Zone N18 age is based on the occurrence of these species, and on the absence of the genus *Sphaeroidinella* in the presence of abundant *Sphaeroidinellopsis*. The occurrence of *Pulleniatina obliquiloculata praecursor* in four of the samples suggests a late Zone N18 age; in one of the samples many specimens are transitional to *P. obliquiloculata obliquiloculata*. The remaining two samples, containing only *P. primalis*, may be slightly older.

DR6A, 84640022.

Seven samples were examined; these consisted of small pieces of burrowed brown to grey calcareous mudstone and grey marl contained in abundant Holocene ooze. The foraminiferal specimens in some

Table 1. Data on dredge hauls: cruise L7-84-SP.

Number	Location	Latitude (S)	Longitude (E)	Water Depth (m)	Recovery
DR 1	W. Manus	2°03.87′	146°04.41′	600-800	None
DR 1A	W. Manus	2°03.73′	146°04.93′	550-600	None
DR 1B	W. Manus	2°03.25′	146°04.64′	600-750	None
DR 2	W. Manus	1°52.39′	146°04.26′	1100-1400	1 cobble
DR 3	W. Manus	1°53.92′	145°56.89′	1100-1250	3 kg
DR 4	NW. Manus	1°37.44′	146°34.29′	1300-2000	None
DR 5	NE. Manus	1°20.94′	147°18.26′	2000-2300	40 kg
DR 6	SW. Rambutyo	2°28.70′	147°40.10′	1200-1750	None
DR 6A	SW. Rambutyo	2°27.68′	147°41.16′	1000-1500	2 kg
DR 7	NE. Rambutyo	2°05.63′	148°12.25	1000-1250	10 kg
DR 7A	NE. Rambutyo	2°02.91′	148°12.61′	800-900	40 kg
DR 8	W. New Hanover	2°23.46′	149°16.34′	1400-1500	20 kg
DR 9	W. New Hanover	2°24.69′	149°37.14′	950-1020	15 kg
DR 10	W. Tabar	2°40.64′	151°48.67′	900-1350	50 kg
DR 11	NW. Tanga	3°20.71′	153°09.01′	900-950	100 kg
DR 12	S. Feni	4°10.36′	153°33.70′	2000-2250	250 kg

samples are heavily manganese stained.

Six of the samples are given a late Miocene (late Zone N18) age. The fauna includes *Globorotalia (G.) tumida tumida, G.(G.) tumida flexuosa, G. (G.) multicamerata, Pulleniatina spectabilis* and *Sphaeroidinellopsis seminulina seminulina*; the late Zone N18 age is based on the occurrence of *Pulleniatina spectabilis* and on the absence of *Sphaeroidinella* in the presence of *Sphaeroidinellopsis*.

One sample contains a similar fauna but has in addition *Globorotalia (Truncorotalia) tosaensis tenuitheca* and *Sphaeroidinella dehiscens dehiscens*, and is given a late Pliocene (Zone N21) age. The occurrence of *Pulleniatina spectabilis* in this sample suggests a lower Zone N21 age.

DR7, 84640023.

One sample of a buff chalk has been examined, and is of late Pliocene (Zone N21) age; the fauna includes *Globorotalia (Truncorotalia) tosaensis tosaensis, G.(G.) tumida tumida*, the *G.(G.) cultrata* group, *Globigerina calida praecalida* and *Sphaeroidinella dehiscens dehiscens*.

DR7A, 84640024

Four samples have been examined. Two are given a Pliocene, Zone N19/20, age, based on the occurrence of *Globorotalia (G.) tumida tumida, G.(G.) tumida flexuosa* and *Sphaeroidinella dehiscens dehiscens*; *Dentoglobigerina conglomerata* is also present, suggesting that the samples are not older than late Zone N19 as originally defined (Blow, 1969).

The two remaining samples contain *Globorotalia (Truncorotalia) tosaensis tenuitheca* and *G.(T.) tosaensis tosaensis*, indicating a late Pliocene-early Pleistocene, Zone N21 age.

DR8, 84640025.

The one sample examined, a buff calcarenite of late Pliocene, early Zone N21 age, contains *Sphaeroidinella dehiscens dehiscens, Globorotalia (G.) tumida tumida* and pink *Globigerina rubescens*. Early forms of *Globorotalia (Truncorotalia) tosaensis tosaensis*, showing the development from the *G.(T.) crassaformis* group also occur. The specimens have five chambers in the last whorl, but still retain a somewhat quadrate, lobate outline and show differences in the relative proportion of the last chamber; one 4-chambered form with a more circular outline is also present. The occurrence of specimens trending toward *G.(T.) tosaensis tosaensis* leads to the conclusion that the sample is of early Zone N21 age.

Table 2. Data on cores: cruise L7-84-SP.

Number	Location	Latitude (S)	Longitude (E)	Water Depth (m)	Recovery (cm)
G 1	N. Manus	1°39.60′	146°51.12′	1514	227
G 2	E. Manus	1°53.38′	147°35.39′	833	None
G 2A	E. Manus	1°53.90′	147°34.96′	842	None
G 3	NE. Rambutyo	1°50.69′	148°16.17′	1086	85
G 4	W. New Hanover	2°20.94′	149°41.58′	607	None
G 4A	W. New Hanover	2°21.14′	149°40.97′	617	None
G 5	W. New Hanover	2°10.82′	150°18.82′	749	Minor
G 5A	W. New Hanover	2°10.62′	150°18.79′	785	None
G 6	NE. New Hanover	2°18.79′	150°39.16′	847	166
G 7	NE. Kavieng	2°29.21′	151°03.17′	1036	277
G 8	NE. New Ireland	2°48.39′	151°31.28′	1236	151
G 9	NE. New Ireland	3°08.46′	152°05.66′	1320	100

DR9, 84640026.

Three samples from this dredge have been examined -- two of buff chalk, and a small pebble of brown limestone. The chalk is of late Pliocene (Zone N21) age, the fauna including *Globorotalia (Truncorotalia) tosaensis tenuitheca*, pink *Globigerina rubescens*, *Sphaeroidinella dehiscens dehiscens* and *Globorotalia (G.) tumida tumida*. The limestone pebble contains algal, coral and molluscan fragments, but no foraminifera have been seen, and no age can be given.

DR10, 84640027.

Ten samples from this large dredge haul have been examined. The oldest sample is from a white chalk, and is given a late Miocene, Zone N17, age. The fauna includes *Globorotalia (G.) multicamerata*, *G.(G.) tumida plesiotumida*, *G.(G.) cultrata limbata*, *Streptochilus latum* and *Globorotaloides hexagona*. *Globorotalia (G.) tumida tumida* is absent, as is the genus *Pulleniatina*. *Streptochilus latum* was given an upper stratigraphic limit of Zone N17 by Bronnimann and Resig (1971). The absence of *Pulleniatina*, which first occurs in N17B, indicates that this sample is referable to Zone N17A of Srinivasan and Kennett (1981).

Four samples from sandy marl, calcareous mudstone and white chalk are of Pliocene (Zone N19/20) age, the fauna including *Sphaeroidinella dehiscens dehiscens*, *Globorotalia (G.) tumida tumida*, *Pulleniatina obliquiloculata obliquiloculata* and *P. obliquiloculata praecursor*. One sample contains pink *Globigerina rubescens* and is placed high in the Zone N19/20 interval.

The five remaining samples, consisting mainly of marl, are of late Pliocene (Zone N21) age, with a fauna similar to that of previous samples, but containing, in addition, the *Globorotalia (Truncorotalia) tosaensis* group.

CORE SAMPLES

All cores taken are of late(?) Pleistocene-Holocene age (Zone N23), based on the occurrence of *Globigerina calida calida* and *Hastigerina adamsi*. In one sample one specimen of *Hastigerinopsis digitiformans* Saito and Thompson has been found; this species is restricted to Zone N23 by Blow (1969), (included in *Hastigerinella digitata*). Saito, Thompson and Breger (1981) stated that *H. digitiformans* probably appeared in the very late Pleistocene or Recent.

ADDITIONAL DREDGE SAMPLES

Four dredge samples collected during the CCOP/SOPAC Cruise PNG 79-1 have been examined. Locations of three of the stations are shown in Figure 1; the fourth, station 24, was near Buka Island at 4°48.02′S, 154°25.07′E. Sample 84640055, from station 14, contains only poorly preserved smaller benthonic foraminifera. The only species identified is *Calcarina spengleri* (Gmelin), indicating a Pleistocene to Recent age.

The remaining samples, 84640056, 84640057 and 84640058, from stations 16, 22 and 24 respectively, are late Pliocene (Zone N21) in age. The fauna includes *Globorotalia (G.) tumida tumida*, *G.(G.) tumida flexuosa*, *Sphaeroidinella dehiscens dehiscens* and the *Globorotalia (Truncorotalia) tosaensis* group. Specimens from stations 22 and 24 are manganese stained, and in this respect resemble the sample from dredge 6A of the *S.P. Lee* material which is also of Zone N21 age. One difference is that *Pulleniatina spectabilis* has not been recorded from the CCGP/SOPAC samples.

OUTCROP SAMPLES, NEW IRELAND, AND COMPARISON WITH CRUISE SAMPLES

Six outcrop samples from New Ireland have been examined for comparison with dredge and core material. Four contain a fauna similar to that from the dredge samples.

One sample from the Lossuk River beds, 84640044, is of latest early Miocene (Zone N8) age, with a fauna including *Praeorbulina transitoria* and *P. glomerosa curva*. Rare larger foraminifera, *Lepidocyclina (?Nephrolepidina)* sp. and *Miogypsina (M.)* sp., derived from a possibly contemporaneous shallow-water fauna, also occur. Sample 84640045B, also from the Lossuk River beds, contains *Miogypsina (M.)* sp. No comparable samples occur in the dredged material.

One sample from the Punam Limestone, 84640046, yielded a fauna indicating a late Miocene, (Zone N18) age. Species present include *Globorotalia (G.) tumida tumida*, *G.(G.) tumida flexuosa*, *G.(G.) merotumida*, *Sphaeroidinellopsis seminulina* and *Pulleniatina spectabilis praespectabilis*. Banner and Blow (1967), who recorded the taxon, subsequently formally described as *P. praespectabilis*, as an intermediate form between *P. primalis* and *P. spectabilis*, stated that it has been recorded only from late Zone N18 and early Zone N19 in land based sections. They also noted that these intermediate forms have been recorded only from archipelagic sections, but Bronnimann and Resig (1971) recorded *P. praespectabilis* from a deep-sea core.

The fauna from sample 84640046 is comparable to those from the *S.P. Lee* dredges 5 and 6A; the sediments from these dredge samples can be correlated with the Punam Limestone. *Pulleniatina spectabilis*, which was recorded from dredge 6A, is not present in the New Ireland sample, but as noted by Banner and Blow (1967), this species appears to need a fully oceanic environment, being recorded, to that time, only from deep-sea cores.

One other dredged sample, 84640027I, from dredge number 10, of late Zone N17 age, may also be correlatable to the Punam Limestone, which ranges in age from Zone N16 to Zone N18, but no samples from New Ireland containing a comparable fauna have been seen.

Three samples from the Rataman Formation, 84640047, 84640048 and 84640049, were also examined. Sample 84640047 is Zone N21 in age, containing *Globorotalia (G.) tumida tumida*, *G.(G.) tumida flexuosa*, *G.(Truncorotalia) tosaensis tosaensis* and *Sphaeroidinella dehiscens dehiscens*. Sample 84640048 has a meager fauna indicating a Zone N19/20 age, the fauna being similar to that of 84640047, but lacking the *G. tosaensis* group. Sample 84640049 contains a fauna giving only a general Pliocene, (N19-N21) age.

The remaining dredge material, from all sampled dredge hauls except 5 and 6A, can be correlated with the Rataman Formation (Zone N19/20 or Zone N21).

NOTES ON THE FAUNA

Comments are given below for selected planktonic species; species recorded from each sample are shown in Table 3.

Globigerina rubescens

Pink *G. rubescens* has been found in dredge samples assigned to Zone N19/20 and Zone N21, but these occurrences may result from contamination with Recent material during dredging. Parker (1967) raised the possibility that the pink coloration developed in the Pleistocene, and this was supported by Jenkins and Orr (1972). Chaproniere (1985a) did not record pink *G. rubescens* below Zone N22.

The possible occurrence of *G. rubescens* in beds older than Zone N22 requires confirmation from a cored sequence.

Globigerina sp. cf. *G. rubescens* Hofker

Included here are small, white specimens very difficult to differentiate from pink *G. rubescens*, but occurring in beds older than the level of first appear-

Table 3. Distribution of planktonic foraminifera.

LATE TERTIARY AND QUATERNARY FORAMINIFERA

Distribution chart of planktonic foraminifera species against planktonic foraminiferal zones (N.8–N.23).

Genus	Subgenus	Species	Author
Globorotalia	(Globorotalia)	tumida plesiotumida	Blow & Banner, 1965
Globorotalia	(Globorotalia)	tumida flexuosa	(Koch, 1923)
Globorotalia	(Globorotalia)	tumida tumida	(Brady, 1877)
Globorotalia	(Globorotalia)	ungulata	Bermudez, 1960
Globorotalia	(Obandyella)	bermudezi	Rögl & Bolli, 1973
Globorotalia	(Obandyella)	hirsuta hirsuta	(d'Orbigny, 1839)
Globorotalia	(Obandyella)	margaritae	Bolli & Bermudez, 1965
Globorotalia	(Obandyella)	scitula scitula	(Brady, 1882)
Globorotalia	(Obandyella)	sp. 1	
Globorotalia	(Truncorotalia)	cf. tosaensis	Takayanagi & Saito, 1962
Globorotalia	(Truncorotalia)	crassaformis crassaformis	(Galloway & Wissler, 1927)
Globorotalia	(Truncorotalia)	crassaformis ronda	Blow, 1969
Globorotalia	(Truncorotalia)	tosaensis tosaensis	Takayanagi & Saito, 1962
Globorotalia	(Truncorotalia)	tosaensis tenuitheca	Blow, 1969
Globorotalia	(Truncorotalia)	truncatulinoides truncatulinoides	(d'Orbigny, 1839)
Globorotaloides		hexagona	(Natland, 1938)
Globorotaloides		variabilis	Bolli, 1957
Hastigerina		adamsi	Banner & Blow, 1959
Hastigerina		siphonifera	(d'Orbigny, 1839)
Hastigerinopsis		digitiformans	Saito & Thompson, 1976
Neogloboquadrina		dutertrei	(d'Orbigny, 1839)
Orbulina		universa	d'Orbigny, 1839
Praeorbulina		glomerosa curva	(Blow, 1956)
Praeorbulina		transitoria	(Blow, 1956)
Pulleniatina		obliquiloculata finalis	Banner & Blow, 1967
Pulleniatina		obliquiloculata obliquiloculata	(Parker & Jones, 1865)
Pulleniatina		obliquiloculata praecursor	Banner & Blow, 1967
Pulleniatina		primalis	Banner & Blow, 1967
Pulleniatina		sp. 1	
Pulleniatina		spectabilis spectabilis	Parker, 1965
Pulleniatina		spectabilis praespectabilis	Bronnimann & Resig, 1971
Sphaeroidinella		dehiscens dehiscens	(Parker & Jones, 1865)
Sphaeroidinellopsis		seminulina seminulina	(Schwager, 1866)
Streptochilus		globigerum	(Schwager, 1866)
Streptochilus		latum	Bronnimann & Resig, 1971
Tenuitella		tokelauae	(Boersma, 1969)
Turborotalia		anfracta	(Parker, 1967)
Turborotalia		acostaensis	Blow, 1959
Turborotalia		continuosa	(Blow, 1959)
Turborotalia		humerosa	(Takayanagi & Saito, 1962)
Turborotalia		obesa	Bolli, 1957
Turborotalia		pseudopima	(Blow, 1969)
Turborotalita		humilis	(Brady, 1884)

AGE OF SAMPLE (Planktonic foraminiferal zone)

ance of *G. rubescens* given by Blow (1969), that is, base of Zone N20. Specimens of both *G. rubescens* and cf. *G. rubescens* are four-chambered, with a low to high trochospiral test, a small arched to rounded aperture, and an identical reticulate shell texture. The only observed difference is the slightly tighter coiling of some specimens of *G.* cf. *rubescens*.

The occurrence in beds of Zone N18 age of small specimens very difficult to distinguish from *G. rubescens* affects the use of this species as an indication of the base of Zone N20. This does not apply to pink specimens of *G. rubescens*, but in the absence of pink forms there is a degree of uncertainty introduced by the demonstrated occurrence of very similar forms in older beds. The older specimens may be young forms of *Globigerina decoraperta*, but it is not possible to demonstrate this. If this could be shown it would support Blow's (1969) conclusion that *G. decoraperta* and *G. rubescens* are probably only subspecifically distinct.

Globigerinoides quadrilobatus ?fistulosus (Schubert)

Included here are weakly fistulose forms from Recent deposits. Blow (1969) considered that the very weakly fistulose forms sometimes seen in Recent beds are not synonymous with *G. quadrilobatus fistulosus*. Banner and Blow (1960) noted the development of narrow digitate extensions on the last, or last two, chambers of *G. quadrilobatus sacculifer*, and regarded these forms as transitional to *G. quadrilobatus fistulosus*. The present specimens are of this type. Kennett (1973) illustrated specimens very similar to those from the New Ireland Basin and regarded them as simpler forms at the end of the evolutionary bioseries.

Globorotalia (Globorotalia)

Specimens from the present material referred to several species groups within the subgenus *Globorotalia (Globorotalia)* show the development of flexuose final chambers. The development of flexuose forms in this subgenus is well-known in *G.(G.) tumida flexuosa* and *G.(G.) neoflexuosa*. In addition, specimens here placed in *G.(G.) multicamerata, G.(G.) cultrata cultrata, G. (G.) cultrata limbata* and *G.(G.) ?miocenica* have flexuose final chambers (see Plates 1 and 2). Several of the flexuose specimens here placed in *G.(G.) cultrata limbata* are close to *G.(G.) multicamerata*. Parker (1973) used the presence of 8 or more chambers in the final whorl to separate *G. multicamerata* from *G. limbata*, as did Kennett and Srinivasan (1983). The specimens here referred to as *G.(G.) ?miocenica* have the flat dorsal surface, the convex ventral surface, radial depressed ventral sutures and the almost non-lobate outline of this species, but have a more strongly developed peripheral keel and coarsely pustulose ornament on the early chambers of the last whorl. Very similar specimens were figured by Boltovskoy (1974) as *Globorotalia ?miocenica* Palmer, from the Indian Ocean, DSDP Leg 26. The specimens figured as *G. miocenica* by Krasheninnikov and Hoskins (1973) are more compressed than the present specimens, have a smooth test surface and a less strongly developed keel. *G.(G.) pseudomiocenica* Bolli and Bermudez also has a flat dorsal surface, but has a more lobate outline than specimens referred to *G. ?miocenica*, and also has a thinner peripheral keel. Flexuose forms are rare in specimens of *G. ?miocenica*.

Kennett (1976) discussed the flexed condition of the final chamber in *G. (G.) tumida flexuosa* and *G.(G.) neoflexuosa*; he noted that it is unclear if this condition resulted from phenotypic variation or geographic isolation. Boltovskoy (1974) considered that flexuose forms resulted from environmental influences. The occasional occurrence of this feature in several morphotypes within the present material, in deep-sea samples from different areas and of different ages, suggests that it is environmentally controlled rather than resulting from geographic isolation or genetic factors.

Specimens of the *Globorotalia (G.) tumida plesiotumida* morphotype have been recorded from the present material in samples ranging up to Zone N21 in age, although Blow (1969) gave the upper limit as Zone N18. These specimens are not sufficiently tumid to be referable to *G. tumida tumida*, but may be rare less tumid variants of the *G. tumida tumida* morphotype, which also occurs in the samples.

G.(G.) cultrata fimbriata is rare in the samples; the specimens show well-developed peripheral spines. Other specimens with only slightly developed peripheral pustules or spines are referred to *G.* cf. *fimbriata*.

Two specimens found cannot be referred to any described species. The first, here referred to as *Globorotalia (G.)* sp. 1 is a tightly coiled biconvex form, with a small umbilicus, a circular lobate outline, five chambers in the last whorl, increasing slowly in size, curved limbate dorsal sutures, radial depressed ventral sutures, a finely perforate test wall and a strongly developed peripheral keel. The second form, *G.(G.)* sp. 2, is also biconvex, with the ventral surface the more strongly convex, a small umbilicus, a lobate oval outline, five chambers in the last whorl, increasing quickly in size, curved limbate dorsal sutures, depressed radial or slightly curved ventral sutures, a very finely perforate smooth test wall, and a well-developed peripheral keel.

Pulleniatina

Involute forms approaching the morphotype of *Pulleniatina obliquiloculata finalis* occur in our samples as early as Zone N18. This is noteworthy, because Blow (1969) reported the first appearance of *P. obliquiloculata finalis* in mid-zone N22. Strongly involute specimens occur in samples assigned to Zones N18, N19/20 and N21; most seem to be variants of *P. obliquiloculata obliquiloculata*, but in sample 846400274 involute forms here referred to as *Pulleniatina* sp. 1 occur with *P. obliquiloculata praecursor*. Involute specimens are illustrated on Plate 3. The innermost end of the dorsal surface of the last chamber does not cover the innermost end of the penultimate chamber to the extent seen in *P. obliquiloculata finalis*. However, the presence of involute (advanced?) forms of *Pulleniatina* in beds down to Zone N18 does not seem to have been recorded previously.

Coiling direction data of specimens of *Pulleniatina* are given in Table 4. *Pulleniatina obliquiloculata obliquiloculata* and *P. obliquiloculata praecursor* are mainly dextrally coiled, each being dominantly sinistral in only two samples. *P. primalis* and *P. spectabilis spectabilis* are wholly or dominantly sinistral in all samples in which they occur.

In three samples, different forms are coiled in different directions. In sample 84640020A, *P. primalis* is wholly sinistral, while *P. obliquiloculata praecursor* is wholly dextral. In 84640020D, *P. primalis* is wholly sinistral, while *P. obliquiloculata praecursor* is dominantly (94%) dextral. In 84640022G, *P. spectabilis* is wholly sinistral, and *P. obliquiloculata obliquiloculata* wholly dextral. These observations are not in agreement with those of Saito (1976), who noted that species differences appear to play no significant role in selection of the preferred coiling direction.

Two samples are of Zone N18 age and are older than the sequence examined by Saito (1976). However, he noted a left coiling interval from the Gilbert Epoch and earlier; some of the present samples from this level contain dominantly dextrally coiled specimens.

Globorotalia (Turborotalia) pseudopima (Blow)

Blow (1969) used the first occurrence of this species to define the base of Zone N20. However, Bronnimann and Resig (1971) recorded *G. pseudopima* to near the base of Zone N18, and therefore included Zone N19 within the lower range of Zone N20, suggesting that another species be sought to define the Zone N19/Zone N20 boundary. Chaproniere (1985b) also found *G. pseudopima* to be an unreliable index for the base of Zone N20. In the present samples *G. (T.) pseudopima* has been recorded from samples given a Zone N18 age (see Plate 3, Figures 23-27). The specimens agree well with Blow's (1969) description, having 4 chambers visible ventrally, with a small umbilicus and a low arched interiomarginal umbilical-extraumbilical aperture.

Globigerina calida group

Blow (1969) restricted the definition of *Globigerina calida* as originally defined by Parker (1962) and proposed the subspecies *G. praecalida* to accommodate forms lacking radially elongate chambers, with a less widely open umbilicus, 4-4½ chambers in the last whorl and a more restricted aperture. It is to be noted that the holotype of *G. calida* has 4½ chambers in the final whorl, and does not have marked radial elongation of the chambers.

Bronnimann and Resig (1971) considered that the distinction of *G. calida* from *G. praecalida* involved a subjective interpretation, and suggested that *G. calida* should not be used for formal definition of Zone N23. Specimens of the type figured by Bronnimann and Resig have not been observed in beds older than Zone N23, and to this extent *G. calida calida* can be used as an indication of this zone. Transitional forms between *G. calida calida* and *G. calida praecalida* have been observed; in this paper specimens with a 4½-5 chambered loosely coiled test and radially elongate chambers are referred to *G. calida calida*.

The specimens here referred to *G. calida praecalida* have a greater extraumbilical development of the aperture than does the holotype figured by Blow (1969), and in this respect are closer to the specimen figured by Parker (1962, Plate 4, Figure 12), and placed by Blow in the synonymy of *G. calida praecalida*. The four-chambered specimen figured by Jenkins and Orr (1972) as a juvenile of *G. calida* appears to be of the type here referred to *G. calida praecalida*.

Table 4. Coiling Direction of *Pulleniatina*.

SPECIES	SAMPLE	COILING	
Pulleniatina obliquiloculata obliquiloculata	84644022G	100.0%	d*
” ” ”	84640023A	68.2%	d
” ” ”	84640024D	83.3%	d
” ” ”	84640026A	100.0%	d
” ” ”	84640026B	84.1%	s
” ” ”	84640027A	73.3%	d
” ” ”	84640027E	73.1%	d
” ” ”	84640027G	69.7%	d
” ” ”	84640027J	56.0%	s
” ” ”	84640030A	95.8%	d
” ” ”	84640030D	88.0%	d
” ” ”	84640030E	100.0%	d
” ” ”	84640038	100.0%	d
” ” ”	84640039	97.0%	d
” ” ”	84640040	95.5%	d
” ” ”	84640041	100.0%	d
” ” ”	84640042	98.3%	d
” ” ”	84640043	95.8%	d
Pulleniatina obliquiloculata praecursor	84640020A	100.0%	d
” ” ”	84640020D	94.1%	d
” ” ”	84640024A	100.0%	d
” ” ”	84640024B	97.1%	d
” ” ”	84640024C	95.8%	d
” ” ”	84640025A	59.6%	d
” ” ”	84640027B	73.2%	s
” ” ”	84640027F	57.4%	d
” ” ”	84640027G	50.0%	d
” ” ”	84640027H	97.7%	s
Pulleniatina primalis	84640020A	100.0%	s
” ”	84640020B	100.0%	s
” ”	84640020C	100.0%	s
” ”	84640020D	100.0%	s
” ”	84640020E	100.0%	s
” ”	84640020F	100.0%	s
Pulleniatina spectabilis spectabilis	84640022A	100.0%	s
” ” ”	84640022B	92.7%	s
” ” ”	84640022C	100.0%	s
” ” ”	84640022D	85.7%	s
” ” ”	84640022E	100.0%	s
” ” ”	84640022F	97.0%	s
” ” ”	84640022G	100.0%	s

* d - dextral, s - sinistral

Globorotalia (Truncorotalia) tosaensis group.

The concept of this group has been modified since the original description by Takayanagi and Saito (1962). Four chambers were said commonly to form the final whorl; however, the holotype shows slightly more than four chambers. Stainforth et al. (1975) noted 4 chambers (in the holotype), or more commonly 5 chambers, in the final whorl. It is now generally accepted that 5 chambers are present in the final whorl, this being one criterion for distinguishing *G. tosaensis* from the *G. crassaformis* group.

A second feature which distinguishes *G. tosaensis* from the *G. crassaformis* group is the rounded outline of *G. tosaensis*. In the present material, specimens have been referred to *G. tosaensis* based on the presence of either or both, of these criteria. Some specimens with a quadrate outline have 5 chambers in the final whorl (Plate 5, Figure 17-22); others with only 4 chambers in the final whorl have a circular outline (Plate 5, Figure 15-16). Specimens of these kinds are possibly transitional from *G. crassaformis* to *G. tosaensis*.

Benthonic foraminifera

The benthonic fauna of the samples has not been fully identified; species recognized are shown in Table 5.

PALEOBATHYMETRY

Statistical information on the planktonic/benthonic ratio of the samples is given in Table 6. These data are the mean for each sample: obtained from separate random counts on three fractions, one from 125 μm to 250 μm, one from 250 μm to 500 μm, and one greater than 500 μm. Murray (1973) noted that in the Mediterranean Sea the planktonic foraminifera form 95%-99% of the fauna in depths greater than 1000 m. Almost all the present samples are from water depths greater than 1000 m and most have a planktonic/benthonic ratio within the range given by Murray (1973).

Resig (1986) gave a compilation of observed depth ranges of benthonic species in the Solomons area. In general, species from the present samples fall within the depth ranges given by Resig, but there are some exceptions. One exception is *Bolivinita quadrilatera*, which Resig did not find below about 600 m, but which occurs in one dredge sample taken between 1000 m and 1500 m, with planktonic species making up 98.6% of the fauna. *B. quadrilatera* also occurs in core 1, taken at 1514 m. Frerichs (1970) recorded *B. quadrilatera* as a rare element of the fauna (less? than 1%) in samples from 1390 m in the Andaman Sea. Another example is *Gyroidina neosoldanii*, which occurs in the present material in samples taken above the approximate 1700 m minimum depth shown by Resig (1986).

The most obvious depth discrepancy is shown by a dredge sample taken at station 14 on CCOP/SOPAC cruise PN79-1, north of New Hanover Island, in a water depth of 830 m to 537 m. The only species identified from this sample is *Calcarina spengleri* (Gmelin), a species characterized as a reef dweller by Adams (1970), and given an intertidal to 10(?) m range by Murray (1973). Either the sediments containing *C. spengleri* were transported down slope following lithification, or there has been considerable downward movement of the sea-bottom. The first alternative is considered to be the more probable; most species occurrences in other samples are within known depth ranges and there is no evidence of any significant movement of the sea-floor since deposition of the sampled sediments. The reef-dwelling species *Baculogypsina sphaerulata* and *Tinoporus hispidus* occur in the otherwise deep-water fauna from sample 84640041; this mixing is also attributed to transport of the shallow-water fauna into deep water.

ACKNOWLEDGMENTS

I wish to thank Dr. Neville Exon and Dr. George Chaproniere of Bureau of Mineral Resources, Australia, Ms. Paula Quinterno of the U.S. Geological Survey, and Dr. David Haig of the University of Western Australia for commenting on the manuscript. Mr. P.W. Davis assisted with photography. The text-figures and tables were drawn by T. Kimber.

REFERENCES

Adams, C.G., 1970, A reconsideration of the East Indian letter classification of the Tertiary: Bulletin of the British Museum (Natural History), v. 19, no. 3, p. 87-137.

Banner, F.T., and Blow, W.H., 1960, Some primary types of species belonging to the superfamily *Globigerinaceae*: Contributions from the Cushman Foundation for Foraminiferal Research, v. 11, no. 1, p. 1-41.

Banner, F.T., and Blow, W.H., 1967, The origin, evolution and taxonomy of the foraminiferal genus *Pulleniatina* Cushman, 1927: Micropaleontology, v. 13, no. 2, p. 133-162.

Blow, W.H., 1969, Late middle Eocene to Recent planktonic foraminiferal biostratigraphy, *in* Bronnimann, P. and Rene, H.H., eds., Proceedings of the First International Conference on Planktonic Microfossils: Leiden, E.J. Brill, p. 199-422.

Boltovskoy, E., 1974, Neogene planktonic foraminifera of the Indian Ocean (DSDP, Leg 26), *in* Davies, T.A., Luyendyk, B.P., et al., Initial Reports of the Deep Sea Drilling Project, v. 26: Washington D.C., U.S. Government Printing Office, p. 675-741.

Bronnimann, P., and Resig, J., 1971, A Neogene globigerinacean biochronologic time-scale of the southwestern Pacific, *in* Winterer, E.L., et al., Initial Reports of the Deep-Sea Drilling Project, v. 7: Washington D.C., U.S. Government Printing Office, p. 1235-1469.

Chaproniere, G.C.H., 1985a, Late Tertiary and Quaternary foraminiferal biostratigraphy and paleobathymetry of cores and dredge samples from cruise KK820316 Leg 2, *in* Brocher, T.M.,

Table 5. Distribution of benthonic foraminifera.

LATE TERTIARY AND QUATERNARY FORAMINIFERA

Genus	Species	Author
Melonis	affinis	(Reuss, 1851)
Melonis	pompilioides	(Fichtel & Moll, 1798)
Miliolinella	subrotunda	(Montagu, 1803)
Miogypsina (Miogypsina)	sp.	
Neoeponides	subornatus	(Cushman, 1921)
Nodosaria	sp.	
Nummoloculina	sp.	
Operculina	sp.	
Oridorsalis	umbonatus	(Reuss, 1851)
Oridorsalis	variapertura	Belford, 1966
Osangularia	bengalensis	(Schwager, 1866)
Osangularia	culter	(Parker & Jones, 1865)
Parrelloides	bradyi	(Trauth, 1918)
Parvicarinina	altocamerata	(Heron-Allen & Earland, 1910)
Peneroplis	pertusus	(Forskål, 1775)
Planulina	wuellerstorfi	(Schwager, 1866)
Pleurostomella	alternans	Schwager, 1866
Pleurostomella	brevis	(Schwager, 1866)
Protoglobobulimina	notovata	(Chapman, 1941)
Protoglobobulimina	ovata	(d'Orbigny, 1846)
Psammosphaera	sp.	
Pseudoclavulina	serventyi	(Chapman & Parr, 1935)
Pseudoclavulina	sp.	
Pullenia	bulloides	(d'Orbigny, 1846)
Pyrgo	cf. serrata	(Bailey, 1862)
Pyrgo	depressa	(d'Orbigny, 1826)
Pyrgo	lucernula	(Schwager, 1866)
Pyrgo	murrhyna	(Schwager, 1885)
Pyrgo	sp.	
Quinqueloculina	auberiana	d'Orbigny, 1839
Quinqueloculina	sp.	
Rectobolivina	dimorpha	(Parker & Jones, 1865)
Rectuvigerina	striata	(Schwager, 1866)
Reophax	nodulosus	Brady, 1879
Reussella	aculeata	Cushman, 1945
Saccammina	sp.	
Sigmoilopsis	schlumbergeri	(Silvestri, 1904)
Siphotextularia	sp.	
Siphouvigerina	proboscidea	(Schwager, 1866)
Sphaeroidina	bulloides	d'Orbigny, 1826
Spiroloculina	sp.	
Spiroplectammina	sp.	
Stilostomella	insecta	(Schwager, 1866)
Stilostomella	sp.	
Textularia	sp.	
Tinoporus	hispidus	(Brady, 1876)
Triloculina	sp.	
Triloculina	tricarinata	d'Orbigny, 1826

Table 6. Percentage of planktonic (P%) and benthonic (B%) species, and percentage of textulariine (T%), milioline (M%) and rotaliine (R%) forms in the benthonic population. NUMID is the number of planktonic specimens counted.

REG.NO.	T%	M%	R%	P%	B%	NUMID	FRACT	STATION NO.	WATER DEPTH (m)
84640020A	14.0	4.0	82.0	94.2	5.8	806	Mean	DR5	2000-2300
84640020B	8.0	8.0	84.0	98.4	1.6	1500	Mean	”	”
84640020C		3.0	97.0	98.1	1.9	1700	Mean	”	”
84640020D	3.6		96.4	98.4	1.6	1700	Mean	”	”
84640020E			100	98.7	1.3	1600	Mean	”	”
84640020F	18.2		81.8	99.3	0.7	1500	Mean	”	”
84640022A	7.1		92.9	98.6	1.4	974	Mean	DR6A	1000-1500
84640022B			100	99.3	0.7	1088	Mean	”	”
84640022C			100			335	Mean	”	”
84640022D	14.3	14.3	71.4	92.0	8.0	80	Mean	”	”
84640022E			100	94.4	5.6	67	Mean	”	”
84640023A			100	98.9	1.1	1209	Mean	DR7	1000-1250
84640024A		7.7	92.3	99.0	1.2	1300	Mean	DR7A	800-900
84640024B	40.0		60.0	99.3	0.7	1500	Mean	”	”
84640024C			100	98.7	1.3	1500	Mean	”	”
84640024D			100	98.2	1.8	1300	Mean	”	”
84640025A			100	98.7	1.3	1300	Mean	DR8	1400-1500
84640026A	37.5		62.5	99.0	1.0	1600	Mean	DR9	950-1020
84640026B			100	99.3	0.7	1500	Mean	”	”
84640027A		2.5	97.5	96.4	3.6	1078	Mean	DR10	900-1350
84640027B			100	99.6	0.4	1500	Mean	”	”
84640027D	7.1	14.3	78.6	99.1	0.9	1500	Mean	”	”
84640027E		5.9	94.1	98.9	1.1	1500	Mean	”	”
84640027F		8.3	91.7	99.2	0.8	1500	Mean	”	”
84640027G		10.0	90.0	99.3	0.7	1500	Mean	”	”
84640027H	25.0		75.0	99.2	0.8	1500	Mean	”	”
84640027I			100	99.7	0.3	1500	Mean	”	”
84640027J		6.3	93.7	98.9	1.1	1500	Mean	”	”
84640030A	8.3	8.3	83.4	99.2	0.8	1500	Mean	G1	1514
84640030D		15.4	84.6	99.1	0.9	1500	Mean	”	”
84640030E	18.2	13.6	68.2	98.5	1.5	1500	Mean	”	”
84640038	50.0	4.5	45.5	98.6	1.4	1500	Mean	G6	847
84640039		26.3	73.7	98.7	1.3	1500	Mean	G7	1037
84640040	11.1	11.1	77.8	99.4	0.6	1500	Mean	G8	1236
84640041	14.3	7.1	78.6	99.1	0.9	1500	Mean	G9	1320
84640042		11.1	88.9	99.4	0.6	1500	Mean	Ooze,DR5	2000-2300
84640043		9.5	90.5	98.6	1.4	1500	Mean	Ooze,DR6A	1000-1500
84640044	1.7	1.1	28.8	68.4	31.6	121	Mean	Outcrop,	
84640046	16.7	8.3	75.0	99.2	0.8	1500	Mean	New	
84640047	11.1	33.3	55.6	99.4	0.6	1500	Mean	Ireland	

ed., Investigations of the northern Melanesian Borderland: Circum-Pacific Council for Energy and Mineral Resources, Earth Science Series, v. 3, p. 103-122.

Chaproniere, G.C.H., 1985b, Late Neogene and Quaternary planktonic foraminiferal biostratigraphy and paleobathymetry of dredge samples from the southern Tonga platform (Cruise L5-82-SP), in Scholl, D.W., and Vallier, T.L., compilers and eds., Geology and offshore resources of Pacific island arcs-Tonga region, Circum-Pacific Council for Energy and Mineral Resources Earth Science Series, v. 2: Houston, Texas, Circum-Pacific Council for Energy and Mineral Resources, p. 131-139.

Frerichs, W.E., 1970, Distribution and ecology of benthonic Foraminifera in the sediments of the Andaman Sea: Contributions from the Cushman Foundation for Foraminiferal Research, v. 21, no. 4, p. 123-147.

Jenkins, D.G., and Orr, W.N., 1972, Planktonic foraminiferal biostratigraphy of the eastern equatorial Pacific-DSDP Leg 9, in Hollister, C.D., Ewing, J.I., et al., Initial Reports of the Deep Sea Drilling Project, v. 9: Washington D.C., U.S. Government Printing Office, p. 1060-1193.

Krasheninnikov, V.A., and Hoskins, R.H., 1973, Late Cretaceous, Paleogene and Neogene planktonic foraminifera, in Heezen, B.C., MacGregor, I.D., et al., Initial Reports of the Deep Sea Drilling Project, v. 20: Washington D.C., U.S. Government Printing Office, p. 105-203.

Kennett, J.P., 1973, Middle and late Cenozoic planktonic foraminiferal biostratigraphy of the southwest Pacific - DSDP Leg 21, in Burns, R.E., Andrews, J.E., et al., Initial Reports of the Deep Sea Drilling Project, v. 21: Washington D.C., U.S. Government Printing Office, p. 576-639.

Kennett, J.P., 1976, Phenotypic variation in some Recent and Late Cenozoic planktonic Foraminifera: Foraminifera, v. 2, p. 111-170.

Kennett, J.P., and Srinivasan, M.S., 1983, Neogene planktonic foraminifera: A phylogenetic Atlas: Pennsylvania, Hutchinson Ross Publishing Company, 265 p.

Murray, J.W., 1973, Distribution and ecology of living benthic foraminiferids: London, Heinemann, 274 p.

Parker, F.L., 1962, Planktonic foraminiferal species in Pacific sediments: Micropaleontology, v. 8, no. 2, p. 219-254.

Parker, F.L., 1967, Late Tertiary biostratigraphy (planktonic foraminifera) of tropical Indo-Pacific deep-sea cores: Bulletins of American Paleontology, v. 52, no. 235, p. 115-108.

Parker, F.L., 1973, Late Cenozoic biostratigraphy (planktonic foraminifera) of tropical Atlantic deep-sea sections: Revista Espanola de Micropaleontologia, v. 5, no. 2, p. 253-289.

Resig, J.M., 1986, Foraminiferal stratigraphy and paleobathymetry of dredged rock, R.V. *S.P. Lee* cruise, Solomon Islands in Vedder, J.G., Pound, K.S., and Bounty, S.Q., eds., Geology and offshore resources of Pacific island arcs--central and western Solomon Islands, Circum-Pacific Council for Energy and Mineral Resources Earth Science Series, v. 4: Houston, Texas, Circum-Pacific Council for Energy and Mineral Resources, p. 255-260.

Saito, T., 1976, Geologic significance of coiling direction in the planktonic foraminifera *Pulleniatina*: Geology, v. 4, no. 5, p. 305-309.

Saito, T., Thompson, P.R., and Breger, D., 1981, Systematic index of Recent and Pleistocene planktonic foraminifera: University of Tokyo Press, 190 p.

Srinivasan, M.S., and Kennett, J.P., 1981, A review of Neogene planktonic foraminiferal biostratigraphy: applications in the equatorial and South Pacific: Society of Economic Paleontologists and Mineralogists, Special Publication, no. 32, p. 395-432.

Stainforth, R.M., Lamb, J.L., Luterbacher, H.P., Beard, J.H., and Jeffords, R.M., 1975, Cenozoic planktonic foraminiferal zonation and characteristics of index forms: University of Kansas Paleontological Contributions, Article 62, Appendix, p. 163-425.

Takayanagi, Y., and Saito, T., 1962, Planktonic foraminifera from the Nobori Formation, Shikoku, Japan: Tohoku University, Science Reports, Geology, v. 5, ser. 2, p. 67-106

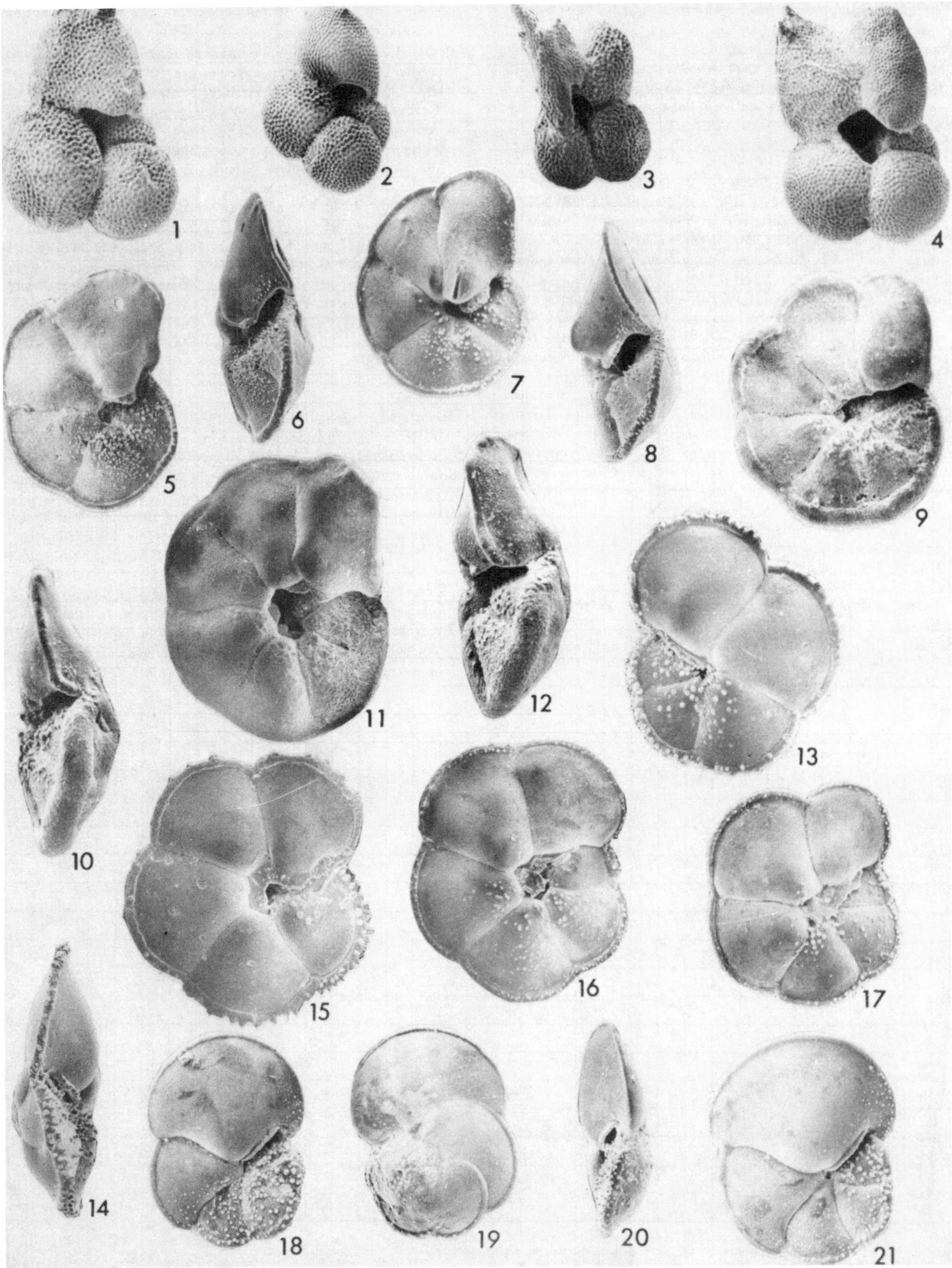
1
2
3
4
5
6
7
8
9
10
11
12
13
14
15
16
17
18
19
20
21

Plate 1 (facing page).

Figures	
1-4	*Globigerinoides quadrilobatus ?fistulosus* Schubert 1, # CPC. 25359, sample 84640030D, x60. 2, # CPC. 25360, sample 84640042, x60. 3, # CPC. 25361, same sample, x60. 4, CPC. 25362, same sample, x60.
5-8	*Globorotalia (G.) cultrata cultrata* (d'Orbigny) 5-6, CPC. 25363, sample 84640024C. 5, ventral view; 6, edge view, both x50. 7-8, CPC. 25364, same sample. 7, ventral view; 8, edge view, showing flexuose tendency of later chambers, both x50.
9-12	*Globorotalia (G.) cultrata limbata* (Fornasini) 9-10, CPC. 25365, sample 84640020D. 9, ventral view; 10, edge view, both x50. 11-12, CPC. 25366, same sample. 11, ventral view; 12, edge view, showing flexuose development of later chambers, both x50.
13-15	*Globorotalia (G.) cultrata fimbriata* (Brady) 13-14, CPC. 25367, sample 84640030. 13, ventral view; 14, edge view, both x50. 15, CPC. 25368, sample 84640030D, ventral view, x50.
16-17	*Globorotalia (G.)* cf.*cultrata fimbriata* (Brady) 16, CPC. 25369, sample 84640025A, ventral view, x50. 17, CPC. 25370, sample 84640027E, ventral view, x50.
18-21	*Globorotalia (Obandyella) margaritae* Bolli and Bermudez 18-20, CPC. 25371, sample 84640024A. 18, ventral view; 19, dorsal view; 20, edge view, all x80. 21, CPC. 25372, same sample, ventral view, x80.

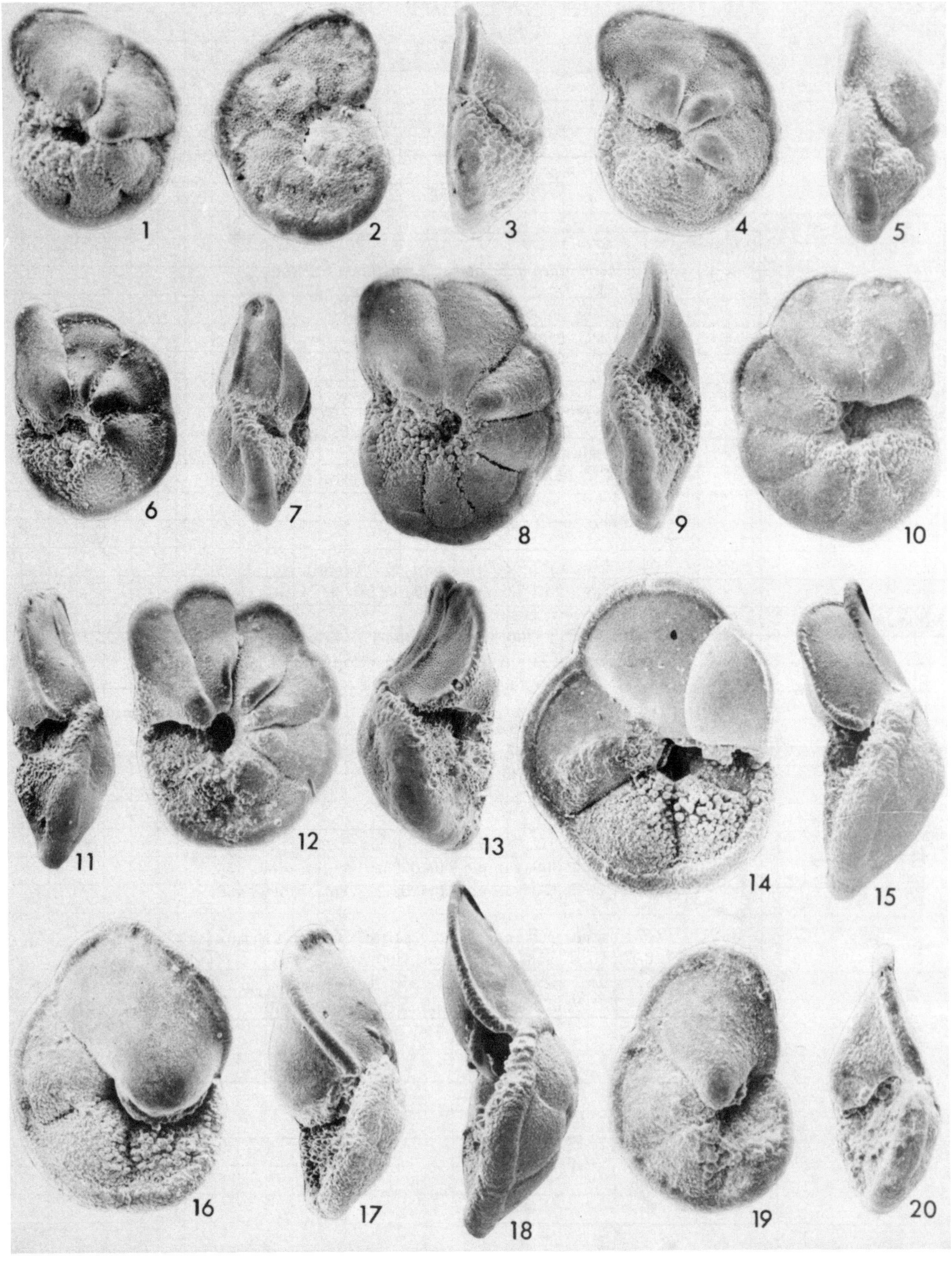
1
2
3
4
5
6
7
8
9
10
11
12
13
14
15
16
17
18
19
20

Plate 2 (facing page).

Figures
1-7   *Globorotalia (G.) ?miocenica* Palmer   1-3, CPC. 25373, sample 84640024A.   1, ventral view; 2, dorsal view; 3, edge view, showing flat dorsal surface, all x50.   4-5, CPC. 25374, same sample. 4, ventral view; 5, edge view, showing only slightly convex dorsal surface, both x50.   6-7, CPC. 25375, sample 84640024B. 6, ventral view; 7, edge view, showing flexuose development of later chambers, both x50.
8-13   *Globorotalia (G.) multicamerata* Cushman and Jarvis   8-9, CPC. 25376, sample 84640020D. 8, ventral view; 9, edge view, both x50.   10-11, CPC. 25377, sample 84640024C. 10, ventral view; 11, edge view, showing slightly flexuose development of later chambers, both x50.   12-13, CPC. 25378, sample 84640020D. 12, ventral view; 13, edge view, showing strongly flexuose development of later chambers, both x50.
14-18 *Globorotalia (G.) cultrata neoflexuosa* Srinivasan, Kennett and Be.   14-15, CPC. 25379, sample 84640030E. 14, ventral view; 15, edge view, both x50.   16-17, CPC. 25380, sample 84640030A. 16, ventral view; 17, edge view, both x50.   18, CPC. 25381, sample 84640030B, edge view, x50.
19-20 *Globorotalia (G.) tumida plesiotumida* Blow and Banner.   CPC. 25382, sample 84640024B.   19, ventral view; 20, edge view, both x50.

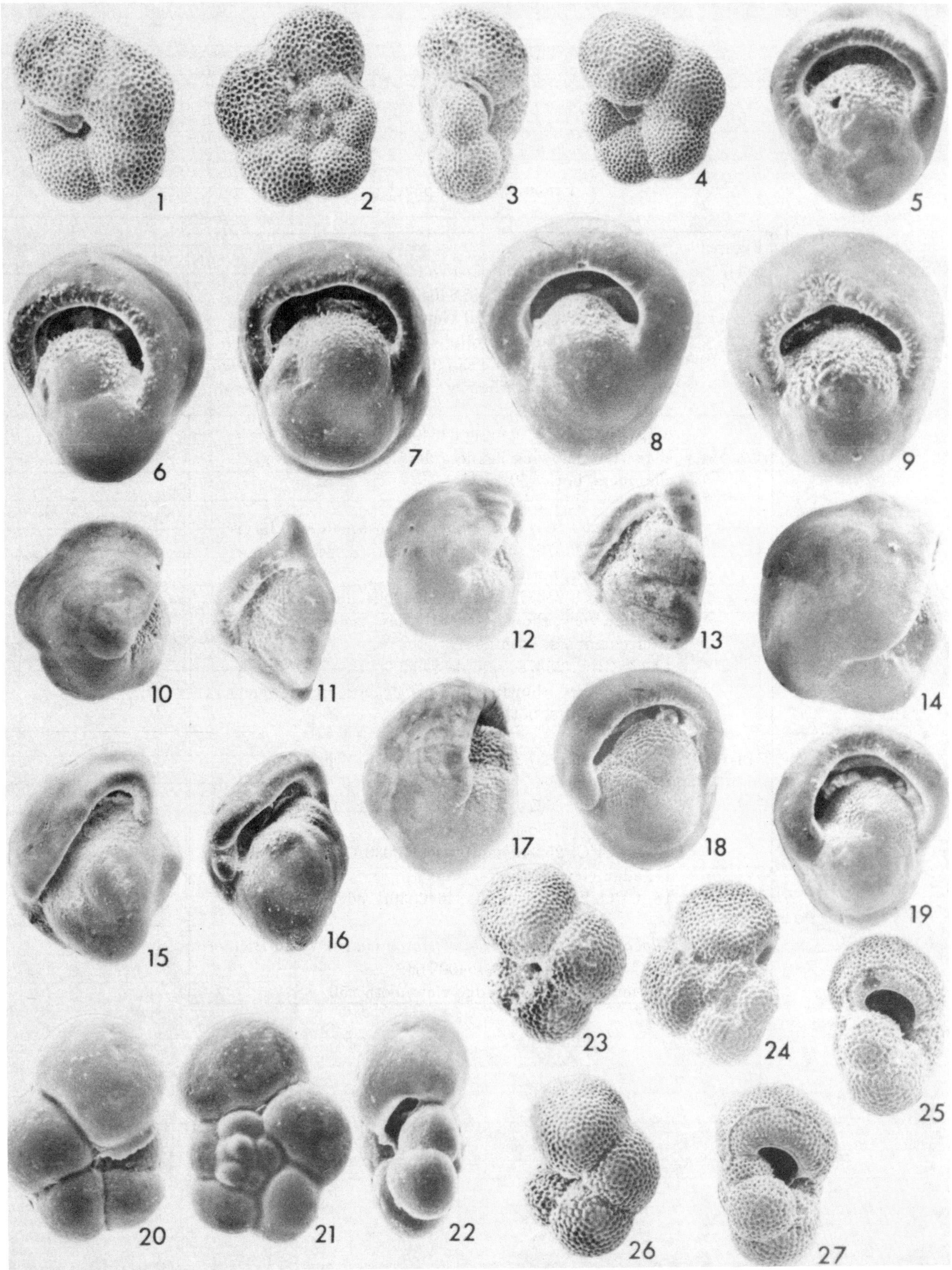
1
2
3
4
5
6
7
8
9
10
11
12
13
14
15
16
17
18
19
20
21
22
23
24
25
26
27

Plate 3 (facing page).

Figures
1-4 *Globorotaloides hexagona* (Natland)
1-3, CPC. 25393, sample 84640024A.
1, ventral view; 2, dorsal view; 3, edge view, all x80.
4, CPC. 25394, same sample, ventral view, x80.

5-8 *Pulleniatina obliquiloculata obliquiloculata* (Parker and Jones), strongly involute forms.
5, CPC. 25395, sample 84640026A, apertural view, x60.
6, CPC. 25396, sample 84640020D, apertural view, x60.
7, CPC. 25397, same sample, apertural view, x60.
8, CPC. 25398, same sample, apertural view, x60.

9 *Pulleniatina obliquiloculata finalis* Banner and Blow.
CPC. 25399, sample 84640030E, apertural view, x60.

10-13 *Pulleniatina spectabilis spectabilis* Parker.
10-11, CPC. 25400, sample 84640022B.
10, ventral view; 11, edge view, both x60.
12-13, CPC. 25401, same sample. 12, ventral view;
13, edge view, both x60.

14-16 *Pulleniatina spectabilis praespectabilis* Bronnimann and Resig.
14-15, CPC. 25402, sample 84640046. 14, ventral view;
15, edge view, both x60.
16, CPC. 25403, same sample, ventral view, x60.

17-19 *Pulleniatina* sp. 1
17-18, CPC. 25404, sample 84640027H. 17, ventral view;
18, edge view showing strongly involute test, both x60.
19, CPC. 25405, same sample, edge view showing strongly involute test, x60.

20-22 *Tenuitella anfracta* (Parker)
CPC. 25406, sample 84640039.
20, ventral view; 21, dorsal view; 22, edge view, all x200.

23-27 *Turborotalia pseudopima* (Blow)
23-25, CPC. 25407, sample 84640020B.
23, ventral view; 24, dorsal view; 25, edge view, all x80.
26-27, CPC. 25408, same sample. 26, ventral view; 27, edge view, both x80.

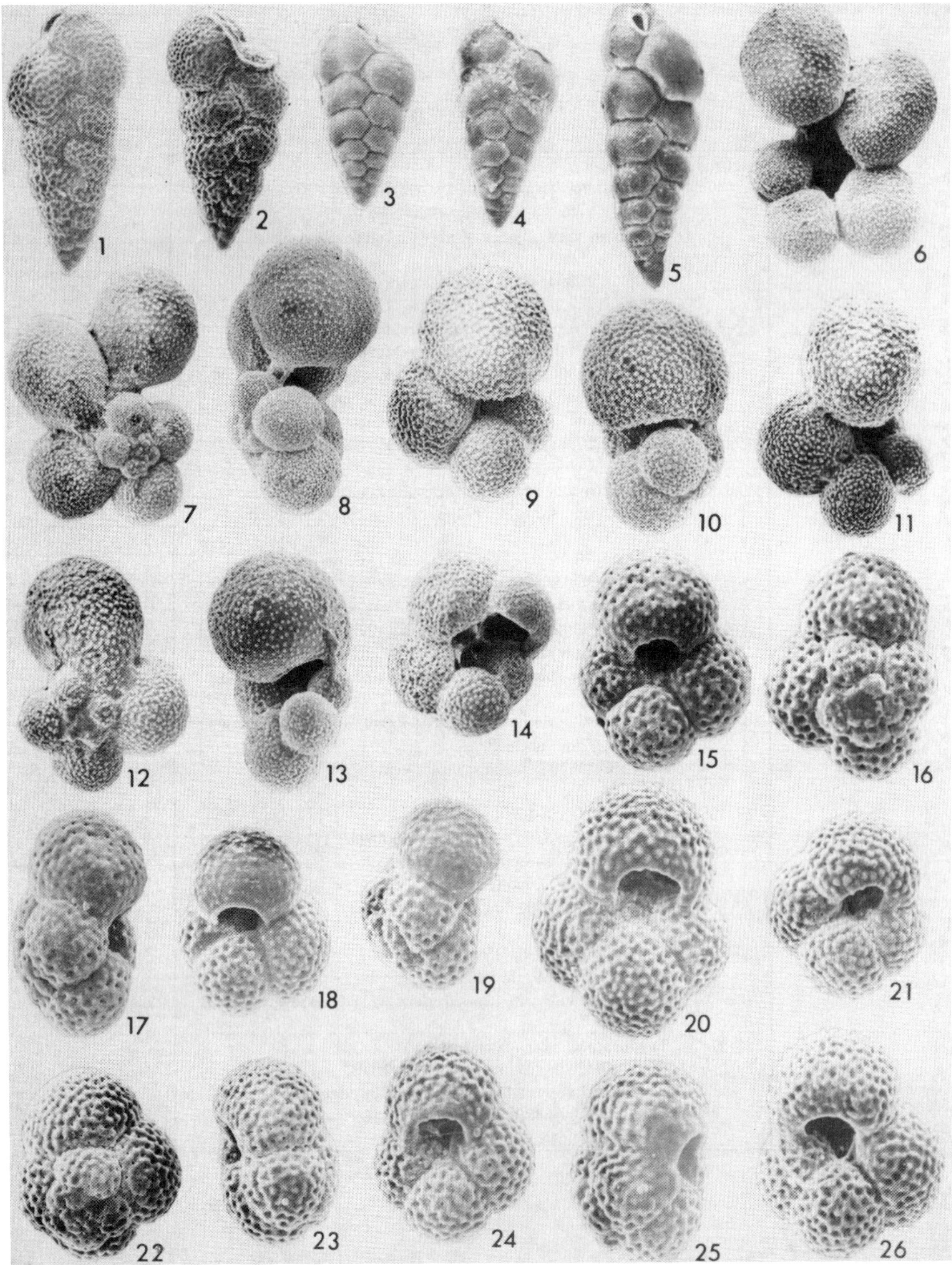
1
2
3
4
5
6
7
8
9
10
11
12
13
14
15
16
17
18
19
20
21
22
23
24
25
26

Plate 4. (facing page).

Figures	
1-2	*Streptochilus globigerum* (Schwager) 1, CPC. 25344, sample 84640022A, side view, x150. 2, CPC. 25345, sample 84640020F, side view, x150.
3-4	*Streptochilus latum* Bronnimann and Resig 3, CPC. 25346, sample 84640027I, side view, x150. 4, CPC. 25347, same sample, side view, x150.
5	*Streptochilus tokelauae* (Boersma) CPC. 25348, sample 84640027F, side view, x150.
6-8	*Globigerina (Globigerina) calida calida* Parker CPC. 25349, sample 84640030E; 6, ventral view; 7, dorsal view; 8, edge view, all x60.
9-13	*Globigerina (Globigerina) calida praecalida* Blow 9-10, CPC. 25350, sample 84640024C. 9, ventral view; 10, edge view, both x80. 11-13, CPC. 25351, sample 84640030E. 11, ventral view; 12, dorsal view; 13, edge view, all x80.
14	*Globigerina (Globigerina) bulloides umbilicata* Orr and Zaitzeff CPC. 25352, sample 84640027D, ventral view, x100.
15-20	*Globigerina (Globoturborotalita) rubescens rubescens* Hofker, pink form 15-17, CPC. 25353, sample 84640022A. 15, ventral view; 16, dorsal view; 17, edge view, all x200. 18-19, CPC. 25354, sample 84640022F. 18, ventral view; 19, edge view, both x200. 20, CPC. 25355, sample 846400438 ventral view, x200.
21-26	*Globigerina (Globoturborotalita)* cf. *rubescens* *rubescens* Hofker 21-23, CPC. 25356, sample 84640020B.21, ventral view; 22, dorsal view; 23, edge view, all x200. 24-25, CPC. 25357, sample 84640020D. 24, ventral view; 25, edge view, both x200. 26, CPC. 25358, sample 84640025A, ventral view, x200.

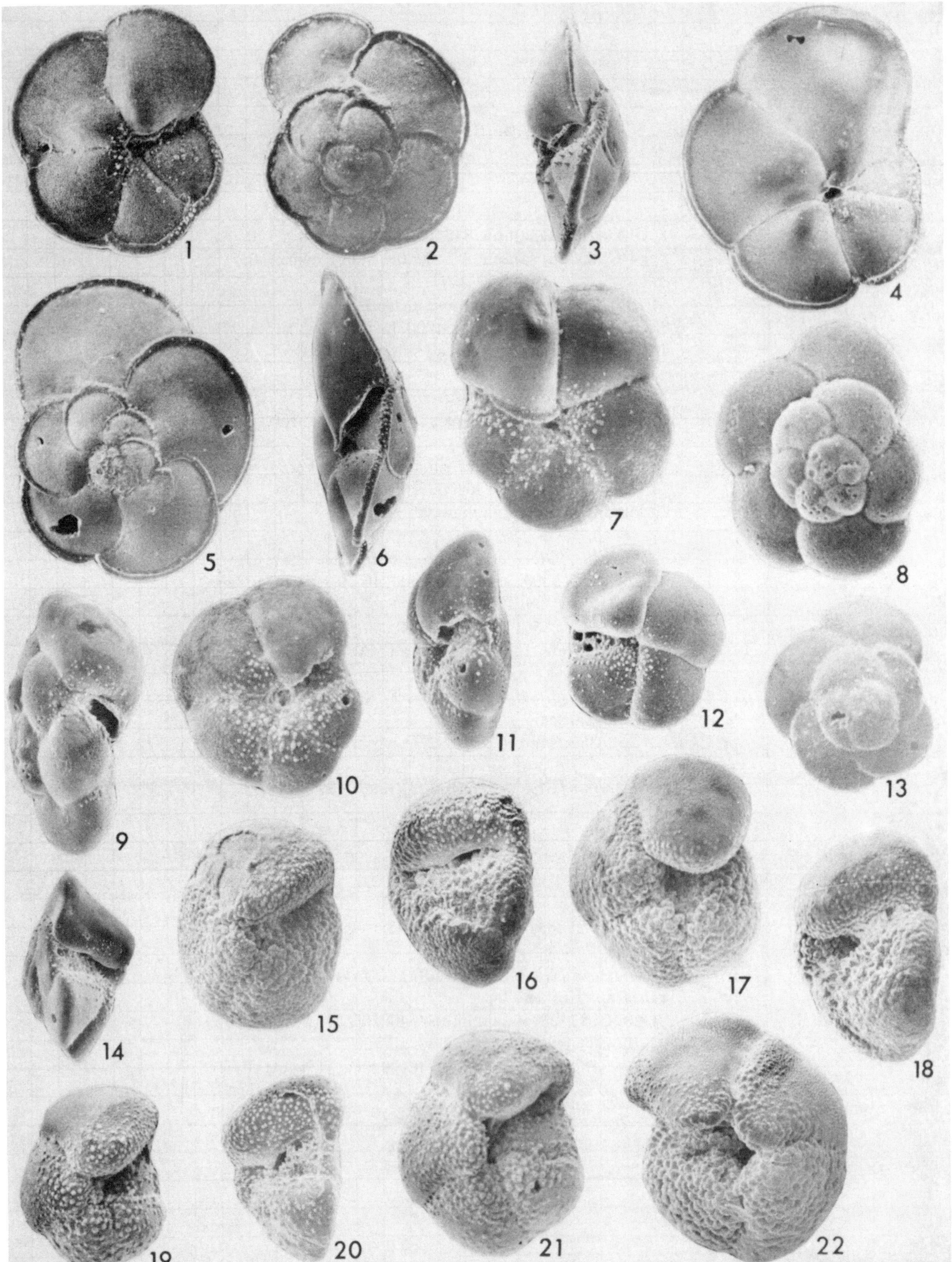
1
2
3
4
5
6
7
8
9
10
11
12
13
14
15
16
17
18
19
20
21
22

Plate 5 (facing page).

Figures	
1-3	*Globorotalia (G.)* sp. 1 CPC. 25383, sample 84640024B. 1, ventral view; 2, dorsal view; 3, edge view, all x50.
4-6	*Globorotalia (G.)* sp. 2 CPC. 25384, sample 84640040. 4, ventral view; 5, dorsal view; 6, edge view, all x50.
7-11	*Globorotalia (Obandyella) bermudezi* Rogl and Bolli 7-9, CPC. 25385, sample 84640039. 7, ventral view; 8, dorsal view; 9, edge view, all x150. 10-11, CPC. 25386, sample 84640038. 10, ventral view; 11, edge view, both x150.
12-14	*Globorotalia (Obandyella)* sp. 1 CPC. 25387, sample 84640024D. 12, ventral view; 13, dorsal view; 14, edge view, all x60.
15-22	*Globorotalia (Truncorotalia)* cf. *tosaensis tosaensis* Takayanagi and Saito. 15-16, CPC. 25388, sample 84640025A, four-chambered form with circular outline. 15, ventral view; 16, edge view, both x80. 17-18, CPC. 25389, same sample, five-chambered form with quadrate outline. 17, ventral view; 18, edge view, both x80. 19-20, CPC. 25390, same sample, five-chambered form with quadrate outline. 19, ventral view; 20, edge view; both x80. 21, CPC. 25391, sample 84640040, five-chambered form with quadrate outline, ventral view, x80. 22, CPC. 25392, same sample, five-chambered form with quadrate outline, ventral view, x80.

Marlow, M.S., Dadisman, S.V., and Exon, N.F., editors, 1988, Geology and offshore resources of Pacific island arcs—New Ireland and Manus region, Papua New Guinea, Circum-Pacific Council for Energy and Mineral Resources Earth Science Series, v. 9: Houston, Texas, Circum-Pacific Council for Energy and Mineral Resources.

NEOGENE FORAMINIFERA AS TIME-SPACE INDICATORS IN NEW IRELAND, PAPUA NEW GUINEA

D.W. Haig, P.J. Coleman
University of Western Australia, Nedlands, Western Australia

ABSTRACT

The oldest known foraminiferal assemblage (early Oligocene) occurs in conglomerate pebbles from the Jaulu Volcanics in New Ireland, Papua New Guinea. Early Miocene inner to mid-neritic assemblages from the Lelet Limestone probably range from the latest Oligocene to the late Miocene. During the middle Miocene, bathyal sedimentation occurred in northern and southern New Ireland, the oldest known deposits being in the north (Lossuk River beds, Langhian, ca. 16.6-15.2 Ma). Foraminifera indicative of middle bathyal environments are recorded from the upper Kapsu unit (Messinian, ca. 6.0-5.3 Ma) in northern New Ireland; the Djaul unit (Piacenzian, ca. 3.1-1.9 Ma) on Djaul Island; the Punam Limestone (Zanclian, ca. 5.3-4.0 Ma) in central-southern New Ireland; the Ramat unit (Zanclian - latest Piacenzian, ca. 5.1-1.9 Ma) in central New Ireland; the Gogo unit (latest Piacenzian - early Pleistocene, ca. 2.0-?1.5 Ma) in central New Ireland; and the Uluputur unit (Piacenzian - early Pleistocene) in central-southern New Ireland. Some of the latest Pliocene and early Pleistocene assemblages indicate shallowing from middle bathyal to outermost neritic environments, and date the final emergence of the New Ireland landmass from approximately 2.0 Ma.

INTRODUCTION

New Ireland is an extensively faulted landmass (Figure 1) on the uplifted southern margin of the New Ireland Basin, containing as much as 5,600 m of Miocene and younger sediment. The basin developed initially, as part of the Outer Melanesian Arc Complex, in the forearc west of the West Melanesian Trench. Later, because of late Miocene arc reversal and mid-Pliocene backarc spreading (associated with the new plate configuration), it moved to the northwest along a strike-slip fault system (Weissel, Taylor, and Karner, 1982; Falvey and Pritchard, 1984; Stewart, Francis, and Pederson, in press). Since exploratory drilling has not been undertaken in the basin, the rock exposures on New Ireland and adjacent islands provide the only direct means for establishing the stratigraphic succession for the region.

Figure 2 illustrates the stratigraphic development of New Ireland, following the revised nomenclature of Stewart and Sandy (1986; and in this volume). The basement rocks belong to the Upper Paleogene Jaulu Volcanics which presumably formed on Upper Cretaceous or Lower Paleogene oceanic crust. Over much of the island, the volcanics are overlain by platform limestones of Miocene age; these are followed by Pliocene bathypelagic limestones and volcaniclastic sediment.

Pervasive faulting makes mapping and correlation of outcrops difficult (Stewart and Sandy, this volume). But the sediment contains abundant foraminifera, including age-diagnostic planktonic and larger benthonic species. High-resolution zonations based on these microfossils are used to reconstruct

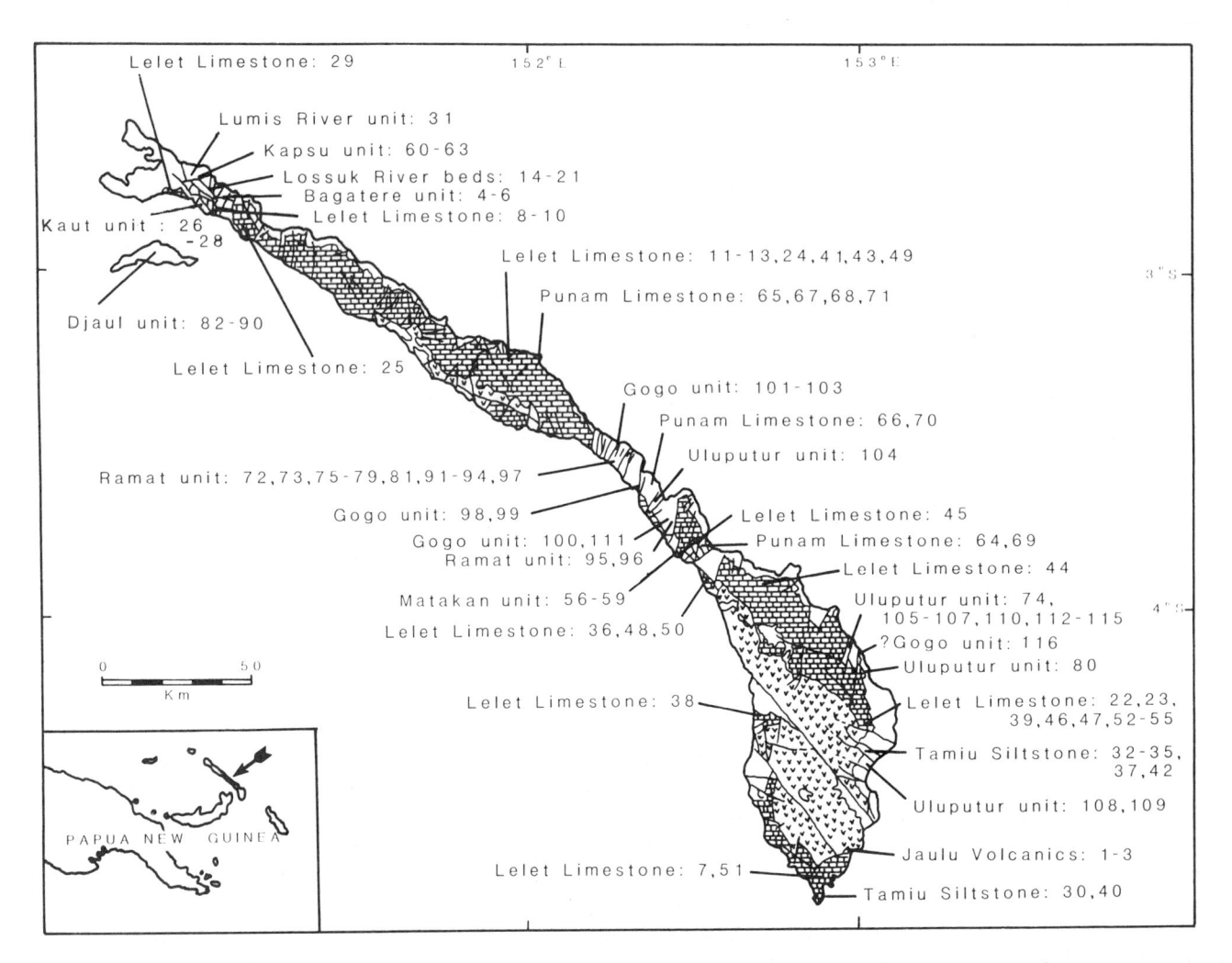

Figure 1. Locality map showing the outcrop extent of the Jaulu Volcanics (v-symbols) and the Lelet Limestone (block-symbols). Sample numbers are cited after the unit name. More precise sample localities are given in Table 1.

the stratigraphy and to interpret depositional environments. The time-space information provided by the foraminifera also contributes to regional tectonic modelling.

Although this is the type area of several important planktonic species, foraminiferal assemblages from New Ireland have received scant documentation (Brady, 1877; Schubert, 1910a,b, 1911; Coleman, 1969; Keston, 1969, Belford and Chaproniere, 1974; Belford, 1976; Hohnen, 1978; Belford, 1981; Brown, 1982; McGowran *in* Mitchell and Weiss, 1982). Brady (1877) obtained the type material of *Globorotalia tumida* (Brady) and *Globigerinoides sacculifer* (Brady) from a carved "chalk" figure collected on a beach on the east coast of New Ireland. The type material of *Globigerinoides fistulosus* (Schubert) came from Djaul Island, off the northwest coast of New Ireland (Schubert, 1910a).

METHODS

Sample Preparation

The material on which the study is based was collected by the Geological Survey of Papua New Guinea during a detailed mapping program carried on between 1983 and 1985. Friable samples were wet-sieved and the > 150 μm sediment fraction was examined microscopically. Random thin sections were prepared from the hard limestone samples. The unprocessed rock material from each sample and the thin sections are housed at the Geological Survey of Papua New Guinea. Washed residues and picked foraminiferal assemblages are in the micropaleontological collection of the University of Western Australia.

Table 1 lists sample locality and lithostrati-

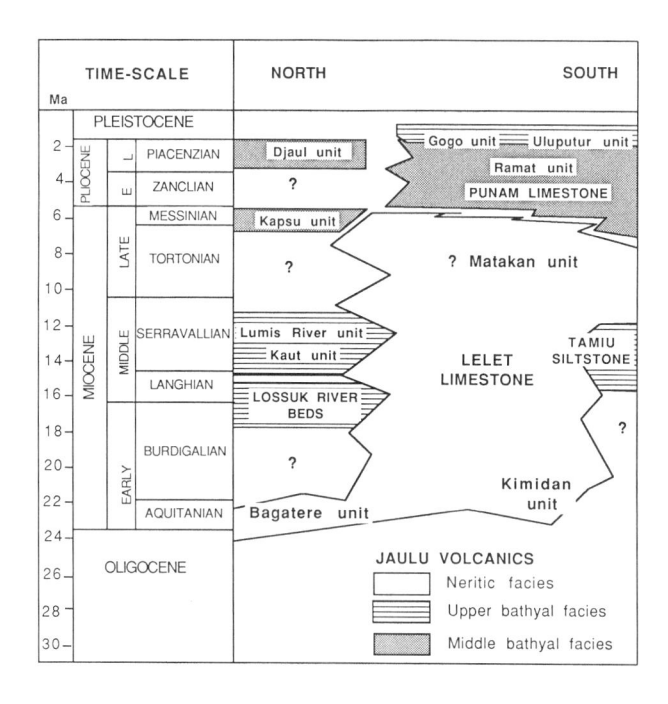

Figure 2. Time-space diagram illustrating the stratigraphic development of New Ireland. Stratigraphic nomenclature follows Stewart and Sandy (1986). The Bagatere unit, Kimidan unit and Matakan unit belong to the Lumis River Volcanics. The Ramat unit, Uluputur unit, Gogo unit, and Djaul unit belong to the Rataman Formation. A detailed description of these units is provided by Stewart and Sandy (1986; and in this volume).

graphic data (following the system used by Stewart and Sandy, 1986). Generalized sample locations are shown on Figure 1. The data contained in Table 1, and other tables, are discussed under the heading of "Results".

Zonation and Age Determination

The planktonic foraminiferal N-zonation devised by Blow (1969; with modifications by Kennett and Srinivasan, 1983; see Table 2) facilitates chronometric correlation of the bathyal sediment found on New Ireland. Zonal boundaries are placed at critical levels within evolving lineages. The "datum levels" are defined in various ways (Table 2). Those marked by a first evolutionary appearance represented by the inception of a distinct morphological feature are the most reliable for correlation because these can be recognized consistently by different workers. The datum levels placed at gradual morphological transitions are positioned subjectively, and are therefore less reliable. The chronometric scale attached to the zonation is based on integrated biostratigraphic, magnetostratigraphic and radiometric data (Berggren et al., 1985).

Tables 3 and 4 show the planktonic foraminiferal zonation of the New Ireland samples.

Because the diversity of planktonic foraminifera decreases markedly in shallow water, the N-zonation usually cannot be applied to neritic sediment. But these shelf deposits often contain larger benthonic foraminifera which form the basis of the Indonesian T-Letter "stages". The larger foraminiferal zonation was redefined by Adams (1984) who established boundary datum levels. Distributions of many of the index species, however, show marked facies control, and their datum levels are not widely applicable. Confusion has also occurred with correlation between T-letter stages and N-zones, but much of this has now been resolved (Coleman and McTavish, 1964; 1967; Cole, 1975; Haak and Postuma, 1975; Coleman, 1978; Belford, 1981; Chaproniere, 1981; Adams, 1984; Adisaputra-Sudinta and Coleman, 1984; Haig, 1985).

Table 5 shows the distribution of larger foraminifera in New Ireland samples of Miocene age. Because many of the index species are absent or cannot be confidently identified, only an approximate T-letter stage designation is made for each sample. Some of the samples contained both larger benthonic and planktonic foraminifera and can be directly correlated to N-zones.

Bathymetric Zonation

Paleobathymetric deductions are made using several criteria based on modern analogies: (a) percentage of planktonic tests in total assemblage (Grimsdale and van Morkhoven, 1955; Keij, 1963; Betjeman, 1969; Murray 1976; Palmieri, 1976); (b) maximum planktonic test size (Murray, 1976); (c) benthonic order ratios (Murray, 1973, modified here using the order classification of Haynes, 1981); (d) Fisher diversity indices for benthonic assemblage (Murray, 1973); (e) upper depth limits for selected benthonic species (derived from Pflum and Frerichs, 1976; and central Indo-Pacific studies of modern faunas including Cushman, 1921; Collins, 1958; Graham and Militante, 1959; Frerichs, 1970; Biswas, 1976; Palmieri, 1976; Hughes, 1977; Hofker, 1978; Burke, 1981).

The paleobathymetric classification adopted here follows Ingle (1980): inner neritic zone, 0-50 m; outer neritic zone, 50-150 m; upper bathyal zone,

Table 1. Sample locality data. Samples are listed in approximate time-stratigraphic order: 1-3, Early Miocene or older; 4-13, Early Miocene; 14-36, Middle Miocene; 37-55, Middle or Late Miocene; 56-69, no younger than Late Miocene; 60-63, Late Miocene; 64-73, Early Pliocene; 74-110, Late Pliocene; 111-115, early Pleistocene; 116, probably early Pleistocene.

SAMPLE	PREPARATION	FIELD NUMBER	GRID REFERENCE	MAP SHEET (1:100,000)	STRATIGRAPHIC UNIT	LITHOLOGY
	wr: washed residue					
	ts: thin section					
1	ts	DK191-272A	MV967785	Cape St. George	Jaulu Volcanics	volcanic arenite
2	ts	DK191-272B	MV967785	Cape St. George	Jaulu Volcanics	algal limestone
3	ts	DK191-272C	MV967785	Cape St. George	Jaulu Volcanics	algal calcarenite
4	wr	DK52-85	KB831884	Fangalawa	Bagatere unit	sandstone
5	wr	DK18-16	KB788915	Fangalawa	Bagatere unit	shelly sandstone
6	wr	DK19-18B	KB787915	Fangalawa	Bagatere unit	siltstone
7	ts	GFK58-111	MV887847	Cape St. George	Lelet Limestone	foram-algal packstone
8	ts	MSK1A	KB814919	Fangalawa	Lelet Limestone	"Halimeda" calcarenite
9	ts	MSK1B	KB814919	Fangalawa	Lelet Limestone	"Halimeda" calcarenite
10	ts	MSK31	KB827924	Fangalawa	Lelet Limestone	"Halimeda" calcarenite (allochthonous block)
11	wr	GFK112-173	LB732424	Konos	Kimidan unit	carbonaceous mudstone
12	ts	GFK112-175	LB732424	Konos	Lelet Limestone	foram-algal calcarenite
13	ts	GFK112-176	LB732424	Konos	Lelet Limestone	foram-algal calcarenite
14	wr	DI4	KB838935	Fangalawa	Lossuk River beds	siltstone
15	wr	MSK158	KB837931	Fangalawa	Lossuk River beds	siltstone
16	wr	MSK30BD	KB826924	Fangalawa	Lossuk River beds	siltstone
17	wr	MSK30B	KB826924	Fangalawa	Lossuk River beds	siltstone
18	wr	BPK86	KB832945	Fangalawa	Lossuk River beds	siltstone
19	wr	MSK14	KB832945	Fangalawa	Lossuk River beds	siltstone
20	wr	MSK5	KB836936	Fangalawa	Lossuk River beds	siltstone
21	wr	DK55-87	KB837932	Fangalawa	Lossuk River beds	siltstone
22	ts	DK222-307	NAO24211	Mimias	Lelet Limestone	foram-algal calcarenite
23	ts	GFK76-129	NAO16167	Mimias	Lelet Limestone	algal limestone
24	ts	MSK73	LB812402	Konos	Lelet Limestone	foram-algal grainstone
25	ts	MSK80	KB946823	Fangalawa	Lelet Limestone	foram-algal mudstone
26	wr	MSK6	KB835938	Fangalawa	Kaut unit	siltstone
27	wr	DK148-202	KB679958	Kavieng	Kaut unit	siltstone
28	wr	DK20-24	KB680956	Kavieng	Kaut unit	siltstone
29	wr	DK32-53A	KB674953	Kavieng	Kaut unit	siltstone
30	wr	BPK19B	NV019995	Siar	Tamiu siltstone	siltstone
31	wr	MSK38	KB736996	Kavieng	Lumis River unit	siltstone
32	wr	GFK79-135	NAO27162	Mimias	Tamiu siltstone	siltstone
33	wr	GFK83-139	NAO36150	Mimias	Tamiu siltstone	siltstone
34	wr	GFK84-141	NAO37149	Mimias	Tamiu siltstone	mudstone
35	wr	GFK85-142	NAO37146	Mimias	Tamiu siltstone	siltstone
36	ts	MSK91	MA466713	Dolomakas	Lelet Limestone	foram-algal packstone/mudstone
37	ts	GFK84-140	NAO37149	Mimias	Tamiu siltstone	fine calcarenite
38	ts	GFK97-157	MA668218	Konogaiang	Lelet Limestone	foram packstone
39	ts	GFK100-162	MA985259	Konogaiang	Lelet Limestone	foram packstone
40	ts	BPK19A	NV019995	Siar	Tamiu siltstone	calcarenite
41	ts	DK153-204	LB821355	Konos	Lelet Limestone	algal packstone
42	wr	GFK75-126	NAO14167	Mimias	Tamiu siltstone	siltstone
43	ts	GFK111-172	LB734425	Konos	Lelet Limestone	foram-algal packstone
44	ts	DK126-148	MA714679	Dolomakas	Lelet Limestone	foram-algal packstone
45	ts	MSK246C	MA389803	Namatanai	Lelet Limestone	foram-algal mudstone/packstone
46	ts	GFK75-127	NAO14167	Mimias	Lelet Limestone	micrite (10% volcanic ash)
47	ts	GFK76-128	NAO16167	Mimias	Lelet Limestone	foram-algal calcarenite
48	ts	MSK89	MA458712	Dolomakas	Lelet Limestone	algal packstone
49	ts	MSK181	LB821355	Konos	Lelet Limestone	algal packstone
50	ts	MSK92	MA471714	Dolomakas	Lelet Limestone	algal mudstone/packstone
51	ts	DK200-280	MV887713	Cape St. George	Lelet Limestone	algal-foram calcarenite
52	ts	DK224-309A	NAO28205	Mimias	Lelet Limestone	bioclastic calcarenite
53	ts	DK226-311-1	NAO36203	Mimias	Lelet Limestone	algal packstone
54	ts	DK226-311-3	NAO36203	Mimias	Lelet Limestone	foram packstone
55	ts	DK226-311-6	NA036203	Mimias	Lelet Limestone - Punam Limestone	foram packstone
56	wr	DK227-314	MA398777	Namatanai	Matakan unit	carbonaceous siltstone
57	wr	DK229-315	MA399784	Namatanai	Matakan unit	siltstone
58	wr	MSK218	MA398779	Namatanai	Matakan unit	siltstone
59	wr	MSK306	MA398778	Namatanai	Matakan unit	siltstone

Table 1. (Continued)

SAMPLE	PREPARATION	FIELD NUMBER	GRID REFERENCE	MAP SHEET	STRATIGRAPHIC UNIT	LITHOLOGY
	wr: washed residue			(1:100,000)		
	ts: thin section					
60	wr	MSK9A	KB826936	Fangalawa	Kapsu unit	siltstone
61	wr	MSK11	KB820952	Fangalawa	Kapsu unit	siltstone
62	wr	DK145-201	KB800985	Fangalawa	Kapsu unit	sandstone
63	wr	MSK15	KB800985	Fangalawa	Kapsu unit	siltstone
64	wr	DK136-182	MA459795	Dolomakas	Punam Limestone	biomicrite
65	wr	DK79-104	LB689454	Konos	Punam Limestone	biomicrite
66	wr	MSK63	MA291942	Namatanai	Punam Limestone	biomicrite
67	wr	DK61-91	LB810442	Konos	Punam Limestone	biomicrite
68	wr	DK67-96	LB817444	Konos	Punam Limestone	biomicrite
69	wr	DK95-114	MA470822	Dolomakas	Punam Limestone	biomicrite
70	wr	DK184-250	MB283017	Namatanai	Punam Limestone	biomicrite
71	wr	MSK75	LB810442	Konos	Punam Limestone	biomicrite
72	wr	MSK242	MB112121	Namatanai	Ramat unit	calcareous sandstone
73	wr	DK182-249	MB282019	Namatanai	Ramat unit	tuffaceous sandstone
74	wr	DK169-221	MA996435	Konogaiang	?Uluputur unit	biomicrite
75	wr	DK121-141	MB249081	Namatanai	Ramat unit	interbedded sandstone/biomicrite
76	wr	MSK44	MB121118	Namatanai	Ramat unit	siltstone
77	wr	MSK45	MB131114	Namatanai	Ramat unit	siltstone
78	wr	MSK194	MB133109	Namatanai	Ramat unit	calcareous sandstone
79	wr	MSK200	MA293899	Namatanai	Ramat unit	tuffaceous biomicrite
80	wr	MSK202	NA015406	Mimias	Uluputur unit	chalky limestone
81	wr	DK113-134	MB224035	Namatanai	Ramat unit	lutite
82	wr	DK8-4	KB670769	Kavieng	Djaul unit	siltstone
83	wr	MSK22	KB621758	Kavieng	Djaul unit	mudstone
84	wr	MSK25A	KB644752	Kavieng	Djaul unit	mudstone
85	wr	MSK20B	KB623762	Kavieng	Djaul unit	mudstone
86	wr	MSK20A	KB623762	Kavieng	Djaul unit	siltstone
87	wr	MSK19	KB627765	Kavieng	Djaul unit	mudstone
88	wr	MSK26	KB626733	Kavieng	Djaul unit	mudstone
89	wr	MSK28	KB626737	Kavieng	Djaul unit	mudstone
90	wr	DK7-3	KB644758	Kavieng	Djaul unit	mudstone
91	wr	DK114-135	MB224035	Namatanai	Ramat unit	lutite
92	wr	DK118-138	MB258091	Namatanai	Ramat unit	biomicrite
93	wr	DK180-247	MB283023	Namatanai	Ramat unit	siltstone
94	wr	DK183-249	MB282019	Namatanai	Ramat unit	tuffaceous biomicrite
95	wr	MSK95	MA332851	Namatanai	Ramat unit	sandstone
96	wr	MSK57	MA372873	Namatanai	Ramat unit	sandstone
97	wr	MSK49	MA285945	Namatanai	Ramat unit	siltstone
98	wr	MSK50	MA305931	Namatanai	Gogo unit	tuffaceous biomicrite
99	wr	MSK52	MA310920	Namatanai	Gogo unit	biomicrite
100	wr	MSK55	MA368898	Namatanai	Gogo unit	biomicrite
101	wr	MSK98	MB204042	Namatanai	Gogo unit	biomicrite
102	wr	DK117-138	MB259091	Namatanai	Gogo unit	biomicrite
103	wr	DK155-205	MB258091	Namatanai	Gogo unit	biomicrite
104	wr	DK143-196	MA339926	Namatanai	Uluputur unit	siltstone
105	wr	DK167-220	MA996435	Konogaiang	Uluputur unit	sandstone matrix in conglomerate
106	wr	DK171-223	MA990436	Konogaiang	Uluputur unit	sandstone
107	wr	DK205-286	MA978425	Konogaiang	Uluputur unit	siltstone
108	wr	MSK211	NA106145	Mimias	Uluputur unit	sandstone
109	wr	MSK212	NA026077	Mimias	Uluputur unit	silty mudstone
110	wr	DK172-224	MA987433	Konogaiang	Uluputur unit	siltstone
111	wr	MSK51	MA307917	Namatanai	Gogo unit	biomicrite
112	wr	DK168-220	MA996435	Konogaiang	Uluputur unit	siltstone
113	wr	GFK1-2	MA990404	Konogaiang	Uluputur unit	silty sandstone
114	wr	GFK3-2	MA989402	Konogaiang	Uluputur unit	shelly siltstone
115	wr	GFK5-4	MA991405	Konogaiang	Uluputur unit	siltstone
116	wr	DK158-209	NA011439	Mimias	? Gogo unit	biomicrite

Table 2. Neogene time scale. Stage correlations follow Berggren, Kent, Flynn, and Van Couvering (1985); planktonic zonation follows Blow (1969) with modifications by Kennett and Srinivasan (1983); age (Ma) follows Berggren, Kent, Flynn, and Van Couvering (1985); T-Letter stage correlation follows Adams (1984). [F.A.D. - first appearance datum; X - evolution represented by inception of a distinct morphological feature; Y - evolution represented by gradual transition in morphology; Z - abrupt appearance of morphotype with uncertain ancestry; L.A.D. - last appearance datum].

150-500 m; middle bathyal zone, 500-2,000 m; lower bathyal zone, 2,000-4,000 m.

Table 6 charts the various criteria listed above, for each sample, and the bathymetric interpretations derived from these data.

RESULTS

Pre-Miocene Assemblages

Limestone pebbles from a conglomerate in the Jaulu Volcanics (at grid reference MV883847 Cape St. George 1:100,000 sheet) contain the only pre-Miocene assemblage found during this study. *Nummulites fichteli* Michelotti occurs in abundance in the samples and indicates a latest Eocene or, more likely, an early Oligocene age (following Adams, 1970, p. 122). Rare examples of a primitive planorbulinid and an unidentified alveolinellid also are present. The pebbles are biomicrite with varying amounts of volcanic detritus. Strong pressure-solution effects are evident. The foraminiferal assemblage indicates an inner neritic depositional environment for the limestone.

Terpstra (*in* Hohnen, 1978) examined a sample from a large limestone "lens" (Hohnen, 1978, p. 3) within the Jaulu Volcanics along the Sapom Fault, and recorded *Nummulites* cf. *fichteli.*

We have examined three other samples (1-3 on Table 1) from the Jaulu Volcanics. These contain no age diagnostic foraminifera, although an abundant assemblage of larger foraminifera (Table 6) indicative of the mid-neritic zone, is present in two of the samples.

Early Miocene Neritic Assemblages

The early Miocene assemblages we recognize (Table 5, samples 4-13) are dated by larger foraminifera and are referred tentatively to the upper Te-letter stage. But index forms such as *Spiroclypeus* and *Lepidocyclina (Eulepidina)* have not been identified in our material (apart from a single doubtful record of *Eulepidina* in sample 9). The present determination is based on the occurrence of primitive *Miogypsinoides dehaartii* and a *Lepidocyclina (Nephrolepidina)* assemblage clearly older than the more diverse one which characterizes the Tf-letter stage in this region.

It is possible that some of the samples included here are of latest Oligocene rather than Miocene age. These may include assemblages (4, 5) from the Bagatere unit of the Lelet Limestone in northern New Ireland which seem to be the oldest in this group (suggested by the presence of primitive *Miogypsinoides dehaartii* and *Miogypsina thecideaeformis*). In contrast, assemblages (12, 13) from the lower Lelet Limestone in the Kimidan region (southern New Ireland) are more firmly positioned near the Te-Tf boundary (equivalent to N6 planktonic zone; Burdigalian; ca. 19.0-17.6 Ma) as indicated by the presence of primitive *Austrotrillina howchini*. We suggest that the base of the Lelet Limestone is diachronous -- a notion consistent with disjunct shoaling of a faulted island arc terrain.

Inner neritic paleobathymetry is inferred for most of these samples with large foraminifera (Table 6), as indicated by their presence, the absence of planktonic forms, and the absence of benthonic species with modern upper depth limits deeper than 50 m. Some of the samples were probably deposited at mid-neritic depths (40-60 m) as suggested by the presence of *Cycloclypeus.*

Table 3. Early Neogene age-diagnostic planktonic foraminifera (interpreted in sense of Kennett and Srinivasan, 1983). Samples are arranged in approximate time-stratigraphic order: 14-21, Langhian; 26-36, Serravallian; 37-41, probably Serravallian; 42, 54 and 55, Middle or Late Miocene. [x: present; cf: comparable form; ?: doubtful record].

SAMPLES (wr: washed residue ts: thin section)	14 wr	15 wr	16 wr	17 wr	18 wr	19 wr	20 wr	21 wr	26 wr	27 wr	28 wr	29 wr	30 wr	31 wr
GLOBIGERINOIDES PRIMORDIUS-ALTIAPERTURA LINEAGE														
G. altiapertura (N5-N7)	cf	cf	cf	cf	cf	cf	cf	cf	cf		cf	cf		cf
G. obliquus (N5-N22)										X				X
GLOBIGERINA (ZEAGLOBIGERINA) LINEAGE														
G. (Z.) woodi (N4A-N21)	X	X	X	X	X	X	X	X	X	X	X	X	X	X
G. (Z.) druryi (N7-N15)														
G. (Z.) decoraperta (N9-N21)														
GLOBIGERINOIDES SUBQUADRATUS-RUBER LINEAGE														
G. subquadratus (N4B-N15)	X	X	X	X	X	X	X	X	X	X	X	X	X	X
G. mitra (N7-N13)	X	X		X	X			X						X
GLOBIGERINOIDES TRILOBA-SACCULIFER LINEAGE														
G. triloba (N4B-N22)	X	X	X	X	X	X	X	X	X	X	X		X	X
G. immaturus (N5-N23)	X	X	X	X	X	X	X	X	X	X	X			X
G. quadrilobatus (N6-N23)	X	X	X	X	X	X	X	X	X	X	X	X	X	X
G. sacculifer (N6-N23)		X	X		X	X		X			X			
PRAEORBULINA-ORBULINA LINEAGE														
P. sicanus (N8-N9)	X	X	X	X		X		X						
P. glomerosa curva (mid N8 - basal N9)	X	X	X	X	X			X						
P. glomerosa glomerosa (mid N8 - lower N9)	X	X				X	X	X						
P. glomerosa circularis (upper N8 - N9)				X	X	X	X							
P. transitoria (mid N8 - mid N9)		X			X	X								
Orbulina suturalis (N9-N23)														X
Orbulina universa (N9-N23)										X				X
Orbulina bilobata (N9-N23)														X
SPHAEROIDINELLOPSIS-SPHAEROIDINELLA LINEAGE														
S. disjuncta (N6-N11)	X	X	X	X	X	X	X	X						
S. seminulina (N7-N21)													X	X
S. kochi (N10-N19)														
GLOBOROTALIA (FOHSELLA) LINEAGE														
G. (F.) birnageae (N7-N8)		X												
G. (F.) peripheroronda (N4B-N10)	X	X		X			X	X						
G. (F.) peripheroacuta (N10-N11)										X			X	X
G. (F.) praefohsi (N11- low N12)														X
G. (F.) fohsi fohsi (N12)														
G. (F.) fohsi lobata (N12)														
G. (F.) fohsi robusta (N12)														
GLOBOROTALIA (JENKINSELLA) LINEAGE														
G. (J.) siakensis (P22-N14)														
G. (J.) mayeri (N4A-N14)	X	X	X	X	X	X	X	X	X	X		X	X	X
GLOBOROTALIA (MENARDELLA) LINEAGE														
G. (M.) archeomenardii (N8-N10)	X	X				X								
G. (M.) praemenardii (N10-N12)										X				
G. (M.) menardii (N12-N23)														
DENTOGLOBIGERINA LINEAGE														
D. altispira altispira (N4B-N21)	X	X	X	X	X	X	X	X	X		X	X	X	X
D. altispira globosa (P22-N19)		X		X	X									
GLOBOQUADRINA LINEAGE														
G. dehiscens (N4B-N18)	X		X						X					X
G. baroemoenensis (N5-N18)	X			X							X			
"G." venezuelana (P22-N19)	X	X	X		X	X		X	X				X	X
GLOBIGERINELLA LINEAGE														
Globorotalia obesa (P22-N23)	X	X	X	X	X						X			
Globigerinella praesiphonifera (N4B-N13)														
GLOBIGERINITA-GLOBIGERINATELLA LINEAGE														
G. glutinata (P22-N23)	X					X								X
Gt. insueta (N6-N9)	X		X	X	X									
PLANKTONIC FORAMINIFERAL ZONE (after Blow, 1969; Kennett & Srinivasan, 1983)	N8	N8	N8	N8	up. N8	up. N8	up. N8	up. N8	N10				N10	N11
APPROXIMATE T-LETTER "STAGE" (see Table 5)		low Tf	low Tf	low Tf				low Tf	low Tf					low Tf

Table 3. (Continued)

SAMPLES (wr: washed residue ts: thin section)	32 wr	33 wr	34 wr	35 wr	36 ts	37 ts	38 ts	39 ts	40 ts	41 ts	42 wr	54 ts	55 ts
GLOBIGERINOIDES PRIMORDIUS-ALTIAPERTURA LINEAGE													
G. altiapertura (N5-N7)													
G. obliquus (N5-N22)		X											
GLOBIGERINA (ZEAGLOBIGERINA) LINEAGE													
G. (Z.) woodi (N4A-N21)	X	X	X	X									
G. (Z.) druryi (N7-N15)		X	X										
G. (Z.) decoraperta (N9-N21)	X	X		X									
GLOBIGERINOIDES SUBQUADRATUS-RUBER LINEAGE													
G. subquadratus (N4B-N15)	X	X	X	X									
G. mitra (N7-N13)				X									
GLOBIGERINOIDES TRILOBA-SACCULIFER LINEAGE													
G. triloba (N4B-N22)	X	X											
G. immaturus (N5-N23)	X	X	X	X									
G. quadrilobatus (N6-N23)	X	X	X	X							X		
G. sacculifer (N6-N23)			X	X									
PRAEORBULINA-ORBULINA LINEAGE													
P. sicanus (N8-N9)													
P. glomerosa curva (mid N8 - basal N9)													
P. glomerosa glomerosa (mid N8 - lower N9)													
P. glomerosa circularis (upper N8 - N9)													
P. transitoria (mid N8 - mid N9)													
Orbulina suturalis (N9-N23)			X										
Orbulina universa (N9-N23)	X	X	X	X		X					X	X	X
Orbulina bilobata (N9-N23)	X		X	X									
SPHAEROIDINELLOPSIS-SPHAEROIDINELLA LINEAGE													
S. disjuncta (N6-N11)													
S. seminulina (N7-N21)	X	X	X	X	X		X	X					X
S. kochi (N10-N19)	X	X	X	X									
GLOBOROTALIA (FOHSELLA) LINEAGE													
G. (F.) birnageae (N7-N8)													
G. (F.) peripheroronda (N4B-N10)													
G. (F.) peripheroacuta (N10-N11)			X	X									
G. (F.) praefohsi (N11- low N12)													
G. (F.) fohsi fohsi (N12)	X	X	X	X	?	?	?	?	?	?			
G. (F.) fohsi lobata (N12)	X	X		X									
G. (F.) fohsi robusta (N12)				X									
GLOBOROTALIA (JENKINSELLA) LINEAGE													
G. (J.) siakensis (P22-N14)			X	X									
G. (J.) mayeri (N4A-N14)	X	X	X	X									
GLOBOROTALIA (MENARDELLA) LINEAGE													
G. (M.) archeomenardii (N8-N10)													
G. (M.) praemenardii (N10-N12)													
G. (M.) menardii (N12-N23)				X									
DENTOGLOBIGERINA LINEAGE													
D. altispira altispira (N4B-N21)	X	X	X	X									
D. altispira globosa (P22-N19)													
GLOBOQUADRINA LINEAGE													
G. dehiscens (N4B-N18)		X											
G. baroemoenensis (N5-N18)													
"G." venezuelana (P22-N19)		X											
GLOBIGERINELLA LINEAGE													
Globorotalia obesa (P22-N23)											X		
Globigerinella praesiphonifera (N4B-N13)			X	X									
GLOBIGERINITA-GLOBIGERINATELLA LINEAGE													
G. glutinata (P22-N23)		X											
Gt. insueta (N6-N9)													
PLANKTONIC FORAMINIFERAL ZONE (after Blow, 1969; Kennett & Srinivasan, 1983)	N12	N12	N12	N12	? N12	? N12	? N12	? N12	? N12	? N12			
APPROXIMATE T-LETTER "STAGE" (see Table 5)					Tf	Tf	Tf		Tf	Tf		Tf	Tf

Table 4. Late Neogene age-diagnostic planktonic foraminifera (interpreted in sense of Kennett and Srinivasan, 1983). Samples are arranged in approximate time-stratigraphic order: 60-73, Zanclian; 74-110, Piacenzian; 111-115, ?116, early Pleistocene. [x: present; D: dextral coiling dominant; S: sinistral coiling dominant; ?: doubtful record].

SAMPLES (washed residues)	60	61	62	63	64	65	66	67	68	69	70	71	72	73	74	75	76	77	78
GLOBIGERINOIDES PRIMORDIUS -ALTIAPERTURA LINEAGE																			
G. obliquus (N5-N22)	X	X	X	X	X	X	X	X	X	X	X	X			X			X	
G. extremus (N16-N21)	X	X	X	X	X	X	X	X	X	X	X				X		X	X	
GLOBIGERINA (ZEAGLOBIGERINA) LINEAGE																			
G. (Z.) woodi (N4A-N21)	X	X	X	X	X	X	X	X	X	X	X				X		X	X	X
G. (Z.) nepenthes (N14-N19)				X	X		X			X	X								
G. (Z.) apertura (N16-N21)			X	X		X		X	X	X	X	X							
G. (Z.) decoraperta (N9-N21)			X	X	X	X	X	X	X	X	X								
GLOBIGERINOIDES SUBQUADRATUS-RUBER LINEAGE																			
G. ruber (N15-N23)															X		X	X	X
GLOBIGERINOIDES TRILOBA-SACCULIFER LINEAGE																			
G. triloba (N4B-N22)	X	X		X	X	X	X	X	X	X		X	X	X					
G. immaturus (N5-N23)	X	X	X	X	X	X	X	X	X	X	X	X			X			X	X
G. quadrilobatus (N6-N23)	X	X	X	X	X	X	X	X	X	X	X	X			X		X	X	X
G. sacculifer (N6-N23)		X	X	X	X	X	X	X	X	X	X	X	X		X		X		
G. fistulosus (N21-low N22)																	X	X	X
ORBULINA LINEAGE																			
O. universa (N9-N23)	X		X	X		X	X	X	X	X	X	X			X	X		X	X
O. bilobata (N9-N23)							X												
SPHAEROIDINELLOPSIS-SPHAEROIDINELLA LINEAGE																			
S. seminulina (N7-N21)	X	X	X	X	X	X	X	X	X	X	X	X	X	X				X	X
S. kochi (N10-N19)					X	X	X	X	X	X	X	X							
S. paenedehiscens (N17B-N20)	X	X	X	X	X	X	X	X	X	X	X	X							
S. dehiscens (N19-N23)											X	X	X	X	X		X	X	X
NEOGLOBOQUADRINA LINEAGE																			
N. acostaensis (N16-N20)	X		D	X	D	D	D	D	D	D	X	D		D	D		D	D	
N. pachyderma (N16-N23)				D			D		D	D					D				
N. humerosa (N18-N22)							D		D					D	D		D	D	D
N. dutertrei (N21-N23)																			
PULLENIATINA LINEAGE																			
P. primalis (N17b-N20)	S		S	S	S	S	S	S	S	S	S	S							
P. spectabilis (upper N18-N19)							S	S	S	S	S								
P. praecursor (N19-N21)								?	S		S	S	S	S	S	S	S	D-S	
P. obliquiloculata (N19-N23)													S	S	S	S	S	D-S	X
GLOBOROTALIA (GLOBOROTALIA) LINEAGE																			
G. (G.) paralenguaensis (N15-N17A)			X	X															
G. (G.) merotumida (N16-N18)	S	?																	
G. (G.) plesiotumida (N17-N19)	S				S	S	S	S	S	S	S	S							
G. (G.) tumida tumida (N18-N23)					S	S	S	S	S	S	S	S	S	S	S	S	S	S	S
G. (G.) tumida flexuosa (upper N18-N21)															S	S	S	S	
G. (G.) ungulata (N21-N23)																			
GLOBOROTALIA (MENARDELLA) LINEAGE																			
G. (M.) menardii (N12-N23)	S																		
G. (M.) limbata (N14-N21)			S	S	S		D-S	D-S	D	D		D	D		S			D	
GLOBOROTALIA (TRUNCOROTALIA) LINEAGE																			
G. (T.) crassula (N18-N22)																			
G. (T.) crassaformis (upper N19-N23)															X	X	X	X	X
G. (T.) tosaensis (N21-N22)															X	X	X	X	X
G. (T.) truncatulinoides (N22-N23)																			
DENTOGLOBIGERINA LINEAGE																			
D. altispira altispira (N4B-N21)	X		X		X	X	X	X	X	X	X	X						X	
D. altispira globosa (P22-N19)		X		X	X		X	X	X	X	X								
GLOBOQUADRINA LINEAGE																			
G. dehiscens (N4B-N18)	X			X		X											X		
G. baroemoenensis (N5-N18)	X			X		X													
"G." venezuelana (P22-N19)	X	X		X	X	X	X	X	X	X	X	X						?	
GLOBIGERINELLA LINEAGE																			
Globorotalia obesa (P22-N23)		X	X	X	X		X	X		X							X		
G. aequilateralis (N12-N23)		X		X	X		X	X	X	X									
GLOBIGERINITA-GLOBIGERINATELLA LINEAGE																			
G. glutinata (P22-N23)				X															
PLANKTONIC ZONE (after Blow, 1969; Kennett & Srinivasan, 1983)	N17B	N17B	N17B	N17B	N18	N18	N18	up. N18	up. N18	up. N18	N19	N19	N19	N19 -20	N21	N21	N21	N21	N21

Middle Miocene Neritic-Upper Bathyal Assemblages

The oldest middle Miocene assemblages recorded here are from the upper part of the Lossuk River beds in northern New Ireland (Table 3, samples 14-21). These beds contain abundant planktonic foraminifera indicative of the N8 zone (Langhian, ca. 16.6-15.2 Ma), and reflect an outermost neritic to uppermost bathyal depositional environment. Slightly younger assemblages characteristic of similar water depths are present in the overlying Lumis River Volcanics within the Kaut unit (sample 26, N10, early Serravallian, ca. 14.9-13.9 Ma) and the Lumis River unit (sample 31, N11, Serravallian, ca. 13.9-12.9 Ma).

A feature of these assemblages is the occurrence of larger foraminifera belonging to the

Table 4. (Continued)

SAMPLES (washed residues)	79	80	81	82	83	84	85	86	87	88	89	90	91	92	93	94	95	96	97
GLOBIGERINOIDES PRIMORDIUS -ALTIAPERTURA LINEAGE																			
G. obliquus (N5-N22)		X																	
G. extremus (N16-N21)		X											X						
GLOBIGERINA (ZEAGLOBIGERINA) LINEAGE																			
G. (Z.) woodi (N4A-N21)		X		X					X		X	X					X		
G. (Z.) nepenthes (N14-N19)																			
G. (Z.) apertura (N16-N21)																			
G. (Z.) decoraperta (N9-N21)		X						X									X		
GLOBIGERINOIDES SUBQUADRATUS-RUBER LINEAGE																			
G. ruber (N15-N23)	X	X		X	X	X	X	X	X	X	X	X	X	X		X	X		
GLOBIGERINOIDES TRILOBA-SACCULIFER LINEAGE																			
G. triloba (N4B-N22)		X																	
G. immaturus (N5-N23)		X		X	X	X	X		X	X	X		X	X	X	X			X
G. quadrilobatus (N6-N23)	X	X		X	X	X	X	X	X	X	X	X	X	X		X	X		X
G. sacculifer (N6-N23)		X		X	X	X	X		X	X			X	X		X	X		
G. fistulosus (N21-low N22)		X		X	X	X	X	X											
ORBULINA LINEAGE																			
O. universa (N9-N23)		X		X	X		X	X	X	X	X	X	X	X		X		X	X
O. bilobata																			
SPHAEROIDINELLOPSIS-SPHAEROIDINELLA LINEAGE																			
S. seminulina (N7-N21)	X	X						X							X	X			
S. kochi (N10-N19)																			
S. paenedehiscens (N17B-N20)																			
S. dehiscens (N19-N23)	X	X	X	X	X	X	X	X	X	X	X	X	X	X	X	X	X	X	X
NEOGLOBOQUADRINA LINEAGE																			
N. acostaensis (N16-N20)					D			D	D	D			D	D	D	D	D	D	D
N. pachyderma (N16-N23)					D			D											
N. humerosa (N18-N22)		D	D	D	D	D	D	D	D	D	D	D	D	D		D	D	D	D
N. dutertrei (N21-N23)															D				
PULLENIATINA LINEAGE																			
P. primalis (N17b-N20)																			
P. spectabilis (upper N18-N19)																X			
P. praecursor (N19-N21)	D-S	D-S	D	S	D-S	D	D-S	D-S	S	S	D-S	S	S	S	D-S	D-S	S	D	S
P. obliquiloculata (N19-N23)	X		D		D			X				S	S	S	D-S	D-S			S
GLOBOROTALIA (GLOBOROTALIA) LINEAGE																			
G. (G.) paralenguaensis (N15-N17A)																			
G. (G.) merotumida (N16-N18)																			
G. (G.) plesiotumida (N17-N19)																			
G. (G.) tumida tumida (N18-N23)		S	S	S	S	S	S	S	S	S	S	S	S	D	S	S	S	S	S
G. (G.) tumida flexuosa (upper N18-N21)	S			S															
G. (G.) ungulata (N21-N23)													X	X					
GLOBOROTALIA (MENARDELLA) LINEAGE																			
G. (M.) menardii (N12-N23)																			
G. (M.) limbata (N14-N21)		D			D			D											
GLOBOROTALIA (TRUNCOROTALIA) LINEAGE																			
G. (T.) crassula (N18-N22)																			
G. (T.) crassaformis (upper N19-N23)	X	X	X	X	X	X			X	X	X		X						X
G. (T.) tosaensis (N21-N22)	X	X	X	X	X	X	X	X	X	X	X	X	X	X	X	X	X	X	X
G. (T.) truncatulinoides (N22-N23)																			
DENTOGLOBIGERINA LINEAGE																			
D. altispira altispira (N4B-N21)															X				
D. altispira globosa (P22-N19)																			
GLOBOQUADRINA LINEAGE																			
G. dehiscens (N4B-N18)																			
G. baroemoenensis (N5-N18)																			
GLOBIGERINELLA LINEAGE																			
Globorotalia obesa (P22-N23)					X	X	X	X	X				X						
G. aequilateralis (N12-N23)					X	X							X						
GLOBIGERINITA-GLOBIGERINATELLA LINEAGE																			
G. glutinata (P22-N23)																			
PLANKTONIC ZONE (after Blow, 1969; Kennett & Srinivasan, 1983)	N21	N21	N21	N21	N21	N21	N21	N21	N21	N21	N21	up. N21	up. N21	up. N21	up. N21	up. N21	up. N21	up. N21	up. N21

lower Tf-letter stage with the predominantly planktonic fauna (samples 15-17, 21, 26, 31). The larger benthonic forms probably hosted symbiotic algae (like their modern counterparts) and thus would have inhabited the photic zone. Their presence here suggests either downslope transport of the tests or exceptionally clear water conditions which allowed light penetration to a depth of approximately 150 m.

No micropaleontological age control exists for the lower Lossuk River beds because of the lack of exposure. The Lumis River Volcanics conformably overlies the Lossuk River beds, but much of the unit is barren of foraminifera and comprehensive biostratigraphic subdivision is not possible.

In southern New Ireland, the oldest known outermost neritic - upper bathyal sample belongs to the N10 zone (sample 30), and was referred to the Tamiu Siltstone by Stewart and Sandy (1986). Other samples from the upper part of this unit (samples 32-35) contain an upper bathyal foraminiferal

Table 4. (Continued)

SAMPLES (washed residues)	98	99	100	101	102	103	104	105	106	107	108	109	110	111	112	113	114	115	116
GLOBIGERINOIDES PRIMORDIUS-ALTIAPERTURA LINEAGE																			
G. obliquus (N5-N22)	X	X																	
G. extremus (N16-N21)																			
GLOBIGERINA (ZEAGLOBIGERINA) LINEAGE																			
G. (Z.) woodi (N4A-N21)								X											
G. (Z.) nepenthes (N14-N19)																			
G. (Z.) apertura (N16-N21)																			
G. (Z.) decoraperta (N9-N21)						X													
GLOBIGERINOIDES SUBQUADRATUS-RUBER LINEAGE																			
G. ruber (N15-N23)	X	X	X	X	X	X	X	X	X	X		X	X	X	X	X	X	X	X
GLOBIGERINOIDES TRILOBA-SACCULIFER LINEAGE																			
G. triloba (N4B-N22)																			
G. immaturus (N5-N23)		X	X	X	X	X	X	X		X				X	X	X	X	X	X
G. quadrilobatus (N6-N23)	X	X	X	X	X	X	X	X	X		X		X	X	X	X	X	X	X
G. sacculifer (N6-N23)	X	X	X	X	X	X	X	X		X	X			X	X	X	X		
G. fistulosus (N21-low N22)				X															
ORBULINA LINEAGE																			
O. universa (N9-N23)	X	X	X		X	X	X							X	X	X			
O. bilobata (N9-N23)		X																	
SPHAEROIDINELLOPSIS-SPHAEROIDINELLA LINEAGE																			
S. seminulina (N7-N21)																			
S. kochi (N10-N19)																			
S. paenedehiscens (N17B-N20)																			
S. dehiscens (N19-N23)	X	X	X	X	X	X	X	X										X	
NEOGLOBOQUADRINA LINEAGE																			
N. acostaensis (N16-N20)	D	D	D	D	D	D	D	D	D	D	D	D	D	D	D	D	D	D	D
N. pachyderma (N16-N23)		D	D																
N. humerosa (N18-N22)	D	D	D	D	D	D	D	D	D	D	D	D	D	D	D	D	D	D	
N. dutertrei (N21-N23)	D	D		D	D	D			D				D		D	D	D	D	
PULLENIATINA LINEAGE																			
P. primalis (N17b-N20)																			
P. spectabilis (upper N18-N19)																			
P. praecursor (N19-N21)	D	D	D	D	D	D	D	D	D	D	D	D	D	D	D	D	D	D	
P. obliquiloculata (N19-N23)	D	D	D	D	D	D	D		D		D		D	D	D	D	D	D	
GLOBOROTALIA (GLOBOROTALIA) LINEAGE																			
G. (G.) paralenguaensis (N15-N17A)																			
G. (G.) merotumida (N16-N18)																			
G. (G.) plesiotumida (N17-N19)																			
G. (G.) tumida tumida (N18-N23)	S	S	S	S	D	S	S	S	X	S	S	S		S	S		S	S	
G. (G.) tumida flexuosa (upper N18-N21)																			
G. (G.) ungulata (N21-N23)																			
GLOBOROTALIA (MENARDELLA) LINEAGE																			
G. (M.) menardii (N12-N23)	S	S	S	S		S	S			S		S	S	S	S	S	S	S	
G. (M.) limbata (N14-N21)																			
GLOBOROTALIA (TRUNCOROTALIA) LINEAGE																			
G. (T.) crassula (N18-N22)								X		X					X		X		
G. (T.) crassaformis (upper N19-N23)	X				X	X	X			X							X		
G. (T.) tosaensis (N21-N22)	X	X	X	X	X	X	X	X	X	X		X		X	X		X	X	
G. (T.) truncatulinoides (N22-N23)														X	X	X	X		
DENTOGLOBIGERINA LINEAGE																			
D. altispira altispira (N4B-N21)		X																	
D. altispira globosa (P22-N19)																			
GLOBOQUADRINA LINEAGE																			
G. dehiscens (N4B-N18)																			
G. baroemoenensis (N5-N18)																			
GLOBIGERINELLA LINEAGE																			
Globorotalia obesa (P22-N23)					X					X							X		
G. aequilateralis (N12-N23)	X		X	X	X	X	X			X	X		X						
GLOBIGERINITA-GLOBIGERINATELLA LINEAGE																			
G. glutinata (P22-N23)																			
PLANKTONIC ZONE (after Blow, 1969; Kennett & Srinivasan, 1983)	up. N21	up. N21	up. N21	up. N21	up. N21	up. N21	up. N21	up. N21	up. N21	up. N21	up. N21	up. N21	up. N21	N22	N22	N22	N22	N22	?

assemblage belonging to the N12 zone (middle Serravallian, ca. 12.9-11.6 Ma). The base of the Tamiu Siltstone has not been located in the field.

Assemblages from the inner to mid-neritic Lelet Limestone cannot be precisely dated because of the absence of age-diagnostic planktonic foraminifera. Some of the samples (22, 24) are designated as lower Tf as indicated by the occurrence of *Miogypsina* with relatively advanced *Lepidocyclina*. Others may range into the late Miocene. The mid-neritic samples usually contain *Cycloclypeus* and rare planktonic foraminifera of small size.

Late Miocene Assemblages

Neritic foraminiferal biostratigraphy for late Miocene deposits is not clear. Among the age-diagnostic larger foraminifera, very few new elements are introduced, and a marked decrease in species diversity occurs. In the New Guinea region, the youngest known *Miogypsina* spp. are associated with planktonic zone N12 of middle Serravallian age (Belford, 1981; Haig, 1987). Most of the lower-Tf *Lepidocyclina* also disappeared during the middle Serravallian, although *L. radiata* and closely related forms

Table 5. Early Neogene age-diagnostic "larger" benthonic foraminifera (interpreted in sense of Coleman, 1963; Belford, 1982a,b; Chaproniere, 1984). Samples are arranged in approximate time-stratigraphic order: 4-13, Early Miocene (4, 5 may be latest Oligocene); 15-26, 31, 36, Middle Miocene; 37-41, probably Middle Miocene; 43-45, 52, 54, 55, Middle or late Miocene. [x: present; p:primitive form; cf: comparable form; ?:doubtful record].

SAMPLES (wr: washed residue ts: thin section)	4 wr	5 wr	7 ts	8 ts	9 ts	10 ts	11 wr	12 ts	13 ts	15 wr	16 wr	17 wr	21 wr	22 ts	23 ts
Alveolinella sp.															
Amphisorus sp.															
Austrotrillina howchini								P	P						
Cycloclypeus eidae															
C. indopacificus															
C. carpenteri															
C. (Katacycloclypeus) martini															X
C. sp.				X	X										
Gypsina vesicularis discus															
Heterostegina borneensis											cf				
Lepidocyclina (Eulepidina) ephippioides					?										
Lepidocyclina (Nephrolepidina) parva								X	X						
L. (N.) sumatrensis	X	P		cf	P		P	X	X						
L. (N.) howchini howchini										X					
L. (N.) ferreroi															
L. (N.) japonica														cf	
L. (N.) martini											X	cf	X		
L. (N.) radiata										X					
L. (N.) sp.					X										
Marginopora sp.															
Miogypsinoides bantamensis								cf	cf						
M. dehaarti	P	P	P				X	cf	cf						
M. new species								X	X						
Miogypsina neodispansa	X				cf			X	X						
M. thecideaeformis	X			cf			X							X	
M. polymorpha						cf		cf	cf	X	X	X	?	X	
Operculina complanata						cf	X								
O. complanata japonica								X	X					X	
O. venosa															X
Planorbulinella solida															
Sporadotrema carpenteri								cf	X						
Sporadotrema cylindricum				cf											
APPROXIMATE T-LETTER "STAGE" (After Adams, 1984)	up. Te	up. Te	up. Te	up. Te	up. Te	up. Te	up. Te	Te-Tf	Te-Tf	low Tf	low Tf	low Tf	low Tf	low Tf	low Tf
PLANKTONIC FORAMINIFERAL ZONE (see Table 3)										N8	N8	N8	up. N8		

continue from the middle Miocene into the late Miocene and, in the Fiji area at least, into earliest Pliocene (Adams, Rodda, and Kiteley, 1979; Adams, 1984).

As noted above, some of the Lelet Limestone samples we have placed in Tf-letter stage (Table 5) may be middle or even late Miocene (e.g. samples 37, 38, 40-55). The Lelet Limestone is conformably overlain by the Punam Limestone of early Pliocene age. Because of its position within the Lelet Limestone, the Matakan unit of carbonaceous siltstone and mudstone which crops out in central New Ireland may be no younger than late Miocene, or older than early Miocene. However, no age-diagnostic foraminifera have been recovered from it. The Matakan unit hosts a poorly diverse assemblage which is indicative of innermost neritic slightly brackish conditions (possibly a mangrove environment).

Bathyal planktonic assemblages (samples 60-63) from the upper part of the volcaniclastic Kapsu unit of the Lumis River Volcanics in northern New Ireland contain the earliest stages of *Pulleniatina* evolution, and belong to the N17B zone (Messinian, ca. 6.0-5.3 Ma). Although poorly preserved, benthonic assemblages of high diversity occur with the planktonic foraminifera and are dominated by buliminids, rotaliids, and nodosariids. Among the benthonic species, *Melonis pompilioides* suggests a depositional environment deeper than 500 m.

The Kapsu unit is the youngest Neogene deposit known from the northern sector of mainland New Ireland. It conformably overlies the Lumis River unit, but the age of the lower beds of the Kapsu unit

Table 5. (Continued)

SAMPLES (wr: washed residue ts: thin section)	24 ts	25 ts	26 wr	31 wr	36 ts	37 ts	38 ts	39 ts	40 ts	41 ts	43 ts	44 ts	45 ts	52 ts	54 ts	55 ts
Alveolinella sp.													X			
Amphisorus sp.													X			
Austrotrillina howchini																
Cycloclypeus eidae														cf		
C. indopacificus		cf			cf		cf		X	cf						
C. carpenteri						cf										
C. (Katacycloclypeus) martini																
C. sp.			X					X			X	X	X		X	
Gypsina vesicularis discus							X						X			
Heterostegina borneensis																
Lepidocyclina (Eulepidina) ephippioides																
Lepidocyclina (Nephrolepidina) parva																
L. (N.) sumatrensis																
L. (N.) howchini howchini																
L. (N.) ferreroi	X		X	X											X	
L. (N.) japonica																
L. (N.) martini			cf	cf			cf	cf		cf				cf		
L. (N.) radiata					cf							X				
L. (N.) sp.													?			
Marginopora sp.									X				X			
Miogypsinoides bantamensis																
M. dehaarti																
M. new species	X				X											
Miogypsina neodispansa																
M. thecideaeformis																
M. polymorpha			cf	cf												
Operculina complanata							X						X			
O. complanata japonica																
O. venosa																
Planorbulinella solida	cf											X		cf		X
Sporadotrema carpenteri																
Sporadotrema cylindricum	cf				X											
APPROXIMATE T-LETTER "STAGE" (After Adams, 1984)	low Tf	low Tf	low Tf	low Tf	low Tf	Tf	Tf		Tf	Tf	Tf	Tf	Tf	Tf	Tf	Tf
PLANKTONIC FORAMINIFERAL ZONE (see Table 3)			N10	N11	?N12	?N12	?N12	?N12	?N12	?N12						

is not known because most of the section is barren of foraminifera.

Pliocene - Pleistocene Bathyal Assemblages

Middle bathyal assemblages of Pliocene age (samples 64-110) have been recovered from the central and southern sectors of New Ireland as well as from Djaul Island off the northwest coast. The oldest Pliocene assemblages come from biomicrites of the Punam Limestone (samples 64-69) and belong to N18 zone (early Zanclian, ca. 5.3-5.1 Ma). We have examined other samples collected from the base of the Punam Limestone which contain very poorly preserved foraminifera (which are not documented on the tables here). This material belongs either to the N17B or N18 zone, but the poor preservation does not allow us to make a confident correlation. The youngest samples (70, 71) of Punam Limestone we have found belong to zone N19 (Zanclian, ca. 5.1-4.0 Ma).

In southern and central New Ireland, the volcaniclastic Ramat unit (samples 72, 73, 75-79, 81, 91-97) of the Rataman Formation contains assemblages which range from lower N19 zone (Zanclian, ca. 5.1 Ma) to uppermost N21 zone (near the Pliocene-Pleistocene boundary, ca. 1.9 Ma). Biomicrites of the Gogo unit (98-103, 111, ?116) of the Rataman Formation span the Pliocene-Pleistocene boundary (from uppermost N21 to lower N22, ca. 2.0 Ma to ?1.5 Ma), as does the clastic sediment of the upper part of the Uluputur unit (104-110, 112-115).

On Djaul Island, the Djaul unit (samples 82-90) of the Rataman Formation contains N21-zone planktonic assemblages (Piacenzian, ca. 3.1-1.9 Ma), including one close to the Pliocene-Pleistocene boundary (sample 90). However, the full stratigraphic range of the Djaul unit is not known.

The Pliocene assemblages are overwhelmingly

Table 6. Bathymetric interpretation. Samples are arranged in approximate time-stratigraphic order: 1-3, Early Miocene or older; 4-13, Early Miocene; 14-36, Middle Miocene; 37-55, Middle or Late Miocene; 56-69, no younger than Late Miocene; 60-63, Late Miocene; 64-73, Early Pliocene; 74-110, Late Pliocene; 111-115, early Pleistocene; 116, probably early Pleistocene. [a: abundant; c: common; r: rare; s: sparse; p: present; o: >5% of systematic count of benthonic foraminifera; x: <5% of systematic count of benthonic foraminifera].

SAMPLE	1	2	3	4	5	6	7	8	9	10	11	12	13	14	15	16	17	18	19
(wr: washed residue ts: thin section)	ts	ts	ts	wr	wr	wr	ts	ts	ts	ts	wr	ts	ts	wr	wr	wr	wr	wr	wr
PLANKTONIC FORAMINIFERAL RATIO (%)	0	0	0	0	0	0	0	0	0	r	0	0	0	76	73	61	82	46	54
MAXIMUM PLANKTONIC TEST SIZE										0.6				0.9	0.9	0.7	0.8	0.7	0.8
RATIO BENTHONIC FORAMINIFERAL ORDERS (%)																			
Astrorhizida														1					
Lituolida	r							r						3	2		1	5	2
Miliolida			r			r					r	r		2	2	2			
Nodosariida														16	10	7	8	24	20
Buliminida					2	8	r			r				30	34	44	33	40	42
Robertinida														3	14	7	6	6	5
Rotaliida	a	r	a	100	98	92	a	c	c	c	100	a	a	47	38	40	50	25	31
BENTHONIC DIVERSITY (Fisher diversity index)				3	6	7				3				14	21	19	27	13	12
"LARGER" FORAMINIFERAL OCCURRENCE (p: present)	p	p	p	p	p	p	p	p	p	p	p	p	p	p	p	p	p		
SELECTED BENTHONIC SPECIES																			
MODERN UPPER DEPTH LIMITS <50m																			
Ammonia convexa (Collins)				x	x	o				x				x	x	x		x	x
Amphistegina spp.				x	x		?			o	c	c		x	x	x	o		
Anomalinella rostrata (Brady)			x		o					x					x	x			
Bolivina karreriana (Brady)																		o	o
Bolivina spathulata (Williamson)																			
Cassidulina subglobosa Brady														x	x	x		x	
Cibicidoides cf. mediocris (Finlay)				x	x									o	o	o	o	x	x
Cibicidoides ungeriana (d'Orbigny)														o	x	x	x	x	x
Cycloclypeus spp. (U.D.L. approx. 40m)	a		a					c	c										
Discorbinella bertheloti (d'Orbigny)														x	o			o	o
Elphidium craticulatum (Fichtel & Moll)										o	cf								
Elphidium multiloculum Cushman & Ellisor					o	x													
Operculina spp.	a	?	a							r	o	a	a						
Rectobolivina cf. dimorpha (Parker & Jones)														o	x		x		
Reussella spinulosa (Reuss)					x	o								x	x	x	o		x
Rosalina concinna (Brady)						o													
Siphonina tubulosa Cushman																			
Uvigerina proboscidea Schwager														x	x	o	o		
MODERN UPPER DEPTH LIMITS 50-150m																			
Ceratobulimina pacifica Cushman & Harris														x	x	o	x	x	
Chilostomella oolina Schwager																			
Globobulimina ovata (d'Orbigny)															x	x	x		
Gyroidina cf. neosoldanii Brotzen														o	x	x	x	x	
Hoeglundina elegans (d'Orbigny)														x	o	x	x	o	o
Karreriella bradyi (Cushman)														x					
Martinottiella sp.																			
Planulina wuellerstorfi (Schwager) [U.D.L. >100m]															x		x	x	
Pleurostomella spp. [U.D.L. >100m]														x	x				
Pullenia bulloides (d'Orbigny)															x				
Pullenia quinqueloba (Reuss)																			
Rectobolivina cf. striata (Schwager)																			
Saracenaria italica Defrance														x	x		x	x	x
Sphaeroidina bulloides d'Orbigny														x	x	x	x	x	x
Stilostomella insecta (Schwager)														o	o	x	x	x	o
Stilostomella cf. lepidula (Schwager)															x	x	x		x
Stilostomella longiscata (d'Orbigny)														x	x		x	x	x
Uvigerina cf. schwageri Brady														x	x	x	x	o	o
MODERN UPPER DEPTH LIMITS >150m																			
Oridorsalis umbonatus [sensu Pflum & Frerichs, 1976]														x	x	x			
Osangularia culter (Parker & Jones)															o		x		
Uvigerina hispida Schwager [U.D.L. >400m]																			
Vulvulina pennatula (Batsch)														x		x			
BATHYMETRIC ZONE	A-B	A	A-B	A	A	A	A	A-B	A-B	A-B	A	A	A	low B	up. C	B-C	up. C	low B	B
A: <50m; B: 50-150m; C: 150-500m; D: >500m																			

dominated by planktonic forms, but also contain highly diverse benthonic foraminifera. The composition and relative abundance of the planktonic and benthonic fauna suggests a middle bathyal depositional environment (Table 6). No evidence of any marked syn-depositional dissolution of calcareous shells is apparent, and the assemblages probably were deposited well above the lysocline.

Some of the latest Pliocene and early Pleistocene assemblages (e.g. 114-115, Uluputur unit of the Rataman Formation) indicate shallowing from middle bathyal to upper bathyal and outermost neritic environments (Table 6). No younger deep-water sediments are known on New Ireland. Therefore, uplift

Table 6. (Continued)

SAMPLE (wr: washed residue ts: thin section)	20 wr	21 wr	22 ts	23 ts	24 ts	25 ts	26 wr	27 wr	28 wr	29 wr	30 wr	31 wr	32 wr	33 wr	34 wr	35 wr	36 ts	37 ts	38 ts	39 ts
PLANKTONIC FORAMINIFERAL RATIO (%)	56	64	0	c	0	r	49	35	35	51	70	67	85	90	98	89	a	a	a	a
MAXIMUM PLANKTONIC TEST SIZE	0.8	0.8		0.7		0.4	0.8	0.8	0.5	0.5	0.8	0.8	0.8	0.8	0.7	0.9	0.8	0.8	0.9	0.8	1
RATIO BENTHONIC FORAMINIFERAL ORDERS (%)																				
Astrorhizida		1					1									1	12			
Lituolida		7	r				1	8					1	6	7	9	12		r	
Miliolida		6	r	r									4	5	7	4	8			
Nodosariida	14	11					5	1		1	7	3	10	16	15	16				
Buliminida	48	24					39	14	3	27	26	33	46	47	32	40				
Robertinida	10	9					7	1		1		3	r	3	3	2				
Rotaliida	28	42	a	a	a	a	47	76	97	72	67	56	33	20	36	10	c	a	a	r
BENTHONIC DIVERSITY (Fisher diversity index)	10	45					20	6	4	6	10	20	18	16	18	18				
"LARGER" FORAMINIFERAL OCCURRENCE (p: present)	p	p	p	p	p	p		p	p		p	p	p	p		p	p	p	p	p
SELECTED BENTHONIC SPECIES																				
MODERN UPPER DEPTH LIMITS <50m																				
Ammonia convexa (Collins)	x	x					O		x		x									
Amphistegina spp.	x	x		a	a	c	x	O	x		O	x	x		O					
Anomalinella rostrata (Brady)							x	O	O											
Bolivina karreriana (Brady)	O						O													
Bolivina spathulata (Williamson)							x													
Cassidulina subglobosa Brady		x									x	x	O	O		x				
Cibicidoides cf. mediocris (Finlay)	O	O					O	O	O	x	O	O	x	x	x	x				
Cibicidoides ungeriana (d'Orbigny)	x	x					x	x							O					
Cycloclypeus spp. (U.D.L. approx. 40m)			a		a												a	c	c	r
Discorbinella bertheloti (d'Orbigny)		x									O		x	x						
Elphidium craticulatum (Fichtel & Moll)											x									
Elphidium multiloculum Cushman & Ellisor																				
Operculina spp.			c	c	a	c		x							x	x				
Rectobolivina cf. dimorpha (Parker & Jones)		x					x			x	x	x								
Reussella spinulosa (Reuss)		x						O				x								
Rosalina concinna (Brady)																				
Siphonina tubulosa Cushman													O		O	x				
Uvigerina proboscidea Schwager		x					x					x	O	x		x				
MODERN UPPER DEPTH LIMITS 50-150m																				
Ceratobulimina pacifica Cushman & Harris	x	x										x	x	x	x	x				
Chilostomella oolina Schwager															x	x				
Globobulimina ovata (d'Orbigny)												x			x	x				
Gyroidina cf. neosoldanii Brotzen	x	x					O		x		x	x	x	x	x					
Hoeglundina elegans (d'Orbigny)	O	O					O			x		x	x	x	x	x				
Karreriella bradyi (Cushman)													x							
Martinottiella sp.													x	x						
Planulina wuellerstorfi (Schwager) [U.D.L. >100m]		x					x						x	x	x					
Pleurostomella spp. [U.D.L. >100m]													x	x	x					
Pullenia bulloides (d'Orbigny)													x		x					
Pullenia quinqueloba (Reuss)													x							
Rectobolivina cf. striata (Schwager)													x	x						
Saracenaria italica Defrance		x									x		x							
Sphaeroidina bulloides d'Orbigny	x	x					x					x	x	x	x					
Stilostomella insecta (Schwager)	O	x					O		x			x	x	x	x	x				
Stilostomella cf. lepidula (Schwager)	x	x					x						x	x	x					
Stilostomella longiscata (d'Orbigny)	x	x					x				x	x	O	O	x					
Uvigerina cf. schwageri Brady	O	x					O				O	O	x	x		O				
MODERN UPPER DEPTH LIMITS >150m																				
Oridorsalis umbonatus [sensu Pflum & Frerichs, 1976]													x							
Osangularia culter (Parker & Jones)		x					x				x		O	x		x				
Uvigerina hispida Schwager [U.D.L. >400m]													x	x						
Vulvulina pennatula (Batsch)		x										x	x		O					
BATHYMETRIC ZONE A: <50m; B: 50-150m; C: 150-500m; D: >500m	B	B-C	A	up. B	A	A-B	B-C	up. B	up. B	B	B-C	B-C	C	C	C	C	B-C	B-C	B-C	B-C

of the southwest margin of the New Ireland Basin seems to have commenced close to the Pliocene-Pleistocene boundary (ca. 2.0 Ma).

DISCUSSION

Our foraminiferal determinations provide the main basis for correlations shown on the time-space diagram (Figure 2). The stratigraphic units shown on this diagram are those of Stewart and Sandy (1986); these units represent a revision of the previous nomenclature.

The early Miocene succession shown on Figure 2 differs from that proposed by Hohnen (1978) in the position of the basal Lelet Limestone with respect to the Lossuk River beds. This difference results from the clarification of the Te-Tf letter-stage boundary (within N6 rather than above N8), which was discussed earlier.

The positions of the Punam Limestone and Rataman Formation also have been reversed from those determined by Hohnen (1978). As now defined by Stewart and Sandy (1986), the Punam Limestone includes the succession of biomicrites that occur between the Lelet Limestone and the Ramat unit in

Table 6. (Continued)

SAMPLE (wr: washed residue ts: thin section)	40 ts	41 ts	42 wr	43 ts	44 ts	45 ts	46 ts	47 ts	48 ts	49 ts	50 ts	51 ts	52 ts	53 ts	54 ts	55 ts	56 wr	57 wr	58 wr	59 wr
PLANKTONIC FORAMINIFERAL RATIO (%)	a	r	a	r	r	0	0	0	0	0	0	0	c	0	a	a	0	0	0	0
MAXIMUM PLANKTONIC TEST SIZE	0.9	0.6	0.5	0.3	0.5								0.5		0.7	0.7				
RATIO BENTHONIC FORAMINIFERAL ORDERS (%)																				
Astrorhizida																				
Lituolida		r	c	1									c							
Miliolida			c	7	r		c		r	c	c				a					1
Nodosariida																				
Buliminida		r	r	5	r	r	r	c	r					r		r	c		2	
Robertinida																				
Rotaliida		c	a	87	a	a	a	a	a	a	c	a	a	a	c	a	c	100	98	100	99
BENTHONIC DIVERSITY (Fisher diversity index)				9														<1	1	1	1
"LARGER" FORAMINIFERAL OCCURRENCE (p: present)	p	p	p	p	p	p	p	p	p	p	p	p	p	p	p	p	p				
SELECTED BENTHONIC SPECIES																					
MODERN UPPER DEPTH LIMITS <50m:																					
Ammonia convexa (Collins)																		O	O	O	O
Amphistegina spp.		a	O		c			a			c	a	c								
Anomalinella rostrata (Brady)			O																		
Alveolinella spp.						r			r	r											
Cassidulina subglobosa Brady																					
Cibicidoides cf. mediocris (Finlay)			O																		
Cibicidoides ungeriana (d'Orbigny)																					
Cycloclypeus spp. (U.D.L. approx. 40m)	r	c		r	c	r							c		r						
Elphidium craticulatum (Fichtel & Moll)			O															x		O	O
Marginopora sp.		r			r		r		r						c						
Operculina spp.		c	O		c	c	c		c	c	c	c		c							
Rectobolivina cf. dimorpha (Parker & Jones)																					
Reussella spinulosa (Reuss)			x																		
Siphonina tubulosa Cushman																					
Uvigerina proboscidea Schwager																					
MODERN UPPER DEPTH LIMITS 50-150m:																					
Ceratobulimina pacifica Cushman & Harris																					
Chilostomella oolina Schwager																					
Eggerella bradyi (Cushman)																					
Globobulimina ovata (d'Orbigny)																					
Gyroidina cf. neosoldanii Brotzen																					
Hoeglundina elegans (d'Orbigny)																					
Karreriella bradyi (Cushman)																					
Martinottiella sp.																					
Oridorsalis stellatus [sensu Pflum & Frerichs, 1976]																					
Planulina wuellerstorfi (Schwager) [U.D.L. >100m]																					
Pleurostomella spp. [U.D.L. >100m]																					
Pullenia bulloides (d'Orbigny)																					
Pullenia quinqueloba (Reuss)																					
Rectobolivina cf. striata (Schwager)																					
Saracenaria italica Defrance																					
Sphaeroidina bulloides d'Orbigny																					
Stilostomella insecta (Schwager)																					
Stilostomella cf. lepidula (Schwager)																					
Stilostomella longiscata (d'Orbigny)																					
Uvigerina cf. schwageri Brady																					
MODERN UPPER DEPTH LIMITS >150m																					
Bulimina rostrata Brady																					
Favocassidulina favus (Brady)																					
Laticarinina pauperata (Parker & Jones) [U.D.L. >400m]																					
Melonis pompilioides (Fichtel & Moll) [U.D.L. >500m]																					
Oridorsalis umbonatus [sensu Pflum & Frerichs, 1976]																					
Osangularia culter (Parker & Jones)																					
Uvigerina hispida Schwager [U.D.L. >400m]																					
Vulvulina pennatula (Batsch)																					
BATHYMETRIC ZONE A: <50m; B: 50-150m; C: 150-500m; D: >500m	B-C	A-B	A	A-B	A-B	A-B	A	A	A	A	A	A	B	A	B	B-C	A	A	A	A

central New Ireland. The Punam Limestone is of early Pliocene (N18-N19, Zanclian) age. The Rataman Formation contains the volcaniclastic Ramat unit of late early Pliocene - late Pliocene age (N19-N21, late Zanclian - Piacenzian), the bathypelagic Gogo unit of latest Pliocene - early Pleistocene age (upper N21 - lower N22); and the clastic Uluputur unit (late Pliocene - early Pleistocene).

The foraminiferal/stratigraphic succession on New Ireland is similar to that found in other parts of the Outer Melanesian Arc. The oldest fauna (probably early Oligocene - with *Nummulites fichteli*) recorded here seems slightly younger than the late Eocene assemblage described from the Baining Volcanics of New Britain by Binnekamp (1973) and noted as widespread throughout Outer Melanesia by Coleman (1978). It correlates with the younger part of the Fijian Eocene/Oligocene in the Mt Koromba area, Nandi district. The neritic early and middle Miocene fauna (upper Te - lower Tf letter-stage) is

Table 6. (Continued)

SAMPLE (wr: washed residue ts: thin section)	60	61	62	63	64	65	66	67	68	69	70	71	72	73	74	75	76	77	78	79	80
	wr	wr	wr	wr	wr	wr	wr	wr	wr	wr	wr	wr	wr	wr	wr	wr	wr	wr	wr	wr	wr
PLANKTONIC FORAMINIFERAL RATIO (%)	73	89	96	98	98	99	98	98	98	99	99	99	99	>95	>90	99	100	99	97	>90	>90	99
MAXIMUM PLANKTONIC TEST SIZE	0.8	0.7	0.9	0.8	1	1.1	1.2	1.1	1.1	1.1	1.2	1.1	1.1	1.1	1.1	1.3	1	1.1	1.3	1.2	1.1	1.3
RATIO BENTHONIC FORAMINIFERAL ORDERS (%)																						
Astrorhizida																						
Lituolida		1	2	1	2	3	4	10	6	3	5	8	3		3			9	7	4	r	1
Miliolida						1		1				1	3						3	15	r	1
Nodosariida	5	8	18	11	11	6	14	13	11	7	10	9	13	r	15			14	8	13		18
Buliminida	67	65	36	38	42	43	47	49	47	41	44	44	29	r	37		9	38	18	r	55
Robertinida		1	8	8																		
Rotaliida	28	25	36	42	45	46	35	27	36	49	40	36	55	r	45			65	47	50	r	25
BENTHONIC DIVERSITY (Fisher diversity index)	7	16	16	12	17	>20	>20	>20	>20	13	>20	>20	15		14			>20	14	>11		>20
"LARGER" FORAMINIFERAL OCCURRENCE (p: present)	p			p								p	p									
SELECTED BENTHONIC SPECIES																						
MODERN UPPER DEPTH LIMITS <50m:																						
Ammonia convexa (Collins)																						
Amphistegina spp.	x			x							x		x									
Anomalinella rostrata (Brady)																						
Alveolinella spp.																						
Cassidulina subglobosa Brady	x	O	x	O	x	O	O	O	O	O	O	O	O	O	s	O		x	x	x		x
Cibicidoides cf. mediocris (Finlay)	O	O	O	O																		
Cibicidoides ungeriana (d'Orbigny)			x	O																		
Cycloclypeus spp. (U.D.L. approx. 40m)																						
Elphidium craticulatum (Fichtel & Moll)																						
Marginopora sp.																						
Operculina spp.																						
Rectobolivina cf. dimorpha (Parker & Jones)				x																		
Reussella spinulosa (Reuss)																						
Siphonina tubulosa Cushman		x	x	x				x	x													
Uvigerina proboscidea Schwager	x	x	O	x	x	x	O	O	O	O	O	O			O				O	x		x
MODERN UPPER DEPTH LIMITS 50-150m:																						
Ceratobulimina pacifica Cushman & Harris		O	O																			
Chilostomella oolina Schwager								x			x							O	x			
Eggerella bradyi (Cushman)					x	x	x	x	x	x		x			x				O	x		
Globobulimina ovata (d'Orbigny)						x	x															
Gyroidina cf. neosoldanii Brotzen	O	x	x	x	x	x	x	x	x	O	x	x	O		x			x	O			x
Hoeglundina elegans (d'Orbigny)						x																
Karreriella bradyi (Cushman)								x	O	x		x	x					x				
Martinottiella sp.				x																		
Oridorsalis stellatus [sensu Pflum & Frerichs, 1976]								x				x	x	s	O			O	O	O		
Planulina wuellerstorfi (Schwager) [U.D.L. >100m]					x	O	x	x	x	x	x	x	O		x			O	O	O	s	x
Pleurostomella spp. [U.D.L. >100m]		x	x		x	x	x	x	x	x	x	x			x			x	x	x	s	O
Pullenia bulloides (d'Orbigny)						x	x	x	x	x	x	x	x		O			x	x	O		x
Pullenia quinqueloba (Reuss)		x	x		x	x	x	x	x	x	x	x	x		x			x	x			
Rectobolivina cf. striata (Schwager)	x	x	O	x																		
Saracenaria italica Defrance			x																			
Sphaeroidina bulloides d'Orbigny							x	x	x		x	x										
Stilostomella insecta (Schwager)	x	O	x	O	O		O	x	x	x		x										x
Stilostomella cf. lepidula (Schwager)				x		x						x	O		x					x		x
Stilostomella longiscata (d'Orbigny)		O	x	x															x			x
Uvigerina cf. schwageri Brady		x	x																			
MODERN UPPER DEPTH LIMITS >150m																						
Bulimina rostrata Brady						x	x	O	x	x	x	x	x					x				O
Favocassidulina favus (Brady)						x	x	x	x	O	O	x		O	s	x		x	x	x		x
Laticarinina pauperata (Parker & Jones) [U.D.L. >400m]					x	x	x	x	x	x	x	x			O		x					x
Melonis pompilioides (Fichtel & Moll) [U.D.L. >500m]	x	x	x	x		x			x			x			x			O	O	s		
Oridorsalis umbonatus [sensu Pflum & Frerichs, 1976]				x	x	x		x	O	x	x	x			x				x	s		x
Osangularia culter (Parker & Jones)								x	x	x		x	x									x
Uvigerina hispida Schwager [U.D.L. >400m]		x				x		x	x													x
Vulvulina pennatula (Batsch)																						
BATHYMETRIC ZONE	D	D	D	D	D	D	D	D	D	D	D	D	D	D	D	D	D	D	D	D	D
A: <50m; B: 50-150m; C: 150-500m; D: >500m																						

also widespread throughout the arc, and marks a regional shoaling and the diminution of volcanism. The oldest known outermost neritic to bathyal deposits on New Ireland (Lossuk River beds, Langhian) are younger than the oldest known beds of corresponding facies on Manus Island in the northwest of the New Ireland Basin (Louwa unit, N5, early Burdigalian; Haig, unpublished data).

The Late Miocene/Pliocene sediments (Kapsu - Punam etc.), upper bathyal or deeper, appear to have partial equivalents in the Bougainville Basin (Mono Siltstone, penetrated by L'Etoile 1; Stewart, Francis and Pederson, in press), and with the Pemba Siltstone on Choiseul, the next island to the east (McTavish, 1968). Santa Isabel has elongated areas of bathyal sediments along the northwestern coast and at both ends, which make up the Tanakau Beds, and also correlate in general terms (the faunas have not been described in detail; Stanton, 1961). These bathyal deposits appear to give way over a short dis-

Table 6. (Continued)

SAMPLE (wr: washed residue ts: thin section)	81 wr	82 wr	83 wr	84 wr	85 wr	86 wr	87 wr	88 wr	89 wr	90 wr	91 wr	92 wr	93 wr	94 wr	95 wr	96 wr	97 wr	98 wr
PLANKTONIC FORAMINIFERAL RATIO (%)	>90	>90	99	99	99	>90	>90	95	>90	>90	99	>90	99	>90	99	>90	99	96
MAXIMUM PLANKTONIC TEST SIZE	1.1	1.3	1.3	1.2	1.2	0.8	1.3	1.2	1.1	1.2	1.3	1.1	1.3	1.1	1.3	1	1	1.2
RATIO BENTHONIC FORAMINIFERAL ORDERS (%)																		
Astrorhizida																		
Lituolida		2	4	1	6	2	4	1	s		s	4						1
Miliolida	r		5	8			1	1		s	s							1
Nodosariida	r	13	26	7	7	12	11	8	s	s	s	8	s	8		s		12
Buliminida	r	27	35	63	71	55	39	66	s	s	s	37	s	54	s	s	s	59
Robertinida																		6
Rotaliida	r	58	30	21	16	31	45	24	s	s	s	51	s	38	s	s	s	21
BENTHONIC DIVERSITY (Fisher diversity index)		>11	15	11	8	>20	>10	9				16		16				>20
"LARGER" FORAMINIFERAL OCCURRENCE (p: present)																		
SELECTED BENTHONIC SPECIES																		
MODERN UPPER DEPTH LIMITS <50m:																		
Amphistegina spp.																		
Bolivina karreriana (Brady)																		
Bolivina spathulata (Williamson)																		
Cassidulina subglobosa Brady		x	x	x		x	x	x		s		O	s	O		s	s	O
Cibicidoides cf. mediocris (Finlay)							x											x
Cycloclypeus spp. (U.D.L. approx. 40m)																		
Rectobolivina cf. dimorpha (Parker & Jones)							x	x										
Reussella spinulosa (Reuss)																		x
Siphonina tubulosa Cushman																		
Uvigerina proboscidea Schwager	x	O	x	x	x	O	O	O	s	s	s	O				s		
MODERN UPPER DEPTH LIMITS 50-150m:																		
Ceratobulimina pacifica Cushman & Harris																		O
Chilostomella oolina Schwager			x	x								s		x				
Eggerella bradyi (Cushman)			x	x	x				s			x						
Globobulimina ovata (d'Orbigny)			x	x			x											
Gyroidina cf. neosoldanii Brotzen		x		x		x		x							s	s		
Hoeglundina elegans (d'Orbigny)																		
Karreriella bradyi (Cushman)																		x
Martinottiella sp.			x	x			O	x										
Oridorsalis stellatus [sensu Pflum & Frerichs, 1976]			x	x	O		x	x	s			x						
Planulina wuellerstorfi (Schwager) [U.D.L. >100m]	x	O	O	x	x	x	O	O	s			x		O	s	s		
Pleurostomella spp. [U.D.L. >100m]		x	x	x	x	x				s		x		x				O
Pullenia bulloides (d'Orbigny)		x				x	x		s		s	O	s	x		s		x
Pullenia quinqueloba (Reuss)			x	x														x
Sphaeroidina bulloides d'Orbigny				x														
Stilostomella insecta (Schwager)			O	x		x		x	s					x				
Stilostomella cf. lepidula (Schwager)			x	x		x		x						x		s		O
Stilostomella longiscata (d'Orbigny)			O		x	O	O	x										x
Uvigerina cf. schwageri Brady																		
MODERN UPPER DEPTH LIMITS >150m:																		
Bulimina rostrata Brady												s	O	s	O	s		
Favocassidulina favus (Brady)																		
Laticarinina pauperata (Parker & Jones) [U.D.L. >400m]						x	x	x				x		x		s		x
Melonis pompilioides (Fichtel & Moll) [U.D.L. >500m]			x								s	x		x				x
Oridorsalis umbonatus [sensu Pflum & Frerichs, 1976]		O	x	x	x		x	O	s	s	s	x		O			s	
Osangularia culter (Parker & Jones)																		
Uvigerina hispida Schwager [U.D.L. >400m]							x											
Vulvulina pennatula (Batsch)																		
BATHYMETRIC ZONE A: <50m; B: 50-150m; C: 150-500m; D: >500m	D	D	D	D	D	D	D	D	D	D	D	D	D	D	D	D	D	D

tance, and over a small stratigraphic interval, to shallow-water sediment. In a similar way on New Ireland, the neritic Lelet Limestone changes to the middle bathyal Punam Limestone (early Pliocene) over a dozen meters vertically.

The explanation for these deep-water strata so closely juxtaposed to shallow-water ones probably lies in the movement of great faults of which the complex Weitin Fault system is a prime example (see Figure 19 in Hohnen, 1978). Such faults are neither simple planes nor are they truly vertical; they are undulatory. Rapid strike-slip movement along such a complex brings about vertical movement with great speed (the so-called "ramp" effect), and removes them just as fast. The portions of crust on either side of the fault complex rise and fall with respect to each other. Large strike-slip faults have shaped the Outer Melanesian islands at least since the mid-Pliocene, and have accommodated the left lateral movements of large slivers of arc crust making up the Outer Melanesian complex (the strike-slip movement is a consequence of the southward drift of the rotational pole for the Indo-Australian/Pacific plates).

As mentioned previously, the type material of *G. tumida* (index species for the N17-N18 boundary) and *G. sacculifer* came from a stone carving found on New Ireland. This material was examined

Table 6. (Continued)

SAMPLE (wr: washed residue ts: thin section)	99	100	101	102	103	104	105	106	107	108	109	110	111	112	113	114	115	116
	wr	wr	wr	wr	wr	wr	wr	wr	wr	wr	wr	wr	wr	wr	wr	wr	wr	wr
PLANKTONIC FORAMINIFERAL RATIO (%)	96	99	99	99	99	97	>90	92	93	89	93	>90	92	96	90	76	80	70
MAXIMUM PLANKTONIC TEST SIZE	1.2	1.1	1.2	1.3	1.5	1.5	1.1	1.1	1	1.2	1.1	1.6	1.1	1.4	1.1	1.2	1	1.1
RATIO BENTHONIC FORAMINIFERAL ORDERS (%)																		
Astrorhizida																		
Lituolida	1	1	5	5	6	1	s			1		8	2	1		6		1
Miliolida	2	1	1	2						1		1		1			5	1
Nodosariida	7	28	17	15	14	9	s			4		1	16	2		3	5	3
Buliminida	47	42	39	44	38	35	s	s		74	s	57	50	55	s	43	37	45
Robertinida						4						2					1	
Rotaliida	43	28	38	34	42	52	s	s	s	20		31	32	41	s	48	52	50
BENTHONIC DIVERSITY (Fisher diversity index)	13	>20	>20	>20	14	>20				>20		>15	>20	13		14	16	>12
"LARGER" FORAMINIFERAL OCCURRENCE (p: present)						p	p											p
SELECTED BENTHONIC SPECIES																		
MODERN UPPER DEPTH LIMITS <50m:																		
Amphistegina spp.						O	s											x
Bolivina karreriana (Brady)		x					s			x	s			x	s	x	x	
Bolivina spathulata (Williamson)																		x
Cassidulina subglobosa Brady	O	x	O	O	O	O	s			x		O	O	O	s	x	O	x
Cibicidoides cf. mediocris (Finlay)						O				x		x	x			O	x	x
Cycloclypeus spp. (U.D.L. approx. 40m)																		x
Rectobolivina cf. dimorpha (Parker & Jones)																		
Reussella spinulosa (Reuss)																		
Siphonina tubulosa Cushman						x												
Uvigerina proboscidea Schwager	O	O	O	O	x	x	s			x		O	x	O		x	x	
MODERN UPPER DEPTH LIMITS 50-150m:																		
Ceratobulimina pacifica Cushman & Harris						x												
Chilostomella oolina Schwager				x														
Eggerella bradyi (Cushman)			x	x		x												
Globobulimina ovata (d'Orbigny)													x					
Gyroidina cf. neosoldanii Brotzen	x		x		x		s		s	x		x	x	x		x		
Hoeglundina elegans (d'Orbigny)						x							x				x	
Karreriella bradyi (Cushman)			x															
Martinottiella sp.																		
Oridorsalis stellatus [sensu Pflum & Frerichs, 1976]										x		x	x		x			
Planulina wuellerstorfi (Schwager) [U.D.L. >100m]	O		O	x	O	x		s	s					x				
Pleurostomella spp. [U.D.L. >100m]	x	x	x	x	x	x							x		s	x		
Pullenia bulloides (d'Orbigny)	x	x	x	x	x	x				x		x			s	x		
Pullenia quinqueloba (Reuss)			x	x	x								x					
Sphaeroidina bulloides d'Orbigny		x		x						O		x	x			x	x	
Stilostomella insecta (Schwager)		x			x								O					
Stilostomella cf. lepidula (Schwager)	x	x	x	O	x	x												
Stilostomella longiscata (d'Orbigny)			x	x														
Uvigerina cf. schwageri Brady																		x
MODERN UPPER DEPTH LIMITS >150m:																		
Bulimina rostrata Brady		x	x	x								x						
Favocassidulina favus (Brady)	O		x	x	x	x	s							x				
Laticarinina pauperata (Parker & Jones) [U.D.L. >400m]		x	O	x	x	x							x	x		x		
Melonis pompilioides (Fichtel & Moll) [U.D.L. >500m]	x	x	x	x									x					
Oridorsalis umbonatus [sensu Pflum & Frerichs, 1976]	O	x	x	O	O											x		
Osangularia culter (Parker & Jones)		x	x															
Uvigerina hispida Schwager [U.D.L. >400m]						x												
Vulvulina pennatula (Batsch)																		
BATHYMETRIC ZONE	D	D	D	D	D	D	D	C-D	C-D	C-D	C-D	D	D	D	D	B-C	C-D	B-C
A: <50m; B: 50-150m; C: 150-500m; D: >500m																		

by Parker (1967, p.157) who recorded an N18 planktonic assemblage associated with the type specimens. It seems, therefore, that the carving originated from the Punam Limestone which crops out on the central east coast of New Ireland. The type locality of *G. fistulosus* apparently lies within the Djaul unit of the Rataman Formation on Djaul Island.

ACKNOWLEDGMENTS

We wish to thank officers of the Geological Survey of Papua New Guinea - W.D. Stewart, M.J. Sandy, G. Francis and B. Pawih - who collected and provided the samples. Thanks are due also to technicians of the Geological Survey for processing many samples as well as Technicians from the Department of Geology, University of Papua New Guinea. Constructive criticism of the manuscript was given by Dr G. Chaproniere and Dr H. Davies of the Bureau of Mineral Resources, Canberra; Dr K. McDougall of the U.S. Geological Survey; and by W.D. Stewart of the Geological Survey of Papua New Guinea.

REFERENCES

Adams, C.G., 1970, A reconsideration of the East Indian Letter Classification of the Tertiary: British Museum (Natural History)

Geology Bulletin, v. 19, 87-137.

Adams, C.G., 1984, Neogene larger foraminifera, evolutionary and geological events in the context of datum planes, in Ikebe, I. and Tsuchi, R., eds., Pacific Neogene datum planes, University of Tokyo Press, p. 93-110.

Adams, C.G., Rodda, P., and Kiteley, R.J., 1979, The extinction of the foraminiferal genus *Lepidocyclina* and the Miocene/Pliocene boundary problem in Fiji: Marine Micropaleontology, v. 4, 319-339.

Adisaputra-Sudinta, M., and Coleman, P.J., 1984, Associations of mid-Tertiary large benthic and small planktonic Foraminifera, Central West Java: Search, v. 15, p. 231-233.

Belford, D.J., 1976, Samples from Kavieng 1:250,000 Sheet area, Papua New Guinea: Geological Survey of Papua New Guinea Technical File 56 (unpublished).

Belford, D.J., 1981, Co-occurrence of Middle Miocene larger and planktic smaller foraminifera, New Ireland, Papua New Guinea: Bureau of Mineral Resources, Geology and Geophysics, Australia, Bulletin, v. 209, p. 1-21.

Belford, D.J., 1982a, Redescription of *Miogypsina neodispansa* (Jones & Chapman), Foraminiferida, Christmas Island, Indian Ocean: Bureau of Mineral Resources Journal of Australian Geology and Geophysics, v. 7, p. 315-320.

Belford, D.J., 1982b, *Planorbulinella solida* sp. nov. (Foraminiferida) from the Miocene of Papua New Guinea: Bureau of Mineral Resources Journal of Australian Geology and Geophysics, v. 7, p. 321-325.

Belford, D.J., and Chaproniere, G.C., 1974, Report on samples from New Hanover, Dyaul and Mussau Islands, Papua New Guinea: Geological Survey of Papua New Guinea Technical File, 56 (unpublished).

Berggren, W.A., Kent, D.V., Flynn, J.J., and Van Couvering, J.A., 1985, Cenozoic geochronology: Geological Society of America Bulletin, v. 96, p. 1407-1418.

Betjeman, K.J., 1969, Recent foraminifera from the western continental shelf of Western Australia: Contributions from the Cushman Foundation for Foraminiferal Research, v. 20, p. 119-138.

Binnekamp, J.G., 1973, Tertiary larger foraminifera from New Britain, PNG: Bureau of Mineral Resources, Geology and Geophysics, Australia, Bulletin, v. 140, p. 1-26.

Biswas, B., 1976, Bathymetry of Holocene foraminifera and Quaternary sea-level changes on the Sunda Shelf: Journal of Foraminiferal Research, v. 6, p. 107-133.

Blow, W.H., 1969, Late middle Eocene to Recent planktonic foraminiferal biostratigraphy, in Bronnimann, P. and Renz, H.H., eds., Proceedings of the First International Conference on Planktonic Microfossils: Leiden, Brill, p. 199-422.

Brady, H.B., 1877, Supplementary note on the foraminifera of the Chalk(?) of the New Britain Group: Geological Magazine, v. 4, Decade 2, p.534-536.

Brown, C.M., 1982, Kavieng, Papua New Guinea - 1:250,000 Geological Series: Bureau of Mineral Resources, Geology and Geophysics, Australia, Explanatory Notes, SA/56-9, 26 p.

Burke, S.C., 1981, Recent benthic foraminifera of the Ontong Java Plateau: Journal of Foraminiferal Research, v. 11, p. 1-19.

Chaproniere, G.C.H., 1981, Australasian mid-Tertiary larger foraminiferal associations and their bearing on the East Indian Letter Classification: Bureau of Mineral Resources Journal of Australian Geology and Geophysics, v. 6, p. 145-151.

Chaproniere, G.C.H., 1984, Oligocene and Miocene larger Foraminiferida from Australia and New Zealand: Bureau of Mineral Resources, Geology and Geophysics, Australia, Bulletin, v. 188, 98 p.

Cole, W.S., 1975, Concordant age determinations by larger and planktonic foraminifera in the Tertiary of the Indo-Pacific region: Journal of Foraminiferal Research, v. 5, p. 21-39.

Coleman, P.J., 1963, Tertiary larger foraminifera of the British Solomon Islands, southwest Pacific: Micropaleontology, v. 9, p. 1-38.

Coleman, P.J., 1969, Report following palaeontological examination of samples from New Ireland. Appendix C in The geology of Permit 48 (Territory of Papua and New Guinea): Continental Oil Co. Australia Ltd. Report (unpublished), Geological Survey of Papua New Guinea Open File.

Coleman, P.J., 1978, Reflections on outer Melanesian Tertiary larger foraminifera: Bureau of Mineral Resources, Geology and Geophysics, Australia, Bulletin, 192, p. 31-36.

Coleman, P.J., and McTavish, R.A., 1964, Associations of larger and planktonic foraminifera in single samples from Middle Miocene sediments, Guadalcanal, Solomon Islands, Southwest Pacific: Royal Society of Western Australia Journal, v. 47, p. 13-24.

Coleman, P.J. and McTavish, R.A., 1967, Association of early Miocene planktonic and larger foraminifera from the Solomon Islands, Southwest Pacific: Australian Journal of Science, v. 29, p. 373-374.

Collins, A.C., 1958, Foraminifera: British Museum (Natural History), Great Barrier Reef Expedition 1928-29, Scientific Reports, v. 6, no. 6, p. 335-437.

Cushman, J.A., 1921. Foraminifera of the Philippine and adjacent seas: United States National Museum Bulletin, v. 100, no. 4, 608 p.

Falvey, D.A. and Pritchard, T., 1984, Preliminary paleomagnetic results from northern Papua New Guinea: evidence for large microplate rotations, in Watson, S.T., ed., Transactions of the Third Circum-Pacific Energy and Mineral Resources Conference, p. 593-599.

Frerichs, W.E., 1970, Distribution and ecology of benthonic foraminifera in sediments of the Andaman Sea: Contributions from the Cushman Foundation for Foraminiferal Research, v. 21, p. 123-147.

Grimsdale, T.F. and van Morkhoven, F.P.C.M., 1955, The ratio between pelagic and benthonic foraminifera as a means of estimating depth of deposition of sedimentary rocks: Proceedings of Fourth World Petroleum Congress, Section 1/D, p. 473-491.

Graham, J.J., and Militante, P.J., 1959, Recent foraminifera from the Puerto Galera area, northern Mindoro, Philippines: Stanford University Publications, Geological Sciences, v. 6, no. 2, 171 p.

Haak, R. and Postuma, J.A., 1975, The relation between the tropical planktonic foraminiferal zonation and the Tertiary Far East Letter Classification: Geologie en Mijnbouw, v. 54, p. 195-198.

Haig, D.W., 1985, *Lepidocyclina* associated with early Miocene planktic foraminiferids from the Fairfax Formation, Papua New Guinea: South Australia, Department of Mines and Energy Special Publication, v. 5, p. 117-131.

Haig, D.W., 1987, Tertiary foraminiferal rock samples from the Western Solomon Sea: Geo-Marine Letters, v. 6, p. 219-228.

Haynes, J.R., 1981, Foraminifera: London, Macmillan, 433 p.

Hofker, J., 1978, Biological results of the Snellius Expedition XXX, the foraminifera collected in 1929 and 1930 in the eastern part of the Indonesian archipelago: Zoologische Verhandelingen, v. 161, 69 p.

Hohnen, P.D., 1978, Geology of New Ireland, Papua New Guinea: Bureau of Mineral Resources, Geology and Geophysics, Australia, Bulletin, v. 194, 39 p.

Hughes, G.W., 1977, Recent foraminifera from the Honiara Bay area, Solomon Islands: Journal of Foraminiferal Research, v. 7, p. 45-57.

Ingle, J.C., 1980, Cenozoic paleobathymetry and depositional history of selected sequences within the southern California continental borderland: Cushman Foundation for Foraminiferal Research Special Publication, v. 19, p. 163-195.

Keij, A.J., 1963, The relative abundance of recent planktonic foraminifera in seabed samples collected offshore Brunei and Sabah: Borneo Region, Malayasia Geological Survey Annual Report, v. 1963, p. 146-152.

Kennett, J.P. and Srinivasan, M.S., 1983, Neogene planktonic foraminifera, a phylogenetic atlas: Stroudsburg, Hutchinson Ross, 265 p.

Keston, S., 1969, Palaeontological notes on the New Ireland samples. Appendix B *in* The geology of Permit 48 (Territory of Papua and New Guinea): Continental Oil Co. Australia Ltd. Report (unpublished), Geological Survey of Papua New Guinea Open File.

McTavish, R.A., 1968, A note on some planktonic foraminifera from the Pemba Siltstones on NW Choiseul: British Solomon Islands Geological Survey Record no. 3, Report 82, p. 71,72.

Mitchell, P.A., and Weiss, T.V., 1982, Prospecting Authority 485, New Ireland, Papua New Guinea, final report: Esso Papua New Guinea Inc. Report (unpublished), Geological Survey of Papua New Guinea Open File.

Murray, J.W., 1973, Distribution and ecology of living benthic foraminiferids: London, Heinemann Educational, 274 p.

Murray, J.W., 1976, A method of determining proximity of marginal seas to an ocean: Marine Geology, v. 22, p. 103-119.

Palmieri, V., 1976, Modern and relict foraminifera from the central Queensland shelf: Queensland Government Mining Journal, v. 77, p. 407-436.

Parker, F.L., 1967, Late Tertiary biostratigraphy (planktonic foraminifera) of tropical Indo-Pacific deep-sea cores: Bulletins of American Paleontology, v. 52, p. 115-208.

Pflum, C.E., and Frerichs, W.E., 1976, Gulf of Mexico deep-water foraminifers: Cushman Foundation for Foraminiferal Research Special Publication, v. 14, 125 p.

Schubert, R.J., 1910a, Uber Foraminiferen und einen Fischotolithen aus dem fossilen Globigerinenschlamm von Neu-Guinea: Austria, Verhandlungen K.K. Geologische Reichsanstalt Wien, v. 14, p. 318-328.

Schubert, R.J., 1910b, Uber das Vorkommen von *Miogypsina* und *Lepidocyclina* in pliocanen Globigerinenschlamm von Neu-Guinea: Austria, Verhandlungen K.K. Geologische Reichsanstalt Wien, v. 14, p. 395-398.

Schubert, R.J., 1911, Die fossilen Foraminiferen des Bismarckarchipels und einiger angrenzender Inseln: Austria, Abhandlungen Geologische Reichsanstalt Wien, v. 20, no. 4, p. 1-130.

Stanton, R.L., 1961, Explanatory notes to accompany a first geological map Santa Isabel, British Solomon Islands Protectorate: Overseas Geology and Mineral Resources, v. 8, no. 2, p. 127-149.

Stewart, W.D., Francis, G. and Pederson, S.L., in press, Hydrocarbon potential of the Bougainville and southeastern New Ireland Basins, Papua New Guinea: Oil and Gas Journal.

Stewart, W.D. and Sandy, M.J., 1986, Cenozoic stratigraphy and structure of New Ireland and Djaul Island, New Ireland Province, Papua New Guinea: Geological Survey of Papua New Guinea Report, 86/12, 36 p.

Weissel, J.K., Taylor, B., and Karner, G.D., 1982, The Opening of the Woodlark Basin, subduction of the Woodlark spreading system, and the evolution of northern Melanesia since mid-Pliocene time: Tectonophysics, v. 87, p. 253-277.

Marlow, M.S., Dadisman, S.V., and Exon, N.F., editors, 1988, Geology and offshore resources of Pacific island arcs—New Ireland and Manus region, Papua New Guinea, Circum-Pacific Council for Energy and Mineral Resources Earth Science Series, v. 9: Houston, Texas, Circum-Pacific Council for Energy and Mineral Resources.

VOLCANISM IN THE NEW IRELAND BASIN AND MANUS ISLAND REGION: NOTES ON THE GEOCHEMISTRY AND PETROLOGY OF SOME DREDGED VOLCANIC ROCKS FROM A RIFTED-ARC REGION

R.W. Johnson
Bureau of Mineral Resources, Geology and Geophysics, Canberra, A.C.T., 2601, Australia

M.R. Perfit
Department of Geology, University of Florida, Gainesville, Florida, 32611, USA

B.W. Chappell
Department of Geology, Australian National University, Canberra, A.C.T., 2601, Australia

A.L. Jaques
Bureau of Mineral Resources, Geology and Geophysics, Canberra, A.C.T., 2601, Australia

R.D. Shuster
Department of Geology, University of Florida, Gainesville, Florida, 32611, USA

W.I. Ridley
U.S. Geological Survey, Denver Federal Center, Denver, Colorado, 80225, USA

ABSTRACT

Geochemical and petrological data for a suite of volcanic rocks dredged from four localities in the New Ireland Basin and Manus Island region serve to illustrate part of the exceptional compositional diversity of magmas erupted in this rifted-arc region. A ferrobasalt from west-northwest of Manus Island is similar in composition to ferrobasalts on Manus itself and to ocean-island basalts in general. Andesite recovered from southwest of Rambutyo Island is geochemically similar to andesites of late Miocene to Pliocene age in eastern Manus Island. Rocks recovered from two dredges near Tanga and Ambitle (Feni) islands are similar in composition to those of the dominantly alkaline arc-type rocks found in the Tabar-to-Feni island chain in the New Ireland Basin. The oceanic ferrobasalts of the Manus region are of particular interest because, like the ocean-island-type rocks of Fiji, they are evidence that ocean-island-type upper mantle may exist beneath an island arc. Lava flows of unknown composition are buried beneath sediment north and northwest of Manus Island. They may represent either forearc volcanism related to subduction at the Manus Trench, or former back-arc volcanism related to the northward-dipping subduction zone now present beneath New Britain.

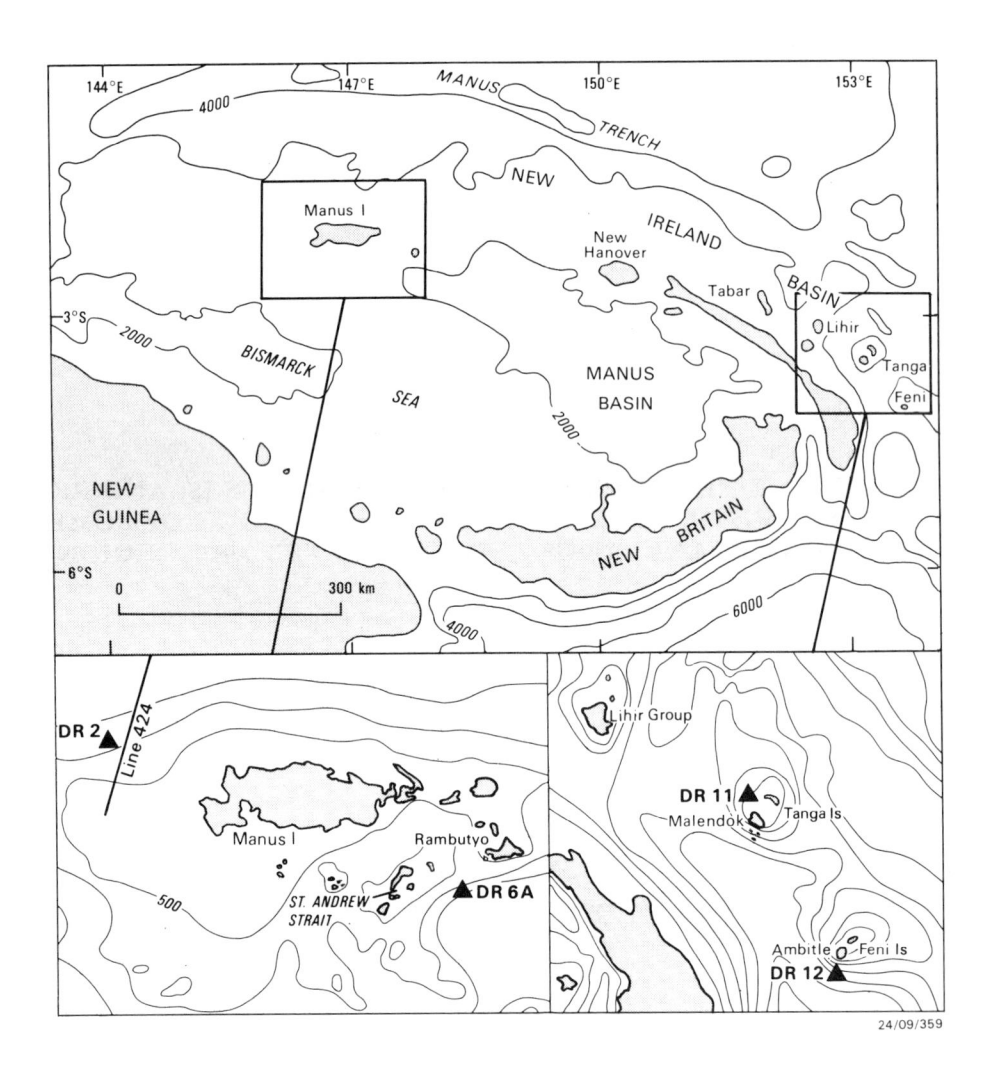

Figure 1. Rock-dredge stations (filled triangles) in the New Ireland Basin and Manus Island region occupied by the *S.P. Lee* in 1984. Isobaths in meters. Line 424 represents the ship's track for the seismic-reflection profile shown in Figure 2.

INTRODUCTION

Subduction-related volcanism has been a major geological process in the evolution of the New Ireland Basin, contributing particularly to the build-up of island-arc-type volcanic sequences on the large islands of New Ireland (Hohnen, 1978), New Hanover (Brown, 1982), and Manus (Jaques, 1980). However, the character of the New Ireland Basin volcanism has not been entirely andesitic, particularly during the late Cenozoic. For example, silica-oversaturated "orogenic" andesites are absent from the dominantly alkaline Tabar-to-Feni volcanic chain that runs down the northeastern side of New Ireland (Wallace et al., 1983; Figure 1). They also appear to be absent from the St. Andrew Strait area, southeast of Manus Island, where the Quaternary volcanism is bimodal, consisting mainly of basalt and rhyolite (Johnson, Smith, and Taylor, 1978). In addition, Quaternary, ocean-island-type volcanic rocks are known on Manus Island (Jaques, 1981; 1983).

The actively spreading Manus Basin impinges on the southwestern side of the New Ireland Basin (Figure 1; B. Taylor, 1979; Taylor et al., 1986), and is known to be floored with a range of volcanic rocks including a low-K_2O tholeiitic-basalt/ferroandesite/dacite suite and a higher K_2O suite of arc-type affinities (Sinton et al., 1986). Formation of the Manus Basin is thought to have begun in the Pliocene and caused the northwestward translation of New Ireland and Manus Island past New Britain, so dislocating this part of the Tertiary island arc from its subduction system (see, for example, Curtis, 1973). The change in compositional character of the

Table 1. Summary of rock-dredge haul characteristics.

Rock Dredge no.	Lat.°S Long.°E Water Depth (m)	Haul Characteristics
DR2	1°52.39′ 146°04.26′ 1100-1400	A small, single fragment of basalt with a 0.5 cm manganese crust on one side, was recovered in this dredge about 55 km west-northwest of Manus Island. The dredge was sited on a scarp over highly reflecting, flat-lying layers that were thought to represent possible, young oceanic crust.
DR6A	2°27.68′ 147°41.16′ 1000-1500	A full haul largely of mud and some cobbles was recovered about 13 km southwest of Rambutyo Island on a steep slope leading down into the Manus Basin. The cobbles included two pieces of volcanic rock.
DR11	3°20.71′ 153°09.01′ 900-950	Seventy percent of this haul, sited about 13 km northwest of the Tanga Islands, consisted of blocks of lava derived from the submarine northwestern flank of Tanga volcano. Agglomerate made up the remainder of the haul.
DR12	4°10.36′ 153°33.70′ 2000-2250	A 250 kg haul of volcanic rocks was obtained from a steep scarp about 7 km southwest of Ambitle Island, the largest of the volcanic Feni Islands. Dark mafic lavas, dark finer-grained porphyritic rocks, and light to dark-grey porphyritic samples were identified on board.

late Cainozoic volcanism in the New Ireland Basin may be a response to this major change in tectonic setting caused by arc dislocation (Johnson, Mutter, and Arculus, 1979). Jaques (1981; 1983) also identified a temporal change in rock composition on Manus Island, from arc type to ocean-island type. This particular change is analogous to the one reported by Gill (1976a,b; 1984) for the Fijian region where arc-rifting also has taken place.

Identification during the 1984 *S.P. Lee* cruise (L7-84-SP) of widespread, flat-lying, strongly reflecting, and strongly magnetic layers north and west of Manus Island (Marlow, Exon, and Tiffin, this volume), represents a major new discovery in our knowledge of the igneous character of the Manus Island forearc. These layers are probably lava flows which can not be interpreted certainly as the eruption products of island-arc volcanoes. An opportunity to dredge igneous rocks for the first time from this and other submarine volcanic parts of the Manus Island region and New Ireland Basin, was afforded by the *Lee* cruise.

This paper is a presentation and discussion of geochemical and petrological data obtained for rocks from the only dredge hauls of the 1984 *S.P. Lee* cruise that recovered igneous rocks (Table 1). Only small samples were available from two of the four dredges (DR2 and DR6A, close to Manus Island), and the other analyzed samples were from hauls on the flanks of Tanga and Ambitle volcanoes in the geochemically well-documented Tabar-to-Feni chain (Wallace et al., 1983).

ANALYTICAL METHODS AND DATA PRESENTATION

Major and trace-element analyses of whole-rock volcanic samples were obtained by X-ray-fluorescence spectrometry (Norrish and Chappell, 1977), spark-source mass spectrography (S.R. Taylor, 1979), and electron-microbe analysis of fused glass beads. All of these data are from analytical laboratories at the Australian National University, Canberra. Mineral

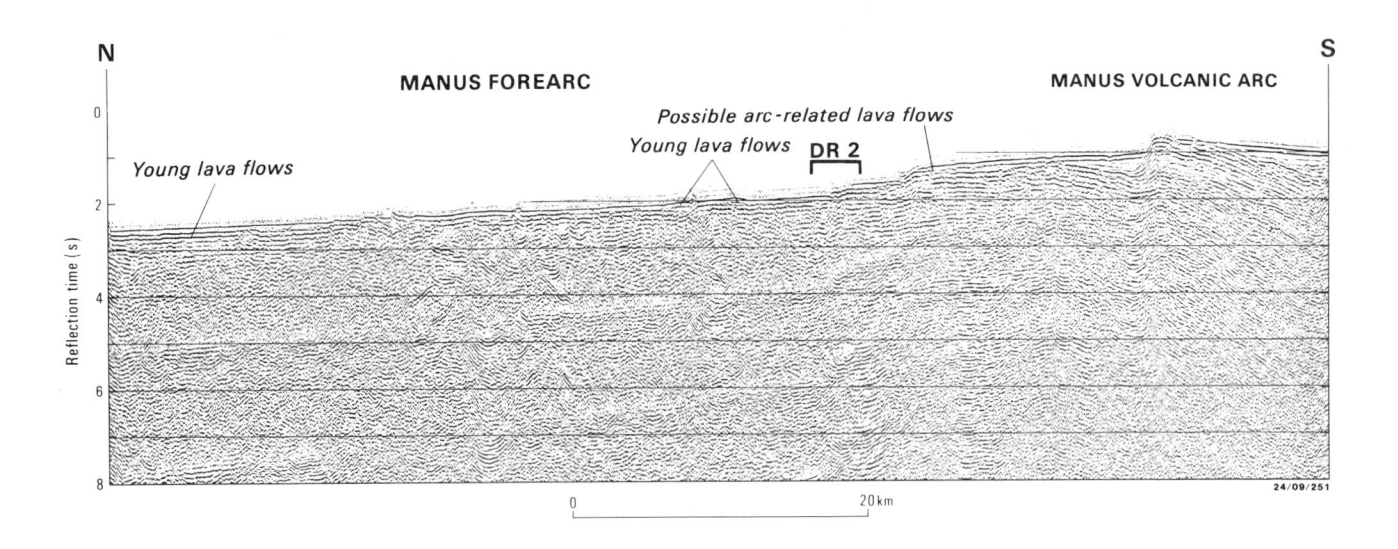

Figure 2. Seismic-reflection profile for Line 424 (Figure 1) showing the approximate projected position for dredge station DR2 (Marlow et al., this volume).

data were obtained by electron-microprobe analysis at the United States Geological Survey laboratories in Denver.

Sr and Pb isotopic compositions were determined on a fully automated VG Isomass 345 thermal-ionization mass spectrometer[1] at the University of Florida, Gainesville (see Table 4 for further details). The first two authors, in conjunction with others, for several years have been collecting isotopic data for a wide range of late Cainozoic volcanic rocks from western Melanesia (Papua New Guinea and the western Solomon Islands), and this data base is used below for general comparative purposes. A compilation of these largely unpublished isotopic data is being prepared currently for publication.

OCEAN-ISLAND FERROBASALT WEST-NORTHWEST OF MANUS ISLAND

Dredge DR2 was targeted on a scarp where exposures of the flat-lying, strongly reflecting, volcanic horizons were thought to exist (Table 1; Figure 1). Its location, however, is only 55 km west-northwest of Manus Island, so there was also the possibility that retrieved samples might have affinities with rocks exposed on Manus itself. Indeed, this seems likely given the approximate position of the dredge shown on the seismic-reflection profile in Figure 2.

Rock sample DR2/1 was a small sliver, less than 10 cm long, and was the only igneous rock in the dredge. It is dark, fine-grained, and vesicular. It lacks visible phenocrysts in hand specimen, and in this respect alone the rock is not typically arc-type. Petrographically identifiable crystals are tabular plagioclase, pale brown-green clinopyroxene, and olivine, some of which is skeletal in habit. Some of the grains are microphenocrysts, and crystal aggregates are common. The groundmass in thin section is extremely dark, almost black, and presumably consists mainly of glass charged with quench granules of titanomagnetite. Plagioclase microphenocrysts have labradorite cores (about An_{60}). They are only slightly zoned, and have negligible Or contents. Groundmass clinopyroxene is augite containing relatively high TiO_2 and Al_2O_3 (Table 2). Ti-Al^{IV} substitution in clinopyroxene is a common feature of ocean-island basalt.

The DR2/1 sample, cleaned of its manganese crust, was too small for routine chemical analysis by X-ray fluorescence spectrometry, so fused beads were prepared from powdered rock and analyzed for major elements using the electron microprobe (Table 3). The sample is a ferrobasalt characterized by high total FeO, low Al_2O_3, and high TiO_2, all of which are non-arc-type geochemical features. It is slightly quartz-normative (setting FeO equal to 0.85 total FeO, and recasting the remaining FeO as Fe_2O_3, before calculating the CIPW norm). Its K_2O content

---

[1] Any use of trade names is for purpose of identification only and does not imply endorsement by the U.S. Geological Survey, the New Zealand Geological Survey, or the Bureau of Mineral Resources, Australia.

is too high for that of a mid-ocean-ridge-type ferrobasalt, but is similar to those of ferrobasalts from oceanic islands such as the Galapagos (Baitis and Lindstrom, 1980).

Other distinctive geochemical features of DR2/1 include enrichments in several incompatible trace elements, including the light rare-earth elements (REE), Zr, Hf, and Nb (Figure 3). In addition, Y/Ti, Y/Zr, Zr/Nb, and Ba/La values are low. The $^{87}Sr/^{86}Sr$ value is 0.70386, but after mild leaching of the sample in hydrochloric acid a reduced value of 0.70379 was obtained (Table 4), so the rock may have been slightly altered by seawater. Both values are within the range known for arc-type rocks from the west Melanesian region, and for ocean-island basalts (see, for example, White and Hoffman, 1982) including those from Fiji (Gill, 1984). $^{207}Pb/^{204}Pb$ is slightly low compared to other values shown in Figure 4, falling outside of the compositional fields for

Table 2. Representative pyroxene analyses for rocks from dredges DR2 and DR6. Oxides in weight percent.

	DR2/1		DR6A/4			
	1	2	3	4	5	6
	G	G	Pc	Pi	Pr	Pc
SiO_2	50.44	49.73	51.19	53.71	51.47	52.93
TiO_2	1.33	1.11	0.57	0.18	0.54	0.25
Al_2O_3	3.55	3.31	2.68	1.14	1.62	1.11
FeO	9.91	10.27	10.00	17.98	9.72	19.12
MnO	0.24	0.24	0.37	0.63	0.32	0.69
MgO	15.72	16.40	14.97	24.39	15.40	23.23
CaO	17.15	17.04	19.02	1.91	19.09	1.86
Na_2O	n.d.	0.25	0.38	0.04	0.32	0.05
Cr_2O_3	0.16	0.17	0.03	n.d.	n.d.	n.d.
Total	98.50	98.52	99.21	99.98	98.47	99.24
Cations per six oxygens
Si	1.900	1.870	1.918	1.968	1.941	1.966
Al^{4+}	0.100	0.130	0.082	0.032	0.059	0.034
Al^{6+}	0.058	0.017	0.361	0.017	0.013	0.014
Ti	0.038	0.031	0.016	0.005	0.015	0.007
Fe^{3+}	0.000	0.063	0.041	0.009	0.039	0.008
Fe^{2+}	0.312	0.259	0.273	0.542	0.267	0.585
Mn	0.008	0.008	0.117	0.019	0.010	0.022
Mg	0.883	0.919	0.836	1.331	0.866	1.286
Ca	0.692	0.687	0.763	0.750	0.771	0.074
Na	0.000	0.018	0.028	0.003	0.023	0.004
Cr	0.005	0.005	0.001	0.000	0.000	0.000
Total	3.995	4.008	4.005	4.001	4.005	4.001
Wo	36.7	36.8	40.8	3.8	40.5	3.8
En	46.8	49.3	44.6	68.3	45.5	66.1
Fs	16.5	13.9	14.6	27.8	14.0	30.1

G: groundmass. P: phenocryst. c: core. i: intermediate. r: rim. n.d.: not detected.

The column-4 orthopyroxene is rimmed by the column-5 clinopyroxene.

Table 3. Chemical analyses of ferrobasalt sample DR2/1. All trace-element data by spark-source mass spectrography. Oxides in weight percent. Trace elements in ppm. Abbreviation e.f. is enrichment factor.

	1.	2.
SiO_2	50.30	50.23
TiO_2	2.84	2.86
Al_2O_3	13.31	13.37
FeO	14.78	14.75
MgO	5.17	5.17
CaO	9.57	9.51
Na_2O	2.86	2.86
K_2O	0.66	0.66
P_2O_5	0.42	0.50

La	16.2	Cs	0.21
Ce	41.2	Ba	90
Pr	5.80	Pb	2.9
Nd	28.4	Y	34
Sm	7.10	U	0.41
Eu	2.34	Th	1.40
Gd	7.04	Zr	190
Tb	1.10	Hf	4.7
Dy	6.62	Nb	18
Ho	1.4		
Er	3.86	[La/Yb]e.f.	
Yb	3.47	= 3.15	

1,2: averages of four electron microprobe analyses for each of two separate fused glass beads.

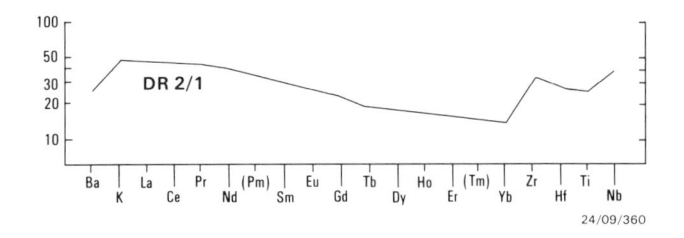

Figure 3. Chondrite-normalized trace-element values for ferrobasalt DR2/1 (Table 3). Normalizing values are those listed by Taylor and McLennon (1981), except for K (120, a primitive-mantle value).

western Melanesia and Fijian arc rocks.

DR2/1, therefore, has the geochemical characteristics of ocean-island "transitional and mildly alkaline" (Coombs, 1963) basalts which, though quartz-normative, have the mineralogy of alkali basalt, including the apparent absence of a Ca-poor pyroxene. Volcanic rocks closely similar in composition to DR2/1 were reported by Jaques (1981) for ferrobasalts from Southwest Bay, Manus Island. A representative major-element analysis of one of these ferrobasalts (73290531) is given in Table 5, together with some new trace-element data obtained specifically for comparison with DR2/1. Both samples are similarly light-REE enriched, lack enrichments in Ba and K relative to the light REE (an island-arc characteristic), and have strong enrichments in the high-field-strength elements (Zr-Nb). $^{87}Sr/^{86}Sr$ values are also similar (Table 4). The implication of these ocean-island-type compositions is discussed below.

ANDESITE SOUTHWEST OF RAMBUTYO ISLAND

Dredge station DR6A was sited 13 km southwest of Rambutyo Island at the southern end of *Lee* line 421, and about 50 km east of the St. Andrew Strait area (Table 1, Figure 1). The site is on the steep slope that leads down into the Manus Basin, so the possibility existed for sampling the deeper parts of the Manus/New-Ireland island arc, or for obtaining volcanic rocks related to the St. Andrew Strait volcanism. The dredge yielded mud containing some cobbles, two of which were pieces of volcanic rock. One of these volcanic samples, DR6A/4, has been chemically analyzed (Tables 4 and 6).

Table 4. Sr and Pb isotope data.

	$^{87}Sr/^{86}Sr$	$^{206}Pb/^{204}Pb$	$^{207}Pb/^{204}Pb$	$^{208}Pb/^{207}Pb$
DR2/1	0.703861±16	18.551	15.487	38.356
	(0.703789±16)			
73290531	0.703708±16			
DR6A/4	0.703543±18	18.743	15.565	38.395
DR11.1/2A	0.704195±18			
DR11.1/4A	0.703975±16	18.729	15.534	38.304
DR12.2/2A	0.704157±22	18.599	15.508	38.224
DR12.1/8A	0.704005±20			
DR12.3/4A	0.704024±18	18.627	15.534	38.303
DR12.3/6A	0.703977±22			

Isotopes were measured on fresh samples ultrasonically washed in 1N hydrochloric acid. The value in parenthesis is for a sample that was leached for 12 hours in 2N HCl. Sr separation was by closed-capsule HF dissolution followed by cation exchange (2N HCl). Sr was analyzed as nitrate on oxidized Ta filaments. Errors are two standard deviations of the mean. Replicate Sr-isotope analysis of standard NBS987 is 0.71023±2.

Whole-rock common-lead analysis was by two-column HBr and 6N HCl separation. Pb was loaded with phosphoric acid and silica gel on a single Re filament. Pb isotopes are corrected for mass fractionation of 0.05 percent AMU based on multiple runs of NBS981 and 982.

Table 5. Chemical analyses of ferrobasalt sample 73290531 from southwestern Manus Island. Oxides in weight percent. Trace elements in ppm. Abbreviation e.f. is enrichment factor.

SiO_2	49.1	Ba	120
TiO_2	2.30	Rb	8
Al_2O_3	14.7	Sr	270
Fe_2O_3	3.70	Zr	171
FeO	9.60	Nb	17
MnO	0.25	Y	37
MgO	4.71	La	18
CaO	9.90	Ce	36
Na_2O	2.84	Cr	30
K_2O	0.49	Ni	29
P_2O_5	0.32	Cu	57
H_2O^+	1.01	Zn	113
H_2O^-	0.57		
Total	99.5		

Oxide and above trace-element analyses from Jaques (1981, Table 1).

La	14.2	Cs	0.12
Ce	37.0	Y	30
Pr	4.71	U	0.43
Nd	21.4	Th	1.17
Sm	5.36	Hf	3.1
Eu	1.72	Nb	16
Gd	5.06		
Tb	0.87	[La/Yb]e.f.	
Dy	5.21	=3.65	
Ho	1.09		
Er	2.91		
Yb	2.63		

Remaining trace-element data by spark-source mass spectrography

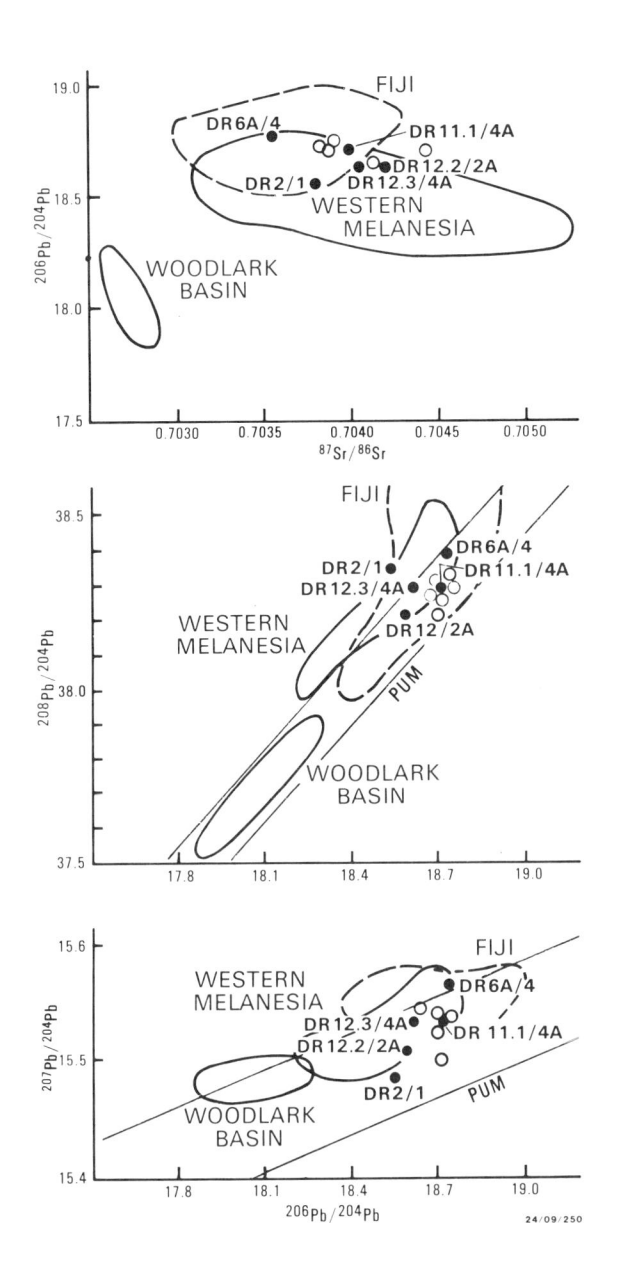

Figure 4. Pb and Sr-isotope data for dredged samples from the New Ireland Basin and Manus Basin region (filled circles; Table 4) compared with data (unpublished) for samples from the Tabar-to-Feni islands (open circles) and generalized compositional fields for arc-type rocks from elsewhere in western Melanesia (excluding the five Tabar-to-Feni samples). The other generalized fields are for Fiji (Gill, 1984) and mid-ocean-ridge-type basalts from the Woodlark Basin marginal basin, Papua New Guinea (Staudigel et al., in press; M.R. Perfit, unpublished data). Generalized PUM (Pb of the Upper Mantle) trend is from Gill (1984).

DR6A/4 is a two-pyroxene "orogenic" andesite. It is highly porphyritic, especially in plagioclase and clinopyroxene, but also has pleochroic orthopyroxene (hypersthene) phenocrysts (some as cores in augite) and Fe-Ti oxide (magnetite or titanomagnetite) microphenocrysts. Some of the plagioclase is oscillatory zoned and many grains have groundmass and mineral inclusions. Glomeroporphyritic aggregates of some or all the phenocryst minerals are also present. The groundmass is vesicular and fine-grained to glassy, and plagioclase is the most readily identifiable

Table 6. Chemical analyses of andesite sample DR6A/4 (1) and average of five late Miocene andesites from Manus Island (2). Abbreviation e.f. is enrichment factor.

	1.	2.		1.	2.
SiO_2	53.58	54.71	Ba	200	257
TiO_2	0.74	0.73	Rb	22.0	23
Al_2O_3	16.81	17.82	Sr	480	757
Fe_2O_3	3.37	3.51	Pb	5	6
FeO	4.21	3.04	Zr	119	175
MnO	0.12	0.12	Nb	3.0	3
MgO	4.05	3.65	Y	24	23
CaO	8.95	7.79	La	11	25
Na_2O	3.51	3.89	Ce	29	51
K_2O	1.59	1.90	V	213	218
P_2O_5	0.40	0.34	Cr	49	42
S	<0.02		Ni	22	18
H_2O^+	1.39	2.02	Cu	95	157
H_2O^-	0.99		Zn	84	
CO_2	0.26		Ga	16.5	20
Total	99.97	99.62			

[Ba/La]e.f. = 1.85   [Rb/La]e.f. = 2.10

[K/La]e.f. = 3.67

1. All data (except FeO, H_2O, and CO_2) by X-ray fluorescence spectrometry.
2. Based on unpublished data (A.L. Jaques)

mineral. The glass is dark brown. Clinopyroxene phenocrysts are characterized by the low TiO_2 and Al_2O_3 and fairly high Na_2O values typical of clinopyroxene in arc-type rocks (Table 2). Some oscillatory zoning is found in the clinopyroxene phenocrysts, and at least one groundmass grain is pigeonite.

Sample DR6A/4 is a low-SiO_2 andesite that is low in both MgO and Ni (and therefore relatively fractionated), and which has all the familiar features of a "normal" island-arc-type andesite. It is a "medium-K" andesite in the sense used by Gill (1978) and has several geochemical features in common with the Bagana (Papua New Guinea) reference andesite composition published by Bultitude, Johnson, and Chappell (1978). Its most significant chemical features contrast markedly with those of the ocean-island-type ferrobasalt DR2/1 described above including, particularly, much higher Al_2O_3 and much lower total Fe. In addition, TiO_2, Zr, and Nb, are all lower than in the ferrobasalt, and Zr/Nb, Y/Ti, and Y/Zr are all higher. Furthermore, the chondrite-normalized value of Ba and the primitive-mantle-normalized values for K and Rb are greater than they are for La (Table 6) and Ce. In other words, these large-ion-lithophile elements plot as positive anomalies relative to the light REE on normalization diagrams - another typical feature of arc-type rocks.

DR6A/4 has a $^{87}Sr/^{86}Sr$ value (0.703543, Table 4) similar to those for other arc rocks from western Melanesia (Figure 4). It has slightly higher $^{207}Pb/^{204}Pb$ (15.565) and $^{208}Pb/^{204}Pb$ (38.395) compared to the Pb-isotope values for other samples from this region, but they are not unusually high.

The Rambutyo Beds of Rambutyo Island consist of well-bedded tuffaceous, calcareous lithic sandstone and siltstone, conglomerate interbeds, and coarse volcanic conglomerate at the base, but no lava flows or pyroclastic deposits have been reported (Jaques, 1980). The beds are lithologically and stratigraphically equivalent to the late Miocene to Pliocene Lauis Formation which is exposed over much of eastern Manus Island and which, in places, is intercalated with basalts and low-SiO_2 andesites known collectively as the Lorengau Basalt. The chemical composition of sample DR6A/4 is similar in composition to the average of five late Miocene to Pliocene andesites from eastern Manus Island, although the Manus rocks have higher abundances of Ba, Sr, Zr, light REE, and Cu (Table 6). Other Tertiary volcanic rocks on Manus Island that have similar SiO_2 contents to that of DR6/4 have distinctly lower K_2O contents (Jaques, 1983; unpublished data).

QUARTZ TRACHYTE AND ALKALINE ROCKS OF TANGA AND AMBITLE VOLCANOES

Tabar-to-Feni Volcanism

The Tabar, Lihir, Tanga, and Feni groups of volcanic islands form a southeast-trending chain 70-80 km northeast of New Ireland, extending down the center of the New Ireland Basin (Figure 1; Wallace et al., 1983; Exon et al., 1986). The volcanism is mainly Pliocene-Pleistocene, but pre-middle Miocene volcanic rocks are known in the Tabar Islands, and some volcanism extended into the Holocene (radiocarbon ages on tephras; W.R. Ambrose and P.S. Licence, personal communications, 1986). Present-day thermal activity is found in all four island groups, but especially on Lihir Island where epithermal gold is being deposited by an active geothermal system, and on Ambitle Island, the largest of the Feni Islands.

Most of the analyzed rocks from the Tabar-to-Feni islands are alkaline - that is, they have nepheline in the CIPW norm. Mafic rocks are basanite, tephrite, transitional basalt, and lavas in which clinopyroxene crystals have accumulated forming ankaramitic rocks (rock classification of Wallace et al., 1983). Intermediate lavas are phonolitic tephrite, trachybasalt (both nepheline and hypersthene normative), tephritic phonolite, and trachyandesite (both ne and hy normative). Felsic rocks are nepheline trachyte and quartz trachyte. The nepheline-normative character of the Tabar-to-Feni rocks is manifest modally in the more fractionated rocks by the appearance of phenocrysts and microphenocrysts of a feldspathoid belonging to the hauyne-sodalite series. Leucite is also present. The quartz trachyte appears to represent the youngest volcanism in all four island groups - in Tanga, for example, forming the intracaldera islands of Bitlik and Bitbok, and on Ambitle forming a large coulee or cumulodome within the breached central crater.

Most analyzed rocks from the islands have high alkali contents (mainly 5.5-11.0 weight percent) and K_2O/Na_2O values between 0.5 and 1.1. Island-arc geochemical features include low TiO_2 (less than 1.2 weight percent), high Al_2O_3, low Th/U, high Zr/Nb, low abundances of Zr, Hf, and Nb, and enrichments in some large-ion/low-valency elements relative to the light REE. Enrichments in incompatible elements are similar to those in alkaline mafic rocks from other island-arc areas, but the high abundances of Sr (about 1000-2195 ppm), Pb (up to 45 ppm), and K_2O (some greater than 4.5 weight percent) are rather unusual.

Tabar-to-Feni rocks are isotopically similar to rocks from other west Melanesian arcs. Sr-isotope ratios range 0.70365-0.70452 and correlate poorly, if at all, with major and trace-element compositions. Nd-isotopic ratios (as ϵ_{Nd} values) range 5.4-8.6, and have the familiar, negative, "mantle-array" correlation where plotted against $^{87}Sr/^{86}Sr$. However, some fall slightly to the right of the mantle array in common with values for other intra-oceanic arc rocks. Samples having the most fractionated REE patterns (high La/Yb) - as well as the quartz trachytes - have, in general, the most radiogenically enriched Sr and

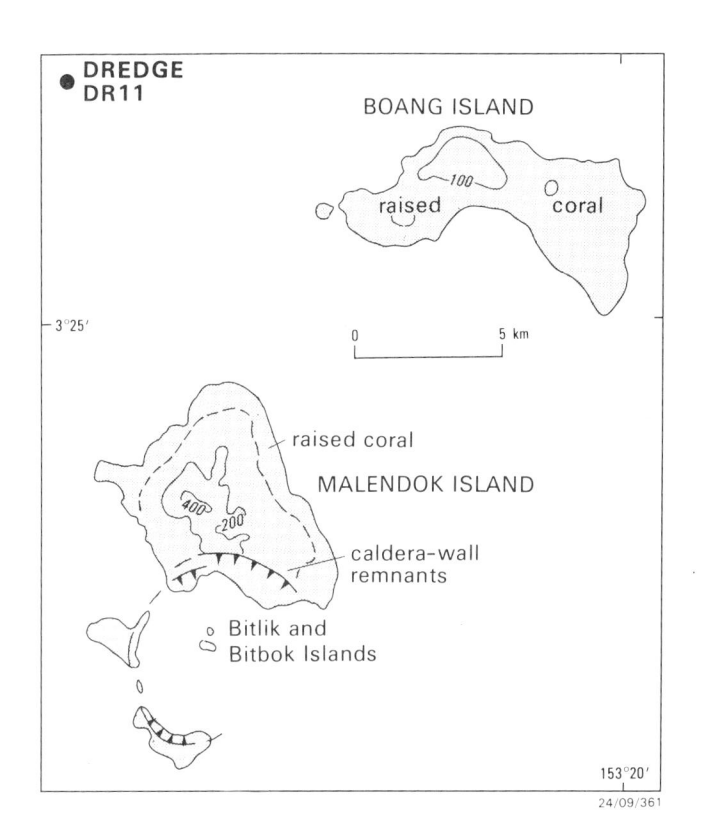

Figure 5. Tanga Islands and approximate location of DR11.

the lowest Nd-isotopic values. The range of Pb-isotopic values is quite narrow compared to the ranges measured for samples from other arcs and from western Melanesia as a whole. Samples have slightly lower $^{207}Pb/^{204}Pb$ and $^{208}Pb/^{204}Pb$ values relative to $^{206}Pb/^{204}Pb$, compared to values for some arc rocks from, for example, Fiji (Gill, 1984).

Dredge Samples of Tanga Volcano

The Tanga Islands represent the smallest of the four island groups in the Tabar-to-Feni chain. One of the two largest of the islands, Boang, consists only of limestone (Figure 5). The volcanic Tanga islands represent the subaerial remnants of a stratovolcano whose summit was largely destroyed by caldera formation (Wallace et al., 1983). Raised reef limestone partially encircles Malendok, the largest of the volcanic islands. The analyzed samples from the Tanga group are mainly phonolitic tephrite and ne-trachybasalt, together with some trachyte (both nepheline and quartz-normative) and transitional basalt (Figure 6).

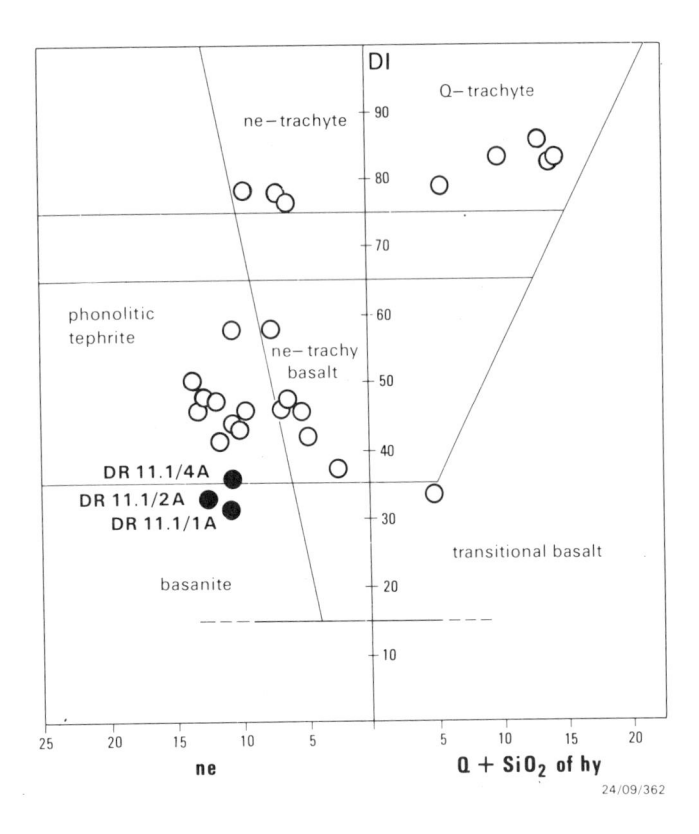

Figure 6. CIPW normative plot of Differentiation Index (DI; Thornton and Tuttle, 1960) against normative nepheline (ne) and normative quartz (Q) plus the silica of normative hypersthene (hy), for rocks from Tanga volcano. Open circles represent analyses of rocks from the islands (Wallace et al., 1983). Filled circles represent dredge samples (DR).

Table 7. Chemical analyses of three rocks from dredge DR11 at Tanga volcano. All data (except FeO, H_2O, and CO_2) by X-ray fluorescence spectrometry. Oxides in weight percent. Trace elements in ppm.

	DR11.1/1A	DR11.1/2A	DR11.1/4A
SiO_2	46.46	46.68	47.08
TiO_2	0.91	0.91	0.83
Al_2O_3	13.88	14.04	14.34
Fe_2O_3	4.53	3.58	6.05
FeO	6.16	7.03	4.84
MnO	0.17	0.18	0.16
MgO	7.17	6.98	6.31
CaO	12.65	12.55	12.35
Na_2O	3.20	3.38	4.59
K_2O	2.35	2.44	0.96
P_2O_5	0.48	0.49	0.55
S	<0.02	<0.02	<0.02
H_2O^+	1.39	1.32	1.85
H_2O^-	0.51	0.37	0.38
CO_2	0.20	0.17	0.16
Total	100.06	100.12	100.45
Ba	125	130	150
Rb	41.5	38.0	12.0
Sr	1070	1080	1300
Pb	6	5	6
Zr	67	66	66
Nb	2.5	2.5	2.5
Y	19	19	18
La	13	14	15
Ce	31	33	35
V	323	313	320
Cr	70	65	28
Ni	43	40	28
Cu	143	155	244
Zn	99	98	96
Ga	18.0	17.5	17.0

1A, 2A: basanites.
4A: phonolitic tephrite.

The lava blocks retrieved in Dredge DR11 (Table 1, Figures 1 and 5) on the northwestern shoulder of Tanga volcano appear to be compositionally similar. Six of them examined petrographically are fine-grained mafic rocks that are porphyritic mainly in euhedral olivine, abundant clinopyroxene, and minor Fe-Ti oxides. Plagioclase is restricted to the groundmass. Sample DR11.1/4A, however, is a little less mafic and contains conspicuous microphenocrysts of an isotropic feldspathoid almost certainly of hauyne-sodalite composition. Three of the six rocks have been chemically analyzed (Tables 4, 7, and 8). Analyzed clinopyroxene is augite containing low TiO_2 and only moderate amounts of Al_2O_3 and Na_2O (Table 9), whereas the clinopyroxene of most other Tabar-to-Feni rocks is salite.

Samples DR11.1/1A and 2A are both basanite characterized by a mafic mineralogy and containing more than 10 percent normative nepheline. They represent the first basanites to be described from Tanga, although similar rocks are known from the Tabar, Lihir, and Feni island groups (Wallace et al., 1983). Sample DR11.1/4A is slightly more felsic than the two basanites. It classifies as a phonolitic tephrite and is similar to others containing hauyne-sodalite from the Tanga Islands.

Table 8. Additional trace-element data for two samples from dredges DR11 and DR12. All data by spark-source mass spectrography. Abbreviation e.f. is enrichment factor.

	DR11.1/4A	DR12.3/3A		DR11.1/4A	DR12.3/3A
La	14.0	1.34	Cs	1.4	2.1
Ce	31.9	2.79	Pb	6.7	11.8
Pr	4.55	0.41	U	1.41	0.43
Nd	21.8	1.78	Th	1.33	0.24
Sm	5.31	0.43	Hf	1.5	1.5
Eu	1.62	0.13			
Gd	4.30	0.43			
Tb	0.57	0.07	[La/Yb]e.f. :		
Dy	3.20	0.34			
Ho	0.59	0.06		6.86	15.1
Er	1.46	0.14			
Yb	1.38	0.06			

Table 9. Representative clinopyroxene and amphibole analyses for rocks from dredges DR11 and DR12. The first two analyses are for clinopyroxene phenocrysts, and the third for a kaersutitic amphibole.

	DR11.1/4A	DR12.1/8A	
SiO_2	51.60	49.86	39.88
TiO_2	0.34	0.65	7.02
Al_2O_3	3.17	2.91	12.61
FeO	8.60	13.62	12.49
MnO	0.14	0.74	0.14
MgO	16.12	8.95	11.32
CaO	19.06	20.90	10.96
Na_2O	0.27	1.71	3.36
K_2O	n.d.	n.d.	1.14
Total	99.30	99.34	98.92
O =	6	6	23
Si	1.915	1.909	5.799
Al^{4+}	0.085	0.091	2.201
Al^{6+}	0.054	0.040	0.028
Ti	0.009	0.019	0.792
Fe^{3+}	0.031	0.140	0.000
Fe^{2+}	0.236	0.296	1.567
Mn	0.004	0.024	0.018
Mg	0.892	0.511	2.531
Ca	0.758	0.857	1.761
Na	0.019	0.127	0.977
Cr	0.000	0.000	0.218
Total	4.004	4.015	15.892
Wo	40.2	51.5	30.0
En	47.3	30.7	43.2
Fs	12.5	17.8	26.8

Island-arc characteristics of the two basanites are low TiO_2, Zr, and Nb, high Zr/Nb, and enrichments in Rb, K, and Sr (but not Ba) relative to the light REE (La and Ce). The phonolitic tephrite, DR11.1/4A, has similar geochemical features, except for its K and Rb values which are exceptionally low for Tabar-to-Feni rocks. The K_2O/Na_2O value of this sample, which appears to be fresh, is only 0.21 which is well outside the normal range of 0.5-1.1. Sr contents, however, are high causing a typical arc-type positive anomaly on a chondrite-normalization trace-element diagram (Figure 7). The REE pattern for DR11.1/4A is only moderately light-REE enriched, despite the strongly alkaline character of the rock, and is somewhat sigmoidal in form. Rather similar trace-element patterns (excluding the low Rb and K) are known for rocks from the Tabar, Lihir, and Feni islands, but this is the first recognition of them in the Tanga group.

The Sr-isotopic ratios of samples DR11.1/2A and 4A (Table 4) are similar to those measured in samples from Tanga (0.70382-0.70428; average 0.7040±2) and from the Tabar-to-Feni islands as a whole (average of 0.70399±38 for 28 samples; Wal-

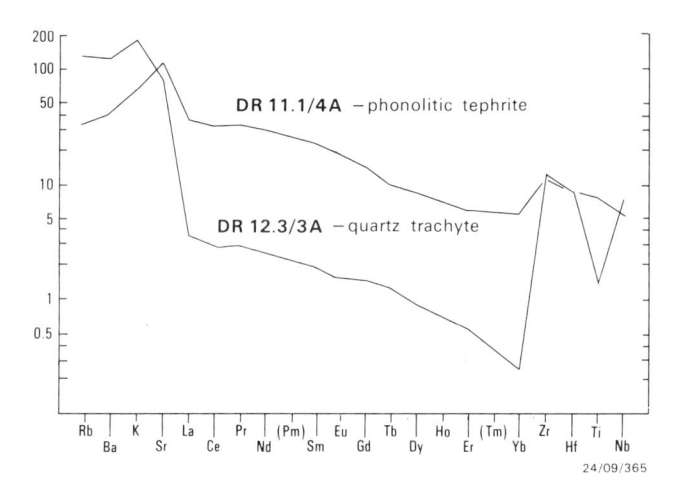

Figure 7. Chondrite-normalized trace-element values for two rocks from DR11 and DR12 (Tables 7, 8, and 11). Normalizing values as in Figure 3 (a primitive-mantle value of 0.35 is used for Rb).

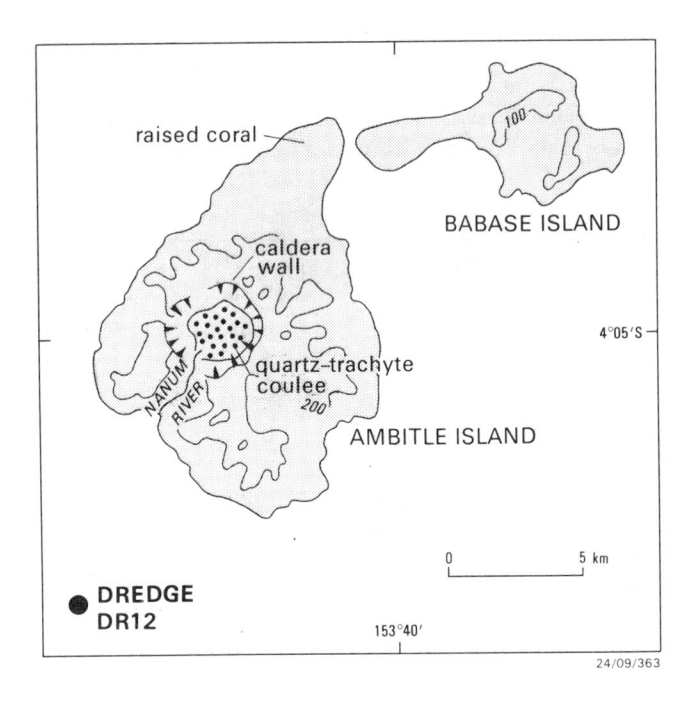

Figure 8. Feni Islands and approximate location of DR12.

Table 10. Chemical analyses of four alkaline rocks from dredge DR12 at Ambitle volcano. All data (except FeO, H_2O, and CO_2) by X-ray fluorescence spectrometry. Oxides in weight percent. Trace elements in ppm.

	DR12.1/7A	DR12.3/7A	DR12.1/1A	DR12.2/2A
SiO_2	45.88	53.18	55.52	58.45
TiO_2	0.69	0.58	0.70	0.35
Al_2O_3	8.01	19.36	17.43	19.32
Fe_2O_3	5.64	3.70	4.20	1.97
FeO	4.30	2.22	2.11	0.84
MnO	0.16	0.15	0.10	0.10
MgO	10.12	2.14	2.90	0.64
CaO	17.71	5.59	5.63	3.42
Na_2O	1.08	5.56	6.64	6.91
K_2O	1.62	2.94	3.29	5.17
P_2O_5	0.28	0.39	0.42	0.11
S	0.02	<0.02	0.07	0.07
H_2O^+	1.01	2.64	0.64	1.83
H_2O^-	1.06	1.37	0.32	0.43
CO_2	2.39	0.15	0.23	1.20
O=S	0.01		0.03	0.03
Total	99.96	99.97	100.17	100.78
Ba	94	390	425	370
Rb	28.5	19.5	53	96
Sr	830	1940	1670	1550
Pb	4	13	13	20
Zr	39	90	90	124
Nb	1.5	3.5	4.0	7.5
Y	12	16	14	12
La	9	18	12	22
Ce	21	40	28	40
V	270	210	247	150
Cr	145	4	4	3
Ni	85	7	9	3
Cu	84	59	91	87
Zn	62	66	47	45
Ga	12.0	23.0	22.0	27.5

12.1/7A: clinopyroxene-cumulate lava.
12.3/7A: ne-trachybasalt.
12.1/1A: ne-trachyandesite.
12.2/2A: ne-trachyte.

lace et al., 1983). The Pb-isotopic composition of sample DR11/4A is also closely similar to several samples from the other islands (Figure 4).

Nepheline-Normative Rocks Dredged
From Ambitle Volcano

Ambitle is a mainly Quaternary stratovolcano built on a basement of poorly exposed Oligocene limestone (Wallace et al., 1983). It has a 3 km-wide eroded caldera, the southwestern side of which is breached by the Nanum River (Figure 8). Raised reef crops out at the northern end of the island, and scattered thermal areas within the caldera and on the western flank are evidence that heat (probably from a cooling magma body) is being channelled to the surface. The main edifice of Ambitle volcano consists of a wide range of volcanic rock types - especially phonolitic tephrite, trachybasalt, and trachyandesite, but also tephritic phonolite, tephrite, basanite, and alkali basalt (Figure 9; Heming, 1979; Wallace et al., 1983). Lavas rich in accumulated clinopyroxene crystals are also present, and the floor of the caldera is covered by the quartz-trachyte coulee.

A wide range of volcanic rock types was obtained in dredge haul DR12 stationed about 7 km southwest of Ambitle (Table 1, Figures 1 and 8). Most of the samples are similar to rocks previously described from the island itself, and ten were selected for chemical analysis (Tables 4, 8, 10, and 11). Four are nepheline-normative, and the remainder are quartz trachyte.

Table 11. Chemical analyses of six quartz trachytes from dredge DR12 at Ambitle volcano. All data (except FeO, H$_2$O, and CO$_2$) by X-ray fluorescence spectrometry.

	DR12.1/8A	DR12.2/1A	DR12.2/6A	DR12.3/3A	DR12.3/4A	DR12.3/6A
SiO$_2$	66.32	66.85	67.85	69.00	69.49	66.67
TiO$_2$	0.23	0.23	0.20	0.15	0.15	0.23
Al$_2$O$_3$	16.92	16.78	16.69	15.88	16.05	17.09
Fe$_2$O$_3$	1.62	1.91	1.47	0.85	0.79	1.93
FeO	0.48	0.23	0.27	0.43	0.46	0.20
MnO	0.02	0.03	0.02	0.02	0.01	0.03
MgO	0.87	1.08	1.42	1.01	0.95	0.85
CaO	1.34	1.55	1.65	0.33	0.31	1.54
Na$_2$O	7.88	8.11	7.90	8.33	8.42	8.14
K$_2$O	3.22	2.82	2.53	2.63	2.60	3.03
P$_2$O$_5$	0.13	0.13	0.12	0.09	0.06	0.12
S	0.06	0.02	<0.02	0.03	0.04	<0.02
H$_2$O$^+$	0.49	0.29	0.41	0.95	0.88	0.26
H$_2$O$^-$	0.35	0.11	0.18	0.42	0.25	0.14
CO$_2$	0.58	0.15	0.19	0.29	0.25	0.11
O=S	0.03	0.01		0.01	0.02	
Total	100.48	100.28	100.90	100.40	100.69	100.34
Ba	540	460	455	440	460	450
Rb	57	47.5	48.5	45.0	45.0	53
Sr	1720	1760	1680	920	970	1950
Pb	14	24	11	14	14	9
Zr	79	75	75	71	70	77
Nb	3.5	4.0	3.5	3.5	3.5	3.5
Y	3	3	2	2	1	3
La	3	<2	3	<2	<2	2
Ce	8	6	7	4	3	5
V	58	58	38	23	24	55
Cr	5	10	23	19	15	5
Ni	5	8	21	16	14	5
Cu	5	3	3	2	2	3
Zn	21	19	23	23	23	20
Ga	20.0	19.0	18.0	19.0	20.0	20.5

Sample DR12.1/7A is the most mafic of all the analyzed dredge samples from the Tabar-to-Feni chain, containing 10.12 percent MgO and 17.71 percent CaO (Table 10). Its phenocrysts assemblage is characterized by an abundance of clinopyroxene (some crystals 1 cm across), olivine which is completely pseudomorphed by serpentine-like minerals, and magnetite. Alteration also affects the groundmass, and the total volatile content of the rock is almost 4.5 percent. Clinopyroxene accumulation is a fairly common feature in the Tabar-to-Feni chain. Ankaramites are known on Lihir Island, and some lava flows also contain cumulate xenoliths, some up to several tens of centimeters in diameter.

Samples DR12.3/7A and DR12.1/1A are representative of nepheline-normative trachybasalt and trachyandesite, respectively, found on Ambitle Island. They both contain phenocrysts of clinopyroxene, a green amphibole (probably magnesiohastingsite), plagioclase, magnetite, and apatite. Hauyne-sodalite phenocrysts are common in the trachyandesite, and biotite is well developed in the trachybasalt which is somewhat altered. Nepheline-normative trachyte has not been recognized previously from Ambitle, and sample DR12.2/2A is the first recorded example of one. The sample is not especially fresh, as it contains secondary carbonate (CO_2 1.20 percent). Plagioclase is the most abundant phenocryst in the trachyte, but clinopyroxene, amphibole, apatite, magnetite, and hauyne-sodalite are also present.

These four samples are good examples of the wide range of alkaline lava compositions that characterize the Feni Islands. Their most notable arc-type chemical features are low TiO_2 (0.70 percent or less), high Al_2O_3 (19.36 percent in the trachybasalt), enrichments in Rb, Ba, K, and especially Sr (1940 ppm in the trachybasalt), relative to La and Ce, and low Nb and high Zr/Nb. Nepheline-trachyte DR12.2/2A is isotopically similar to other analyzed rocks from the Tabar-to-Feni chain (Table 4), but has the lowest $^{206}Pb/^{204}Pb$ value of any measured sample from the island groups (Figure 4).

Quartz Trachyte Dredged From Ambitle Volcano

A large suite of Ambitle quartz-trachyte samples was obtained from dredge DR12. Six of the trachytes have been chemically analyzed (Tables 8 and 11).

The Tabar-to-Feni quartz trachytes are one of the more striking petrological features of the volcanic chain, because they are in marked compositional contrast to the more alkaline, mafic and intermediate rocks that make up the bulk of each island group. The notably light-colored trachytes appear to represent the youngest eruptions in each island group. All are probably at least as young as Middle to Upper Pleistocene and may be Holocene on Ambitle Island. The quartz trachyte may represent a dramatic and synchronous change in magma composition throughout the island chain, perhaps marking the final stages of Tabar-to-Feni volcanism. Similar striking contrasts in composition are known in other alkaline volcanoes where strongly silica-oversaturated trachyte is a part of sequences consisting mainly of silica-undersaturated lavas - for example, Dunedin volcano (New Zealand; Price and Taylor, 1973) and Mount Kenya (Price et al., 1985). The Tabar-to-Feni trachytes are also of some economic importance because they (and related intrusive rocks located by company drilling) are the host for epithermal gold being deposited by the active geothermal system on Lihir Island.

Tabar-to-Feni quartz trachytes have a fairly uniform and simple mineralogy. By far the most abundant phenocryst mineral is alkali feldspar, in grains up to about 5 mm across. Examples of feldspar-composition ranges determined by electron microprobe for Tabar and Tanga quartz trachytes are: Ab_{95}-Or_{91}, Ab_{87}-Or_{43}, and Ab_{71}-Or_{36}. Some of the feldspar is polysynthetically twinned, and many crystals are strained or fractured as if they had been deformed during lava emplacement. Biotite and exsolved magnetite are ubiquitous microphenocrysts or phenocrysts, and amphibole and clinopyroxene are present in small quantities. Some of the amphibole is relatively alkali rich, reaching magnesioarfvedsonite and magnesioriebeckite compositions. The groundmass is fine-grained, consisting mainly of alkali feldspar, magnetite, and a high-temperature silica mineral (probably cristobalite or tridymite).

The six quartz-trachyte samples from DR12 are all fairly fresh and are representative of the range of Ambitle quartz-trachyte compositions (Table 11). The samples may have been plucked by the dredge from outcrops on the southwestern flank of Ambitle volcano, but more likely they represent boulders and cobbles derived from the island coulee and transported downslope from the Nanum River outlet (Figure 8). All six samples are mineralogically and geochemically similar to previously analyzed Tabar-to-Feni quartz trachytes.

Analyzed clinopyroxene grains in DR12.1/8A are relatively rich in Ca and Fe (Table 9) and their compositions plot along the same trends reported for phenocryst-groundmass pairs in less fractionated rocks from the Tabar-to-Feni chain (Ellis, 1975). The Fe-rich nature of mafic minerals in this rock is evident also by the presence of a mica crystal of annite composition coexisting with a salite pyroxene. In addition, a kaersutitic amphibole containing high TiO_2 has been analyzed in DR12.1/8A (Table 9). Kaersutite has not been identified previously in Tabar-to-Feni rocks, and is being investigated further in order to establish whether it is an equilibrium or xenocryst mineral.

The Ambitle-dredge quartz-trachyte samples have high Differentiation Indices (greater than 90) and are therefore strongly felsic (Figure 9). Amounts of normative quartz are mostly less than 10-15 percent, but one sample (DR12.3/4A) has more than 20 percent and is, strictly speaking, an alkali-rich dacite. Alkali contents exceed 10 percent, and Na_2O is at least twice as abundant as K_2O. TiO_2 and P_2O_5 values are low, as are most incompatible trace-element abundances - excepting Rb, Ba, and Sr.

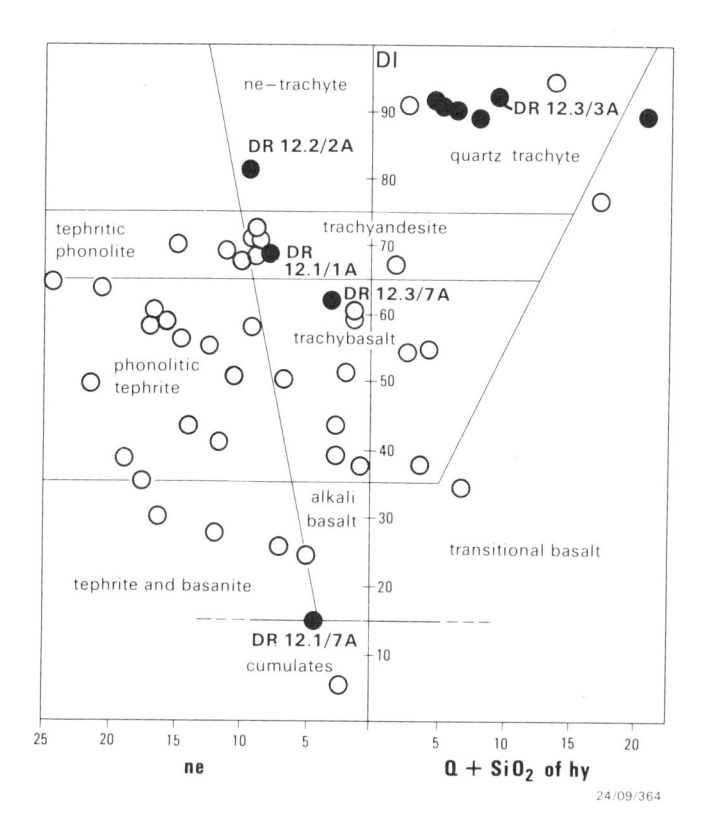

Figure 9. CIPW normative plot (see Figure 6) for rocks of Ambitle volcano. Open circles represent analyses of rocks from Ambitle and Babase islands (Heming, 1979; Wallace et al., 1983). Filled circles represent dredge samples.

REE abundances, for example, are extremely low in sample DR12.3/3A (Table 8) - especially the heavy-REE contents some of which are below those of chondritic values (Figure 7). The REE pattern for this sample is light-REE enriched and there is no Eu anomaly. These are not the trace-element features expected for felsic magmas that represent the product of a long line of fractional crystallization involving common rock-forming minerals such as plagioclase. They could be regarded as consistent with magma derivation by partial melting of an alkali-feldspar-rich crustal source, but the absence of negative Eu anomalies does not appear to support this.

Sr-isotope ratios for three quartz-trachyte samples from Ambitle dredge DR12 are nearly identical (Table 4), and two are only slightly less radiogenically enriched than are quartz trachytes from Tabar and Tanga (0.70412 and 0.70428). The Pb-isotope composition of DR12.3/4A is closely similar to those of other samples from the Tabar-to-Feni islands, as well as of volcanic rocks elsewhere in Papua New Guinea. The similarity in isotopic composition between the more mafic Tabar-to-Feni rocks and the highly fractionated quartz trachytes may mean that if the trachytes were derived by crustal anatexis, then the crustal source is isotopically similar to that of the more mafic volcanic rocks in the island chain.

Discussion

The petrology and geochemistry of the Tabar-to-Feni volcanism is the subject of reports currently being prepared in collaboration with R.J. Arculus, A. Kennedy, and others, and the limited results presented here for the dredge rocks are regarded as additions to a much larger data base which will be discussed at length elsewhere. The petrogenesis of these unusual rocks remains enigmatic, but an interpretation involving a relationship with subduction processes seems unavoidable as the rocks have clear arc-type signatures.

An active Wadati-Benioff zone is absent from beneath the entire length of the Tabar-to-Feni chain, although the deeper parts of the New Britain seismic zone sweep around southeastwards beneath the southeastern end of the chain. In addition, a downgoing slab (or slabs of different polarities at different times) could have existed beneath what is now the Tabar-to-Feni chain in the Tertiary. The relationship between the composition of the Tabar-to-Feni magmas and a previous subduction system (or systems) is still uncertain, especially as no known andesites were produced from it (at least in the Tabar-to-Feni chain).

An alternative view is that the extensive Pliocene and Pleistocene volcanism represented on the islands may be related to faulting caused by the opening of the Manus Basin west of New Ireland and to melting of a previously modified source (Johnson, Mutter, and Arculus, 1979). However, some Tabar-to-Feni volcanism probably began well before the formation of the Manus Basin - in the Tabar Islands, at least before the Middle Miocene - and therefore can not be attributed to Manus Basin opening.

Tabar-to-Feni volcanism has been long-lived, and volcanoes have existed in the New Ireland Basin at least since the mid-Tertiary. The presence of the volcanoes, therefore, must have influenced patterns of sedimentation for much of the history of the New Ireland Basin. In addition, the existence of the volcanism for that length of time implies that the central axis of the eastern New Ireland Basin, which presumably was one of higher-than-normal heatflow, may have influenced maturation temperatures of any organic sediment in the basin.

EFFECTS OF REMOVING AN ARC FROM A SUBDUCTION REGIME

The analytical data obtained for ferrobasalt DR2/1 and andesite DR6/4 are sufficient to illustrate an important feature of the petrotectonic evolution of the New Ireland Basin - namely, the temporal transition from arc-type volcanism in the Tertiary to ocean-island-type in the Quaternary. These new data supplement the preliminary results for Manus Island rocks presented and discussed by Jaques (1981, 1983).

Bimodal volcanism in the St. Andrew Strait area - including rhyolitic volcanism in 1953-57 (Reynolds, Best, and Johnson, 1980) - has been interpreted to be the result of an intraplate melting anomaly involving partial melting of both crust and upper mantle (Johnson and Smith, 1974; Johnson, Smith, and Taylor, 1978). Jaques (1981, 1983) extended this concept by speculating that the Quaternary volcanic rocks from Southwest Bay, on the M'Buke and Johnstone Islands, and in the St. Andrew Strait area, represent the trace of a southwest-tracking "hot spot" or mantle plume. Hot-spot volcanism at, or close to, ocean-basin spreading axes has been taken as evidence that the positions of spreading axes are controlled by deep-mantle plumes (for example, Burke and Dewey, 1973; Hey, 1977). This concept may apply also to St. Andrew Strait, as the volcanism there is about only 50 km from, and on the northwestern extension of, an important, active transform fault defining part of the southwestern side of the Manus Basin (Johnson, Mutter, and Arculus, 1979). However, the question of whether the ocean-island-type volcanism represented by the DR2/1 and Southwest Bay ferrobasalts is related to Manus Basin spreading is debatable, because there is no supporting evidence that they are necessarily part of a St. Andrew Strait hot-spot trace. No radiometric ages for rocks from the different volcanoes are available. In addition, the basalts of St. Andrew Strait (including marginal-basin types on Baluan Island) are distinctly different from the ocean-island-type ferrobasalts, so their sources are probably different too.

The established view (Wilson, 1963; Morgan, 1972; Ringwood, 1982) that volcanoes defining hot-spot tracks in ocean basins represent the existence world-wide of deep-mantle plumes that rise to the base of the lithosphere, has been questioned. Gill (1976a, 1984) pointed out that such a model is difficult to accept for the ocean-island-type rocks of Fiji because a deep-mantle plume would have to produce, fortuitously, ocean-island-type volcanism precisely in an area where previously only arc-type magmas had been erupted in a narrow zone. The arc to ocean-island transition in the Manus region is a direct analogue of the Fijian example, and it too must be regarded as coincidental if a deep-mantle plume now exists beneath the Manus Island region where formerly only spatially restricted arc-type volcanism had taken place.

An alternative view is that upper mantle capable of yielding ocean-island-type basalts exists also in the wedge beneath island arcs and above downgoing slabs (Gill, 1976a,b; 1984; Morris and Hart, 1983). This interpretation was debated recently by Perfit and Kay (1986) who, in contrast, favored the hypothesis that the mantle wedge beneath arcs has the composition of the depleted mantle source thought to exist beneath mid-ocean ridges, but modified by components derived from the downgoing slab. On the other hand, Zindler, Staudigel, and Batiza (1984) proposed that both ocean-island-type and mid-ocean-ridge-type magmas may be produced by different degrees of partial melting from a world-wide mantle source that is heterogeneous on a small scale. Smaller degrees of melting involve a greater proportion of a low-melting-point source component, producing basaltic magmas that are more ocean-island like in composition.

Gill (1984) and Morris and Hart (1983; 1986) argued that a component derived from downgoing slabs exists beneath island arcs as well as both ocean-island and mid-ocean-ridge type sources, and therefore that arcs too have the potential for producing ocean-island-type basalts. The fact that Fiji, and now the Manus region, are the only known examples of the arc to ocean-island transition, may be taken as a weakness of this interpretation if it is to be applied world-wide (Perfit and Kay, 1986). Alternatively, ocean-island basalt generation in arcs may require special tectonic circumstances - ones in which sources activated by arc splitting have not been modified by a slab component and which perhaps control magma generation by degrees of partial melting that are smaller than normal for island arcs.

Vigorous attempts have been made in recent years to recognize the components that contribute to arc-magma genesis (for example, Kay, Sun, and Lee-Hu, 1978; Arculus and Johnson, 1981; McCulloch and Perfit, 1981; Gill, 1984; Johnson et al., 1985). The limited Sr and Pb-isotope values presented above for the New Ireland Basin and Manus region dredge samples are similar to those for arc rocks elsewhere in western Melanesia, corresponding to similar, although almost certainly heterogeneous, mantle sources. Slightly higher $^{87}Sr/^{86}Sr$ values distinguish them and other west Melanesian arc rocks from many ocean-island basalts, where comparing rocks that have the same ^{206}Pb/^{204}Pb (Figure 4) or ^{143}Nd/^{144}Nd, but this does not necessarily exclude the possibility of widespread ocean-island-type sources beneath the region. More particularly, ocean-island-type sample DR2/1 has only slightly different isotopic ratios from indisputable arc-type rocks from western Melanesia.

A further complication in the Manus Island region is represented by the extensive, flat-lying lava flows north and northwest of Manus itself. Sediment cover prevented the sampling of these flows during *S.P. Lee* cruise, and a drilling program is required in order to determine their compositions. The age of the flows is also unknown, but most likely the lavas were produced before the presumed Pliocene opening of the Manus Basin because their sediment cover is much thicker than that in the basin. Lower to middle Miocene limestone is widespread throughout the Bismarck Archipelago, apparently corresponding to a regional hiatus in volcanic and tectonic activity. The volcanism represented by the buried flows north and northwest of Manus Island therefore may either predate this hiatus, in which case the flows are pre-Miocene, or it post-dates the hiatus and the flows are late Miocene.

The buried flows are described as Manus "forearc" rocks and correlated with southwards subduction at the Manus Trench (Marlow, Exon, and Tiffin, this volume). The trench today has no known Quaternary arc-type or forearc volcanism, which may be an indication of low rates of subduction that are reflected at the present-day by the low levels of seismicity (McCue, this volume), or alternatively the subduction is intermittent. However, another interpretation is that the flows north and northwest of Manus were produced when Manus was underlain by the northward-dipping subduction system that now underlies New Britain. If so, the lavas may represent volcanism behind the Manus arc rather than in front of it. Perhaps they are an aborted phase of marginal-basin volcanism before formation of the Manus Basin. Nevertheless, there still remains the tantalizing prospect that ocean-island-type compositions may be represented in these buried flows.

ACKNOWLEDGMENTS

No volcanic petrologists were aboard the *S.P. Lee* when it surveyed the New Ireland Basin, so we are especially indebted to our non-volcanic colleagues for collecting the samples described here - particularly the Co-Chief Scientists, M.S. Marlow and N.F. Exon. Constructive reviews of the draft manuscript were provided by J.B. Gill, R.C. Price, and S.-S. Sun. Spark-source mass spectrographic data was obtained on contract from the Research School of Earth Sciences, and we are grateful to S.R. Taylor for agreeing to undertake this work. We also thank D. Glicksburg for his analytical assistance at the University of Florida. This research was partially supported by the National Science Foundation (OCE-8315725). RWJ and ALJ publish with the permission of the Director of the Bureau of Mineral Resources, Canberra.

REFERENCES

Arculus, R.J., and Johnson, R.W., 1981, Island-arc magma sources: a geochemical assessment of the roles of slab-derived components and crustal contamination: Geochemical Journal, v. 15, p. 109-133.

Baitis, H.W., and Lindstrom, M.M., 1980, Geology, petrography and petrology of Pinyan Island, Galapagos Archipelago: Contributions to Mineralogy and Petrology, v. 72, p. 367-386.

Brown, C.M., 1982, Kavieng, Papua New Guinea: Geological Survey of Papua New Guinea, 1:250,000 Geological Series Explanatory Notes, SA/56-9, 26 p.

Bultitude, R.J., Johnson, R.W., and Chappell, B.W., 1978, Andesites of Bagana volcano, Papua New Guinea: chemical stratigraphy, and a reference andesite composition: Bureau of Mineral Resources Journal of Australian Geology and Geophysics, v. 3, p. 281- 295.

Burke, K., and Dewey, J.F., 1973, Plume-generated triple junctions: key indicators in applying plate tectonics to old rocks: Journal of Geology, v. 81, p. 406-433.

Coombs, D.S., 1963, Trends and affinities of basaltic magmas and pyroxenes as illustrated on the diopside-olivine-silica diagram: Mineralogical Society of America Special Paper 1, p. 227-250.

Curtis, J.W., 1973, Plate tectonics of the Papua - New Guinea - Solomon Islands region: Journal of the Geological Society of Australia, v. 20, p. 21-36.

Ellis, D.J., 1975, A preliminary report on the petrography and mineralogy of the feldspathoid-bearing potassic lavas from the Tabar, Lihir, Tanga, and Feni Islands, off the coast of New Ireland, PNG: Bureau of Mineral Resources, Australia - Record 1975/29 (unpublished), 24 p.

Exon, N.F., Stewart, W.D., Sandy, M.J., and Tiffin, D.L., 1986, Geology and offshore petroleum prospects of the eastern New Ireland Basin, Papua New Guinea: Bureau of Mineral Resources Journal of Australian Geology and Geophysics, v. 10, p. 39-51.

Gill, J.B., 1976a, Evolution of the mantle: geochemical evidence from alkali basalt: comment and reply: Geology, v. 4, p. 625-626.

Gill, J.B., 1976b, From island arc to oceanic islands: Fiji, southwestern Pacific: Geology, v. 4, p. 123-126.

Gill, J.B., 1978, Role of trace element partition coefficients in models of andesite genesis: Geochimica et Cosmochimica Acta, v. 42, p. 709-724.

Gill, J.B., 1984, Sr-Pb-Nd isotopic evidence that both MORB and OIB sources contribute to oceanic island arc magmas in Fiji: Earth and Planetary Science Letters, v. 68, p. 443-458.

Heming, R.F., 1979, Undersaturated lavas from Ambittle [sic] Island, Papua New Guinea: Lithos, v. 12, p. 173-186.

Hey, R., 1977, Tectonic evolution of the Cocos-Nazca spreading center: Geological Society of America Bulletin, v. 88, p. 1404-1420.

Hohnen, P.D., 1978, Geology of New Ireland, Papua New Guinea: Bureau of Mineral Resources, Australia - Bulletin 194, 39 p.

Jaques, A.L., 1980, Admiralty Islands, Papua New Guinea: Geological Survey of Papua New Guinea, 1:250 000 Geological Series Explanatory Notes, SA/55-10, SA/55-11, 25 p.

Jaques, A.L., 1981, Quaternary volcanism on Manus and M'Buke Islands, in Johnson, R.W., ed., Cooke-Ravian Volume of Volcanological Papers: Geological Survey of Papua New Guinea Memoir 10, p. 213-219.

Jaques, A.L., 1983, Manus Island, Papua New Guinea: geochemical evolution from island-arc to oceanic island: Geological Society of Australia Abstracts, v. 9, p. 142-143.

Johnson, R.W., and Smith, I.E.M., 1974, Volcanoes and rocks of St. Andrew Strait, Papua New Guinea: Journal of the Geological Society of Australia, v. 21, p. 333-351.

Johnson, R.W., Mutter, J.C., and Arculus, R.J., 1979, Origin of the Willaumez-Manus Rise, Papua New Guinea: Earth and Planetary Science Letters, v. 44, p. 247-260.

Johnson, R.W., Smith, I.E.M., and Taylor, S.R., 1978, Hot-spot volcanism in St. Andrew Strait, Papua New Guinea: geochemistry of a bimodal rock suite: Bureau of Mineral Resources Journal of Australian Geology and Geophysics, v. 3, p. 55-69.

Johnson, R.W., Jaques, A.L., Hickey, R.L., McKee, C.O., and Chappell, B.W., 1985, Manam Island, Papua New Guinea: petrology and geochemistry of a low-TiO_2 basaltic island-arc volcano: Journal of Petrology, v. 26, p. 283-323.

Kay, R.W., Sun, S.-S., and Lee-Hu, C.-N., 1978, Pb and Sr isotopes in volcanic rocks from the Aleutian Islands and Pribilof Islands, Alaska: Geochimica et Cosmochimica Acta, v. 42, p. 263-273.

McCulloch, M.T., and Perfit, M.R., 1981, $^{143}Nd/^{144}Nd$, $^{87}Sr/^{86}Sr$ and trace element constraints on the petrogenesis of Aleutian island arc magmas: Earth and Planetary Science Letters, v. 56, p. 167-179.

Morgan, W.J., 1972, Deep mantle convection plumes and plate motions: American Association of Petroleum Geologists Bulletin, v. 56, p. 203-213.

Morris, J.D., and Hart, S.R., 1983, Isotopic and incompatible element constraints on the genesis of island arc volcanics from Cold Bay and Amak Island, Aleutians, and implications for mantle structure: Geochimica et Cosmochimica Acta, v. 47, p. 2015-2030.

Morris, J.D., and Hart, S.R., 1986, Isotopic and incompatible element constraints on the genesis of island arc volcanics from Cold Bay and Amak Island, Aleutians, and implications for mantle structure: reply to a critical comment by M.R. Perfit and R.W. Kay: Geochimica et Cosmochimica Acta, v. 50, p. 483-487.

Norrish, K., and Chappell, B.W., 1977, X-ray fluorescence spectrometry, in Zussman, J., ed., Physical Methods in Determinative Mineralogy: London, Academic Press, p. 161-214.

Perfit, M.R., and Kay, R.W., 1986, Comment on "Isotopic and incompatible element constraints on the genesis of island arc volcanics from Cold Bay and Amak Island, Aleutians, and implications for mantle structure" by J.D. Morris and S.R. Hart: Geochimica et Cosmochimica Acta, v. 50, p. 477-481.

Price, R.C., Johnson, R.W., Gray, C.M., and Frey, F.A., 1985, Geochemistry of phonolites and trachytes from the summit region of Mt. Kenya: Contributions to Mineralogy and Petrology, v. 89, p. 394-409.

Price, R.C. and Taylor, S.R., 1973, The geochemistry of the Dunedin volcano, east Otago, New Zealand: rare earth elements: Contributions to Mineralogy and Petrology, v. 40, p. 195-205.

Reynolds, M.A., Best, J.G., and Johnson, R.W., 1980, 1953-57 eruption of Tuluman volcano: rhyolitic volcanic activity in the northern Bismarck Sea. Geological Survey of Papua New Guinea Memoir 7, 44 p.

Ringwood, A.E., 1982, Phase transformations and differentiation in subducted lithosphere: implications for mantle dynamics, basalt petrogenesis, and crustal evolution: Journal of Geology, v. 90, p. 611-643.

Sinton, J.M., Liu, L., Taylor, B., Chappell, B.W., and Shipboard Party, 1986, Petrology, magmatic budget and tectonic setting of Manus Back-arc Basin lavas: EOS, Transactions of the American Geophysical Union, v. 67, p. 377-378.

Staudigel, H., McCulloch, M., Zindler, A., and Perfit, M.R., in press, Complex ridge subduction and island arc magmatism: an isotopic study of the New Georgia forearc and the Woodlark Basin, in Taylor B., and Exon, N.F., eds., Marine Geology and Geophysics of the Woodlark Basin - Solomon Islands region: Circum-Pacific Council for Energy and Mineral Resources Earth Science Series.

Taylor, B., 1979, Bismarck Sea: evolution of a back-arc basin: Geology, v. 7, p. 171-174.

Taylor, B., Crook, K., Sinton, J., and Shipboard Party, 1986, Fast spreading and sulphide deposition in Manus back-arc basin: Program and Abstracts of Papers, Circum-Pacific Conference, Singapore, p. 35-36.

Taylor, S.R., 1979, Trace element analysis of rare earth elements by spark source mass spectrometry, in Gscheider, K.A., Jr., and Eyring, L., eds., Handbook on the Physics and Chemistry of Rare Earths: Amsterdam, North Holland, p. 359-376.

Taylor, S.R., and McLennon, S.M., 1981, The composition and evolution of the continental crust: rare earth element evidence from sedimentary rocks: Philosophical Transactions of the Royal Society of London, v. A301, 381-399.

Thornton, C.P., and Tuttle, O.F., 1960, Chemistry of igneous rocks I. Differentiation Index: American Journal of Science, v. 258, p. 664-684.

Wallace, D.A., Johnson, R.W., Chappell, B.W., Arculus, R.J., Perfit, M.R., and Crick, I.H., 1983, Cainozoic volcanism of the Tabar, Lihir, Tanga, and Feni Islands, Papua New Guinea: geology, whole-rock analyses, and rock-forming mineral compositions: Bureau of Mineral Resources, Australia - Report 243, 62 p.

White, W.M., and Hoffman, A.W., 1982, Sr and Nd isotope geochemistry of oceanic basalts and mantle evolution: Nature, v. 296, p. 821-825.

Wilson, J.T., 1963, A possible origin of the Hawaiian Islands: Canadian Journal of Physics, v. 41, p. 863-870.

Zindler, A., Staudigel, H., and Batiza, R., 1984, Isotope and trace element geochemistry of young Pacific seamounts - implications for the scale of upper mantle heterogeneity: Earth and Planetary Science Letters, v. 70, p. 175-195.

Marlow, M.S., Dadisman, S.V., and Exon, N.F., editors, 1988, Geology and offshore resources of Pacific island arcs—New Ireland and Manus region, Papua New Guinea, Circum-Pacific Council for Energy and Mineral Resources Earth Science Series, v. 9: Houston, Texas, Circum-Pacific Council for Energy and Mineral Resources.

GEOCHEMISTRY OF BATHYAL FERROMANGANESE DEPOSITS FROM THE NEW IRELAND REGION IN THE SOUTHWEST PACIFIC OCEAN

B.R. Bolton[1]
Department of Geology, La Trobe University, Bundoora, Victoria, 3083, Australia

N.F. Exon
Bureau of Mineral Resources, Geology and Geophysics, Canberra, A.C.T., 2601, Australia

ABSTRACT

Five ferromanganese crusts and one nodule from the bathyal New Ireland region in the Southwest Pacific Ocean have been analyzed for Mn, Fe, Ni, Cu, Zn, and Co. On the basis of these data, which show a maximum grade of 0.73% combined Ni + Cu + Co and 0.3% Co, it would appear unlikely that crusts or nodules of economic grade are present. However, a low sampling density, together with one occurrence of manganese oxide which is apparently hydrothermally precipitated, suggest that further sampling in selected areas is warranted, especially near the apparent hydrothermal deposit northwest of Manus island.

INTRODUCTION

This paper reports the results of a geochemical investigation of six dredged ferromanganese samples from the New Ireland region of the Southwest Pacific Ocean (Figure 1). Ferromanganese nodules and crusts on seamounts and other raised areas of the sea-floor are known to contain higher concentrations of cobalt than other marine deposits. This enrichment has in recent years been the focus of increasing study, particularly with regard to the potential of such material for economic exploitation (Commeau et al., 1984; Halbach and Manheim, 1984; Manheim and Hein, 1986; Hein et al., 1987). The main objective of the current investigation was to assess the region in terms of potentially economic Co-rich ferromanganese crusts or high-grade polymetallic ferromanganese nodules.

New Ireland and the nearby islands - Manus and New Hanover - form an arcuate, northwest-trending island chain on the southwestern margin of the Pacific Plate, south of the Manus and Kilinailau Trenches and bounded to the south by the Manus-Williaumez Rise, Manus Basin, and New Britain Trench (Figure 1). Detailed accounts of the stratigraphic and tectonic setting are given by Marlow et al. (this volume) and Exon and Marlow (this volume).

Three of the samples (DR2-1, DR3-2, DR6A-4) were collected during the 1984 R/V *S.P. Lee* cruise (Marlow et al., 1984); the remaining samples (PN79-1(22A), PN79-1(24A) and PN79-1(24An) are from CCOP/SOPAC stations from cruise PN79-1 (Eade, 1980). The samples were dredged from water depths

---

[1] Present address: Department of Geology and Geophysics, University of Adelaide, South Australia, 5001, Australia

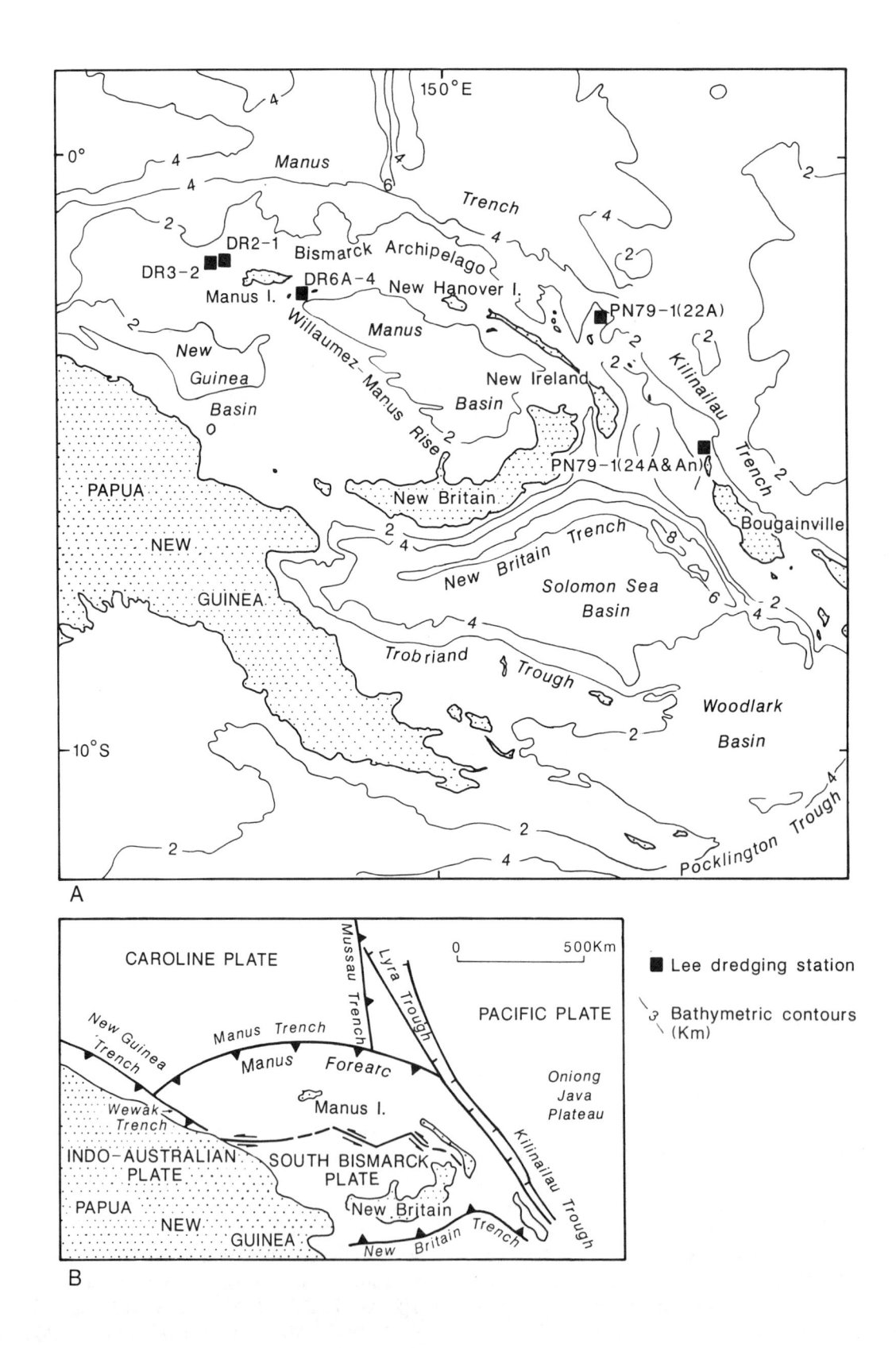

Figure 1. A) Map showing location of dredge sites and major bathymetric features. B) Map showing major tectonic features for the study area.

ranging from 800 to 1500 m (Figure 1).

Previous marine surveys in the region had revealed little about ferromanganese materials on the sea-floor. Cronan and Thompson (1978) conducted a regional geochemical study of Southwest Pacific sediments, and identified several areas of anomalous, high concentrations of Mn in the Manus Basin. Eade (1980) noted the presence of ferromanganese deposits and his samples, as mentioned above, were included in this investigation. Both et al. (1986), reporting on a recent camera survey, note the presence of hydrothermal chimneys along the spreading center of the Manus Basin.

This report also presents a comparison with other selected ferromanganese deposits, and a brief discussion of their origin. A detailed investigation including geochemical, mineralogical, and petrological analysis of all samples is presently underway and will be reported at a later date.

DESCRIPTION OF SAMPLES

Crusts analyzed in this study are of two types. Samples DR2-1, DR6A-4, PN79-1(22A), and PN79-1(24A) consist of thin (1-5 mm), hard, dense, and generally smooth to slightly gritty black crusts. Samples DR2-1 and DR6A-4 formed on altered basalt and andesite respectively, while PN79-1(22A) and PN79-1(24A) developed on late Pliocene-early Pleistocene (N21) foraminifera-bearing sediment, and a highly altered, gray-green volcanic siltstone, respectively.

Sample DR3-2 differs markedly from other samples in that it consists of a slightly gritty surface layer, less than 1 mm thick, beneath which is 15-20 mm of hard, shiny, bluish-black, finely laminated crust with individual laminae usually less than 1 mm thick. This material overlies, in turn, a 5-15 mm thick basal layer of grayish-black crust, showing no internal lamination but having a microgranular surface texture and very irregular porous structure.

The single nodule analyzed (PN79-1(24An)) consists of a smooth, 3 mm thick envelope of ferromanganese material formed on a core of Mn-impregnated volcanic sandstone. Associated materials include gray-green, highly-burrowed, very fine volcanic sandstone, fine marl, coarse foraminiferal sandstone and glauconitic foraminiferal sand.

RESULTS

The results of chemical analyses are presented in Table 1. The most significant feature is the very high Mn/Fe ratio (65:1) in DR3-2. Crust samples DR2-1, DR6A-4, PN79-1(22A) and PN79-1(24A), have a mean Mn/Fe ratio of 0.76, and a mean grade (Ni+Cu+Co) of 0.38% where Ni>Co>Cu. There are insufficient data to indicate any relationship between grade and water depth, but there appears to be a direct relationship between grade and Mn/Fe ratio (Figure 2), with all samples but DR3-2 defining a straight line. The ternary diagram of Ni versus Co versus Cu (Figure 3) indicates that, for all samples, the proportion of Cu remains nearly constant whereas the proportions of Ni and Co vary.

Analysis of crust DR3-2 shows a Mn/Fe ratio of 65.33 and a mean grade of 0.03% (Ni>Cu=Co). The nodule has a Mn/Fe ratio of 1.21 and a grade of 0.73% (Ni>Co>>Cu).

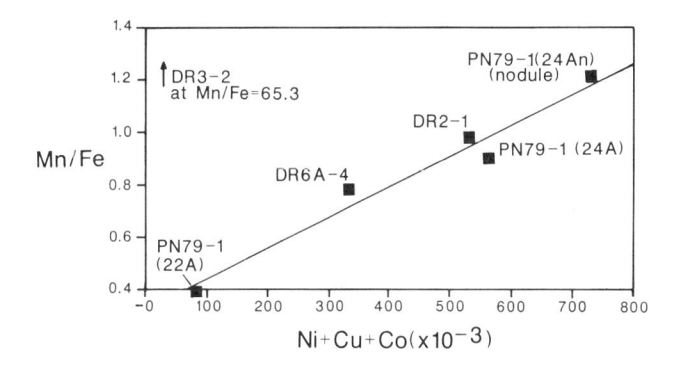

Figure 2. Plot of percent Ni+Cu+Co against Mn/Fe ratio in the New Ireland region. Line of best fit is also shown.

DISCUSSION

Economic evaluation of cobalt-rich ferromanganese crusts requires, at a minimum, detailed studies of grade, thickness and areal extent. As noted by Hein et al. (1987), most prospective areas have undergone only limited evaluation, with resource estimates largely based on a calculation of price per ton of metal (Halbach, Manheim and Otten, 1982; Clark, Johnson and Chinn, 1984; Halbach and Manheim, 1984). As was also pointed out by Hein and co-workers, few studies have assessed the economic and engineering significance of such factors as the roughness of the small-scale topography, and the amount of sediment covering crusts.

For the purposes of this report we assume that a minimum requirement in an economic assessment

Table 1. Metal concentrations in ferromanganese crusts and a nodule from the New Ireland region.

Station	Latitude (S)	Longitude (E)	Water Depth (m)	Mn	Fe	Ni	Cu	Zn	Co	Ni+Cu+Co	Mn:Fe	Sample Type
DR2-1	1°52.39'	146°04.26'	1100-1400	17.00	17.60	0.21	0.02	0.06	0.30	0.53	0.97	3mm crust on basalt
DR3-2	1°53.92'	145°56.89'	1100-1250	49.00	0.75	0.02	0.005	0.01	0.005	0.03	65.33	30 mm crust on volcanics
DR6A-4	2°27.68'	147°41.16'	1000-1500	15.40	19.70	0.23	0.03	0.07	0.07	0.33	0.78	4 mm crust on andesite
PN79-1 (22A)	2°43.06'	152°50.08'	1000-1100	2.83	8.88	0.03	0.01	0.02	0.04	0.08	0.40	5 mm crust on foram-bearing sediments
PN79-1 (24A)	4°48.02'	154°25.07'	838-900	13.30	14.80	0.29	0.04	0.07	0.23	0.56	0.90	1-2 mm crust on volcanic sandstone
PN79-1 (24An)	4°48.02'	154°25.07'	838-900	17.30	14.30	0.38	0.04	0.07	0.31	0.73	1.21	Smooth nodule (7x5x3 cm)

The analyses for Ni, Cu and Zn were performed at La Trobe University using XRF; the analyses for Mn, Fe and Co were carried out by the Australian Mineral Development Laboratories using atomic absorption methods.

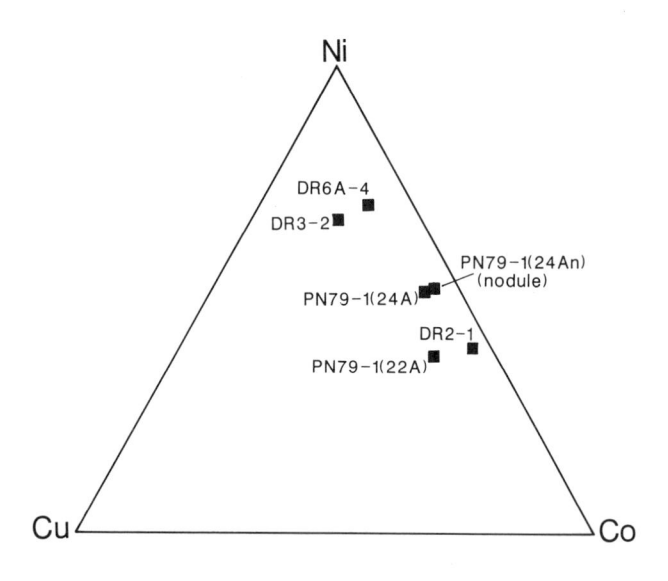

Figure 3. Triangular plot of variations of proportions of Ni, Cu and Co. Values are percentages where Ni+Cu+Co = 100 %.

is that crusts show an average crustal thickness of 2 cm or greater, an areal coverage of 40%, and an average cobalt content of 0.9% and combined Ni+Cu+Co content of 1.46% (Clark, Johnson and Chinn, 1984).

Criteria for assessing the economic potential of a ferromanganese nodule field are more clearly defined, and generally include a prerequisite minimum grade (Ni+Cu+Co) of more than 2% and a minimum concentration of about 10 kg "wet" nodules/m^2 (Seibold, 1978).

No information is available about crust coverage or nodule abundance, but the six samples analyzed do provide some information on grades in the region. Only one crust sample (DR3-2) meets the required 2 cm average thickness. This overall thinness, together with a maximum grade of 0.73% for combined Ni+Cu+Co for the single nodule, and the 0.3% Co and 0.56% combined Ni+Cu+Co maximum for crust samples, suggest little or no economic potential based on this small amount of sampling.

The analyses of metals from four of the samples (DR2-1, DR6A-4, PN79-1(24A), PN79-1(24An)) are similar to one another, and quite different from the other two samples (Table 1). PN79-1(22A) has very low metal values, which can be ascribed to a major admixture of pelagic sediment in the crust.

When the average metal analyses from the four samples are compared with selected regional averages, they can be seen to be typical hydrogenetic bathyal deposits (Table 2). Metal contents are almost identical to average concentrations in ferromanganese deposits from similar water depths on

Table 2. Metal concentrations in ferromanganese crusts and nodules from selected locations.

Reference	Area	Latitude	Water depth (m)	No. of samples	Mn	Fe	Ni	Cu	Co
This study[1]	New Ireland	2-5°S	840-1500	4	15.7	15.9	0.28	0.03	0.03
Hinz et al. (1978)	Scott Plateau	13°S	2100-2300	3	18.7	18.6	0.38	0.04	0.31
von Stackelberg et al.(1980)	Exmouth & Wallaby Plateau crusts	16-25°S	2500-5300	30	14.1	15.4	0.30	0.12	0.18
Cronan (1977)	Seamounts	-	-	15	14.6	15.8	0.35	0.06	1.15
Cronan (1972)	W. Pacific crusts	-	abyssal	-	16.9	13.3	0.56	0.39	0.40

[1] Average of four samples : DR2-1, DR6A-4, PN79-1(24A), PN79-1(24An)

Sample DR3-2 excluded as hydrothermal, and PN79-1(22A) excluded as heavily contaminated with sediment.

the Scott Plateau off northwest Australia, and to averages from deeper water deposits on the Scott Plateau and on the Exmouth Plateau further south. In abyssal areas of the western Pacific, Ni values in nodules average about twice as much, and Cu values about ten times as much, as in the New Ireland region. These differences accord with the changes in water depth.

When the New Ireland analyses are compared with those from seamounts worldwide, it is apparent that Co values are much lower. Halbach, Manheim and Otten (1982) have suggested that high Co values for crusts of the central Pacific appear to be related to the oxygen minimum zone developed at bathyal depths. The low Co values in bathyal crusts and nodules from the New Ireland and western Australian regions may be related to the lack of an oxygen minimum zone.

Sample DR3-2 shows extreme fractionation of Fe from Mn and a low concentration of trace metals (Table 1), and is similar in bulk chemistry to some crusts sampled from the TAG region of the Mid-Atlantic Ridge. These features, together with this sample's remarkably close physical resemblance to hydrothermal crusts recently described by Cronan et al. (1982) from the flanks of the Tonga-Kermadec Ridge in the Southwest Pacific Ocean, suggest a similar origin by precipitation of manganese oxide from a submarine hydrothermal source. The tentative identification of hydrothermal manganese oxide crusts is particularly significant, in that similar deposits in the Red Sea and the Galapagos Rift are associated with earlier-formed polymetallic sulfides of possible economic value (Cronan, 1980). On this basis, the area lying to the northwest of Manus Island warrants further exploration for submarine polymetallic sulfide deposits.

CONCLUSIONS

1. Although the density of sampling is low, the metal grades of hydrogenetic material determined by this study are low, and comparable to those from other bathyal hydrogenetic crusts and nodules in the Australasian region. Thus, it is unlikely that economic crust or nodule fields will be present.
2. The presence of material of probable hydrothermal origin does, however, suggest that further sampling is warranted, particularly in the region northwest of Manus Island, in the search for submarine polymetallic sulfide deposits.

ACKNOWLEDGMENTS

M. Marlow is thanked for his comments on an earlier version of this manuscript. The comments of R. Koski and J. Hein greatly improved the manuscript and are gratefully acknowledged. The help of Ian McCabe (sample preparation and analysis) and Judi Bolton and Joan Brushett (typing) is also acknowledged. Exon publishes with the permission of the Director, Bureau of Mineral Resources.

REFERENCES

Both, R., Crook, K., Taylor, B., Brogan, S., Chappell, B., Frankel, E., Liu, L., Sinton, J., and Tiffin, D., 1986, Hydrothermal chimneys and associated fauna in the Manus back-arc basin, Papua New Guinea: EOS, v. 67, no. 21, p. 489-490.

Clark, A., Johnson, C.J., and Chinn, P.J., 1984, Assessment of cobalt-rich manganese crusts in the Hawaiian, Johnston and Palmyra Islands Exclusive Economic Zones: United Nations, New York, National Resources Forum, v. 8, no. 2, p. 163-174.

Commeau, R.F., Clark, A., Johnson, C., Manheim, F.T., Aruscavage, P.J., and Lane, C.M., 1984, Ferromanganese crust resources in the Pacific and Atlantic Oceans: Oceans '84, Proceedings, p. 421-430.

Cronan, D.S., 1972, Regional geochemistry of ferromanganese nodules in the world ocean, in Horn, D.R., ed., Papers from a Conference on ferromanganese deposits on the ocean floor: Office of the International Decade of Ocean Exploration, National Science Foundation, Washington, D.C., p. 19-29.

Cronan, D.S., 1977, Deep-sea nodules, distribution and geochemistry, in Glasby, G.P., ed., Marine Manganese deposits: Amsterdam, Elsevier, p. 11-44.

Cronan, D.S., 1980, Underwater minerals: London, Academic Press, 362 p.

Cronan, D.S. and Thompson, B., 1978, Regional geochemical reconnaissance survey for submarine metalliferous sediments in the SW Pacific: Transactions Institute Mining Metallurgy Section B., v. 87, p. 87-89.

Cronan, D.S., Glasby, G.P., Moorby, S.A., Thomson, J., Knedler, K.E., and McDougall, J.C., 1982, A submarine hydrothermal manganese deposit from the south-west Pacific island arc: Nature, v. 298, no. 5873, p. 456-458.

Eade, J.V., 1980, Cruise Report, Papua New Guinea Offshore Survey Cruise PN-79(1), 10th-16th March, 1979: Cruise Report No. 25, CCOP/SOPAC, Suva, 10 p.

Halbach, P., Manheim, F.T., and Otten, P., 1982, Co-rich ferromanganese deposits in the marginal seamount regions of the Central Pacific Basin - results of the Midpac '81: Erzmetall, v. 35, no. 9, p. 447-453.

Halbach, P., and Manheim, F.T., 1984, Potential of cobalt and other metals in ferromanganese crusts on seamounts of the Central Pacific Basin: Marine Mining, v. 4, no. 4, p. 319-336.

Hein, J.R., Morgenson, L.A., Clague, D.A. and Koski, R.A., 1987, Cobalt-rich ferromanganese crusts from the Exclusive Economic Zone of the United States and nodules from the oceanic Pacific, in Scholl, D.W., Grantz, A., and Vedder J.G., eds., Geology and resource potential of the continental margin of western North America and adjacent ocean basins - Beaufort Sea to Baja California, Circum-Pacific Council for Energy and Mineral Resources Earth Science Series, v. 6: Houston, Texas, Circum-Pacific Council for Energy and Mineral Resources, [in press].

Hinz, K., Beiersdorf, H., Exon, N.F., Roeser, H.A., Stagg, H.M.J. and von Stackelberg, U., 1978, Geoscientific investigations from the Scott Plateau off northwest Australia to the Java Trench: Bureau of Mineral Resources Journal of Australian Geology and Geophysics, v. 3, p. 319-340.

Manheim, F.T., and Hein, J.R., 1986, Ferromanganese crusts: Exclusive Economic Zone Symposium Exploring the New Ocean Frontier, Oct. 2nd-3rd, 1985, Proceedings, p. 79-95.

Marlow, M., Exon, N.F., and Shipboard Party, 1984, Initial Report on 1984 R/V *S.P. Lee* Cruise L7-84-SP in Northern Papua New Guinea: CCOP/SOPAC Cruise Report No. 95, Suva, 20 p.

Seibold, E., 1978, Deep sea manganese nodules - the challenge since "Challenger": Episodes, no. 4, p. 3-8.

von Stackelberg, U., Exon, N.F., von Rad, U., Quilty, P., Shafik, S., Beiersdorf, H., Seibertz, E., and Veevers, J.J., 1980, Geology of the Exmouth and Wallaby Plateaus off northwest Australia: sampling of seismic sequences: Bureau of Mineral Resources Journal of Australian Geology and Geophysics v. 5, p. 113-140.

Marlow, M.S., Dadisman, S.V., and Exon, N.F., editors, 1988, Geology and offshore resources of Pacific island arcs—New Ireland and Manus region, Papua New Guinea, Circum-Pacific Council for Energy and Mineral Resources Earth Science Series, v. 9: Houston, Texas, Circum-Pacific Council for Energy and Mineral Resources.

OFFSHORE STRUCTURE AND STRATIGRAPHY OF NEW IRELAND BASIN IN NORTHERN PAPUA NEW GUINEA

M. S. Marlow,
U.S. Geological Survey, Menlo Park, California 94025, USA

N. F. Exon,
Bureau of Mineral Resources, Geology and Geophysics, Canberra, A.C.T. 2601, Australia

H. F. Ryan, S. V. Dadisman
U.S. Geological Survey, Menlo Park, California 94025, USA

ABSTRACT

The arcuate, northwest-trending New Ireland Basin lies offshore northeast of the islands of New Hanover and New Ireland and is 600 km long and about 150 km wide. Multichannel seismic-reflection and refraction data show that much of the basin is a simple structural downwarp, which formed as a fore-arc basin northeast of an Eocene to early Miocene volcanic arc. The basin contains about 7 km of sedimentary rocks consisting of lower Miocene and possibly Eocene volcaniclastic rocks, lower to upper Miocene shelf carbonate, upper Miocene and Pliocene bathyal chalks and volcaniclastic sediment, and Pleistocene and Holocene sediment ranging from terrestrial conglomerates to hemipelagic oozes. In the eastern part of the basin, Pliocene and Quaternary volcanism has formed islands and has greatly disrupted the older strata.

New Ireland Basin is part of a tectonically complex zone between the Pacific and Australia-India plates. Small microplates have been proposed in the Bismarck Sea and one model would place the New Ireland Basin within the so-called North Bismarck plate, although other models suggest that the basin is now part of the Pacific plate. New Ireland Basin formed as a forearc basin in the early Tertiary and was filled in part by sediment eroded from a nearby volcanic island arc that today includes the islands of New Ireland and New Hanover. Sedimentation and carbonate formation have continued to fill the basin through the Neogene and Quaternary. Volcanism on New Ireland stopped after the Pliocene but picked up elsewhere in the Pliocene in the Tabar-Feni Islands, uplifting the northeastern section of the New Ireland Basin into a series of horsts capped by islands.

No hydrocarbon exploration wells have been drilled into the basin. A possible lack of source rocks and low thermal gradients may nullify the basin's petroleum potential. However, the thick shelf limestones onshore and the widespread and deeply-buried Miocene shelf limestones and carbonate build-ups offshore, suggest that the basin warrants further exploration. The shelf carbonate deposits are up to 2,000 m thick and could be both petroleum source and reservoir rocks. At least one anticlinal structure exists offshore and the anticline could be a structural trap for hydrocarbons. Fine-grained volcanogenic silts may also exist offshore and their impermeability could provide seals to trap migrating hydrocarbons.

Multichannel seismic-reflection profiles collected in 1984 by the Tripartite (Australia, New Zealand, and USA) CCOP/SOPAC cruise of the R/V *S.P. Lee* traversed the axis of the New Ireland Basin east of New Ireland. Line 401, collected just southeast of Namatanai, New Ireland, revealed a flat, high-amplitude reflector or "bright spot" within the core of an anticline some 20 km east of New Ireland. On line 401, the bright spot is about 2 km wide and occurs 1.2 s (1,700 to 1,800 m) beneath the sea-floor in water depths of 2,500 to 2,600 meters. Thus, the anomalous section may have been buried deeply enough to generate hydrocarbons, but in water too deep at present for commercial development. The western extent of the anticline and associated bright spot in the shallower waters of New Ireland is unknown.

INTRODUCTION

New Ireland Basin is a 600 km long, 150 km wide sedimentary basin that is exposed on the islands of New Ireland and New Hanover and that is flanked to the north and east by the Manus and Kilinailau Trenches and to the south by the 2,500 m deep Manus Basin adjacent to island of New Ireland (Figure 1). Exon and Tiffin (1984) originally defined the basin as extending from the Feni Islands in the east to Manus Island in the west; however, our seismic-reflection data to the west near Manus Island (Marlow, Exon, and Tiffin this volume) show that the basin probably does not extend much farther west than the island of New Hanover (Figure 1). The basin is a gentle downwarp, sloping to the northeast away from exposures on New Ireland, and rising again toward an outer arc high - the Emirau-Feni Ridge - a volcanic arc that extends parallel and north of New Ireland and includes the Mussau, Tabar, Lihir, and Feni island groups (Figure 1; Exon and Tiffin, 1984). Axial water depths of New Ireland Basin increase from around 1,400 m in the northwest near New Hanover Island to 3,000 m in the southeast near the Feni Islands. The southwest flank of the basin is truncated by coastal and offshore faults on the southwest side of New Ireland that step down more than 2,000 m into Manus Basin, a back-arc basin (Connelly, 1976; Taylor, 1979). Uplifted basinal strata are exposed on both New Ireland and New Hanover islands.

Based on offshore sonobuoy refraction measurements and correlation to units exposed on New Ireland, New Ireland Basin contains about 7 km of sedimentary strata of Eocene(?) and younger age (Childs and Marlow, this volume). Exon and Tiffin (1984) state that the basin formed as a forearc basin initially in Eocene time in response to underthrusting along a trench that lay north of the basin. The underthrusting created an ancient island arc, the remains of which underlie the islands of New Ireland and New Hanover. The northeastern side of the basin was intruded and uplifted mainly in the Pliocene and Quaternary by volcanoes in the Tabar to Feni groups of islands along the Emirau-Feni Ridge, a ridge corresponding roughly to the 2,000 m isobath in Figure 1. The ridge uplifts the older basin fill and has confined deposition in the basin since the Pliocene to the axial trough between the Emirau-Feni Ridge and New Ireland.

Earlier studies of New Ireland Basin are summarized by Exon and Tiffin (1984) and Exon et al. (1986). Their work was based on seismic-reflection studies carried out by CCOP/SOPAC, the IFP-CEPM-ORSTOM group (de Broin et al., 1977), and unpublished surveys by Gulf Research and Development Company. This report extends their work and includes the analysis of seismic-reflection profiles collected in 1984 by the Tripartite CCOP/SOPAC cruise to offshore Papua New Guinea of the U.S. Geological Survey's R/V *S.P. Lee*. We present these new data and updated interpretations of older data to help assess the hydrocarbon potential of New Ireland Basin, a frontier basin that is undrilled.

SEISMIC-REFLECTION PROFILES AND STRATIGRAPHY

Seismic Interpretation

In our study of New Ireland Basin, we interpreted all the seismic-reflection profiles east and north of New Ireland shown in Figure 1, and representative sections of these profiles are shown in Figures 2-9. Distinguishing characteristics of major seismic horizons and sequences on the profiles are shown in Figure 2 and are listed in Table 1, both of which are modified from Exon et al. (1986). The sequences listed in Table 1 were first defined by Exon and Tiffin (1984) and consist largely of carbonate and volcaniclastic sediment, except for the lowest (E-V) sequence, which contains mainly volcanic rocks. The upper 2-3 meters of sequence SF-A have been cored and found to consist of calcareous ooze (Carlson et. al, this volume). The A-B and B-C sequences have been sampled by dredging just west of the Tabar Islands and found to consist of upper Miocene, lower Pliocene, and upper Pliocene and lower Pleistocene sedimentary rocks, including white to gray marls, indurated white calcilutite and interbedded calcareous mudstone and volcanogenic sandstone. The A-B and B-C sequences prograde and thin to the northeast from New Ireland and New Hanover (Figures 10 and 11). Based on the seismic-reflection characteristics described in Table 1 and the

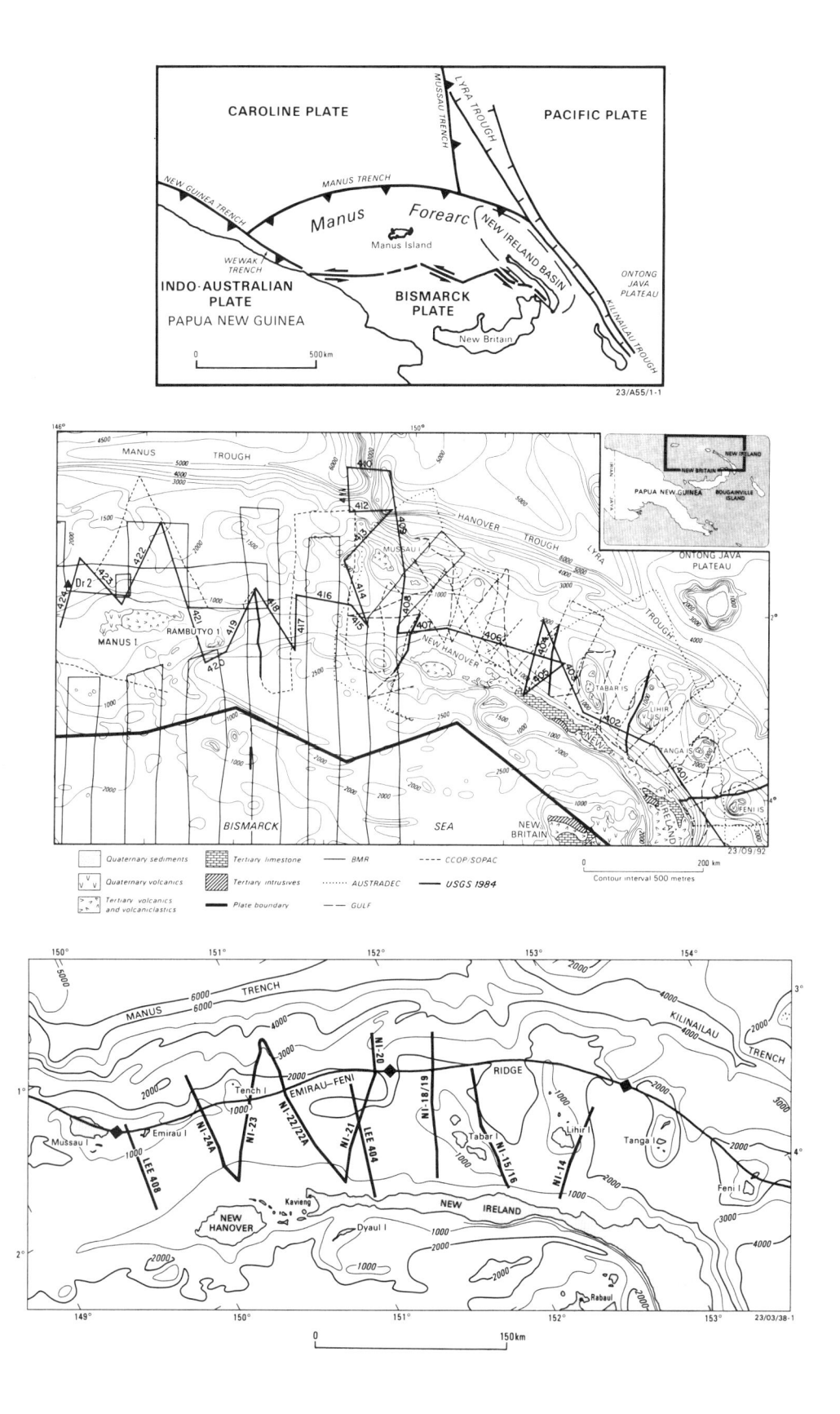

Figure 1. Index map of New Ireland Basin region, Papua New Guinea, and trackline map showing location of geophysical lines; bathymetry from Taylor (1979).

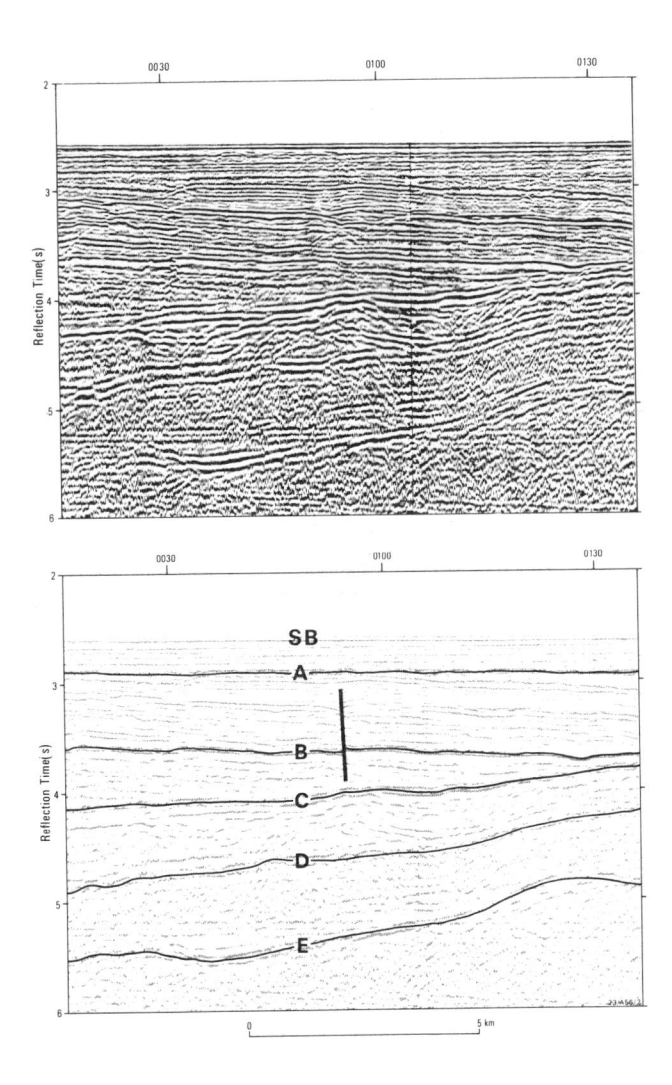

Figure 2. Part of multichannel seismic-reflection profile NI-20 with interpretation in lower panel. Location of line is shown in Figure 1. From Exon et al. (1986).

dredging results, the A-B sequence may be predominantly volcaniclastics equivalent to the Rataman Formation on New Ireland and the B-C sequence may be mainly marls and chalks equivalent to the Punam Limestone of New Ireland.

According to Exon and Tiffin (1984) and Exon et al. (1986), the C-D sequence has an interval velocity of 3,600 m/s based on stacking velocities from multichannel seismic-reflection data. They describe the seismic-reflection characteristics of the C-D unit as mixed, with weak to strong parallel reflectors, dipping reflectors, and what they term buildups or areas of complex and curved reflection events (Figure 2, Table 1). They interpret the unit as a carbonate platform sequence equivalent to the lower and upper Miocene Lelet Limestone exposed on New Ireland.

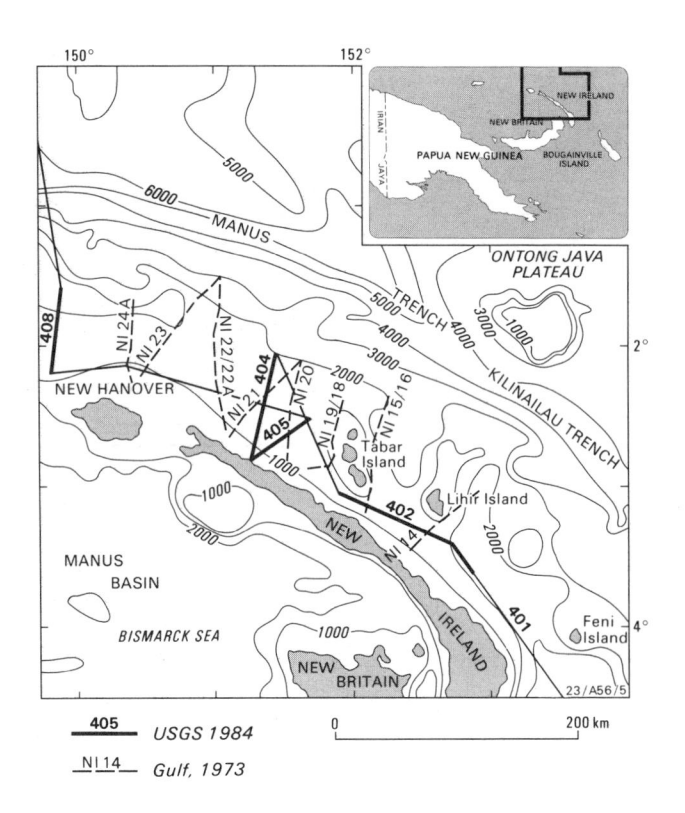

Figure 3. Index map showing location of *Lee* seismic-reflection profiles used in this report.

The buildup features are interpreted by Exon and Tiffin (1984) and Exon et al. (1986) as reefs, platform and back-reef deposits, and fore-reef rubble. They suggest that the carbonate deposits could be prospective source and reservoir beds for hydrocarbons (see Exon and Marlow, this volume). Onshore, the base of the Lelet Limestone is time-transgressive (Exon et al., 1986). Refraction velocities within the C-D sequence average 3,500 m/s over a range of 2,700 to 4,700 m/s for 10 refraction stations over the New Ireland Basin (Childs and Marlow, this volume). The average is within the range measured for compressional velocities in limestone (Clark, 1966).

The deepest resolvable sequence, D-E, is variably reflecting, moderately well-bedded and locally contains dipping reflectors, channels, and buildups (Table 1). Exon and Tiffin (1984) and Exon et al. (1986) correlate the sequence with the marine Lossuk River beds of New Ireland, which stratigraphically underlie the Lelet Limestone (Stewart and Sandy, this volume).

The section below reflector E, listed as E-V in Table 1, is variably reflecting, poorly defined and at the limit of acoustic penetration, with no discernible lower boundary in most places (Figure 2). The reflector E is a low frequency event exhibiting many

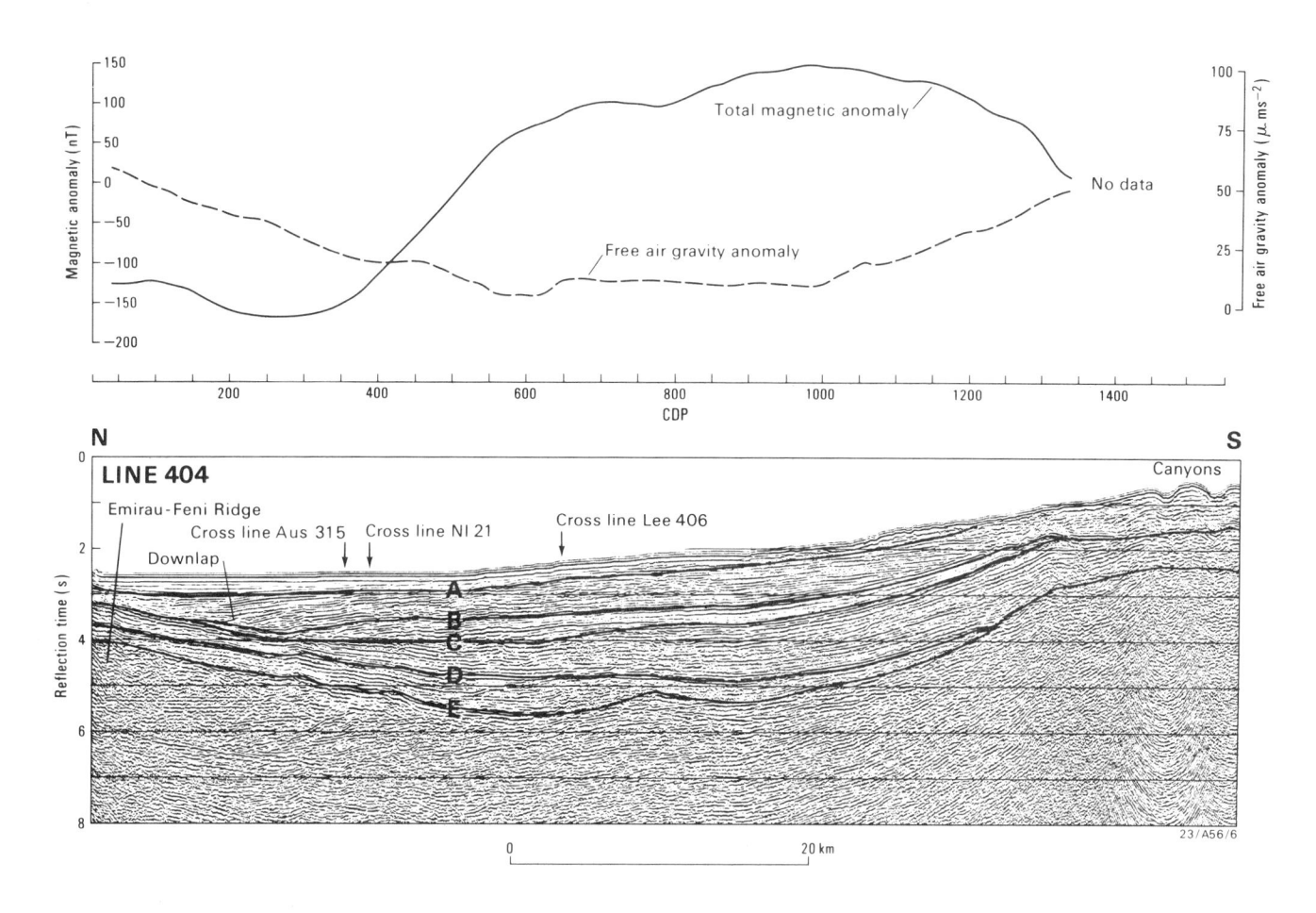

Figure 4. Migrated multichannel seismic-reflection, magnetic, and gravity profiles on line 404 across New Ireland Basin. See Figure 1 for location. Travel time is two-way in seconds. CDP refers to Common Depth Points spaced at 50 m along the line. Reflector B downlaps onto reflector C at the northern end of the line.

diffractions and overlain unconformably by the D-E sequence. Exon and Tiffin (1984) and Exon et al. (1986) correlate the section below reflector E to the Eocene to lower Miocene volcanic and volcaniclastic rocks of the Jaulu Volcanics exposed on nearby islands.

Lines 404 and 405

Lines 404 and 405 traverse the central part of New Ireland Basin near the northern end of New Ireland (Figure 1) and exhibit features typical of the basin. The sea floor slopes gently seaward from New Ireland, and the basin fill thickens near the center of both lines and thins toward the Emirau-Feni Ridge to the northeast (Figures 4 and 5).

The sea floor is incised by two canyons at the southern end of line 404, which are probably related to erosion on nearby New Ireland (Figure 4). On this line, the A-B sequence downlaps onto reflector B near the northern end, suggesting progradation of this sequence to the northeast away from New Ireland. The underlying B-C sequence is lenticular in shape and also progrades away from New Ireland. The C reflector flattens at the southern end of the line, perhaps delineating the top of an eroded and now-buried carbonate platform. The C-D section is thickest near the center of the profile, and this sequence remains thick near New Ireland, whereas the underlying section D-E thins and is truncated toward New Ireland. We interpret the geometry of these units as showing that the C-D sequence was derived mainly from New Ireland and progrades away from the island, whereas the older, underlying D-E section was derived from the north, probably from an ancient outer forearc high.

Along line 404 the gravity profile decreases to a minimum near the center of the profile where the low density sedimentary section is thickest (Figure 4). The broad magnetic anomaly profile suggests that the anomaly sources are deep and beneath the basin

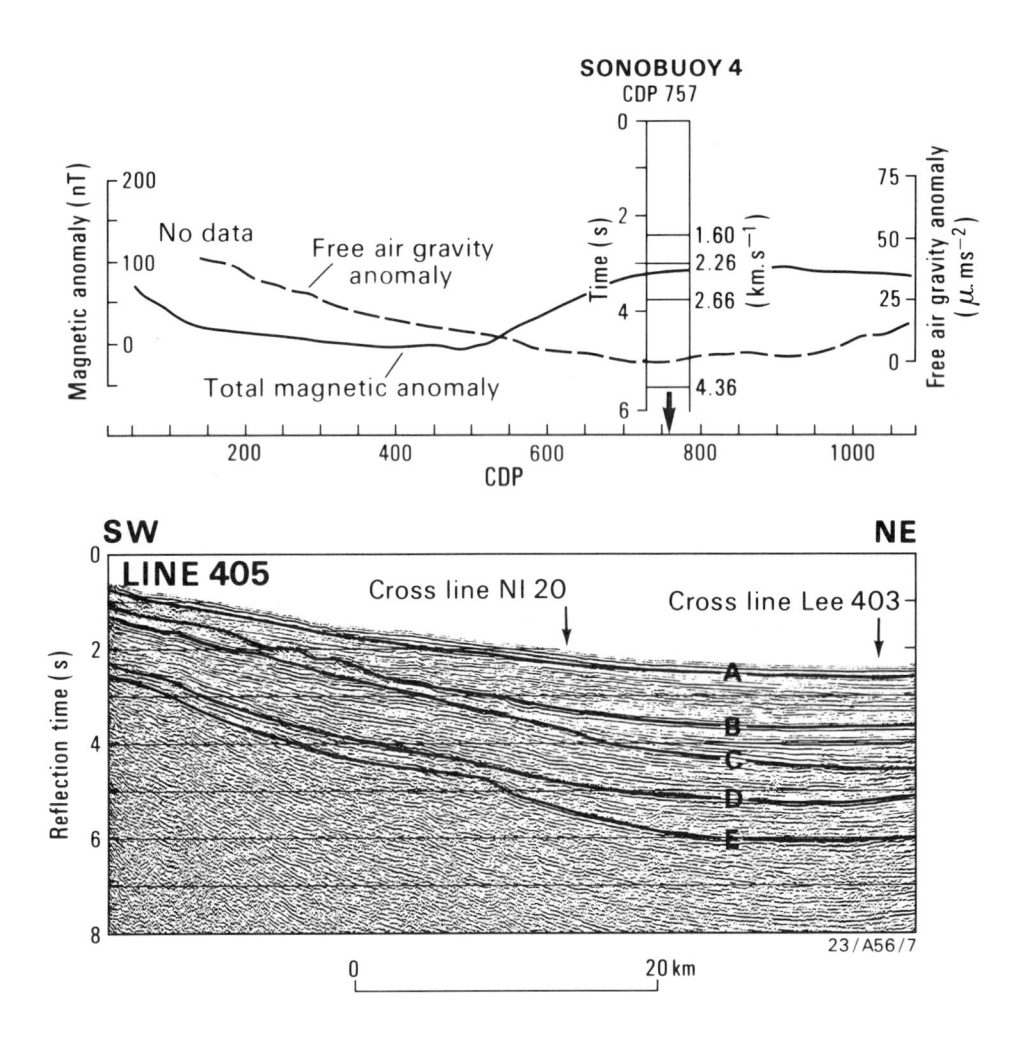

Figure 5. Migrated multichannel seismic-reflection, magnetic, and gravity profiles on line 405 across New Ireland Basin. Location and annotation as in Figure 3.

fill, possibly related to a magnetic basement beneath the basin.

The seismic-reflection sequences on line 405 (Figure 5) are similar to those on the adjacent line 404, except that sequence A-B thickens markedly away from New Ireland and into the basin (Figure 5). Near the southwestern end of the line, reflector B shows hummocky relief, which could be the tops of buried, relict slump structures or the tops of carbonate buildups. The gravity profile on line 405 reaches a minimum over the thickest basin section. Here sonobuoy refraction data show thick sediment bodies with velocities of 2,260 m/s in sequence A-B, 2,660 m/s in sequence B-C, and 4,360 m/s in sequence D-E (Figure 5; see Childs and Marlow, this volume). The magnetic profile on line 405 is characterized by low frequency and long wavelength anomalies. Both lines 404 and 405 are similar to the Gulf line NI 20 (Figures 1 and 9) described by Exon and Tiffin (1984).

Lines 401 and 402

Lines 401 and 402 were run as tie lines to connect the Gulf lines previously interpreted by Exon and Tiffin in 1984 (Figure 1). On both lines, the topography of the sea floor reflects underlying structures in the basin fill (Figures 6 and 7). The gravity profile mimics the underlying basement structure outlined by reflector E. Near CDP's 2,100 and 3,000 on line 401 the basin fill is broadly folded into anticlines some 20 km across that bow up the sea floor. Reflector B laps onto reflector C on the flanks of the western anticline (beneath CDP 3,000), suggesting that this structure has existed as a high since the late Miocene or earliest Pliocene (Table 1). The magnetic anomaly centered near CDP 2,700 on line

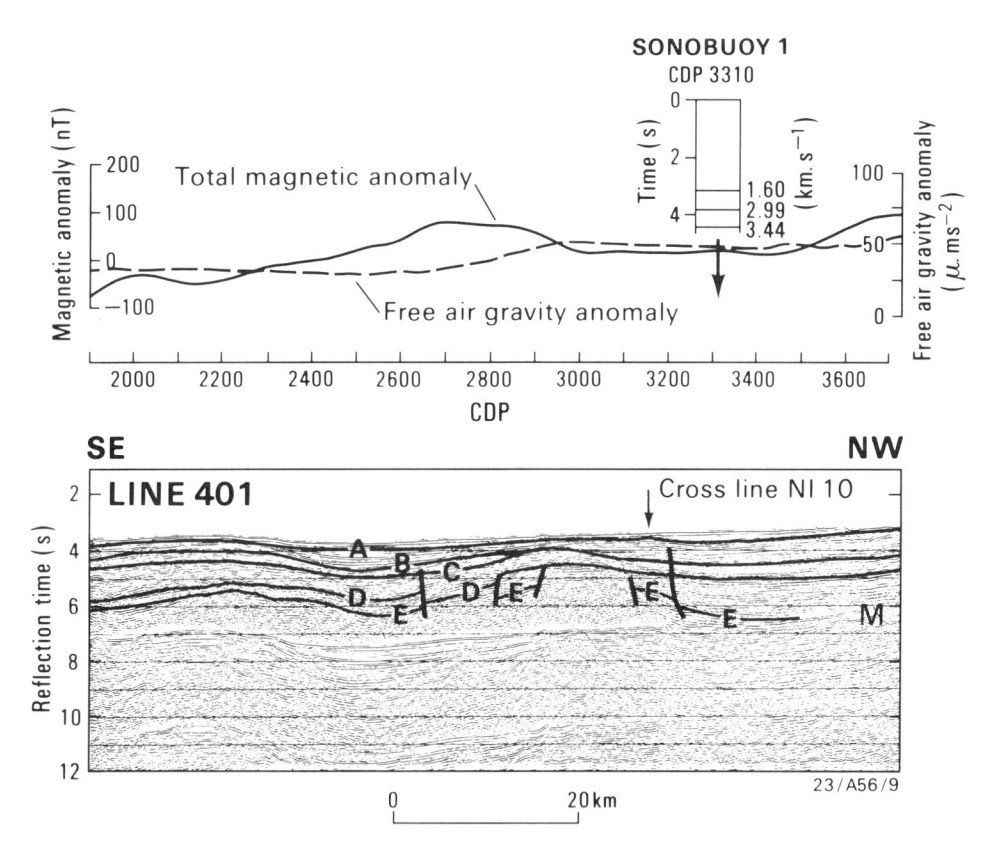

Figure 6. Migrated multichannel seismic-reflection, magnetic, and gravity profiles on line 401 over the New Ireland Basin. Location and annotation as in Figure 3. The figure displays only the northwestern portion of line 401. A prominent anticline occurs near CDP 3,000.

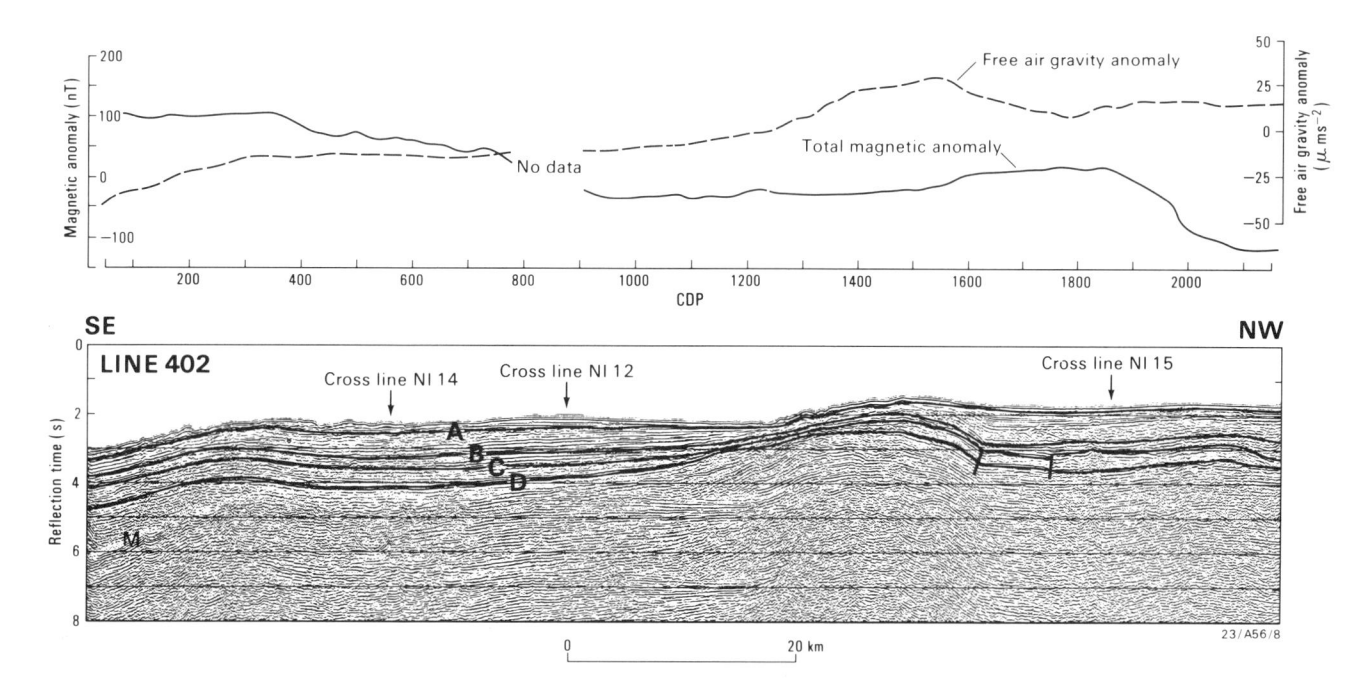

Figure 7. Migrated multichannel seismic-reflection, magnetic, and gravity profiles on line 402 across New Ireland Basin. Location and annotation as in Figure 3. A broad anticline arches the sea floor near CDP 1,400.

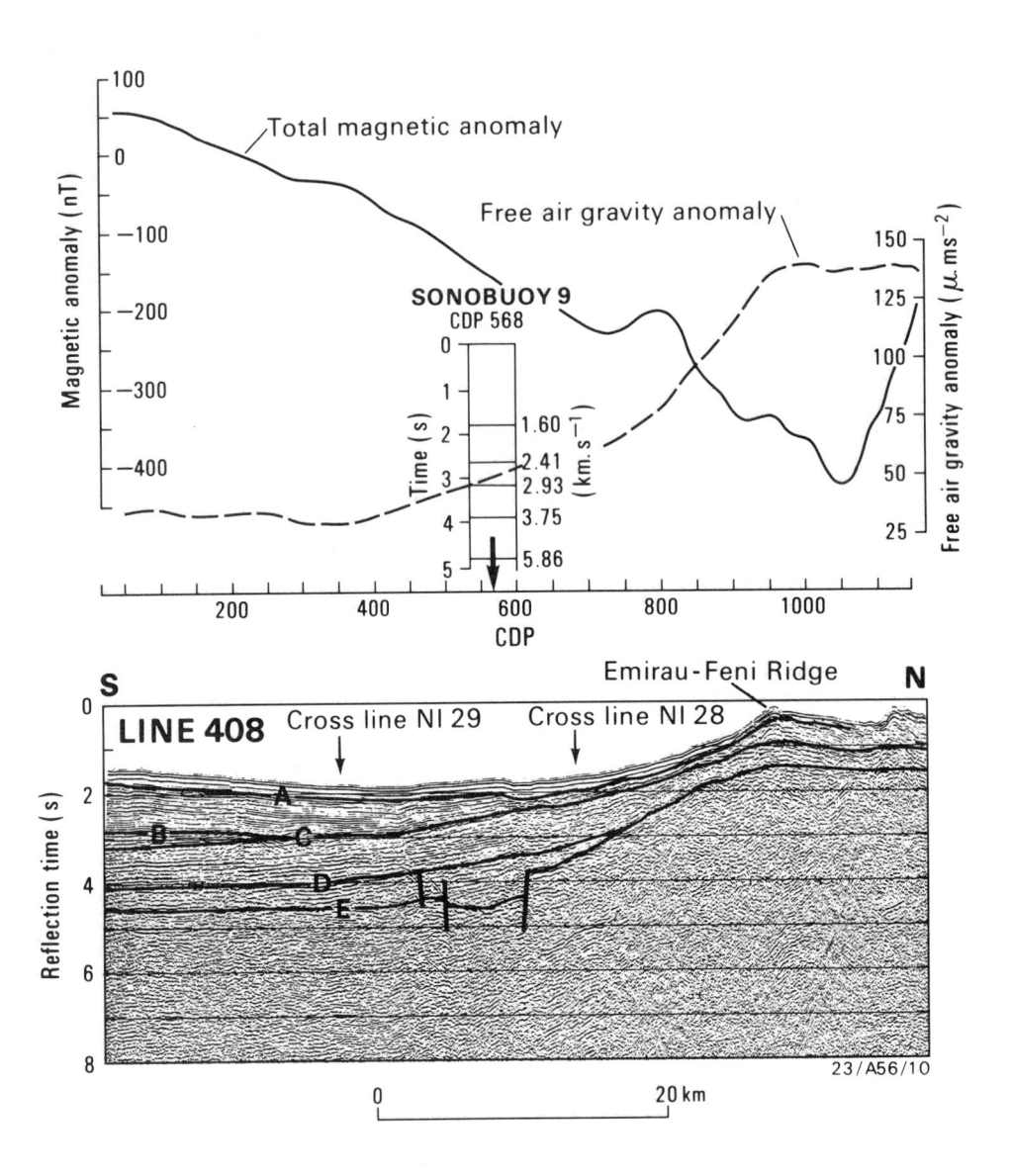

Figure 8. Migrated multichannel seismic-reflection, magnetic, and gravity profiles on line 408 across New Ireland Basin. Location and annotation as in Figure 3. Most of the basin fill thins and onlaps to the north onto the basement of the Emirau-Feni Ridge.

401 is offset from the western anticline, indicating that the anticline probably doesn't have a volcanic core. A flat, high-amplitude reflector or "bright spot" occurs within the core of the anticline at a depth of 4.6 seconds (two-way time). The bright spot is about 2 km wide and occurs 1.2 s (1,700 to 1,800 m) beneath the sea floor in water depths of 2,500 to 2,600 meters. The anomaly is in water too deep for commercial development at present (see Exon and Marlow, this volume). The western extent of the anticline and associated bright spot in the shallower waters off New Ireland is unknown.

The reflecting sequences C-D and D-E are offset by high angle faults on line 401 (Figure 6), indicating that the early tectonic history of New Ireland Basin involved extensive faulting. Sonobuoy results on line 401 yielded refraction velocities of 2,990 m/s along a refractor within sequence A-B and 3,440 m/s along the top of the C-D layer (Childs and Marlow, this volume).

Line 402 traverses the axis of the central part of New Ireland Basin (Figures 1 and 7). The entire basin fill is broadly folded into a faulted, anticlinal structure centered at CDP 1,500. The sequences A-B, B-C, and C-D thin over the southern flank of the anticline, indicating that the structure is at least as old as Miocene in age (Table 1). The anticline is marked by a positive free-air gravity anomaly of

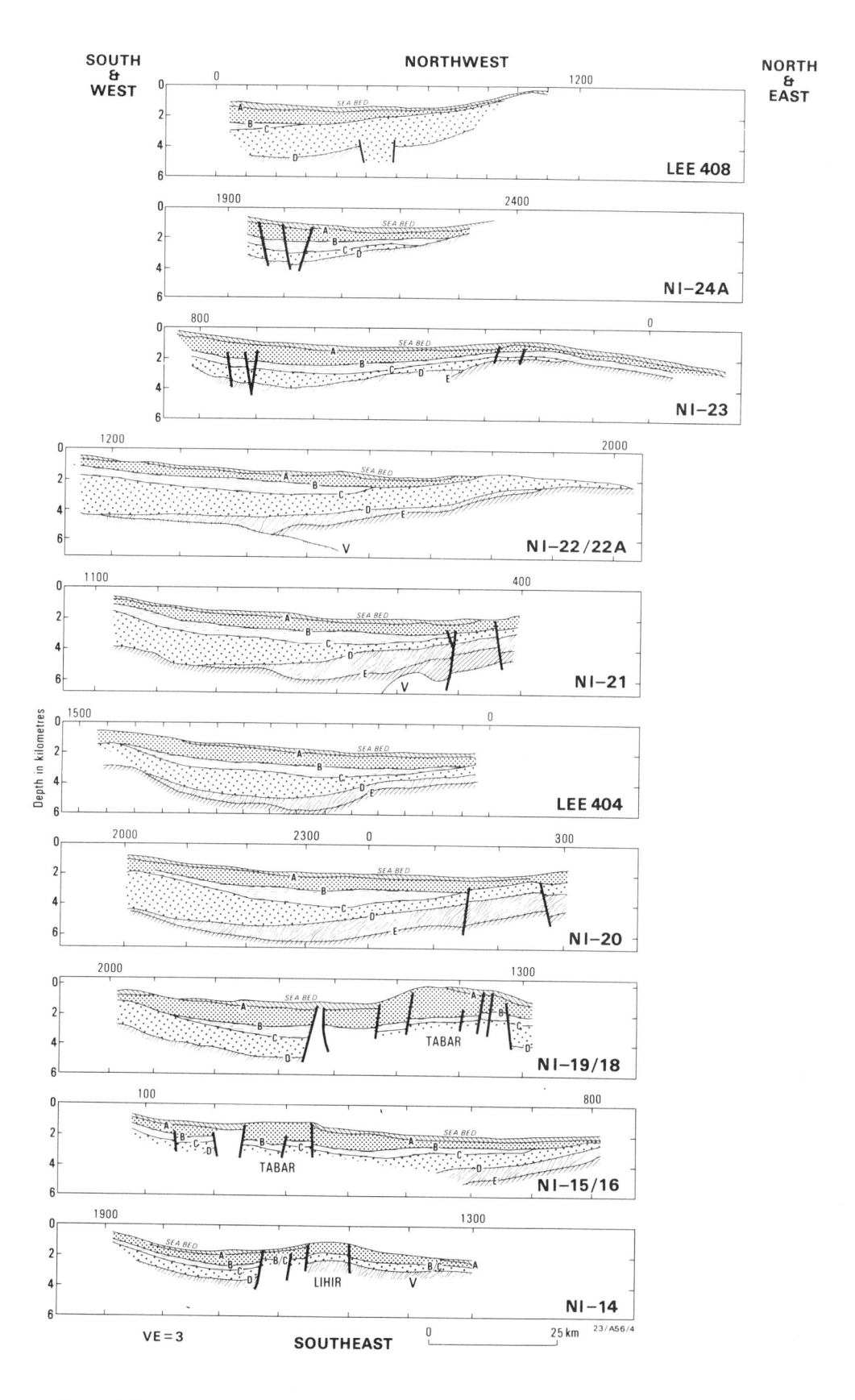

Figure 9. Composite of interpretative drawings of true depth sections based on seismic-reflection profiles across New Ireland Basin. See Figure 1 for location and Table 1 for description of the seismic units.

Table 1. Seismic sequences

Sequence (Velocity*)	Character	Interpretation
SF-A	Semi-transparent, well-bedded sequence immediately beneath sea bed (SB). Conformably overlies high-frequency reflector A, which in turn is unconformable on underlying sequence in places.	Pleistocene and Holocene hemipelagic oozes.
A-B (2270)	Well-bedded sequence with interbedding of strong reflectors and semi-transparent intervals. Progrades to the northeast. Conformably overlies shallowest continuous low-frequency reflector B that is conformable on underlying sequence.	Pliocene volcaniclastic turbidites interbedded with marls and chalks. Equivalent to the Rataman Formation on New Ireland.
B-C (2560)	Well-bedded sequence like A-B. Progrades northeastward and is unconformable on underlying low-frequency reflector (C).	Upper Miocene to lower Pliocene chalks and marls, which accumulated as Globigerina oozes, interbedded with volcaniclastic turbidites. Equivalent to Punam Limestone on New Ireland.
C-D (3520)	Mixed seismic character. Weak to strong parallel reflectors, dipping reflectors, and curved reflectors suggestive of carbonate buildups. Unconformably overlies continuous low-frequency reflector D, which in turn is generally conformable on underlying sequence.	Lower to upper Miocene platform and upper slope limestones equivalent to thick Lelet Limestone of New Ireland. Potential petroleum source and reservoir rocks.
D-E (4330)	Reflecting sequence with variable amplitude, locally containing dipping reflectors, and containing curved reflectors suggestive of channels and carbonate buildups. Unconformably overlies strong, low-frequency reflector E.	Lower Miocene outer shelf and slope sediment, largely volcaniclastic, with some channels and possible carbonate banks. Equivalent to marine Lossuk River beds on New Ireland; potential petroleum source rocks.
E-V (5560)	Variably reflecting, poorly-defined sequence at limit of acoustic penetration. Unconformably overlies high amplitude, diffracting, and irregular reflector.	Probably Eocene to lower Miocene volcanics and volcaniclastics equivalent to Jaulu Volcanics on nearby islands. Diffracting surface beneath E-V sequence may either be oceanic basement or an internal reflector within the Jaulu Volcanics.

* Velocity is average sonobuoy refraction velocity (see Childs and Marlow, this volume).

about 50 mgals, indicating uplift of basement rocks in the core of the anticline. The basin section is also broadly folded near the southeast end of the line. Amplitude anomalies or "bright spots" are not evident on true amplitude plots of these fold structures.

Line 408

Line 408 extends across the northwestern end of New Ireland Basin northwest of New Hanover Island (Figure 1). Our seismic-reflection lines recorded sedimentary sequences west of New Hanover Island, but the sequences B-C and D-E thin and pinchout to the north along line 408 (Figure 8). The reflection characteristics of the sedimentary sequences also change west of New Hanover Island on lines 414-416 (Figure 1), and the reflection signatures that define the sequences A-B, B-C, C-D, and D-E in the New Ireland Basin (Table 1) are no longer evident. Hence, the northwestern end of the New Ireland Basin lies just west of line 408, or west of New Hanover Island. Along line 408 we ran a sonobuoy over the basin that yielded refraction velocities of 2,670 m/s and 3,060 m/s in the C-D sequence, 3,750 m/s in the D-E sequence, and 5,860 m/s in the E-V or presumed volcanic basement beneath the basin (see Childs and Marlow, this volume). The top of the volcanic basement, reflector E, is offset in several places by high angle, normal faults.

At the northern end of line 408, the sea floor and the basement beneath reflector E shoals to form the Emirau-Feni Ridge. The volcanic nature of the

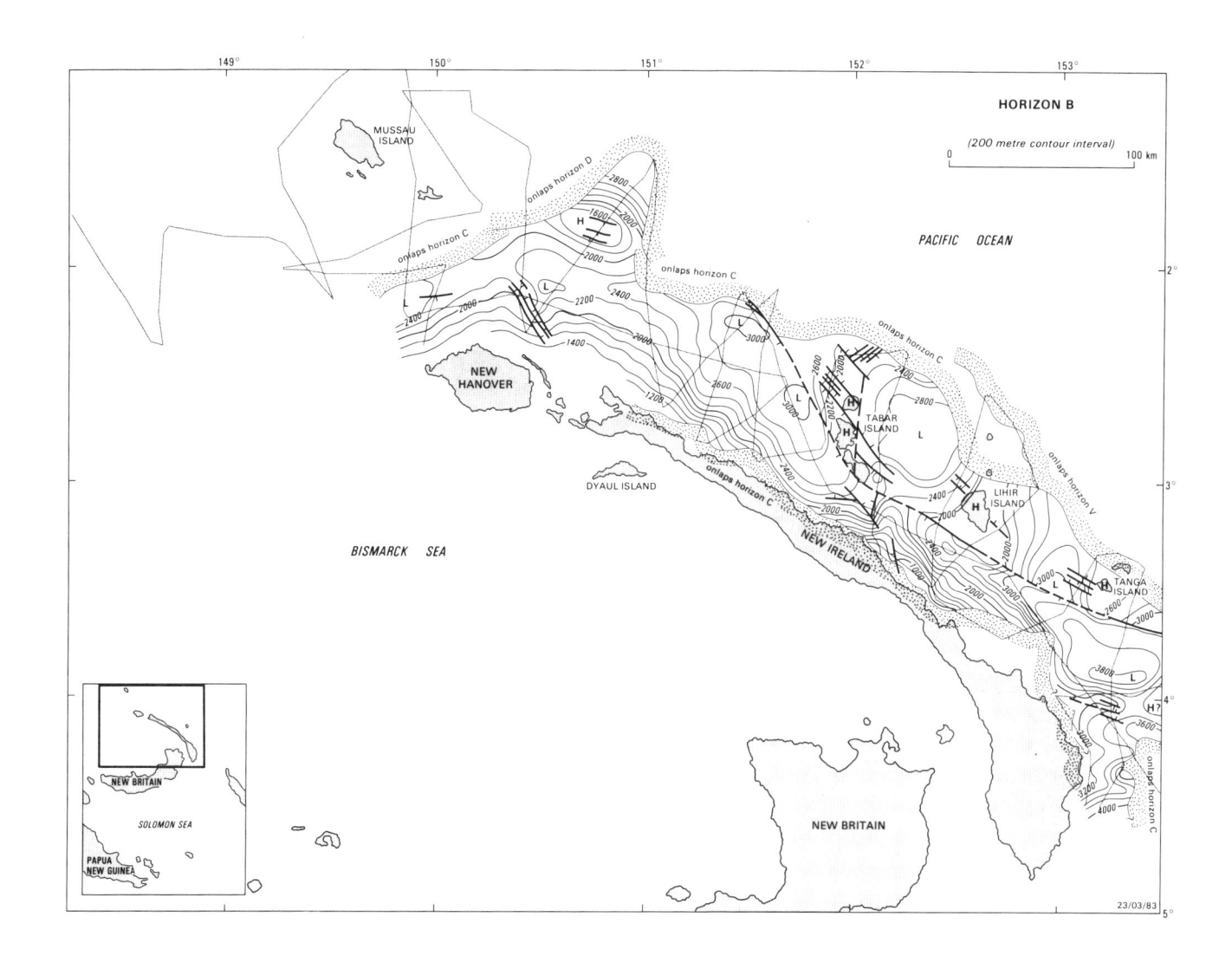

Figure 10. Structure contours in meters on Horizon B, New Ireland Basin, equivalent to top of early Pliocene to late Miocene Punam Limestone exposed on New Ireland. See text for derivation and discussion of contours in this and Figures 10-14. L = low, H = high.

ridge and the ridge's underpinnings is evident in the 75 nanoteslas (nT) negative magnetic anomaly and the positive 100 mgal gravity anomaly near the northern end of line 408 (Figure 8).

Composite Profiles

To show the variation in sediment thickness and structural style in the New Ireland Basin, we constructed a composite of depth-converted and interpreted cross sections. These sections are derived from seismic-reflection data and are shown from the northwest to the southeast (Figure 9). The stratified section thickens toward the center of the basin from the northwest and from the southeast, reaching a maximum of more than 5 km on lines 404 and NI-20 near Tabar Island. Most sequences slope gently to the northeast away from New Ireland. The overall thickening of the sequences occurs toward the basin axis, and the thinning of all the units but sequence D-E is toward the Emirau-Feni Ridge to the northeast.

The southeast profiles, NI-14 to NI-19, show the effects of Pliocene and Pleistocene volcanism, where the sections have been pushed up and exposed in the nearby Tabar-Lihir group of islands. Older sequences, such as C-D, have been faulted and uplifted, especially on Tabar Island, where Miocene carbonates are exposed (Wallace et al., 1983). The uplift of the older C horizon reaches 3,500 m around the volcanic islands of Tanga and Feni (Exon et al.,

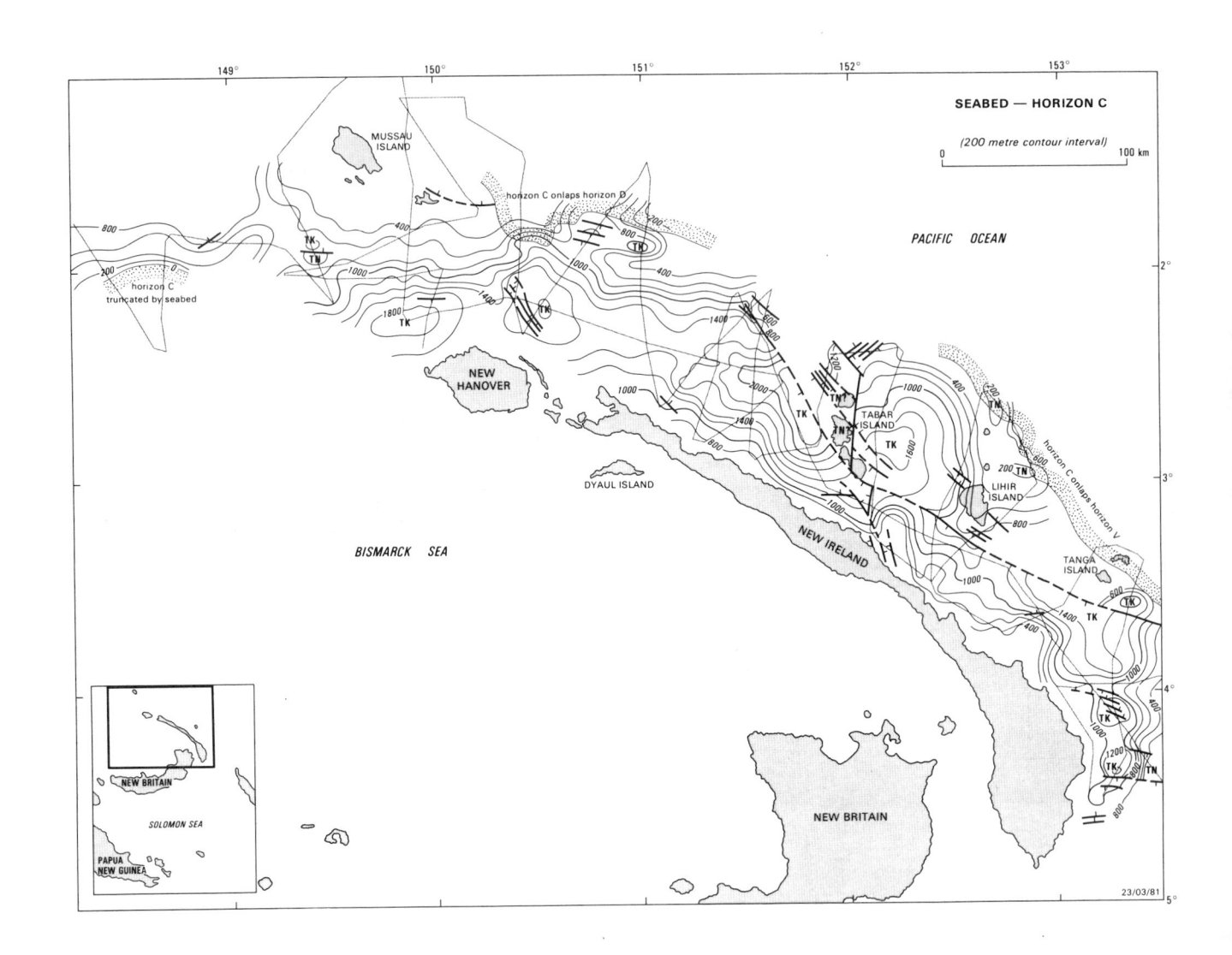

Figure 11. Isopach contours in meters for unit sea floor-Horizon C, New Ireland Basin, equivalent to sea floor to top of early to late Miocene Lelet Limestone. TK = thick, TN = thin.

1986).

Gulf line NI 20 extends north from New Ireland away from any volcanic islands in the Tabar-Feni chain, and the profile extends up on to the flanks of the Emirau-Feni Ridge (Figure 9). The basin fill is more than 5 km (4 s) thick and water depths reach a maximum of just over 2,000 m. The B-C sequence is lenticular in shape and the C-D sequence thickens toward New Ireland, whereas the D-E section thickens away from New Ireland. Exon and Tiffin (1984) reported suspected carbonate buildups in the C-D sequence at both ends of the line and possible buildups in the D-E sequence beneath the Emirau-Feni Ridge.

Gulf Profile NI 21 runs northeastward from the northern end of New Ireland and extends across New Ireland Basin (Figures 1 and 9). Along this line the New Ireland Basin sequence is more than 4 km (3 s) thick. The lower sequences C-D and D-E are lenticular and thin and thicken respectively, away from New Ireland. The entire basin sequence is tilted up to the northeast on Emirau-Feni Ridge. Gulf line NI 22/22A shows the same basin fill sequence as line NI 21, except that on this line, the basin strata onlap onto the Emirau-Feni Ridge, whereas the sequence on line NI 21 is clearly uplifted and disrupted by faulting (Figure 9). This relationship, of faulting in some areas and onlap in other areas along the ridge, suggests that the ridge is locally broken into a series of horsts, some of which are capped by islands.

Gulf lines NI 23 and 24 extend from near New Hanover Island northeast and north, respectively, across the narrow, western end of New Ireland Basin (Figures 1 and 9). Here, the basin fill is only about 3 km (2.5 s) thick, as opposed to a maximum thickness of 5 km (4 s) on Gulf line NI 20. The basin sequence near New Hanover is cut by several high angle normal faults that slightly offset almost all the basin fill. As on line NI-21, line NI 23 shows arching and faulting of the New Ireland Basin strata across the uplifted Emirau-Feni Ridge near the northeastern end of the line.

OFFSHORE STRUCTURE

Structure Contour and Isopach Maps

Seismic-reflection profiles from the 1984 *Lee* cruise as well as the lines reported by de Broin et al. (1977), Exon and Tiffin (1984), and Exon et al. (1986) were interpreted using the horizons and sequences described in Table 1. The horizons were then digitized by Worldwide Exploration Consultants, Inc. as part of a hydrocarbon evaluation of New Ireland Basin that was commissioned by the government of Papua New Guinea (Stewart et al., 1986). The digitized time horizons were converted to depth using the following interval velocities:

Layer	Interval Velocity (m/s)
Sea Level - Seabed	1,496
Seabed - A	1,800
A - B	2,000
B - C	2,500
C - D	3,600
D - E	3,900
E - V	5,000

These interval velocities were derived from stacking velocities of the multichannel seismic-reflection data and from sonobuoy refraction velocities (Childs and Marlow, this volume) acquired near the axis of New Ireland Basin. The converted depth values were contoured by hand to produce the isopach and structure contour maps shown in Figures 10-15.

Horizon B is equivalent to the top of the lower Pliocene Punam Limestone exposed on New Ireland (Table 1), and the horizon is unconformable with underlying horizons (Figure 4). In the northern, southern, and western parts of New Ireland Basin, the B horizon onlaps Horizons C or D (Figure 10). Horizon B is 3,800 m deep in the southeast sector, the deepest part of the basin, and 3,000 m deep in the central depocenter of the basin west of Tabar Island (Figure 10).

Horizon C is interpreted as the top of the upper Miocene Lelet Limestone, a potential source and reservoir for hydrocarbons in New Ireland Basin (Table 1; Exon and Tiffin, 1984; Exon et al., 1986). Because of the possible resource potential of the C-D sequence, we have constructed isopach contours between the sea floor and Horizon C. Isopach contours show that seabed-Horizon C is thickest, more than 2,200 m thick, west of Tabar Island (Figure 11). To the north, Horizon C onlaps Horizon D against the flank of the Emirau-Feni Ridge (Figure 8), and to the west, the C horizon is close to and often truncated at the sea floor (Figure 11).

Isopachs between Horizons C and D, the equivalent of the onshore Lelet Limestone, are shown on Figure 12. The thickest part of the C-D section is in the far western part of the basin, where the section exceeds 3,000 m, and also in the southern portion of the basin near New Ireland, where the sequence is 2,600 m thick. A depocenter containing 2,000 m of strata fills the southeastern end of the basin.

Structure contours on Horizon D, equivalent to the top of the lower Miocene Lossuk River beds on New Ireland (Table 1), show two deeps on this horizon: one in the southeast, where the horizon is 6,200 m deep and the other in the west, where the unit is 5,200 m deep (Figure 13). Horizon D laps onto basement along the northern and southern edges of New Ireland Basin.

The E-V section at the base of the seismic-reflection profiles is probably Eocene to lower Miocene volcanic rocks and volcaniclastic strata that are equivalent to the Jaulu Volcanics exposed on the islands (Table 1). This section also forms the last acoustic unit resolved on the records and is the economic basement of New Ireland Basin. We can recognize Horizon E only in the central portion of New Ireland Basin and Horizon V only around the Emirau-Feni Ridge that forms the northern margin of the basin (Figure 14). Horizon V contours are shown Near Lihir and Taya Islands, where Horizon E is absent. Both Horizons E and V are considered volcanic basement flooring New Ireland Basin. The deepest part of the basin is 20 km west of Tabar Island, where the basement is 7,500 m deep. A second deep is between New Hanover and Mussau Island, where the basement reaches depths of 6,500 m.

The total sedimentary section in New Ireland Basin is shown by isopach contours on Figure 15. This section is measured from Horizons E-V to the

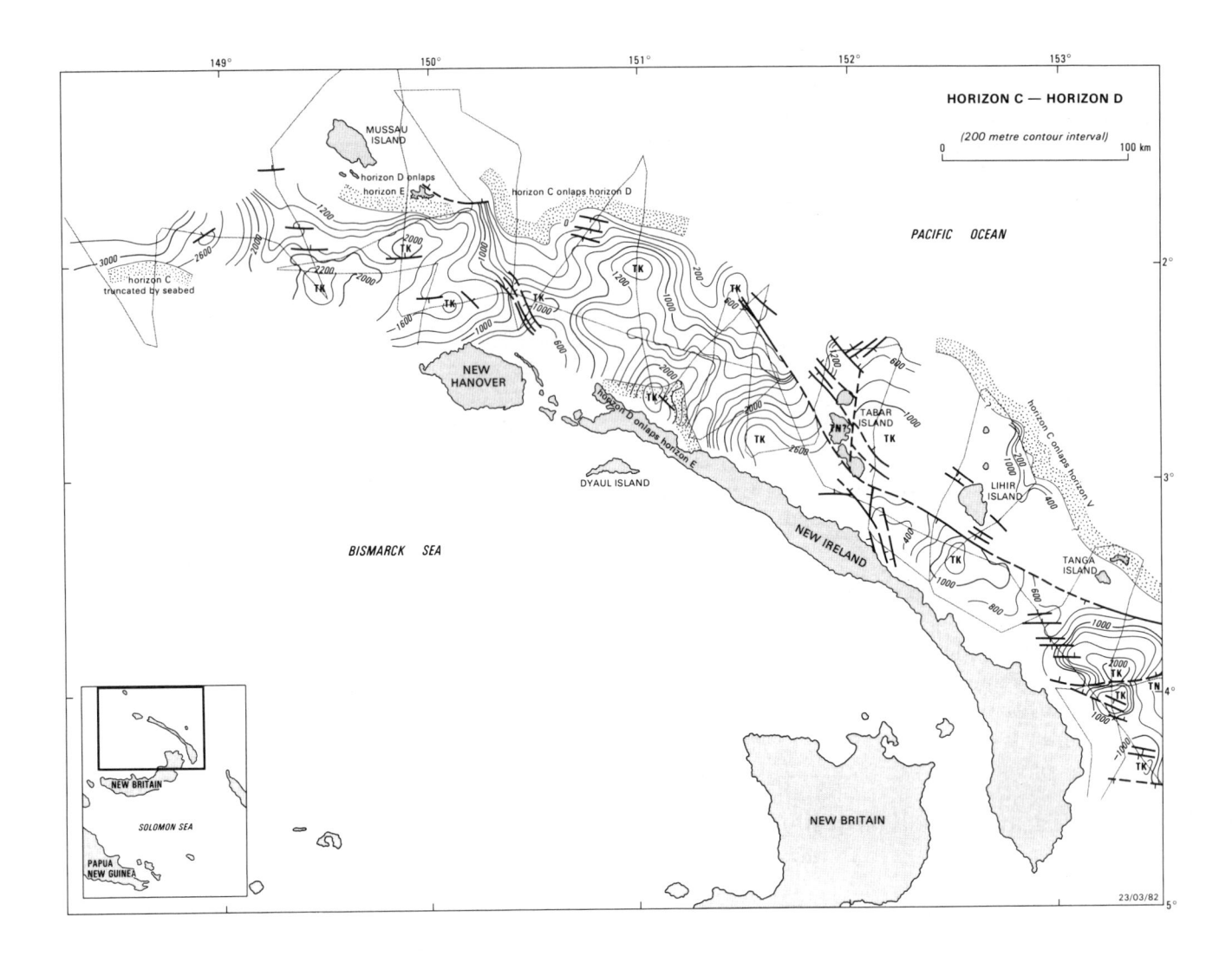

Figure 12. Isopach contours in meters for unit C-D, New Ireland Basin, equivalent to early to late Miocene Lelet Limestone exposed on New Ireland. TK = thick, TN = thin.

sea floor. Three major depocenters in the basin are evident from the contours: a thick section of more than 5,000 m between New Hanover and Mussau Islands, a second, thickest section (> 6,000 m) west of Tabar Island, and a third section more than 5,000 m thick south of Tanga Island. The sediment fill in the basin is truncated by faulting along the eastern edge of the basin near Tabar, Lihir, and Tanga Islands (Figure 15). The basin fill thins over volcanic basement to the north near Mussau Island. The southwestern flank of basin fill is exposed on New Ireland.

Faulting and Tectonics

According to Falvey and Pritchard (1984), New Ireland Island and New Ireland Basin were part of a continuous, linear island arc that included the northern Solomon Islands, and New Ireland, Manus, and New Britain Islands. Back-arc spreading and opening of the Manus Basin in the Pliocene caused New Britain to separate from the region of the present day Manus Island and slide southeastward past New Ireland (Taylor, 1979; see also Figure 2 of Marlow, Exon, and Tiffin this volume). Exon and Tiffin (1984) suggest that heating related to the opening of Manus Basin caused late Miocene to Pliocene uplift of the southern margin of New Ireland Basin, which is now exposed on the island of New Ireland. Uplift has continued to the present. Exon and Tiffin (1984) also note that compressional structures do not occur in the New Ireland Basin, although we did find two anticlinal structures in the southeastern part of the

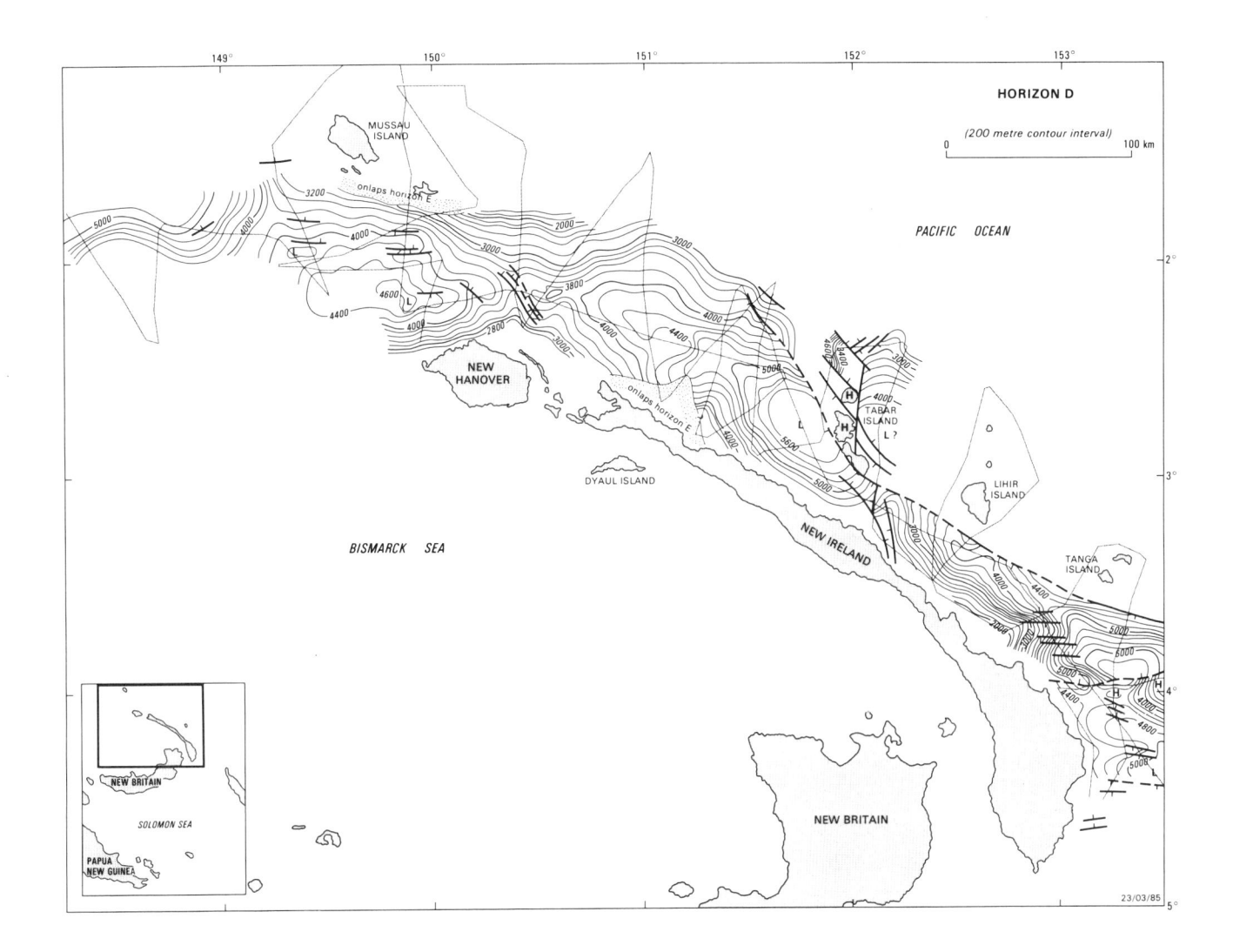

Figure 13. Structure contours in meters on Horizon D, New Ireland Basin, equivalent to top of marine strata in the early Miocene Lossuk River beds exposed on New Ireland. L = low, H = high.

basin (Figures 6 and 7). The Emirau-Feni Ridge (Figure 1) forms the northeastern margin of the New Ireland Basin, and was uplifted originally in the Oligocene (Exon and Tiffin, 1984). For a discussion of the regional tectonics of the Manus Basin, see Johnson (1979) and Marlow, Exon, and Tiffin (this volume).

Extensive faulting in the New Ireland Basin occurs southwest of the island of New Ireland, where the basement flank of the basin is downfaulted into Manus Basin and the total vertical displacement exceeds 2,000 m (Exon et al., 1986). Other large displacements include uplifts of about 2,000 m of the basin sequences in association with the emplacement of the Tabar to Feni Island groups (Exon and Tiffin, 1984). The origin of the volcanic rocks on these islands is enigmatic in that the alkaline Quaternary volcanism on the islands is of arc-trench type but the volcanism is not associated with any clearly-defined, present day Benioff zone (Johnson, 1979). Volcanism along the Tabar-Feni chain may have been related to a northeast-dipping subduction system that existed before the opening of Manus basin (see Figure 2 of Marlow, Exon, and Tiffin, this volume). Opening of Manus basin may have transported the arc, which includes the Tabar-Feni group as well as New Ireland, to the northwest, forcing the destruction of a former trench by faulting along the southwestern side of New Ireland (Taylor, 1979; Hamilton, 1979; Johnson, 1979). Johnson alternatively suggested that faulting and extension related to the uplift of the northern New Ireland Basin may be responsible for

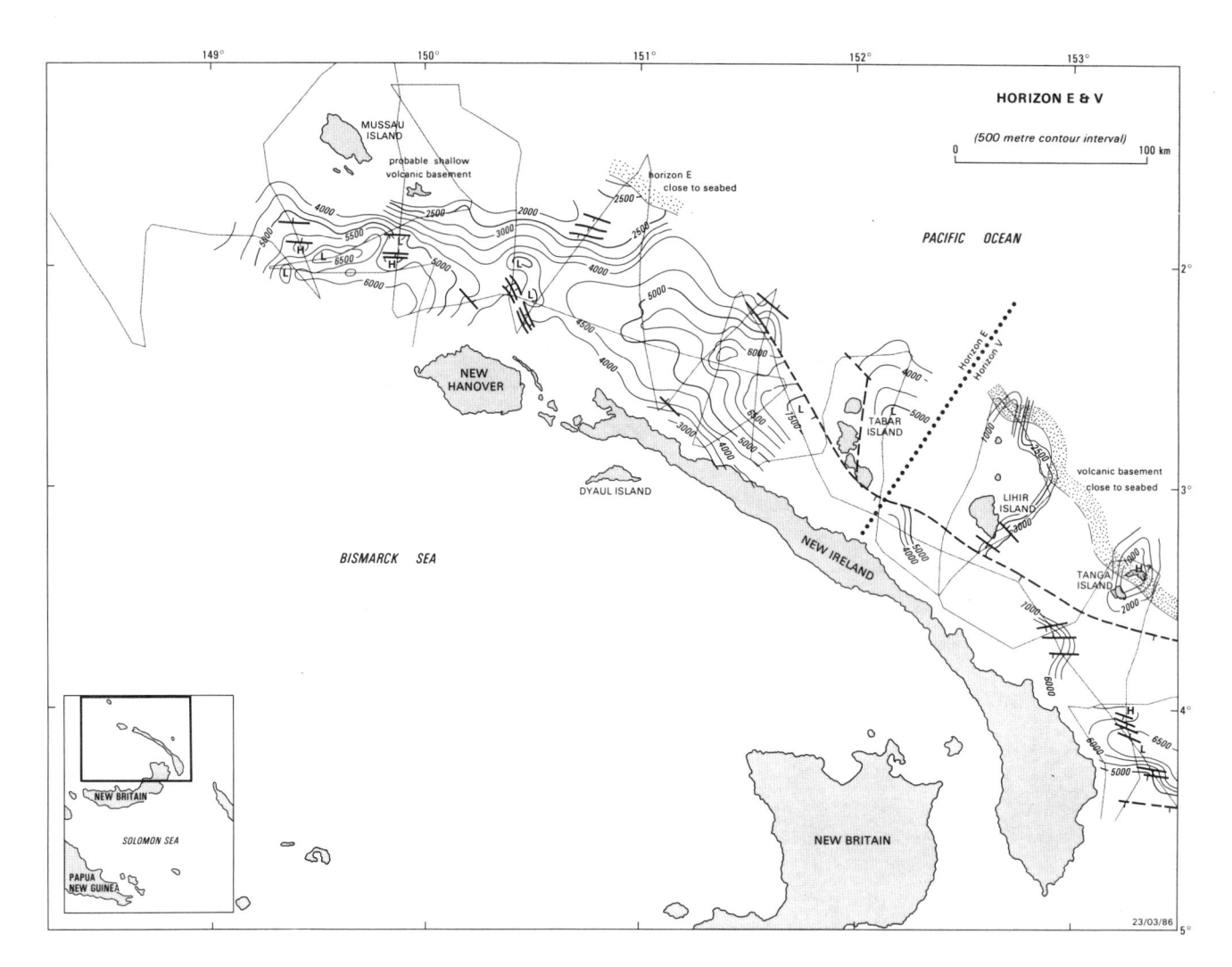

Figure 14. Structure contours in meters on Horizons E and V, New Ireland Basin, equivalent to the Eocene to early Miocene volcanic basement exposed on New Ireland. L = low, H = high.

SUMMARY

New Ireland Basin formed initially in the early Tertiary as a forearc basin adjacent to the Manus Trench. In Figure 16, we have outlined a tentative development history of New Ireland Basin based on seismic-reflection profiles near NI-20 (Figure 1). The basin was originally laid down between an Eocene to early Miocene volcanic arc in the southwest and an outer-arc high, the precursor to the Emirau-Feni Ridge, in the northeast. The lower Miocene basin sequence D-E progrades and thins to the southwest on seismic-reflection profiles and this sequence shows that a source area for the basin fill lay to the

the uplift and formation of the volcanoes in the Tabar-Feni group.

northeast. This source area or outer-arc high may have grown by accretion and uplift induced by subduction along the adjacent Manus Trench. By the early and middle Miocene, subduction ceased, allowing the deposition of carbonate reefs and lagoonal sediment over the sinking volcanic pile around New Ireland. Up to 750 m of the Lelet Limestone is preserved on New Ireland (Stewart and Sandy, this volume) and is stratigraphically equivalent to the C-D sequence mapped offshore. Volcanism recommenced on New Hanover in the late early Miocene and persisted through to latest Miocene times (Lavongai Volcanics and Matanalaua Formation - locally derived and not recognized on offshore seismic-reflection profiles). On northwestern New Ireland volcanism recommenced in the early middle Miocene (Lumis River volcanics - locally derived and not recognized offshore) and persisted into the

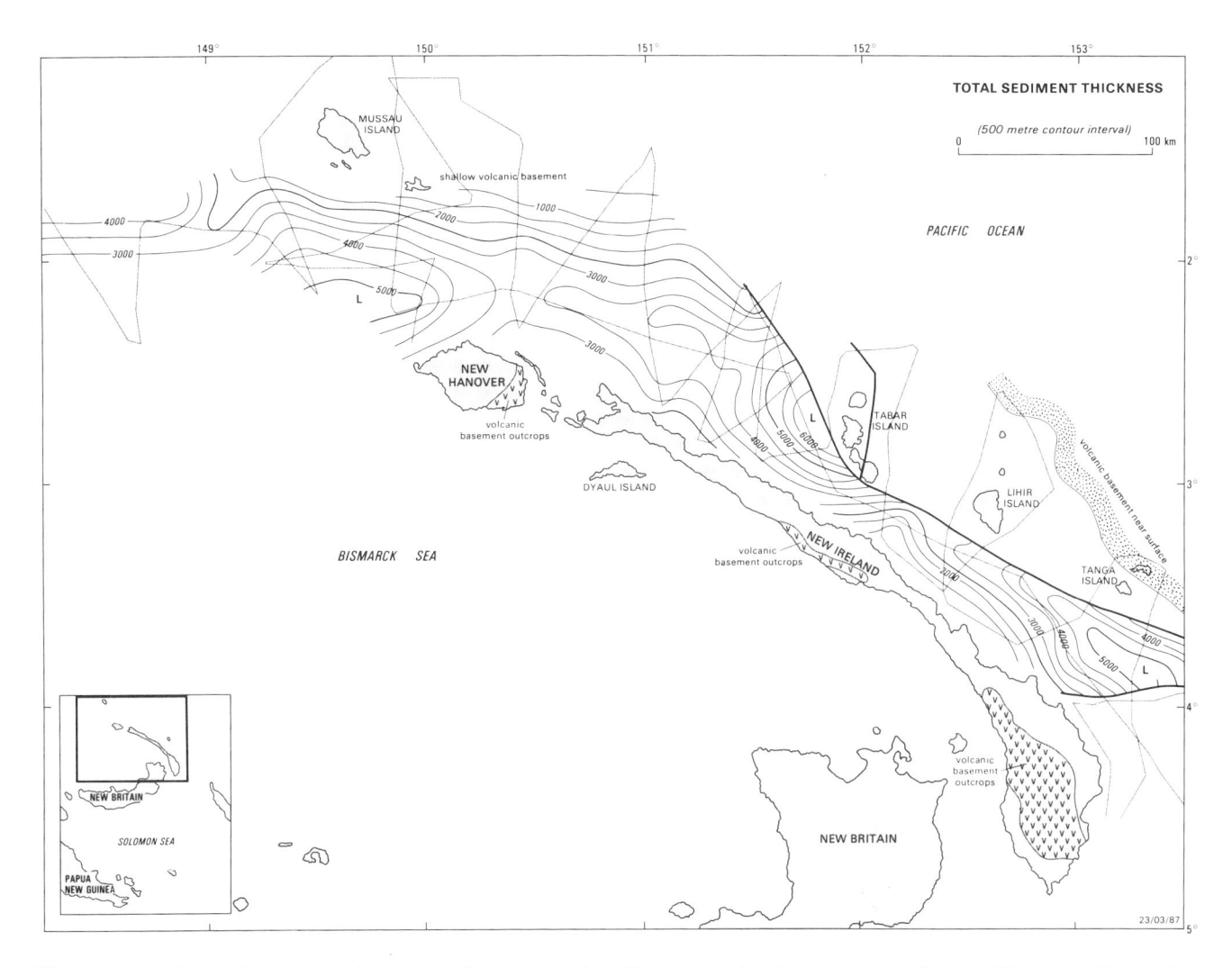

Figure 15. Isopach contours in meters for the total sediment section from the sea floor to Horizons E and V in New Ireland Basin. L = low.

Pliocene (Rataman Formation - offshore A-B sequence). Volcanism did not occur elsewhere on New Ireland until the Pliocene with the deposition of the Rataman Formation.

Manus Basin began to form in the Pliocene, shifting New Britain from west of Manus Island to New Britain's present position south of New Ireland. At the same time volcanism began in the Tabar-Feni Islands, uplifting the northeastern strata in New Ireland Basin into a series of horsts topped by islands along the Emirau-Feni Ridge. Offshore deposition in the basin consisted mainly of a volcaniclastic sequence, A-B, derived from the emergent Tabar-Feni volcanic islands to the northeast as well as from the young volcanoes on New Britain, which developed in response to new, northward-directed subduction beneath New Britain (see Figure 2 of Marlow, Exon, and Tiffin, this volume). Pelagic carbonate deposition dominated the Pleistocene and Holocene sedimentation, laying down the last sequence (SB-A) in New Ireland Basin.

A seismic-reflection profile collected in 1984 across the southeastern part of New Ireland Basin exposed a flat, high-amplitude reflector or "bright spot" within the core of an anticline about 20 km east of New Ireland Island. The bright spot or DHI (direct hydrocarbon indicator) is in water 2,500 to 2,600 m deep, but the anticline may continue upslope into shallower water. No exploration wells have been drilled into New Ireland Basin, but a thick sedimentary section, including shelf limestones, reefs, and carbonate build-ups, are suspected offshore in the basin, warranting further hydrocarbon exploration of this frontier basin.

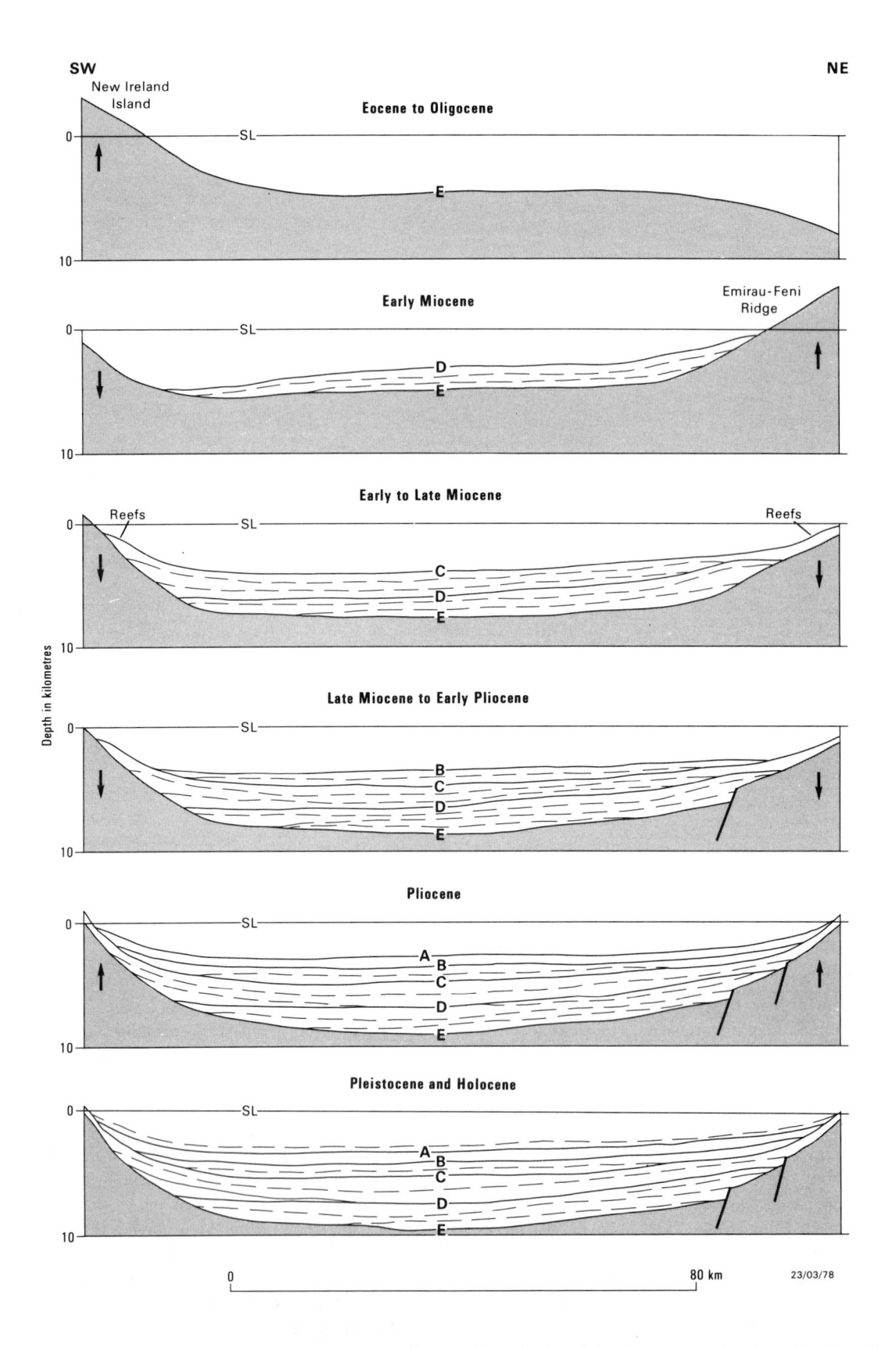

Figure 16. Schematic of the developmental history of New Ireland Basin near seismic-reflection line NI-20 (Figure 1). Seismic units or sequences as defined in Figure 2 and as listed in Table 1. Scales are approximate. Arrows show relative uplift and subsidence directions during basin development. See text for discussion.

ACKNOWLEDGMENTS

We would like to thank the Captain and crew of the R/V *S.P. Lee* for their expert help in the last SOPAC cruise to Papua New Guinea in 1984. We thank the government of Papua New Guinea in their support of the work and logistics involved in field work on New Ireland, and we acknowledge the help of John Pinchin of Flower Doery Buchan (Australia) Ltd. and Worldwide Exploration Consultants Inc. in the preparation of the structure contour and isopach maps. We also thank Terry Bruns and Alan Cooper for their thoughtful reviews.

REFERENCES

Clark, S.P., Jr., 1966, Handbook of Physical Constants: Geological Society of America Memoir 97, 583 p.

Connelly, J.B., 1976, Tectonic development of the Bismarck Sea based on gravity and magnetic modeling: Royal Astronomical Society Geophysical Journal, v. 46, p. 23-40.

de Broin, C.E., Aubertin, F., and Ravenne, C., 1977, Structure and history of the Solomon-New Ireland region: *in* International Symposium on Geodynamics of the Southwest Pacific, New Caledonia, August-September 1976: Editions Technip, Paris. 413p: p. 37-49.

Exon, N.F. and Tiffin, D.L., 1984, Geology of offshore New Ireland basin in northern Papua New Guinea, and its petroleum prospects, *in* Watson, S. T., ed., Transactions of the Third Circum-Pacific Energy and Mineral Resources Conference, 22-28 August, 1982, Honolulu, Hawaii, p. 623-630.

Exon, N.F., Stewart, W.D., Sandy, M.J., and Tiffin, D.L., 1986, Geology and offshore petroleum prospects of the eastern New Ireland Basin, northeastern Papua New Guinea: Bureau of Mineral Resources Journal of Australian Geology and Geophysics, v. 10, p. 39-51.

Falvey, D.A. and Pritchard, T., 1984, Preliminary paleomagnetic results from northern Papua New Guinea: evidence for large microplate rotations *in* Watson, S.T., ed., Transactions of the Third Circum-Pacific Energy and Mineral Resources Conference, 22-28 August, 1982, Honolulu, Hawaii, p. 593-599.

Hamilton, W., 1979, Tectonics of the Indonesian region: U.S. Geological Survey Professional Paper 1078, 345 p.

Johnson, R.W., 1979, Geotectonics and volcanism in Papua New Guinea: a review of the late Cainozoic: Bureau of Mineral Resources Journal of Australian Geology and Geophysics, v. 4, p. 181-207.

Stewart, W.D., Francis, G., and Deibert, D.H., 1986, Cape Vogel basin hydrocarbon potential: Oil and Gas Journal, Nov. 17, p. 67-71.

Taylor, B., 1979, Bismarck Sea: evolution of a back-arc basin: Geology, v. 7, p. 171-174.

Wallace, D.A., Chappell, B.W., Arculus, R.J., Johnson, R.W., Perfit, M.R., Crick, I.H., Taylor, G.A.M., and Taylor, S.R., 1983, Cainozoic volcanism in the Tabar, Lihir, Tanga and Feni Islands, Papua New Guinea: Geology, whole-rock analyses, and rock-forming mineral composition: Bureau of Mineral Resources Report 243, BMR Microfilm MF 197.

Marlow, M.S., Dadisman, S.V., and Exon, N.F., editors, 1988, Geology and offshore resources of Pacific island arcs—New Ireland and Manus region, Papua New Guinea, Circum-Pacific Council for Energy and Mineral Resources Earth Science Series, v. 9: Houston, Texas, Circum-Pacific Council for Energy and Mineral Resources.

HYDROCARBON GAS IN BOTTOM SEDIMENT FROM OFFSHORE THE NORTHERN ISLANDS OF PAPUA NEW GUINEA

K.A. Kvenvolden

U.S. Geological Survey, Menlo Park, California 94025, USA

ABSTRACT

Small concentrations of hydrocarbon gases were found in eleven samples from six gravity-core stations in bottom sediment of the New Ireland Basin. Methane was the most abundant hydrocarbon gas ranging in concentration from 260 to 1,200 nanoliters/liter (nL/L) of wet sediment. The next most abundant gases were ethene and propene that had maximum combined concentrations of 30 nL/L. Ethane and propane, when detected, had combined concentrations of less than 10 nL/L. Butanes were not detected. Coring stations were targeted relative to geophysical anomalies to enhance the chances of sampling hydrocarbon gases from deep sources. The compounds found, however, represent the *in-situ* population of hydrocarbon gases in near-surface sediment; these gases are probably derived from low-level microbial activity and early thermal diagenesis. No evidence was found for petroleum-related hydrocarbons.

INTRODUCTION

The search for petroleum in frontier areas often incorporates geochemical prospecting as an exploration tool (Hunt, 1979). Thus, as part of the 1984 R/V *S.P. Lee* petroleum-oriented cruise in the frontier region of the New Ireland Basin of northern Papua New Guinea, a study of the hydrocarbon gas content of bottom sediment was included. This study was not comprehensive, but it did apply principles of geochemical prospecting in a search for surficial evidence of migrating hydrocarbon gases by using an approach similar to that described by Kvenvolden, Vogel, and Gardner (1981). Although only minimal evidence for petroleum hydrocarbons was found in that survey, previous work in Norton Sound, Alaska (Kvenvolden et al, 1979), and offshore northern California (Kvenvolden and Field, 1981) has shown that the seepage of petroleum hydrocarbons into oceanic sediment can be identified by collecting and analyzing bottom sediment.

The rarity of natural seeps in the marine environment (Kvenvolden and Harbaugh, 1983) makes difficult the finding of petroleum-related hydrocarbons in most bottom sediment. In addition, most oceanic sediment contains a background of nonpetroleum-related hydrocarbons. Oceanic sediment worldwide usually contains low concentrations of the hydrocarbon gases, methane (C_1) ethane (C_2), propane (C_3), isobutane (i-C_4), and n-butane (n-C_4) as well as the unsaturated hydrocarbon gases, ethene ($C_{2:1}$) and propene ($C_{3:1}$). C_1 is most abundant with amounts usually less than about 10,000 nanoliters/liter (nL/L) of wet sediment. The other hydrocarbons (C_2, $C_{2:1}$, C_3, $C_{3:1}$, i-C_4, and n-C_4) generally occur in amounts at least two or three orders of magnitude lower. Most of these hydrocarbons, and also C_1, are dissolved in the interstitial water and probably result from low-level microbiologic and early diagenetic processes (Claypool and Kvenvolden, 1983).

Table 1. Gravity core stations at which sediment samples were recovered for hydrocarbon gas analysis (New Ireland Basin, northern Papua New Guinea).

Station	Water depth (m)	Core length (m)	Latitude	Longitude	Area and description
G-1	1517	2.37	01°39.60′S	146°51.12′E	North of Manus Island. Cored in small basin. Grey, olive-green, hemipelagic foraminiferal mud with silt and minor sand.
G-3	1089	0.85	01°50.69′S	148°16.17′E	Between Manus Island and New Hanover and northeast of Rambutyo. Cored at base of abrupt scarp of a bathymetric high. Grayish-tan, foraminiferal sandy mud.
G-6	848	1.60	02°18.79′S	150°39.17′E	Northeast of New Ireland and northwest of New Hanover. Cored in canyon associated with a fault. Olive-gray, foraminiferal sandy mud.
G-7	1033	2.77	02°29.21′S	151°03.17′E	North of the north end of New Ireland. Cored in canyon. Olive-gray, foraminiferal ooze and calcareous mud.
G-8	1234	1.51	02°48.39′S	151°31.78′E	North of New Ireland. Cored in a minor canyon. Olive-gray, foraminiferal mud.
G-9	1320	1.00	03°08.46′S	152°05.66′E	North of New Ireland and south of Tabar Island. Cored into sediment above dipping reflectors. Gray slightly shelly mud.

Superimposed on this background of hydrocarbons are occasional anomalous occurrences of hydrocarbon gases, consisting principally of C_1, resulting from high levels of microbial activity in anoxic sediment depleted in sulfate (Claypool and Kaplan, 1974). More important to geochemical prospecting are petroleum-related gases composed of C_1 and the higher molecular weight, saturated, hydrocarbon gases, which can be superimposed on this background where conduits, such as faults and fractures, permit migration from depth to surface.

Geophysical surveys can aid in geochemical prospecting by providing targets where petroleum-related hydrocarbons might be expected near the surface due to migration. In our survey of the New Ireland Basin, we attempted to locate some of our sampling stations on features observed on multichannel seismic-reflection records obtained earlier in the cruise (Table 1). Such features included small basins, the base of a scarp, a canyon associated with a fault, and sediment overlying dipping reflectors.

PROCEDURES

Our geochemical survey sought hydrocarbon gases in eleven sediment samples recovered from gravity cores taken at six stations in slope sediment of the New Ireland Basin north of New Ireland and north and east of Manus Island of northern Papua New Guinea (Table 1; Figure 1). Depths of water ranged from 848 to 1,517 m, and the sediment type was mainly olive-gray, fine-grained, silty, and sandy mud, the sand-size material being principally forams. The sample stations were located, wherever possible, near geophysical anomalies (Table 1) to enhance the chance of finding petroleum-related hydrocarbons which might have migrated to the surface from deep sources.

Sediment sampling involved gravity coring to subbottom depths less than 3 m. Ten-centimeter long segments of sediment core (8-cm in diameter) were removed for gas extraction and analysis, using the methods developed by Kvenvolden and Redden (1980). Each segment was placed in a metal can that had previously been prepared with septa to provide for removal of gas after extraction. Enough degassed water was added to establish a 100-ml headspace within the container. The can was sealed, the headspace was purged with helium (100 cc/min for 10 min), and the can was shaken for 10 min to extract gas from sediment into the helium-filled headspace.

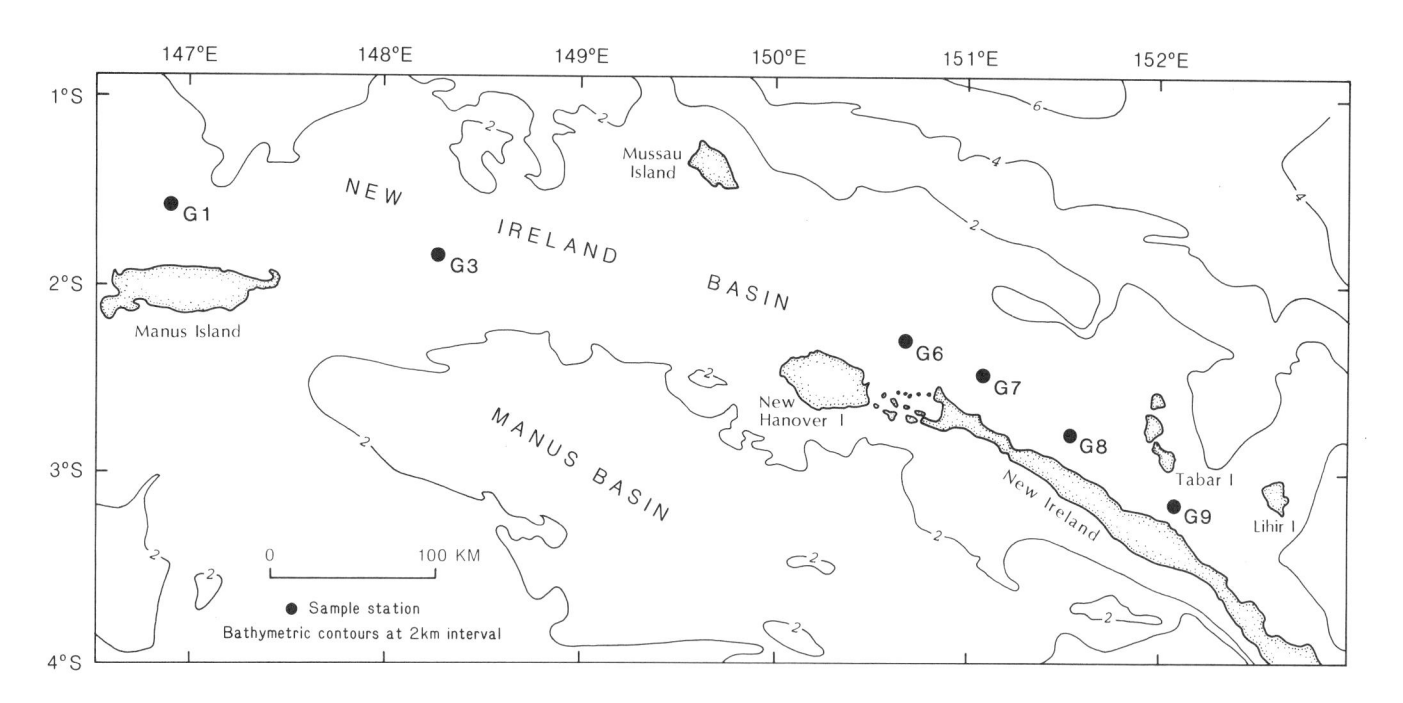

Figure 1. Locations of sediment sampling sites in the New Ireland Basin.

One milliliter of the headspace mixture was analyzed by gas chromatography, utilizing an instrument designed especially for identifying the components of natural gas. The following partition coefficients were used to correct for gas solubilities: C_1=0.8; C_2 and C_3=0.7; and $C_{2:1}$ and $C_{3:1}$=0.6. Concentrations are reported as nanoliters of gas per liter of wet sediment (nL/L).

RESULTS AND DISCUSSION

Table 2 lists the results obtained. The concentrations of all gases were disappointingly low. C_1, the most abundant hydrocarbon gas, ranges in concentration from 260 to 1,200 nL/L. The other hydrocarbons are orders of magnitude lower in amounts. In many samples, the non-C_1 hydrocarbons occurred only in trace amounts or could not be detected above the instrument noise level caused by the movement of the ship. When detected, the non-C_1 hydrocarbons are reported only to one significant figure. The second most abundant hydrocarbons are $C_{2:1}$ and $C_{3:1}$, whose total concentration never exceeded 30 nL/L. C_2 was never measured, and C_3 was determined in only 5 of 11 samples at concentrations of 9 nL/L or less. CO_2 was also present in 10 samples, and its concentration ranged from 60 to 170 nL/L.

Besides C_1, the other hydrocarbon gases were not detected often enough to permit the application of criteria for assessing possible sources. These criteria include measurement of $C_1/(C_2+C_3)$ and $C_2/C_{2:1}$ (see Kvenvolden and Redden (1980) for a discussion of the use of these ratios). Thus, C_1 is the most significant hydrocarbon gas observed in this survey, and its occurrence in each core is considered next.

Core G-1 came from a small basin north of Manus Island. Two samples from this core, separated stratigraphically by 100 cm, were extracted for hydrocarbons. C_1 concentrations approximately doubled with depth (520 to 1,200 nL/L). Core G-3 came from the base of an abrupt scarp between Manus Island and New Hanover, and one sample contained 410 nL/L of C_1. Two samples, separated stratigraphically by 50 cm, were taken from core G-6 located in a canyon associated with a fault northeast of New Ireland and northwest of New Hanover. C_1 concentrations of 700 and 560 nL/L suggest that in the near-surface sediment there is no gradient of increasing C_1 with depth, and thus no evidence for a leaking fault. Three samples were collected from core G-7 taken in a canyon north of New Ireland. The C_1 content with depth was uniform (510 to 600 nL/L). From core G-8, also located in a canyon north of New Ireland, came two samples that had only 320 and 390 nL/L of C_1. The lowest amount of C_1 (260 nL/L) was found in one sample from core G-9 from

Table 2. Concentration (nL/L of wet sediment) of hydrocarbon gases and CO_2 extracted from sediment of the New Ireland Basin, northern Papua New Guinea. (nd=not detected)

Station	Interval (cm)	C_1	C_2	$C_{2:1}$	C_3	$C_{3:1}$	CO_2
G-1	90-100	520	nd	8	nd	10	100
"	190-200	1200	nd	8	4	10	170
G-3	75-85	410	nd	nd	nd	5	20
G-6	90-100	700	nd	6	4	21	80
"	150-160	560	nd	12	9	11	nd
G-7	90-100	510	nd	6	4	16	60
"	190-200	570	nd	12	9	16	60
"	258-268	600	nd	12	9	16	60
G-8	90-100	320	nd	nd	nd	nd	60
"	132-141	390	nd	nd	nd	5	80
G-9	82-92	260	nd	nd	nd	nd	100

near Tabar Island where the surface sediment overlies dipping reflectors. Therefore, the results show that the gases in sediments at these coring stations have not been influenced by the situations defined by the geophysically interpreted features that guided the sampling.

The low concentrations of C_1 and other saturated hydrocarbons indicate that seepage of petroleum-related gases is not taking place in these sediments. In fact, the low amounts of hydrocarbons in all samples suggest that they represent the *in-situ* background for this area. The source of the C_1 is probably from low-level microbial and early diagenetic processes. The olive-gray color of the sediment suggests oxidation, and the C_1 may be partially oxidized causing the low concentrations. The source of C_2 and C_3 may be microbial or the product of very early thermal diagenesis (Claypool and Kvenvolden, 1983). The unsaturated hydrocarbons $C_{2:1}$ and $C_{3:1}$ are generally thought to represent biologically derived substances (Davis and Squires, 1954; Primrose and Dilworth, 1976). The amount of $C_{3:1}$ always exceeds the amount of C_3 -- a relationship that is expected where nonpetroleum-related hydrocarbons are present.

Of course, not finding petroleum-related hydrocarbons on a cruise in which the thrust was a regional assessment of petroleum potential was disappointing. However, negative results from only six cores of sediment from the entire New Ireland Basin should hardly diminish exploration interest. On the other hand, positive results at even one site would have greatly increased the prospects of finding petroleum in the area.

This study has demonstrated that low concentrations of hydrocarbons gases are present in bottom sediment of the New Ireland Basin. These gases probably result from low-level biologic activity and early thermal diagenetic processes in the sediment and pore water. Petroleum-related hydrocarbons are not present in any of the samples.

REFERENCES

Claypool, G.E., and Kaplan, I.R., 1974, The origin and distribution of methane in marine sediments, *in* I.R. Kaplan, Natural Gases in Marine Sediments: New York, Plenum Press, p. 99-139.

Claypool, G.E., and Kvenvolden, K.A., 1983, Methane and other hydrocarbon gases in marine sediment: Annual Reviews of Earth and Planetary Sciences, v. 11, p. 299-327.

Davis, J.B., and Squires, R.M., 1954, Detection of microbially produced gaseous hydrocarbons other than methane: Science, v. 119, p. 381-382.

Hunt, J.M., 1979, Petroleum geochemistry and geology: San Francisco, W.H. Freeman, 617 p.

Kvenvolden, K.A., and Field, M.E., 1981, Thermogenic hydrocarbons in unconsolidated sediment of Eel River Basin, offshore northern California: American Association of Petroleum Geologists Bulletin, v. 65, p. 1642-1646.

Kvenvolden, K.A., and Harbaugh, J.W., 1983, Reassessment of the rates at which oil from natural sources enters the marine environment: Marine Environmental Research, v. 10, p. 223-243.

Kvenvolden, K.A., and Redden, G.D., 1980, Hydrocarbon gas in sediment from the shelf, slope, and basin of the Bering Sea: Geochimica et Cosmochimica Acta, v. 44, p. 1145-1150.

Kvenvolden, K.A., Vogel, T.M., and Gardner, J.V., 1981, Geochemical prospecting for hydrocarbons in the outer continental shelf, southern Bering Sea, Alaska: Journal of Geochemical Exploration, v. 14, p. 209-219.

Kvenvolden, K.A., Weliky, K., Nelson, H., and Des Marais, D.J., 1979, Submarine seep of carbon dioxide in Norton Sound, Alaska: Science, v. 205, p. 1264-1266.

Primrose, S.B., and Dilworth, M.J., 1976, Ethylene production by bacteria: Journal of General Microbiology, v. 93, p. 177-181.

Marlow, M.S., Dadisman, S.V., and Exon, N.F., editors, 1988, Geology and offshore resources of Pacific island arcs—New Ireland and Manus region, Papua New Guinea, Circum-Pacific Council for Energy and Mineral Resources Earth Science Series, v. 9: Houston, Texas, Circum-Pacific Council for Energy and Mineral Resources.

PETROLEUM SOURCE ROCK STUDY, MIOCENE ROCKS OF NEW IRELAND, PAPUA NEW GUINEA

Miryam Glikson
Centre for Resource and Environmental Studies,
Australia National University, Canberra, Australia

ABSTRACT

A study of petroleum source rocks from New Ireland was performed on samples from the Lossuk River beds, Lumis River volcanics and the Lelet Limestone. The hydrocarbon potential of the Lossuk River beds, which contain up to 2.74% total organic carbon (TOC) by weight, is moderate as is that of the Lumis River volcanics, which contains up to 2.68% TOC. The diversity of organic matter, including land plant material and marine cyanobacterial material, throughout these beds, cause difficulties in assessing whether these rock units are oil or gas prone. On the other hand, the extent and thickness of the Lelet Limestone, if it is proven by further work to contain sufficient organic matter, make this formation a potential petroleum source. The limestones contain organic matter only in small quantities (less than 0.5% TOC), but if the sequence is sufficiently thick, the overall amount of organic matter may be high enough to constitute a source rock. Pyrite in some of the limestones indicates reducing conditions, and initial organic matter concentrations were sufficient for bacterial sulphate reduction. Higher concentrations of organic matter may be present in the Lelet Limestone sequence below the main weathering zone. The Matakan unit within the Lelet Limestone contains transported coal beds in a marine organic-rich calcareous-pyritic sequence. The Matakan unit has a high petroleum potential and early generation of hydrocarbons from pollen exines in the Matakan coal is evident.

INTRODUCTION

The objective of this study was to assess the source rock potential of certain strata within the New Ireland Basin. The rocks, collected from outcrops, covered the entire sedimentary sequence from lower Miocene to Pleistocene. Details of location are filed with the Geological Survey of Papua New Guinea (GSPNG), but general locations are shown in Figure 1. Emphasis was placed on the Miocene sequence because they are more likely to contain source rocks than the younger sequences. This study attempts to 1) determine the amount of organic matter in individual samples to enable extrapolation to entire formations, 2) identify the type of organic matter to determine source rock potential, 3) assess hydrocarbon potential based on quantity and quality of organic matter, and 4) assess maturity of organic matter based on vitrinite reflectance.

SAMPLES AND METHODS

Sampling

Rock samples were collected during a 1984 field trip to New Ireland (Figure 1). The Lelet Limestone is represented by a section sampled from the east coast along the road climbing to the Lelet Plateau in the northwest. Another representative section of the

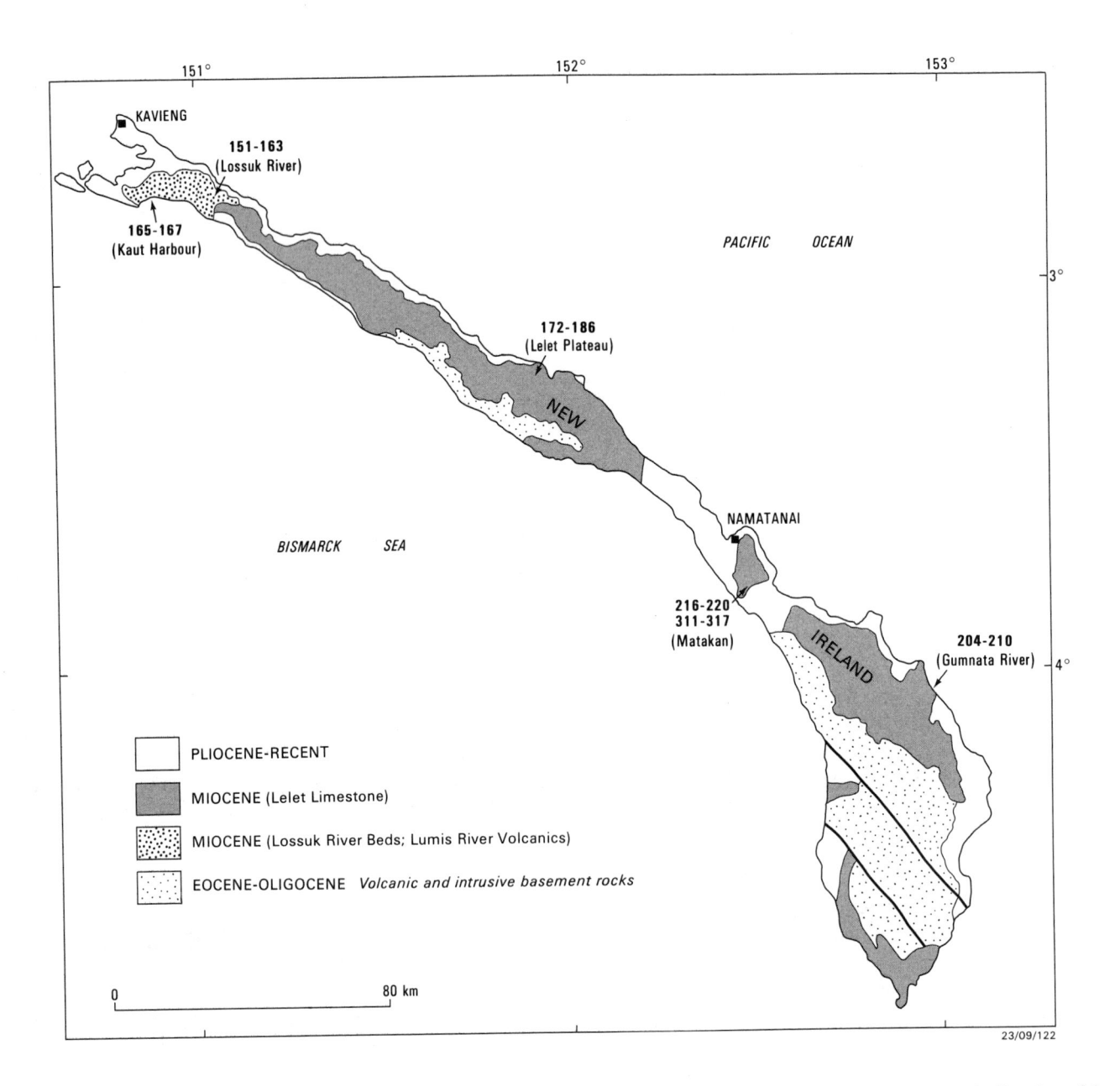

Figure 1. Sample locations on simplified geological map of New Ireland (after Stewart and Sandy, this volume).

Lelet Limestone, was collected along the Gumnata River north of the Nans Meyer Range in the east. Coal samples were collected from the Matakan unit of the Lelet Limestone on the west coast, south of Namatanai. Additional samples from this area were collected by a GSPNG field party in 1985 (M.J. Sandy, written comm.).

All but two samples were collected from surface outcrops, and hence have been affected by weathering and oxidation processes. The Lossuk River beds, in particular, are weathered and oxidized because the sampled outcrops occur along river banks that are commonly submerged during heavy rainfall. Samples 311 and 312 from the Matakan area were collected by auger drilling to depths of 3.0 and 3.4 meters, respectively.

Methods

An optical microscope was used to examine polished blocks in reflected white light as well as ultra violet (UV) light combined with a blue filter. In addition, dispersed (term as used by Burgess,

Table 1. Organic matter types in concentrates (transmitted light, by point count expressed as %).

Formation/Unit	Sample no.	Amorphous OM	Wood/Tissue	Resin	Fibrils	Pollen	Semi-Degraded Wood
Lossuk River beds	152	66	8	-	-	1.0	25
Lossuk River beds	153	65	8	-	-	2.0	25
Lossuk River beds	154	46	32.5	-	1	0.5	20
Lossuk River beds	158	94	5	-	-	1.0	-
Lumis River volcanics	161	-	51	5	-	44	-
Lumis River volcanics	164	-	68	-	-	32	-
Lumis River volcanics	167	34	39	-	13	14	-
Lelet Limestone	178	100	-	-	-	-	-

1974) samples of organic concentrates and thin sections were examined in transmitted light; the percentages of different types of organic matter in certain concentrates are listed in Table 1. Random reflectance measurements (Teichmuller and Teichmuller, 1981) of vitrinite and vitrinite-like (V-L) material were carried out using a Zeiss[1] photometer microscope and reported as mean random reflectance (Rm).

Transmission electron microscopy (TEM) was carried out on ultra-thin sections (0.03 - 0.05μm) of resin-embedded organic concentrates (Glikson and Taylor, 1986) using JEOL 100C and H-500 electron microscopes. Organic matter has been concentrated by digestion of rock samples in hydrochloric acid and hydrofluoric acid. Selected samples were treated with sodium hydroxyide to separate humic acids. The latter were precipitated with dilute HCl and dried for C-isotope analyses. Total organic carbon (TOC) and Rock-Eval analysis of representative samples were carried out by C.J. Boreham (Australian Bureau of Mineral Resources) using a GORDEL LECO carbon analyzer and a ROCK EVAL II Pyroliser. Stable carbon isotope analyses were run using a VG 602D isogas spectrometer after combustion in a isoprep 13 VG.

RESULTS

Lossuk River Beds

Depositional Environment

The Lossuk River beds consist of a variety of marine sediments, including conglomerate, sandstone, siltstone and limestone, deposited in an upper bathyal environment (Stewart and Sandy, this volume). The age of the sequence has been dated by foraminiferal remains as earliest to early middle Miocene (Stewart and Sandy, this volume).

Organic-rich black shale with plant remains, dark grey mudstones and calcareous siltstones were sampled. The beds outcrop only in the northern part of the island and are more than 150 m thick in the type section of the unit. The samples collected have been shown by geochemical and TEM results to be intensely oxidized.

The beds are considered to be of marine origin and were deposited largely in water depths of 150-500 m (Stewart and Sandy, this volume). Planktonic foraminifera are present in most samples examined. Cyanobacteria, bacteria, abundant pyrite, and some pollen and spores from land plants, vary in abundance however, all of which suggest original deposition in a shallow water environment less than 80 m deep, possibly a semi-enclosed basin connected to the open sea. The preservation of substantial organic matter suggests oxygen-poor conditions existed in the bottom sediments. Some samples have a high woody component, whereas others are dominated by amorphous organic matter; the former indicates a greater terrigenous influence than the latter. The land plant material is clearly not *in situ* and its diversity indicates a variety of sources. The variable influx of land plant remains is shown by extreme fluctuations in organic carbon content and organic matter concentrations. This variation is typical of Miocene turbidites in sedimentary basins of Papua New Guinea (G. Francis written comm.). A recent study of the dispersion of terrestrially-derived organic matter in reef environments showed leaf litter from mangrove vegetation to be the most important single source of terrestrial organic carbon (Chivas and Torgensen,

---

[1] Any use of trade names is for purposes of identification only and does not imply endorsement by the U.S. Geological Survey, the New Zealand Geological Survey, or the Bureau of Mineral Resources, Australia.

1985). Such leaf litter can be transported hundreds of kilometers out to sea, and down into the bathyal zone.

Light Microscopy and Transmission Electron Microscopy (TEM)

The organic matter observed in reflected light varies from a very low reflectance (0.1% Rm) vitrinite-like (V-L) component of probable bacterial origin and various degradation products, to vitrinite (0.3-0.6% Rm) representing terrestrial plant remains, and to inertinite - sclerotinite in some samples with reflectances higher than 0.6%Rm. ("Vitrinite-like" corresponds to amorphous organic matter in transmitted light microscopy and is synonymous with "bituminite" as discussed by Stach et al., 1975). The proportions of V-L and true vitrinite vary from sample to sample. Sample 156, when viewed in reflected light, has predominant organic matter reflectances of 0.3 and 0.4%Rm (Figure 2), but higher reflectance (0.6% Rm) organic matter is also present. The latter possibly represents reworked land plant remains. Filaments of probable cyanobacterial origin contribute to the calcareous, as well as organic remains in some samples (Plate 1, figures 5 and 6). Framboidal pyrite is abundant (Plate 1, figures 1, 2, 4), and is present within the organic matter and foraminiferal tests (Plate 1, figure 4). However, a finely granulated pyrite is also present (Plate 2, figure 7). Pyrite apparently predominates over marcasite. Inertinite is rare but recognizable (Plate 1, figure 3). Vitrinite occasionally (sample 158) displays two reflectances (Figure 2), one of 0.2 and 0.3% Rm and another of 0.4 and 0.5% Rm. Ultra-thin sections observed in TEM showed higher electron-dense material, corresponding to the higher reflectance vitrinite, and low electron-dense material corresponding to the lower reflectance obtained in light microscopy (Plate 1, figure 2). As electron density is a function of the mass number, carbon-rich organic matter will display higher electron density than carbon-poor, hydrogen-rich material.

Fluorescence is displayed by some of the organic matter (exinite and sporinite, Plate 2, figure 8). Partly oxidized pyrite is visible. Examination of organic matter in transmitted light (samples 152, 153, 154, and 158) reveals 3 main types: 1) amorphous yellow-orange material, 2) wood remains, and 3) semi-degraded wood with a minor contribution from pollen and conifer fibrils. The proportions of these components vary (Table 1), but amorphous organic matter is dominant overall. TEM work on selected samples has shown humic acids and degraded organic matter to be the major contributors, with lipid-rich components present in some samples.

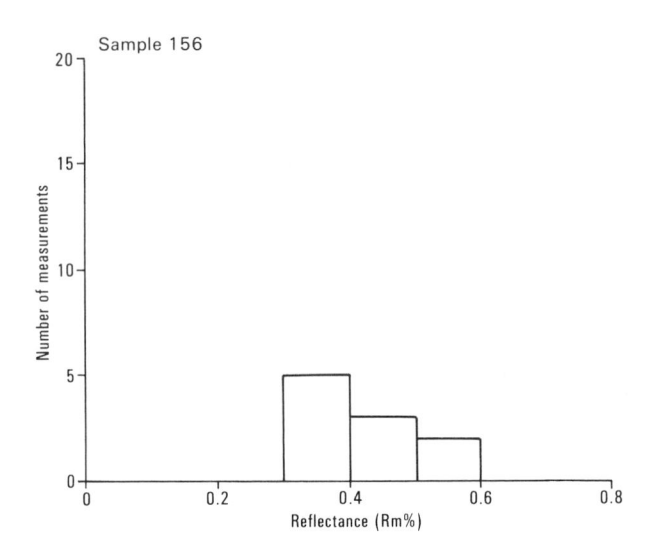

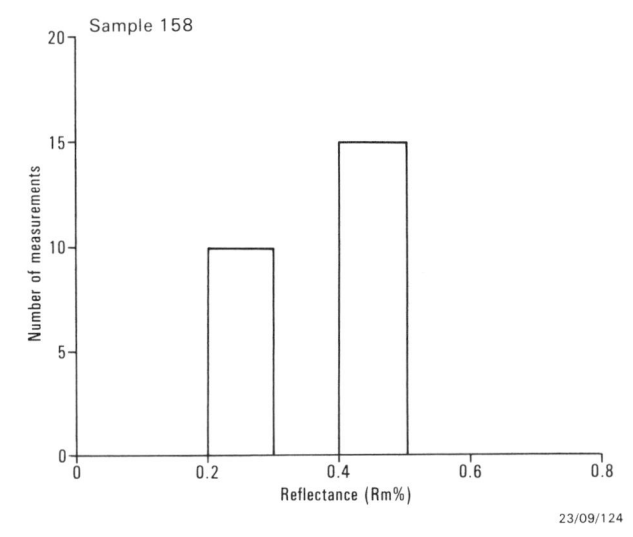

Figure 2. Reflectance measurements in the Lossuk River beds.

Transmitted light microscopy of organic concentrates shows degradation products, which are seen to be low reflectance vitrinite-like organic matter (less than 0.3%Rm) when observed in reflected light. Thin sections of sample 158 also display calcareous remains of filamentous cyanobacteria (Plate 1, figure 5) whose organic remains may form part of the lower reflectance organic matter observed by optical microscopy. There is a scarcity of pollen and spores of higher plants, which suggests a stronger marine influence in these sediments (Table 1).

Table 2. New Ireland Rock Eval data (analysis by C.J. Boreham, Bureau of Mineral Resources, Canberra).

Sample No.	Tmax °C	S_1	S_2	S_3	S_1+S_2	S_2/S_3	Production Index PI	TOC wt. %	Hydrogen Index HI	Oxygen Index OI
			---mg/g---							
LOSSUK RIVER BEDS										
152	432	0.22	1.30	4.76	1.52	0.27	0.14	2.74	47	174
153	422	0.12	0.70	2.92	0.82	0.24	0.15	1.86	38	157
154	431	0.21	1.19	4.28	1.40	0.28	0.15	2.51	47	171
155	426	0.18	0.93	3.40	1.11	0.27	0.16	2.19	42	155
157	307	0.04	0.01	1.20	0.05	0.01	0.80	0.21	5	571
LUMIS RIVER VOLCANICS										
159	407	0.19	0.52	2.71	0.71	0.19	0.27	0.90	58	301
162	403	0.13	0.25	1.90	0.38	0.13	0.34	1.08	23	176
163	423	0.03	0.90	0.85	0.12	0.11	0.25	0.30	30	283
165	422	0.03	0.28	0.82	0.31	0.34	0.10	0.68	41	121
167	379	0.17	0.28	7.01	0.45	0.40	0.38	2.68	10	262
LELET LIMESTONE										
179	394	0.02	0.04	1.00	0.60	0.04	0.33	0.05	80	2000
180	390	0.01	0.05	0.55	0.06	0.09	0.17	0.07	71	786
183	277	0.00	0.00	0.20	0.00	0.00	0.00	0.40	0	500
218	408	0.16	0.22	2.39	0.38	0.09	0.42	0.65	34	368
MATAKAN UNIT										
216	378	1.28	8.69	9.59	9.97	0.91	0.13	10.92	80	88
217	397	14.59	60.92	46.94	75.51	1.30	0.19	43.60	140	108
220	401	12.85	44.07	47.95	56.92	0.92	0.23	43.12	102	111
221	414	0.16	0.50	0.32	0.66	1.58	0.24	0.97	52	33

Hydrocarbon Potential

Lossuk River bed samples display amorphous organic matter and semi-degraded wood as predominant carbonaceous components; both lipids (as shown from TEM observations) and macerals of the liptinite group are poorly preserved in all samples. Thus the samples are gas prone.

Total organic carbon ranges from 0.21 to 2.74% (Table 2). Tmax values indicate low maturation (> 435° C as defined by Espitalie, Madec, and Tissot, 1977) which is also evident from the vitrinite reflectance data (Table 2). Cyanobacterial mats have been shown to degrade and convert into amorphous (bituminite) organic matter (Peters, Rohrback, and Kaplan, 1981; Glikson and Taylor, 1986). Part of the amorphous matter in Lossuk River bed samples may have its origin in cyanobacterial mats, the filamentous remains of which have been observed in the present study. Low hydrogen index (HI) is often indicates gas-prone organic matter but may also be the result of intense oxidation. The former is supported by C-Isotope composition of Lossuk River bed total organic matter (Table 3) which displays values typical of land-sourced material. The significantly "lighter" humic acid fraction may support low maturation, as demonstrated in earlier studies (Peters, Rohrback, and Kaplan, 1981). The microscopy and geochemistry of the outcrop samples of the Lossuk River beds clearly show organic matter of mixed source namely terrigenous and marine, of low maturation and predominantly gas prone.

Lumis River Volcanics

Depositional Environment

A strong terrigenous influence is evident in the middle to late Miocene Lumis River volcanics samples, by the abundance, and type of organic matter. Well-preserved plant tissue, probably derived from mangrove vegetation, is present in some samples (161, 167), and is accompanied by an abundance of mangrove pollen (J.M.K. Owen, Australian National

Table 3. C-isotope compositions of organic concentrates.

Sample	Unit	Lithology	^{13}C ppt relative to PDB — total organic matter	humic acids
152	Lossuk River beds	dark grey silty mudstone	-25.41	
153	Lossuk River beds	dark grey silty mudstone	-25.70	-27.10
164	Lumis River volcanics	fine grained highly carbonaceous mudstone	-25.72	-27.11
166	Lumis River volcanics	dark grey shelly siltstone	-25.78	
167	Lumis River volcanics	black coaly shale	-28.62	
177	Lelet Lst	creamy fine limestone, possible algal reef	-18.96	
178	Lelet Lst	creamy limestone with gastropods, corals and foraminifera	-19.22	
180	Lelet Lst	creamy pitted limestone with gastropods, foraminiferas and algae	-18.94	
218	Lelet Lst (Matakan unit)	black calcareous siltstone with fossils	-22.83	
221	Lelet Lst (Matakan unit)	black calcareous siltstone with fossils	-25.08	
225	Lelet Lst	grey hard limestone	-22.66	
309	Lelet Lst (Matakan unit)	coal	-24.55	-25.45
316	Lelet Lst (Matakan unit)	coal	-26.75	-27.92

University, oral comm.). Probable red algae and benthic foraminifera (sample 161) may indicate derivation from a shelf environment with mangroves in the vicinity. The high pyrite content and excellent preservation of organic matter (including inner linings of foraminifera) point to anoxic conditions in the sediments' pore waters. However, as pointed out earlier, mangrove leaf litter can be transported a long way out to sea, and down into the bathyal zone.

The abundance of the diatom *Navicula*, (sample 165), may indicate shallow intertidal pools with occasional estuarine influence (Hanic and Lobban, 1979). The resinite and other land plant contributors in this sample indicate input from highland vegetation, via rivers. Some of the primary organic matter, except wood, pollen and spores, is degraded and amorphous. The microbial remains of cf. *Flectobacillus marinus* are evidence of bacterial degradation. This bacterium, isolated from marine algae and mangrove leaves in the Florida Keys (Sieburth, 1979), is an aerobe and could be responsible for the degradation of primary organic matter in the aqueous oxygenated environment.

A strong terrigenous element, possibly from stream valleys and swamps, is observed in all samples. Although, the organic components are highly variable, a marine influence is evident. However, the type of marine environment is unclear. Certain components in some samples, such as red algae, suggest water depths not less than 80 m. Stewart and Sandy (this volume) suggest that water depths were variable, ranging from 50 to 500 m.

Light Microscopy and Transmission Electron Microscopy (TEM)

In terms of organic matter content, samples from the middle to late Miocene Lumis River volcanics resemble Lossuk River bed samples. However, sediment from the Lumis River volcanics (sample 159) show a higher diversity in organic components. Bright yellow fluorescing alginite or resinite are apparent in reflected UV light. In incident light, the vitrinite usually has a reflectance of 0.3 to 0.5% Rm, with some particles of 0.6% Rm (Figure 3, samples 159 and 160). Vitrinite-like organic matter, with reflectances of 0.2% Rm, is often associated with calcite. Little evidence of recycled vitrinite was found. Similar reflectance values were obtained for most Lumis River volcanics samples. Non-fluorescing macerals are common (Figure 3, sample 160); their vitrinite reflectance being 0.2 and 0.4% Rm, possibly representing two different vitrinites. Organic matter displaying slightly higher reflectance of 0.6% Rm (sample 159) is probably semi-fusinite (inertinite), the result of oxidation. Pyrite, in framboidal form, as well as in finely disseminated grains, occurs in significant concentrations (Plate 1, figure 7). Framboidal pyrite is abundant within organic matter, as well as within the carbonate, and occasionally shows oxidation rims (orange-red). Remains of foraminifera are often found embedded in the organic matter. TEM observations of the organic concentrate reveal humic acids to be predominant (Plate 2, figure 9, sample 160).

Sample 161 consists of interbedded black and light grey bands. Organic matter dominates in the

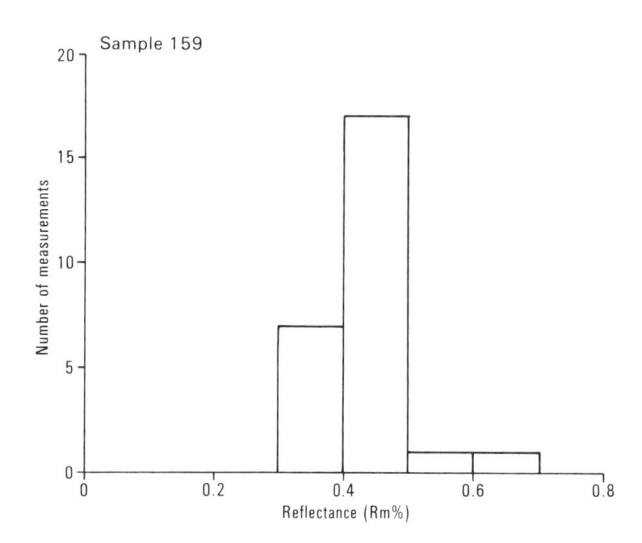

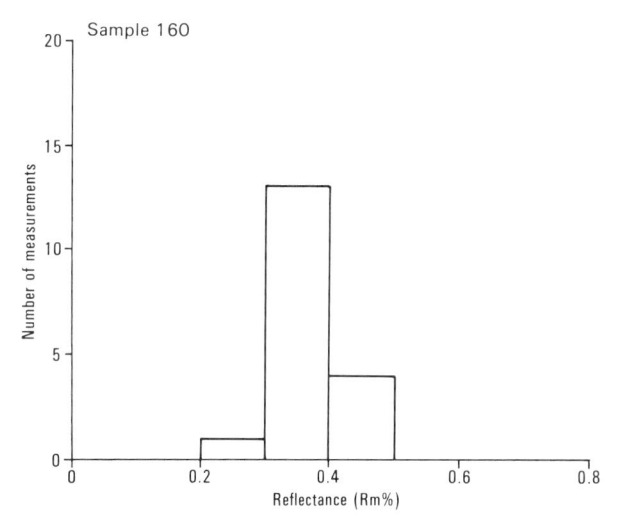

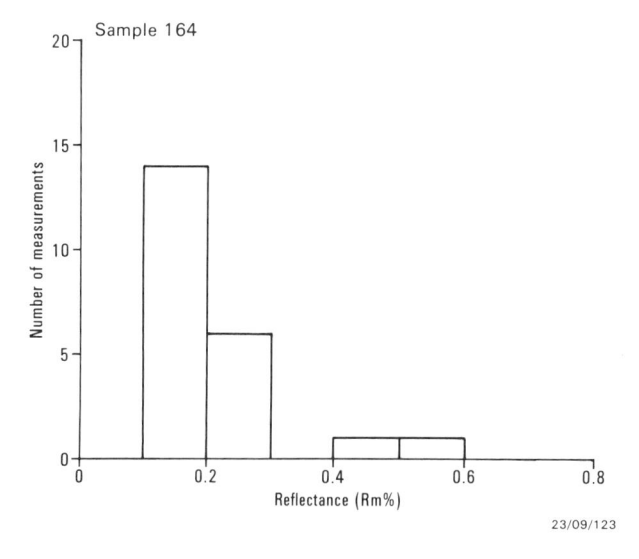

Figure 3. Reflectance measurements in the Lumis River volcanics.

black bands and $CaCO_3$ in the light grey bands. Fine laminations can be recognized within the broader bands (Plate 2, figure 10). In thin section, some of the organic matter within the organic-rich bands displays structures similar to polysporangia of red algae. Organic remains resembling root cells, as well as other cells of *Rhizophora*, are common (Plate 2, figure 11). Pollen of *Rhizophora* have been identified in the organic concentrates of sample 161 (J.M.K. Owen, oral comm.). Yellow organic matter of algal origin is seen in thin sections dispersed throughout the calcareous matrix. Organic matter is also occasionally present as foraminiferal inner linings.

The organic matter, in samples 164 and 165, observed in reflected light, is dominated by a low reflectance vitrinite-like component (Figure 3, sample 164) and pyrite. Resinite is occasionally present and displays bright yellow fluorescence in UV light. Pyritized rod-like structures about 6μm long are abundant and may be sponge spicules. Scanning electron microscope (SEM) images of whole rock fragments (Plate 2, figure 12; sample 165) show a considerable contribution from the diatom *Navicula* sp. and the bacterium cf. *Flectobacillus marinus*. The latter is known to degrade algae in present day marine environments (Sieburth, 1979). The SEM images also show aragonite crystals filling voids (Plate 3, figure 13).

In hand specimen, sample 167 is an almost coaly black shale, finely laminated and highly compacted, with a strong sulphurous smell. A dispersed organic concentrate, when viewed in transmitted light, displays a wealth of pollen and spores of various kinds - including a bisaccate pollen - possibly a *Podocarpus*, and spores of Lycopod affinity and a *Selaginella*-like type. Remains of wood are abundant, particularly conifer rodlets or microfibrils similar to those described by Camier and Siemon (1978) from Victorian (Australia) brown coals. The organic matter in reflected light reveals vitrinite and vitrinite-like components with reflectances of 0.5 and 0.2% Rm respectively, as well as some alginite(?), exinite, and micrinite. Small amounts of vitrinite, with reflectances over 0.5% Rm, are possibly reworked.

The organic matter in these rocks consists almost solely of the remains of well-preserved land plants, including pollen and spores (Table 1). The rest of the organic matter is composed of wood and tissue, and little or no amorphous or degraded organic matter is visible.

Hydrocarbon Potential

Total organic carbon percentages range from 0.30 to 2.68% (Table 2), the average being slightly lower than in Lossuk River bed samples. The organic matter is highly immature and has been partly oxidized. The high input from pollen/spores and cuticles in most Lumis River volcanics organic concentrations may indicate high condensate/oil potential, although the lower TOC may affect the overall hydrocarbon potential. Isotopically the organic matter is of the same composition as that of the Lossuk River beds. The C-isotope values generally reflect the predominant organic components. Four of the five samples analyzed from the Lossuk River beds and Lumis River volcanics, had isotopic compositions in the range of terrestrially-dominated organic matter (Table 3). The humic acid fraction analyzed separately for some samples (153 and 164), had C-isotope composition of -27.10 and -27.11 ppt respectively (Table 3), "lighter" by almost 1.5 ppt, which suggests immature material. Nissenbaum and Kaplan (1972) showed that humic acids formed *in situ* are of similar isotopic composition to their source material. In the Lossuk River bed samples, the possibility of allochthonous sources contributing to these humates should be considered. Mangrove rootlets, visible in thin sections, might be the source for the humic acids.

Rock Eval analyses of Lumis River volcanics samples give low S_2 yields, like the Lossuk River beds. On the standard Rock Eval plot the latter are located in type III kerogen, no doubt due to the high content of humic acids; these being oxidation products may not reflect the nature of deeper buried non-oxidized organic matter. The Tmax values of Lumis River volcanics are high. A few samples with Tmax values lower than Lossuk River beds values, probably reflecting the organic matter compositions; cuticles, pollen and spores are more abundant in Lumis River volcanics. Overall, the abundant exinite-sporinite and cutinite, which can generate hydrocarbons at lower maturation than other macerals, is a favorable characteristic.

Lelet Limestone

General Geology and Lithology

The Lelet Limestone is a massive early to late Miocene limestone, containing much biogenic material, up to 750 m in thickness (Stewart and Sandy, 1985), consisting of biogenic detritus with varying contributions from corals, algae and other organisms. Field observations suggest that certain "terraces" or beds are dominated by algae and corals, and foraminifera and gastropods are subordinate. However, the algae are absent from beds dominated by coral. A few coral heads in the latter indicate *in situ* formation of local patch reefs, but most corals are broken and transported. The two distinct occurrences implies different depositional environments and conditions for organic matter accumulation and preservation.

The Matakan unit of the Lelet Limestone, which includes coal pods wedged within marine sediment, has been assigned to the Lelet Limestone sequence (Stewart and Sandy, this volume). The coal was first described by Noakes (1939) as small seams and lenses within a sequence of folded Tertiary limestone, calcareous shale and mudstone. The coal seams were described as having the same strike as the surrounding rock strata, but a lower dip. The outcrops sampled for the present study are located in various tributaries of the Tahalabar River. From field observations, the coal is discordant with the limestone and closely associated with a highly calcareous, pyritic carbonaceous mudstone containing coral remains, shells, foraminifera, encrusting coralline algae, plant remains and other organic matter. Examination of the microfloras suggests a general Miocene age for the lagoonal sediment (Stewart and Sandy, 1985) and possibly a late Miocene age for the coal (Glikson and Owen, 1987).

In hand specimen, the coal (samples 216, 217, 220, 309, 313, and 316) consists of vitrinitic friable bands, as well as massive dull coal with conchoidal fracture reminiscent of torbanite. The associated sedimentary rocks are grey to black, calcareous, organic-rich, friable to hard, lithified siltstones with an abundance of coral remains (samples 218, 221, 312, and 317).

Depositional Environment

Three distinct depositional environments can be recognized in the massive Lelet Limestone:

1) An outer shelf environment giving an ooze dominated by the foraminifer *Globorotalia*, for example samples 209 and 210. These sedimentary rocks are devoid of carbonaceous matter and are exclusively composed of the tests of planktonic foraminifera and calcareous cement. The lack of organic remains and sulphides points to a well-oxygenated environment.

2) Algal-dominated deposits. Red calcareous algae of the Corallinaceae are important reef forming organisms in the ridges and edges of seaward reefs (Glynn, 1973). Organic matter varies in concentration from one sample to another, but is generally low and mainly preserved in voids within the algal structures, and in hollows created by borers.

3) A mixed environment of coral, foraminifera and algal-bearing sediment (samples 170, 173, 177, and 179).

The preservation of organic matter and pyrite in environmental group 3 favors a back-reef lagoon environment. Some samples (173 and 179) of group 3 are particularly rich (1-3%) in organic matter, for limestones. Anoxic conditions need not have been widespread in the water column during deposition; anoxicity in sediment pore-waters would have been sufficient for organic matter preservation.

In the Matakan unit, the abundance of exines and exinal fragments in some parts of the coal make it a cannel coal. However, vitrinitic coal alternates with the cannel coal, suggesting that the Matakan coals are partly autochthonous and partly allochthonous. The maceral corpocollinite, which is abundant in parts of the coal, is believed to be the product of strongly decomposed tissues of aquatic plants (Stach et al., 1975) and may indicate an aquatic source for some of the coal-forming plants. Mangrove rootlets and leaves may have contributed to the vitrinite.

The field relationships indicate that the coal has been carried into the depositional area (Stewart and Sandy, this volume). The well preserved plant remains, low humic acid content and few bacterial remains indicate that an anaerobic environment of deposition originally prevailed. Resinite and woody rodlets, possibly from conifers (Camier and Siemon, 1978), may indicate that conifers from the hinterland were contributors. Mangrove forests may have given rise to some of the vitrinitic coal; pollen and spores carried through channels and trapped in certain areas are the source of the cannel coal. The sedimentary rocks (samples 218, 221, 312, and 317) associated with the coal are possibly of a back reef lagoonal environment. Coral stems are probably not *in situ*, but have been washed in from a more oxygenated part of the lagoon.

Wiens (1962) noted that restricted lagoons accumulate bottom muds, in contrast to lagoons with channels to the open sea, where fine particles are carried out to sea by flushing. Organic matter concentrations may be low in sediment of the open lagoon

(Wiens, 1962). The opposite would apply to closed lagoons that have higher organic matter content. The high organic matter content in the sediment associated with the Matakan coal suggests that they were deposited in a closed lagoon, rather than one that was open to the ocean. A significant quantity of organic matter can be seen within and around organisms. The exceptionally high concentration of pyrite occurs mostly in framboids, but also as fine granules and solid. The diversity of the fauna excludes a restricted environment. However, tectonic activity may have brought about slumping of peat and its deposition in the lagoon, which may account for the diversity of fauna. Higher sulphur concentrations are obtained in coals subjected to long term marine influence (Smith and Batts, 1974).

Light Microscopy and Transmission Electron Microscopy (TEM)

Thin sections of the Lelet Limestone reveal reef organisms, coralline algae and foraminifera, manganese and/or pyrite and organic matter in varying amounts, often filling in cavities formed by borers (Plate 4, figures 19-21). Organic matter is sparse and is of an amber-yellow or brown color. Pyrite, if present, is in a granular dispersed form.

The organic matter concentrate, when observed in transmitted light, is amorphous and dark yellow in color. Numerous black rods, possibly magnetite, make part of the organic concentrate.

TEM studies of sample 179 revealed ultra structures as minute spicules of the coralline alga *Corallina officinalis* (Plate 4, figure 19). Humic acids and algal lipids are the main organic components. The organic matter in the limestones generally consists of lipids, bacterial metabolites/degradation products, humic acids and some semi-degraded wood. Degraded algae and other organisms, as well as the humic acids, have adsorbed metals, which are displayed as electron dense particles when observed in TEM (Glikson et al., 1985). Natural metal staining of the microbial metabolites and degraded organisms is widely observed in degraded organic matter of an aquatic source, and the preservation of the degraded organic remains in the marine environment is believed to be enhanced by such metal staining (Degens and Ittekkot, 1982).

Some samples (181 to 187) are dominated by coralline algae, and bryozoa. Most of these algae-dominated sedimentary rocks contain organic matter when observed in thin sections. The organic matter is within and around the calcareous remains of organisms and in voids within the calcareous matrix. It is usually of dark yellow coloration and occasionally displays microreticulate structure. In samples dominated by foraminifera (188 and 189), the organic matter is often preserved in their cavities, as well as in spaces between the organisms.

In hand specimen, samples 204 to 210 vary from silty carbonaceous brown limestone to white marls, coral and algal limestones and shelly sandstones. Thin sections show all of these samples are devoid of organic matter, and consist of cemented foraminiferal ooze. The foraminifera were mainly planktonic, and primarily *Globorotalia tumida*, although the assemblage also includes *Lepidocyclina* and *Globigerina*.

The Matakan Unit of the Lelet Limestone Sequence

Polished surfaces of the coal in reflected incident light display vitrinite, the reflectance of which may vary within bands. Some variation can be attributed to oxidation, but most seems to be inherent, as shown by TEM. Cuticular structures are common (Plate 3, figure 18). Other organic components are inertinite/fusinite and the macerals sporinite and resinite. Vitrinite reflectance ranges from 0.3 to 0.5%Rm (Figure 4); observations in blue fluorescent light reveal the predominant organic component in the cannel coal to be exines of pollen and spores of land plants (Plate 4, figures 19 and 20) and resinite. Exsudatinite fills some microfissures leading from concentrations of exines, but more commonly fills the cavities of the pollen grains themselves. The fluorescence colors of the exines range from bright yellow-green through orange to brown. The vitrinite appears as phlobaphinite in some coal samples and as corpocollinite (Plate 3, figure 17) in others.

The sedimentary rocks associated with the coal (samples 218, 221, 312, and 317) are dark grey to black, calcareous and highly fossiliferous silty biomicrites. Polished surfaces of sample 218 observed in reflected light, as well as thin sections viewed in transmitted light, show unusually abundant pyrite, replacing or infiltrating organic matter and calcite in organisms (Plate 3, figure 15). Vitrinite-like organic matter (bituminite, sample 218) displays low reflectances (0.1 to 0.2% Rm) and commonly occurs interstitially between calcareous particles or within cavities of organisms. Organic concentrates observed in transmitted light contain opaque rods that could be sponge spicules where the calcite has been replaced by pyrite. Wood fragments abound, as well as thin-walled trilete and monolete spores and amor-

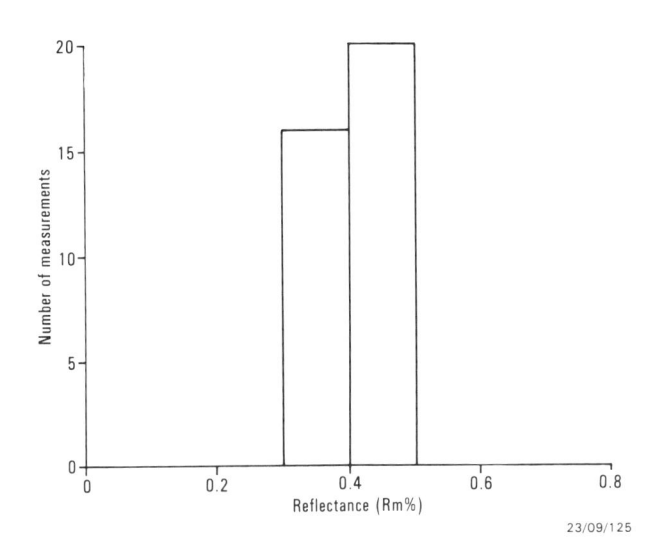

Figure 4. Reflectance measurements of sample 217 in the Matakan coal unit of the Lelet Limestone.

phous organic matter. Rod-shaped woody particles, similar to those described from Victorian brown coals, are common; the latter are attributed to microfibrils of conifers (Camier and Siemon, 1978). In contrast, conifer pollen was sparsely represented in the microfloras analyzed from these samples (Glikson and Owen, 1987). In UV light with blue light excitation, green-yellow fluorescence is displayed by the exines of spores and resin bodies. The predominant organic components (vitrinite and bituminite) in these samples do not fluoresce.

In hand specimen, the marine sedimentary rocks associated with the coal (samples 218, 221, and 317), are highly fossiliferous with planktonic foraminifera, gastropods, encrusting algae and the stems of a coral. The organisms are embedded in a matrix of silt, clay and organic matter. The organic concentrate (sample 218) observed in transmitted light, displays an abundance of conifer microfibrils (Plate 3, figure 14), amorphous organic matter, pollen of land plants and abundant pyrite. In reflected light the organic matter appears to have a high contribution from vitrinite.

Thin sections (samples 221 and 317) show organic matter of algal and terrigenous source, framboidal pyrite within both organic and inorganic components, coral remains and abundant planktonic foraminifera. The filamentous cyanobacterium *Oscillatoria* is a contributor to the calcareous, as well as organic remains in sample 221. Sample 317 is characterized by predominance of coralline encrusting algae, the organic remains of which are commonly preserved.

Ultra-thin sections of the coal observed in TEM often show the vitrinite to be composed of two types: 1) more or less homogeneous, medium to low electron dense material, and 2) material of higher electron density. The first type of vitrinite corresponds to that part displaying lower reflectance in light microscopy, whereas the second kind exhibits higher reflectances.

Exines of pollen and spores, resinite, alginite and exsudatinite are revealed by TEM (Plate 4, figure 21). Hydrocarbons (exsudatinite), which appear to have generated from pollen, are seen filling the pollen cavity, and tend to aggregate in minute globules, displaying the electron density characteristic of lipoid material (electron density, being a function of mass number, exhibits characteristics in TEM corresponding to H/C ratios). Bacterial lipids are usually more electron-transparent than the above mentioned macerals and mostly retain their morphology. The latter are rare in the coal samples. Humic acids and partly decomposed vitrinite can be observed in places (Plate 4, figure 22) having adsorbed metals (electron-dense particles). Pyrite seen in TEM observations of samples associated with the coals (218, 221 and 317) has infiltrated and replaced organic matter (Plate 3, figure 16). The organic matter in these samples is dominated by algal components, lipoid material, pyrite and humic acids.

Hydrocarbon Potential

The Lelet limestones contain varying amounts of organic matter, which could make this sequence a potential source rock for hydrocarbons although values of TOC are generally less than 0.5%. Due to its great thickness and the type of organic matter, the sequence may contribute to hydrocarbon accumulation in deeper parts of the New Ireland Basin. Gehman (1962) showed that some recent calcareous sediments contain up to 8% by weight organic matter, similar ancient sedimentary rocks mostly have a range of 0.3 to 1% and only a few have yielded values between 1 to 8%. He concluded that the loss of organic matter must be tied to processes of lithification such as the circulation of meteoric water in highly porous limestones, which provides oxygen to oxidize the organic matter. The adsorptive power of lime, as compared to clay, is low (Gehman, 1962), which is a factor contributing to the lower organic content of ancient limestones as compared to shales. Gehman (1962) has also shown that hydrocarbons in particular are found in higher concentrations in limestones as compared to shales and also in recent lime sediments as compared to recent clayey sediment.

The algal reef facies of the Lelet Limestone contains up to 2% organic matter in surface samples, and the unweathered section could contain much higher concentrations. T.G. Powell (oral comm.) pointed out that particular organic-rich beds in an otherwise organic-poor reef complex in Canada were found to be petroleum source rocks. The TOC of selected samples analyzed from the limestones (samples 179, 180, and 183) is low (Table 2). Tmax values for the limestone samples is also low. However, Foster, O'Brien and Watson (1986) clearly demonstrated that oxidation and mixed organic matter precursors affect the hydrogen index (HI) and therefore HI/Tmax plots are not always indicators of maturation. Widespread oxidation in the Lelet limestones is evident from concentrations of humic acids observed in TEM. The C-isotopic composition of the organic concentrates indicates that reef algae were the precursor organisms, the occasional lipidic remains of which are observed in TEM. Therefore, parts of the limestone complex may be less affected by oxidation and have higher concentrations of organic matter and greater petroleum potential.

C-isotope compositions of the Lelet Limestone organic concentrates (samples 177, 178, 180, and 225) are typical for algal and coral reef organic matter (Land, Lang, and Barnes, 1977; Holmes, 1983). A reef, being a highly productive environment where organisms are decomposed and some organic matter is oxidized, may form a suitable environment for humic acid formation by melanoidin-type reactions of carbohydrates and amino acids from decomposing organisms. Sample 225 is somewhat lighter than the other reef-sourced organic matter, possibly because of particulate organic matter from plankton.

Organic concentrates of the Matakan unit within the Lelet sequence vary in their C-isotope compositions from -22.8 ppt in the carbonaceous lagoonal sediments of the unit, to -26.7 ppt in the coal (Table 3). The former having had a higher contribution from marine plankton and red algae, whereas sample 221 from the same sequence displays the strong influence of terrestrial material. Isotope compositions of the two coal samples (309 and 316) vary only by 1 ppt. The humic acid fractions of both coal samples are slightly isotopically lighter than the rest of the organic matter, possibly implying the incorporation of external source material.

The high potential of the Matakan unit to generate oil, as shown by geochemical as well as microscopical observations, is undisputed. Sporinite, and possibly resinite, have generated hydrocarbons. The boundaries of the "oil window", as defined by vitrinite reflectance values, are 0.35 to 0.7% Ro, depending on the types of kerogen (Barnard, Collins, and Cooper, 1982; Tissot, 1984; Tissot and Welte, 1984). Snowdon and Powell, (1982) state that the onset of oil generation in terrestrially-derived organic matter is a function of the relative input from resinite, liptinite and vitrinite, and have shown that heavy oils and condensates in the Beaufort-Mackenzie Basin, Canada, were generated from land plant-derived organic matter at reflectance levels of 0.4 to 0.6% Ro. The source for low temperature oil generation has been attributed to resinite (Snowdon, 1980), whose hydrogen-rich, oxygen-poor nature has long been known (Murchison and Jones, 1963). Resinite in the Matakan coal may be one source for hydrocarbons, but much of the exsudatinite seems to be associated with exines, whose waxy part may have generated hydrocarbons early. TEM observations in the present study suggest that the endexine is responsible for hydrocarbon generation.

Free hydrocarbons in the Matakan coal is shown by Rock Eval data (Table 2): the two coal samples analyzed (217 and 220) indicated early generated hydrocarbons/exsudatinite (S_1) and low Tmax values. Bright yellow fluorescing hydrocarbon accumulations in lenses and thin bands, like those observed in the Matakan coal, have been reported from the Posidonia shale, northwestern Germany (van Gijzel, 1981) in a transition zone from immature to early mature (about 0.45 to 0.55% Ro). According to van Gijzel (1981), the type of oil may be distinguished in fluorescence microscopy. Thus yellow or orange fluorescence is displayed by naphthenic oils, which are generated from terrestrial organic matter rich in resin at vitrinite reflectances of 0.40 to 0.60% Ro (Snowdon and Powell, 1982; Powell and Snowdon, 1983).

CONCLUSIONS

The Lossuk River beds and Lumis River volcanics, the oldest of the sedimentary rocks collected for this study, have variable lithologies and a diversity in type and content of organic matter. This diversity complicates assessments of their petroleum potential, especially as the organic matter is immature, and oxidized in outcrop samples. Both sequences are regarded as of bathyal origin, and their organic matter is mixed and includes remains of land plants, as well as degraded algal and cyanobacterial remains. The latter are regarded as promising petroleum source materials (Dean, Claypool, and Thiede, 1984; Williams, 1984).

The Rock Eval analyses (Table 2) indicate at first glance the dominance of kerogen type III. The humic acid content of these samples suggests a classification of Type III or possibly Type IV kerogen. Humic acids are oxidative biodegradation products that predominate during early stages of diagenesis under oxidizing conditions, and have been shown experimentally to be converted into gas-prone kerogen (Peters et al., 1981). Dean, Claypool and Thiede (1984) indicated that highly oxidized organic matter from a marine-source may give pyrolysis results analogous to those obtained from organic matter of Type III kerogen composition, and this may apply to the Lossuk River beds. On the other hand, the Lumis River volcanics show a definite predominance of land plants and are unlikely to have significantly different constituents in their deeply buried counterparts. The type of land-derived plant material is important in assessing hydrocarbon potential, because exines and cuticles are known to generate liquid hydrocarbons and wood remains (vitrinite) and humic acids are gas-prone. Maceral analysis shows that exines occur in sufficient quantities in some parts of the Lumis River volcanics to provide an oil generation potential. The high pyrite content in most samples suggests an initial high organic matter content; up to 0.4% of total organic carbon is lost from sediments during sulphate reduction (Berner, 1982).

Most major oil accumulations in the world originate from source rocks with TOC of more than 2.5% by weight (Jones, 1980) but the source rocks underlying the reservoirs of the Niger Delta and the Mississippi have 0.3 to 1.0% TOC (Evamy et al., 1977; Jones, 1980). A minimum of 0.4% TOC is needed for oil generation and expulsion (Dow, 1977). The thick, biogenic Lelet Limestone sequence may constitute a potential petroleum source rock under certain conditions. TOC values are variable; the samples analyzed yielded values of 0.65% or less. Shorter chain length n-alkanes are expelled more effectively than their higher homologues (Leythauser et al., 1984). Because marine organisms and many bacteria yield the former, and also provide the hydrocarbon source in the Lelet, then perhaps the hydrocarbons have been mobilized.

The samples studied here have all been collected from within the weathering zone, where a considerable part of the organic matter must have been oxidized. Rock samples from drill cores are necessary to determine the amount of primary organic matter, the degree of oxidation and the levels of maturation of the organic matter with depth.

The Matakan unit within the Lelet Limestone has high organic contents, both in the lagoonal sedimentary rocks and in coal inclusions. Matakan marine strata are regarded as a lagoonal facies of the Lelet Limestone (Stewart and Sandy, this volume). The coal is allochthonous and varied in terms of its vitrinitic-cannel components. Coal inclusions rich in cutinite and exinite have generated hydrocarbons at low maturation (0.4 and 0.5% Rm) as observed in reflected UV fluorescence light microscopy. The latter has also been confirmed by free hydrocarbons detected in Rock Eval pyrolysis and low Tmax; however, the main hydrocarbon generation stage has not been reached as judged by their low yields at low T max values (C.J. Boreham, oral comm.). Based on the organic carbon content, H/C ratios, and Rock Eval pyrolysis results, the hydrocarbon generation potential of the Matakan unit samples is considered to be high.

The proportion of the Matakan unit within the Lelet limestone has not been established. The cannel parts of the coal are potentially oil-generating, whereas the vitrinitic coal would be mostly gas-prone. Maceral analysis indicates that the cannel coal is not a major constituent of the coal as a whole. The fossiliferous marine strata associated with the coal are sufficiently organic-rich to constitute a potential hydrocarbon source. Their high humic acid content indicates that they may be gas prone, although this may be the product of post-depositional oxidation and may not reflect the original source material (as shown by C-isotope composition of sample 218). TEM shows that resinite and algal lipidic organic matter are present in these rocks. The decomposing algae may have produced the humic acids. These lagoonal rocks include remains of marine organisms, as well as spores and semi-degraded wood, so dual-sourced organic matter is present.

Two factors may affect the conclusions of the present study: 1) variations in the thermal gradient, and 2) the nature of the organic matter below the weathering zone. The present study indicates a potential for gas and perhaps oil generation based on the organic matter content in the Lossuk River beds, the Lumis River volcanics, and possibly in parts of the Lelet Limestone sequence.

ACKNOWLEDGMENTS

This study was funded by the Australian

Development Assistance Bureau as part of the Tripartite marine geoscience program of assistance for CCOP/SOPAC countries, the Geological Survey of Papua New Guinea (GSPNG), and the Centre for Resource and Environmental Studies (CRES), Australian National University (ANU), Canberra. Special thanks are due to M.J. Sandy and Upu Kila (GSPNG) for assistance in sample collecting, and C.J. Boreham, Bureau of Mineral Resources (BMR), Canberra for Rock Eval pyrolysis. Thanks are due to the ANU Research School of Biological Sciences' (RSBS) Electron Microscope Unit for use of facilities. C-isotope analysis was carried out by Zarko Roksandic, RSBS, and his assistance is gratefully acknowledged. I am indebted to N.F. Exon and G.W. O'Brien (BMR), L. Magoon, (USGS, Menlo Park) and G. Francis (GSPNG) for reviewing the manuscript and many constructive comments. A. Bayes' (CRES) computer program was used for initial presentation of reflectance histograms, and the final figures were drawn by A. Murray in the BMR drawing office.

REFERENCES

Barnard, P.C., Collins, A.G. and Cooper, B.S., 1982, Generation of hydrocarbons: time, temperature and source rock quality, *in* Brooks, J., ed., Organic maturation studies and fossil fuel exploration: New York, Academic Press, p. 337-342.

Berner, R.A., 1982, Burial of organic carbon and pyrite sulfur in the modern ocean: Its geochemical and environmental significance: Australian Journal Science, v. 282, p. 451-473.

Burgess, J.D., 1974, Microscopic examination of kerogen (dispersed organic matter) in petroleum exploration: Geological Society America, Special Paper 153, p. 19-30.

Camier, R.J., and Siemon, S.R., 1978, Colloidal structure of Victorian brown coals. 2. Rod-shaped particles in brown coal: Fuel, v. 57, p. 693-696.

Chivas, A.R., and Torgensen, T., 1985, Terrestrial organic carbon in marine sediments, a preliminary balance for a mangrove environment [abs.]: 2nd Australian Stable Isotope Conference, Macquarie University, Abstracts of Papers, p. 16.

Dean, W.E., Claypool, G.E., and Thiede, J., 1984, Accumulation of organic matter in Cretaceous oxygen-deficient depositional environments in the central Pacific Ocean: Organic Geochemistry, v. 7, p. 39-51.

Degens, E.T., and Ittekkot, V., 1982, *In situ* metal-staining of biological membranes in sediments: Nature, v. 298, p. 262-268.

Dow, W.G., 1977, Kerogen studies and geological interpretation: Journal Geochemical Exploration, v. 7, p. 79-99.

Espitalie, L., Madec, M., and Tissot, B., 1977, Source rock characterization method for petroleum exploration: Offshore Technology Conference, 9th, Houston, Texas, 1977, Proceedings, p. 439-443.

Evamy, B.D., Harembourne, J., Kamerling, P., Knapp, W.A., Mulloy, F.A., and Rowlands, P.H., 1977, Hydrocarbon habitat of Tertiary Niger delta: American Association of Petroleum Geologists Bulletin, v. 62, p. 1-39.

Foster, C.B., O'Brien, G.W., and Watson, S.T., 1986, Hydrocarbon source potential of the Goldwyer Formation, Barbwire Terrace, Canning Basin, Western Australia: The APEA Journal (Australia), v. 261, no. 1, p. 142-155.

Gehman, J.M., Jr., 1962, Organic matter in limestones: Geochimica Cosmochimica Acta, v. 26, p. 885-897.

Glikson, M., Chappell, B.W., Freeman, R.S., and Webber, E., 1985, Trace elements in oil shales, their source and organic association with particular reference to Australian deposits: Chemical Geology, v. 53, p. 155-174.

Glikson, M., and Oven, J.A.K., 1987, A New Ireland coal and associated sediments: hydrocarbon generation from pollen exines at low maturation: Journal of Southeast Asian Earth Sciences, v. 1, no. 4, p. 221-234.

Glikson, M., and Taylor, G.H., 1986, Cyanobacterial mats; major contributors to Toolebuc Formation oil shales: Geological Society of Australia Special Publications no. 12, p. 273-286.

Glynn, P.W., 1973, Aspects of the ecology of coral reefs in the western Atlantic Region, *in* Jones, O.A., and Endean, R., eds., Biology and geology of coral reefs V. II: London, Biology, Academic Press, p. 271-181.

Hanic, L.A., and Lobban, C.S., 1979, Observations on *Navicula ulvacea*, a range foliose marine diatom: Journal Phycology, v. 15, p. 174-181.

Holmes, C.W., 1983, S^{18}O variations in the *Halimeda* of Virgin Islands sands: Evidence of cool water in the northeast Caribbean, late Holocene: Journal Sedimentary Petrology, v. 53, no. 2, p. 429-439.

Jones, R.W., 1980, Some mass balance and geological constraints on migration mechanisms, *in* Roberts, W.H., and Cordell, R.J., eds, Problems of petroleum migration: American Association of Petroleum Geologists Studies in Geology, v. 10, p. 47-68.

Land, L.S., Lang, J.C., and Barnes, D.J., 1977, On the stable carbon and oxygen isotopic composition of some shallow water, ahermatypic scleractinian coral skeletons: Geochimica et Cosmochimica Acta, v. 41, p. 169-172.

Leythauser, D., McKenzie, A., Schaefer, R.G., and Bjory, M., 1984, A novel approach for recognition and quantification of hydrocarbon migration effects in shale-sandstone sequences: American Association of Petroleum Geologists Bulletin, v. 68, p. 196-219.

Murchison, D.G., and Jones, J.M., 1963, Properties of the coal macerals: elementary composition of resinite: Fuel, v. 42, p. 141-158.

Nissenbaum, A., and Kaplan, I.R., 1972, Chemical and isotopic evidence for the *in situ* origin of marine humic substances: Limnological Oceanography, v. 17, no. 4, p. 570-580.

Noakes, L.C., 1939, Geological report on the occurrence of lignite at Matakan plantation, New Ireland: Bureau of Mineral Resources, Australia, Report, (unpublished), 8 p.

Peters, K.E., Rohrback, B.G., and Kaplan, I.R., 1981, Geochemistry of artificially heated humic and sapropelic sediments--I: Proto-kerogen: American Association of Petroleum Geologists Bulletin, v. 65, p. 688-705.

Powell, T.G., and Snowdon, L.R., 1983, A composite hydrocarbon generation model: Erdol und Kohle, v. 36, p. 163-170.

Smith, J.W., and Batts, B.D., 1974, The distribution and isotopic composition of sulfur in coal: Geochimica et Cosmochimica Acta, v. 38, p. 121-133.

Snowdon, L.R., 1980, Resinite - a potential petroleum source in the Upper Cretaceous/Tertiary of the Beaufort-Mackenzie Basin, *in* Miall, A.D., ed., Facts and principles of world petroleum occurrence: Canadian Society of Petroleum Geology Memoir, v. 6, p. 421-44.

Snowdon, L.R., and Powell, T.G., 1982, Immature oil and condensate - modification of hydrocarbon generation model for terrestrial organic matter: American Association of Petroleum Geologists Bulletin, v. 66, p. 775-788.

Stach, E., Mackowsky, M.-Th., Teichmuller, M., Taylor, G.H., Chandra, D., and Teichmuller, R., 1975, Coal Petrology: Berlin, Gebruder Borntraeger, 428 p.

Stewart, W.D., and Sandy, M.J., 1985, Geology of New Ireland and Djaul Island, Northeastern Papua New Guinea: Geological Survey Papua New Guinea Report, (unpublished), 45 p.

Sieburth, J. McN., 1979, Sea Miocrobes: Oxford, Oxford University Press, 491 p.

Teichmuller, M., and Teichmuller R., 1981, The significance of coalification studies to geology - a review: Bulletin Centres Recherche Exploration Production Elf-Aquitaine, v. 5, p. 491-534.

Tissot, B.P., 1984, Recent advances in petroleum geochemistry applied to hydrocarbon exploration: American Association of Petroleum Geologists Bulletin, v. 68, p. 545-563.

Tissot, B.P., and Welte, D.H., 1984, Petroleum formation and occurrence: Berlin, Springer-Verlag, 699 p.

Van Gijzel, P., 1981, Applications of the geomicrophotometry of kerogen, solid hydrocarbons and crude oils to petroleum exploration, in Brooks, J., ed., Organic maturation studies and fossil fuel exploration: New York, Academic Press, p. 351-377.

Wiens, H.J., 1962, Atoll Environment and Ecology: New Haven, Yale University Press, 532 p.

Williams, L.A., 1984, Subtidal stromatolites in Monterey Formation and other organic-rich rocks as suggested source contributors to petroleum formation: American Association of Petroleum Geologists Bulletin, v. 68, p. 1879-1893.

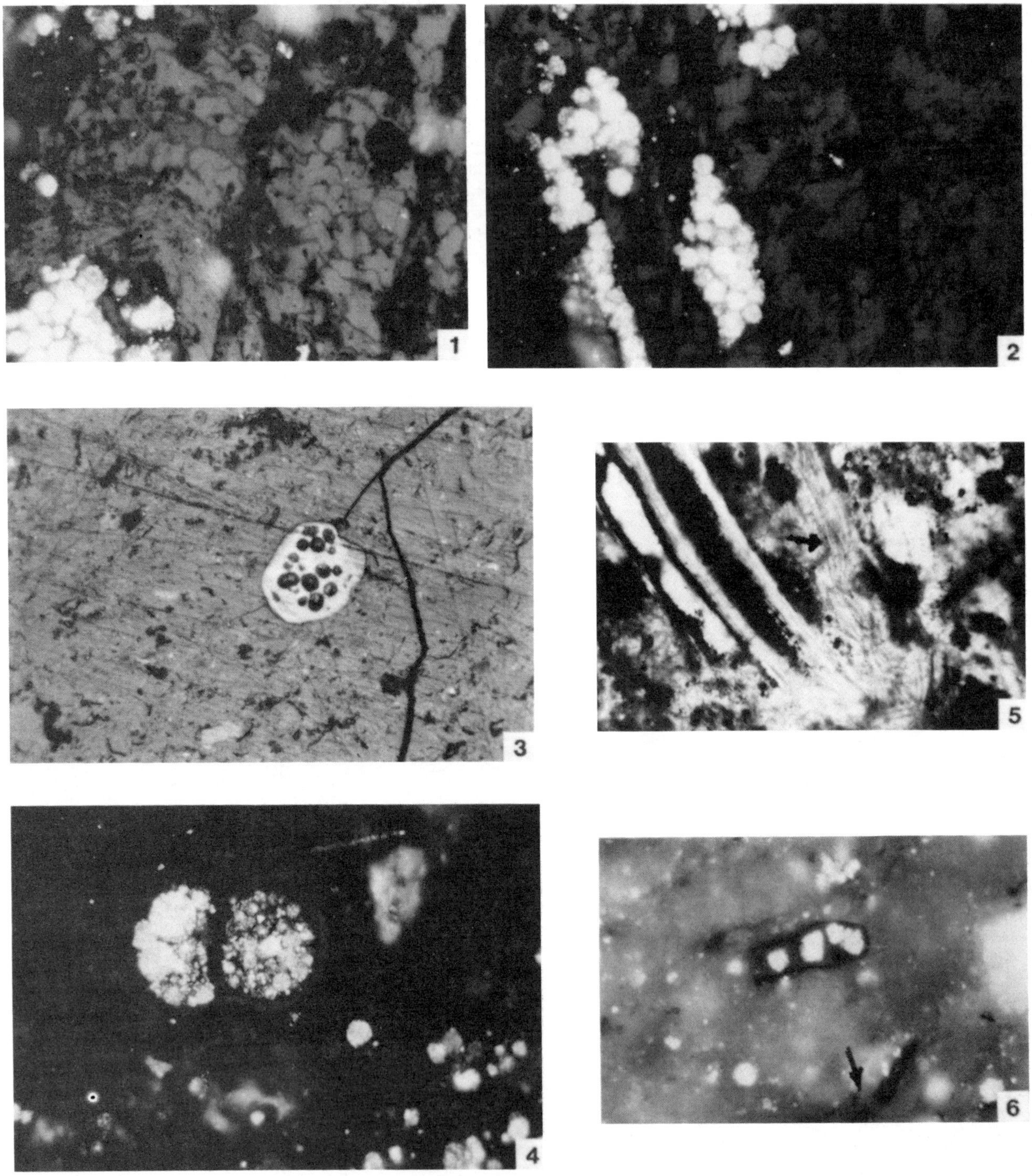
1
2
3
5
4
6

Plate 1 (facing page).

Figure	
1	Sample 158 from Lossuk River beds. Vitrinite (tellinite) showing cell structure and variation in reflectance of cell content (hydrogen rich versus hydrogen poor), framboidal pyrite, non oxidized and oxidized. Very dark spheres are resinite. x 400.
2	Sample 158 from Lossuk River beds. Vitrinite mostly in the form of collinite, framboidal pyrite. x 400.
3	Lossuk River beds sample 158. Inertinite ("sclerotinite" body) in collinitic vitrinite matrix. x 400.
4	Lossuk River beds sample 158. Framboidal pyrite filling foraminiferal test. x 400.
5	Transmitted light, thin section of sample 158 from Lossuk River beds. Cyanobacterial filaments of Girvanella type (arrow). x 250
6	Lossuk River beds sample 156. Organic matter in reflected light shows structures of fungal origin with pyrite and cyanobacterial remains (arrow). x 250.

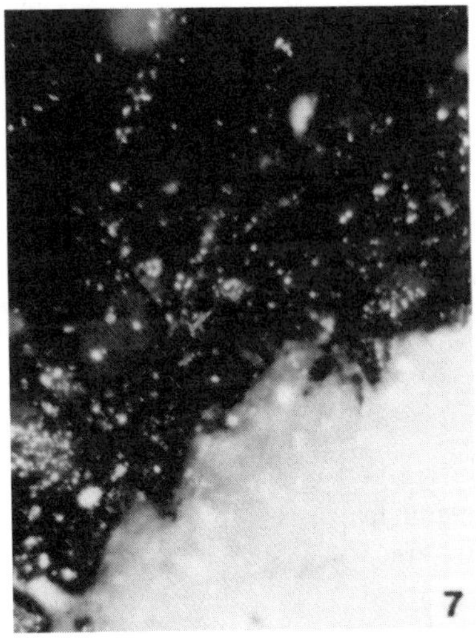
7

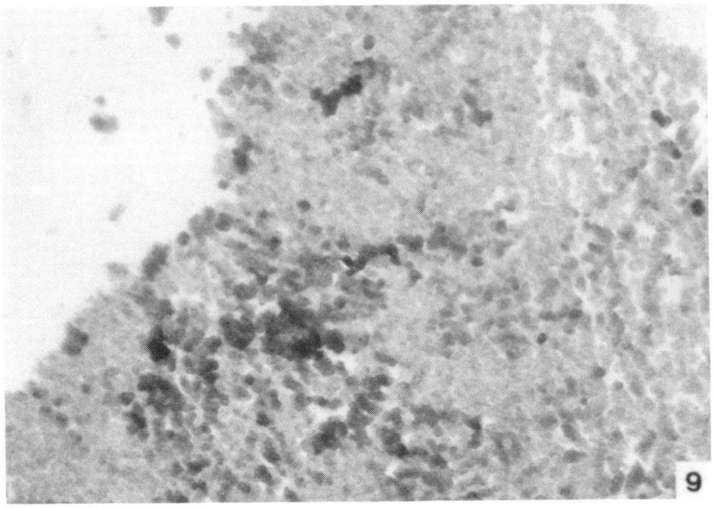
9

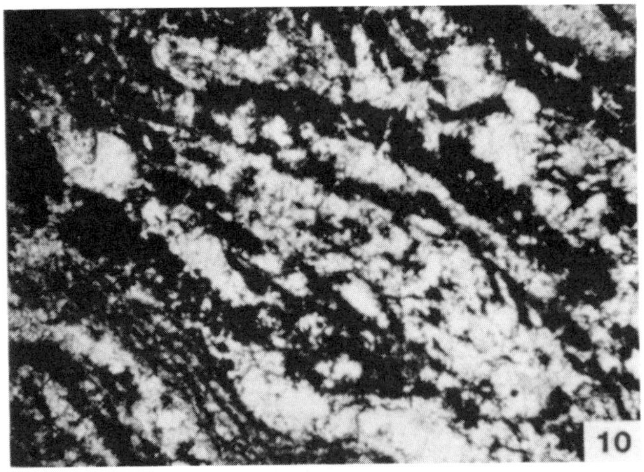
10

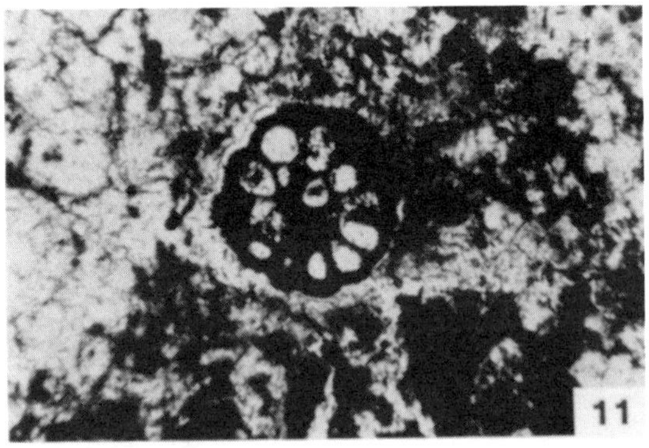
11

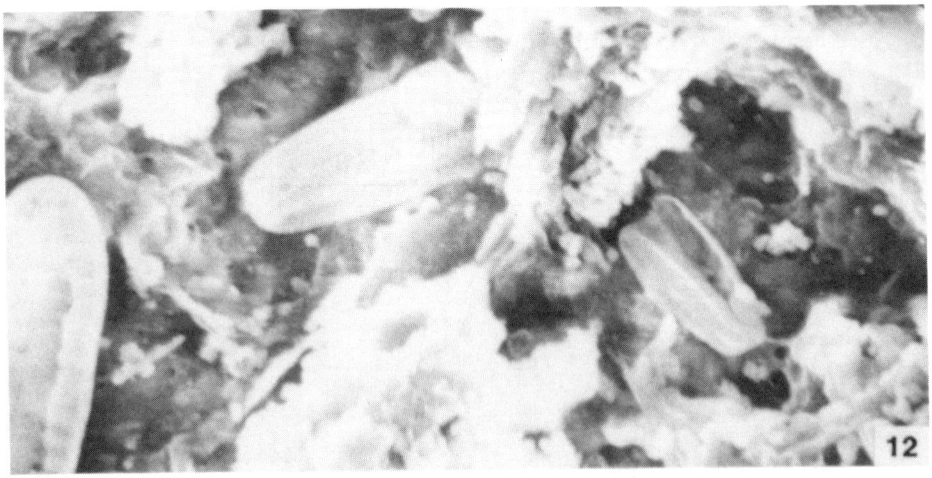
12

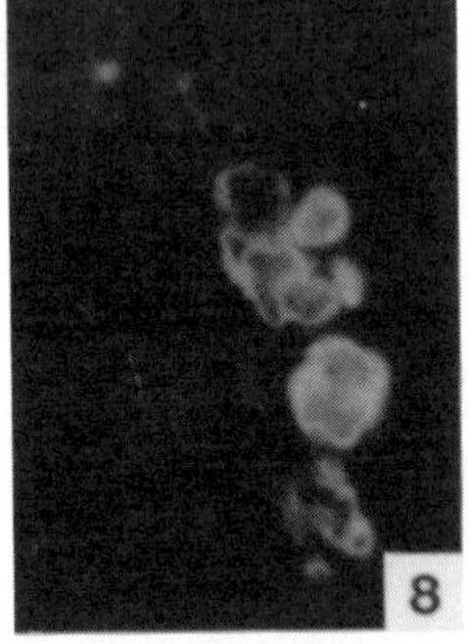
8

Plate 2 (facing page).

Figure

7   Lumis River volcanics. Vitrinite-like (bitumenite) organic matter and finely dispersed pyrite, reflected light, sample 159. x 250.

8   Reflected UV light (blue filter excitation) of sample 158, from Lossuk River beds, showing exines of land plants and exsudatinite filling exinal cavities. x 400.

9   Lumis River volcanics, sample 160 in TEM, displaying humic acid concentrations. x 33,000.

10 Lumis River volcanics, sample 161, viewed in transmitted light. Thin section perpendicular to bedding showing organic laminated matter alternating with calcareous laminae. Pale yellow organic matter dispersed within matrix is probably of algal source. x 250.

11 Lumis River volcanics, sample 161, section parallel to bedding showing pyritized wood and pyritized fructification structure. x 250.

12 Scanning electron micrograph of whole rock sample 165 from Lumis River volcanics, showing the diatom *Navicula*. x 3,000.

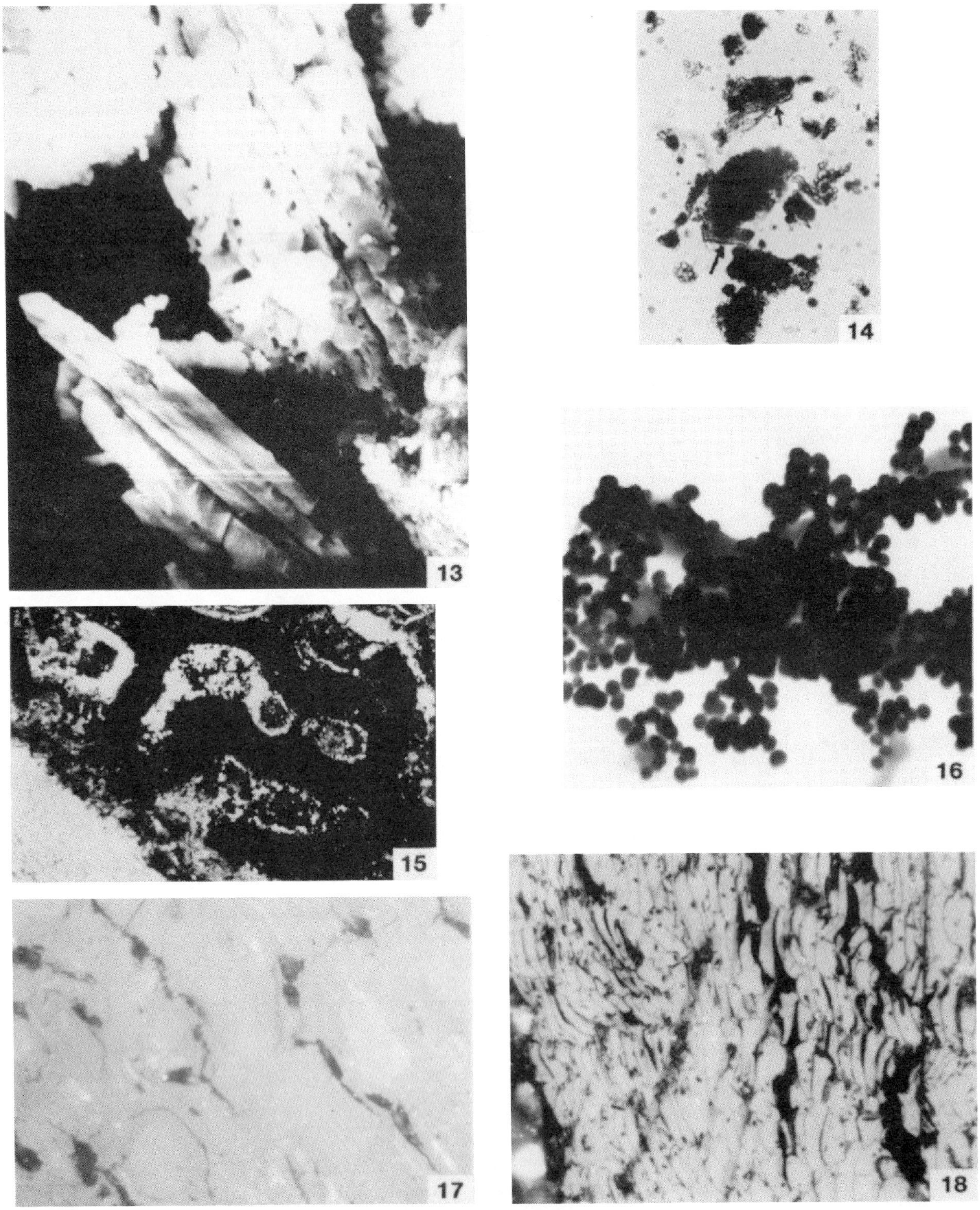
13
14
15
16
17
18

Plate 3 (facing page).

Figure	
13	Scanning electron micrograph of the whole rock sample 165 from Lumis River volcanics. Aragonite crystals filling in voids. x 3,000.
14	Sample 218, Matakan unit of Lelet Limestone. Organic concentrate in transmitted light, showing conifer microfibrils (arrows), amorphous organic matter and pyrite. x 250.
15	Thin section of rock sample 218 (Matakan unit) showing calcareous remains of organisms. The organic remains, as well as other organic matter, have been mostly replaced/infiltrated by pyrite. Transmitted light. x 250.
16	Lelet Limestone. Pyrite concentrations in organic concentrates, TEM, sample 179. x 24,000.
17	Coal sample 307 (Matakan unit) displaying corpocollinite structures on polished surface in reflected light. x 350.
18	Vitrinite (tellinite), sample 217 (Matakan unit) in reflected light, showing exinite and exsudatinite (dark areas). x 350.

182 GLIKSON

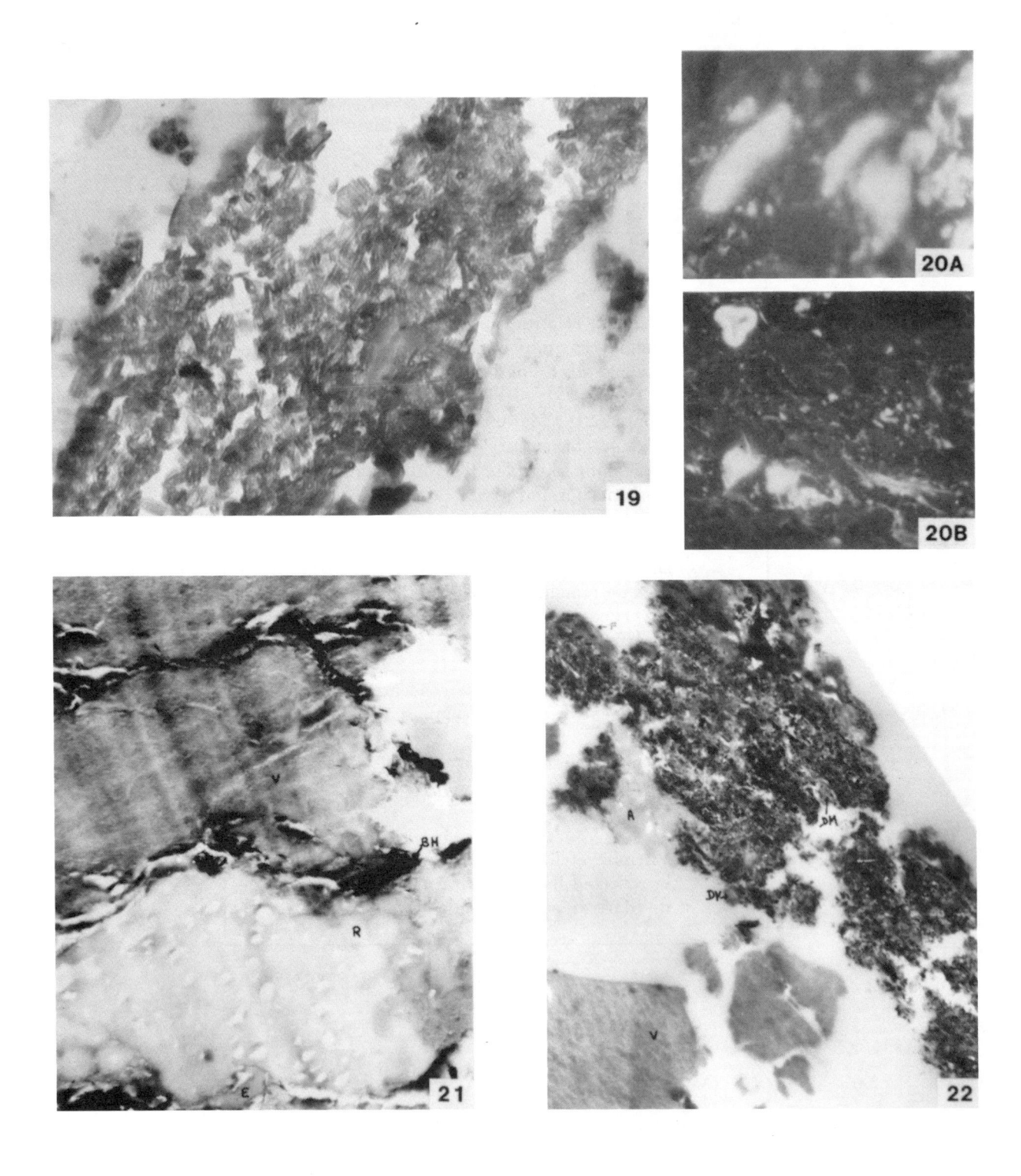

Plate 4 (facing page).

Figure	
19	Lelet Limestone. Ultra structures (spicules) of coralline algae, TEM, sample 179. x 27,000.
20A, B	Exsudatinite in a cannel coal (Matakan unit), mostly filling pollen cavities. Reflected UV light, blue filter, fluorescence mode, sample 217. x 350.
21	Sample 217 (Matakan unit) in TEM. Exsudatinite (R), vitrinite (V), exinite (E). x 26,000.
22	Sample 217, Matakan coal, in TEM. Vitrinite (V), degraded material (DV), filamentous organisms (F), metal adsorbed by degraded organic matter (DM), algal lipids (A). x 20,000.

Marlow, M.S., Dadisman, S.V., and Exon, N.F., editors, 1988, Geology and offshore resources of Pacific island arcs—New Ireland and Manus region, Papua New Guinea, Circum-Pacific Council for Energy and Mineral Resources Earth Science Series, v. 9: Houston, Texas, Circum-Pacific Council for Energy and Mineral Resources.

THE PETROLEUM POTENTIAL OF THE NEW IRELAND BASIN, PAPUA NEW GUINEA

N. F. Exon
Bureau of Mineral Resources, Geology and Geophysics, Canberra, A.C.T. 2601, Australia

M. S. Marlow
U.S. Geological Survey, Menlo Park, California 94025, USA

ABSTRACT

The New Ireland Basin forms a simple arcuate downwarp, mostly offshore, extending 600 km southeastward from Mussau to Feni Island, and is about 150 km wide. The basin is bounded by the slope of the Manus Trench to the north, and the actively spreading 2000 m-deep Manus Basin to the south. It contains as much as 7 km of Eocene and younger strata. The total basin area is about 90,000 km^2 of which about one third is either dry land or is covered by water less than 1000 m deep.

The geology of New Ireland, New Hanover, Mussau, and the Tabar-to-Feni group of islands indicates that the New Ireland Basin contains thick sequences of Eocene to earliest Miocene arc volcanic rocks, Miocene shallow-marine volcaniclastic sedimentary rocks, Miocene shelf limestone, and latest Miocene to Recent pelagic carbonates and volcaniclastic turbidites. The Miocene sedimentary rocks have some potential as source rocks for both oil and gas, but the outcrops are generally immature. The Miocene carbonate sequences would be potential reservoir rocks if deeply buried offshore.

There has been no offshore drilling, but some 2600 km of multichannel seismic-reflection data, supplemented by single-channel seismic-reflection data and dredge samples, allow us to correlate the offshore basin-fill with the land sequences. The correlation is based largely on seismic character, seismic interval velocities, and relative thicknesses. Offshore, depocenters are interpreted to contain more than 1000 m of Miocene clastics, overlain by about 2000 m of Miocene shelf carbonates and 2000 m of younger volcaniclastic rocks and bathyal carbonates. The thickness of the sediment pile suggests that temperatures are adequate to generate hydrocarbons, and a flat "bright spot" on a seismic record near Tanga Island may represent a gas/oil or gas/water contact. Reefal buildups and fore-reef deposits in Miocene limestones are possible offshore petroleum traps, as are numerous normal faults and pinch-outs near the basin margins. Overall, the basin appears to have moderate petroleum potential.

INTRODUCTION

The New Ireland Basin as defined here is some 90,000 km^2 in area, and lies largely north of New Hanover and New Ireland in water depths mostly less than 2000 m (Figure 1). It is the eastern part of what used to be called the New Ireland Basin -- an area extending from north of Manus Island to Feni island (e.g. Exon and Tiffin, 1984). About one third of the basin is dry land or lies beneath less than 1000 m of water. Much of the basin is a structurally simple downwarp, which formed as a forearc basin between an Eocene to early Miocene volcanic arc to the south and an outerarc high to the north. The basin contains as much as 7 km of Cenozoic sedimen-

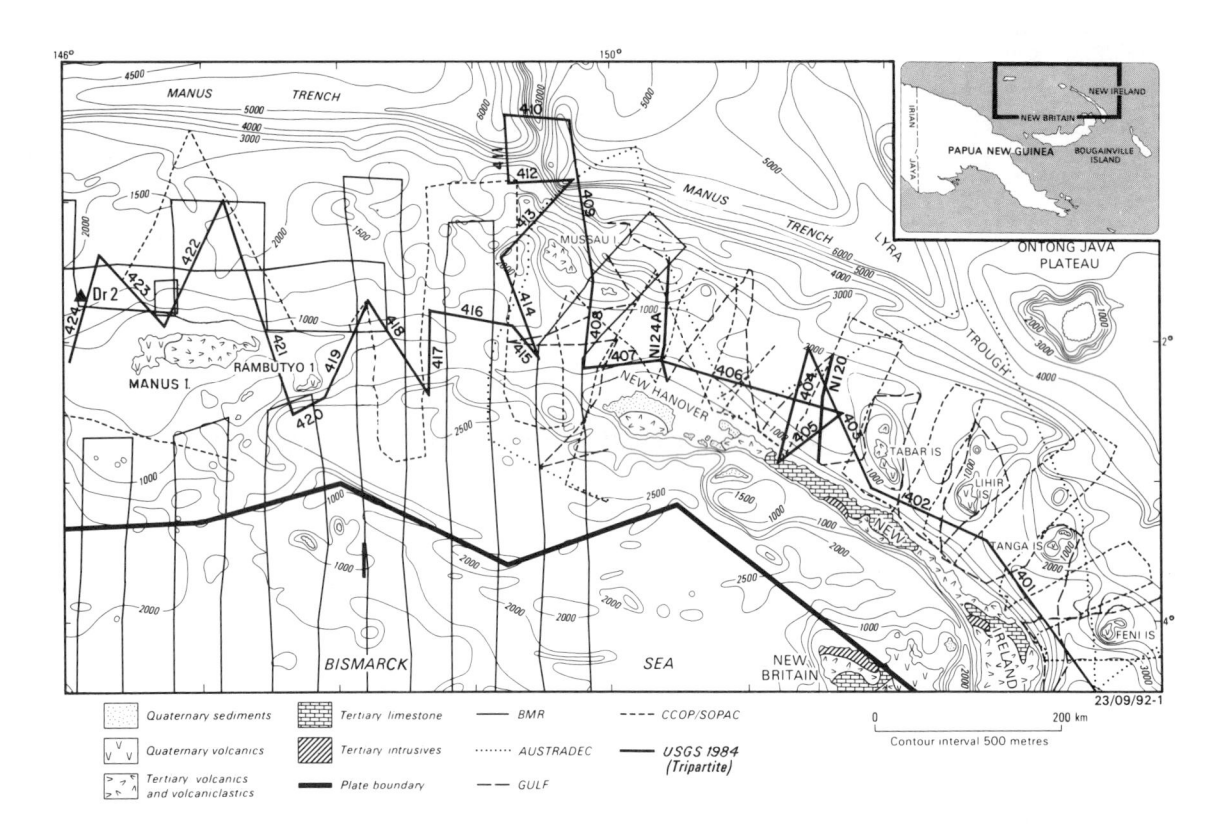

Figure 1. New Ireland Basin showing simplified land geology, bathymetry, and geophysical tracklines.

tary rocks but has not been drilled. This paper deals with the parameters that define the basin's petroleum potential: stratigraphy, structure, source rocks, maturation, reservoirs, seals, and geological history.

A review of the offshore area from north of Manus Island to Feni Island, based on geophysical data recorded by the Australian Bureau of Mineral Resources (BMR), the French IFP-CEPM-ORSTOM group, CCOP/SOPAC, and Gulf Research and Development Company, was published by Exon and Tiffin in 1984. A review of the eastern part of that area (the New Ireland Basin as defined here, where petroleum prospects are better), using only the Gulf geophysical data and the results of recent Geological Survey of Papua New Guinea (GSPNG) mapping on New Ireland, was published by Exon et al. in 1986. A brief review of petroleum prospects was prepared by Stewart, Francis, and Pederson in 1986.

Data

The data set used here comes largely from onshore geological mapping (Francis, this volume; Brown, 1982; Stewart and Sandy; this volume; Wallace et al., 1983) and offshore geophysical and geological surveys. Offshore seismic-reflection surveys used in the present study are those of Gulf (1973) and of the 1984 Tripartite R/V *S.P. Lee* marine geoscience survey, both of which recorded 24-channel data. The seismic data set consists of about 1000 km of unprocessed Gulf data in the Tabar-to-Feni Island area, 1000 km of processed Gulf data in the central basin west of Tabar, and 1600 km of *S.P. Lee* processed data (Figure 1). The *S.P. Lee* data were fully interpreted and tied to the previously interpreted Gulf data by Exon and Marlow. The interpretations were then digitized by Flower Doery Buchan and Robertson Research Australia Pty Ltd, who produced structure and thickness maps for the Papua New Guinea Government. A selection of these maps is illustrated by Marlow et al. (this volume), and they form the basis for our Figures 5 and 6.

GEOLOGICAL SETTING

Onshore geology

Relevant land geology is discussed in detail by Stewart and Sandy (1986) and elsewhere in this volume (Stewart and Sandy; Francis). The onshore stratigraphy of the New Ireland Basin, and Manus Island to the west, is shown in simplified form in Figure 2.

The basin formed as part of an island-arc system in the Eocene, when deposition of 2000 m of Jaulu Volcanics commenced. In the Miocene, silty sequences with some source-rock potential, the Lossuk River beds and Tamiu siltstone, were laid down. Then followed deposition of up to 1000 m of Miocene shelf carbonate of the Lelet Limestone, the main potential reservoir rock. A regional seal is provided by the fine-grained Punam Limestone and the tuffaceous volcaniclastics of the Lumis River volcanics and Rataman Formation.

Farther west, on New Hanover (Brown, 1982), if the identification of "Jaulu Volcanics" is correct, the sequence above them is condensed and lacks Miocene limestones. However, if the volcanics are really part of the Lumis River volcanics (as suggested by G. Francis, written comm.), Miocene limestones could be present at depth. On Manus Island, the sequence above the Tinniwi Volcanics is condensed, but 220 m of Miocene shelf limestone -- the Mundrau Limestone -- are present in central Manus (Francis, this volume). Early Miocene shelf limestones are exposed on Mussau Island (C.M. Brown, pers. comm.), and Miocene forereef limestones are found in the northernmost Tabar Islands (Wallace et al., 1983).

In general, onshore units dip down to the north where their offshore equivalents are thicker. The petroleum potential of the basin probably depends on the presence offshore of the equivalents of three New Ireland sequences: Miocene source rocks equivalent to the Lossuk River beds and the Tamiu siltstone, and Miocene reservoir rocks equivalent to the Lelet Limestone. These three critically important sequences are described in detail by Stewart and Sandy (this volume) and in summary below.

The exposures of Lossuk River beds are 150 m thick, but the whole sequence is thicker. The beds consist of interbedded medium to dark gray, fossiliferous carbonaceous mudstone, siltstone, tuffaceous sandy siltstone and tuffaceous sandstone. Abundant planktonic foraminifera, woody fragments, and shelly fossils are found in places. The beds were laid down at bathyal depths, largely by turbidity currents, the tuffaceous component reflecting airfall input from distant volcanoes.

The Tamiu siltstone is a sequence of unknown thickness in southeastern New Ireland, containing two distinctive units. The lower unit is mostly massively bedded, greenish gray calcareous siltstone containing some tuffaceous biomicrite. Planktonic foraminifera are common, and woody fragments are present in places. The upper unit consists largely of interbedded gray, calcareous tuffaceous siltstone, sandy mudstone, sandstone, and limestone. Pelecypods, benthic foraminifera, and echinoid fragments are common, and some coal lenses are present. The unit appears to be an upper slope to shelf sequence deposited in shallowing water depths.

The Lelet Limestone is estimated to be a maximum of 1000 m thick. The section consists mostly of white to cream-colored, massive to thick-bedded limestone of variable porosity, and is made up of foraminiferal and algal-foraminiferal biomicrite and microsparite. Benthic foraminifera, bivalves, echinoid fragments, gastropods, bryozoa, green and coralline algae, and branching corals are widespread. No framework reefs have been identified. The Kimidan unit of the Lelet Limestone is present low in the formation in central New Ireland and consists of gray, fossiliferous, calcareous muddy sandstone, carbonaceous sandy mudstone, and brown fossiliferous biomicrite. The Matakan unit occurs in the formation on southwest New Ireland, and may possibly be equivalent to the Kimidan unit. The Matakan unit consists of gray, fossiliferous, calcareous carbonaceous silty mudstone, and tuffaceous algal-foraminiferal micrite. Large transported blocks of sub-bituminous coal are incorporated in the mudstone.

Offshore Sequences

The offshore sequences in New Ireland Basin were first described in detail by Exon and Tiffin (1984) and Exon et al. (1986). The addition of the 1984 *S.P. Lee* seismic data led to an even more detailed interpretation by Marlow et al. (this volume). Figure 3 shows typical seismic data in the central basin and Figure 4 shows an interpretation of the data. The seismic reflectors and the characteristic sequences between them can be recognized over most of the area, but cannot be traced into the Manus Island region.

Correlation of the seismic sequences with the onshore geology is hampered by a 5 km gap between the shoreline and the seismic profiles. Correlation therefore depends on interpretation of the seismic character, interval velocities, and relative thicknesses of onshore and offshore sequences (see Table 1). We correlate the D-E sequence (average interval velocity from multichannel stacking and sonobuoy refraction data is 3900 m/s) with the Lossuk River beds and

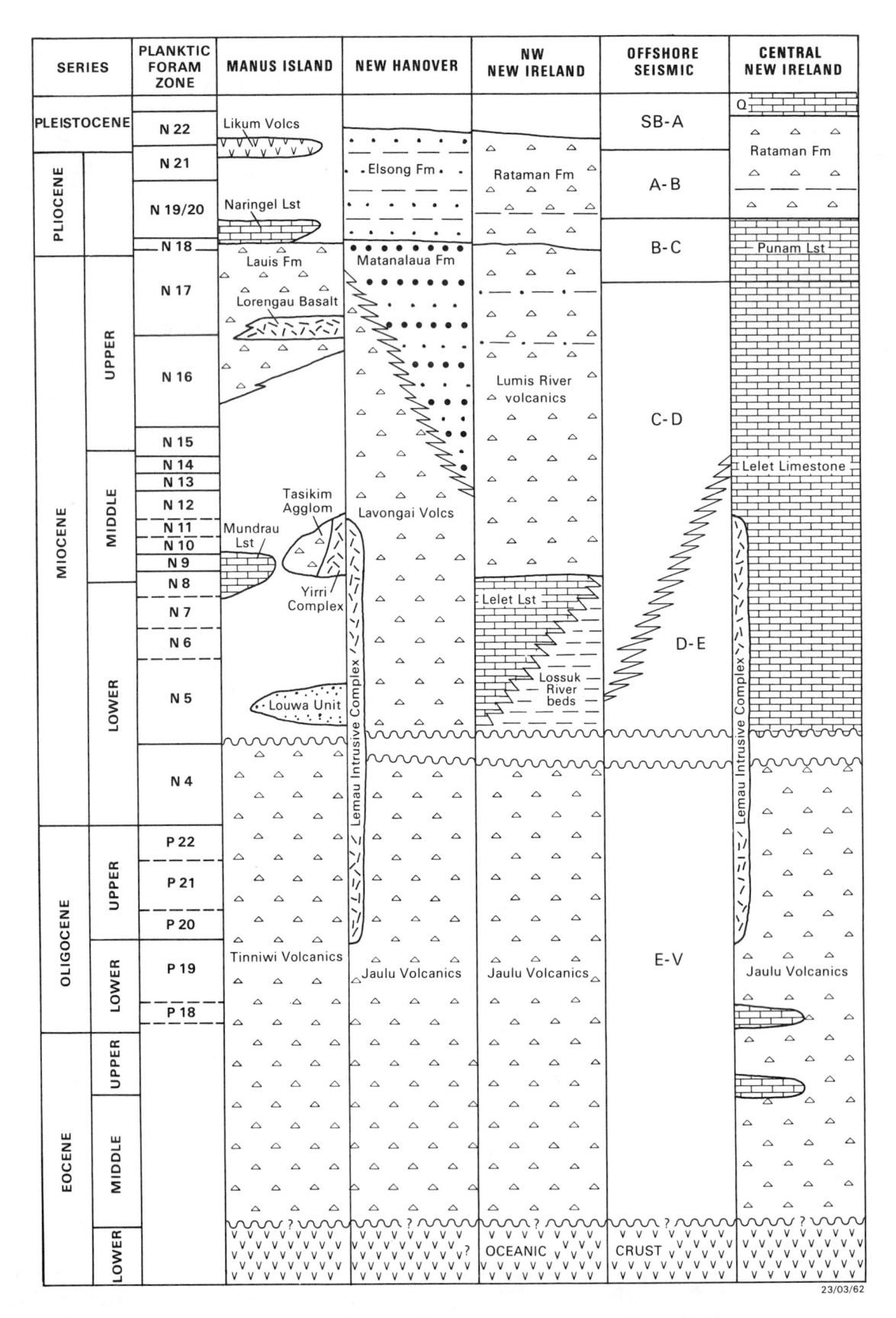

Figure 2. Land stratigraphy related to offshore seismic sequences. The "Jaulu Volcanics" on New Hanover may be younger than shown and equivalent to the Lumis River volcanics on New Ireland.

the Tamiu siltstone, and the C-D sequence (average interval velocity 3600 m/s) with the Lelet Limestone.

The overall shape of the basin is illustrated by Figure 5, which shows the bathymetry and the structure of the C horizon (the presumed upper surface of the Lelet Limestone). Much of the basin is a simple downwarp with an average sediment thickness of 5 km but, in the east, tectonic movement and volcanic activity associated with the Pliocene-Pleistocene emplacement of the Tabar-to-Feni group of islands have formed a series of horsts that complicate the simple structure. The major WNW-trending Tabar fault system, believed to consist of several faults, is thought to separate the islands from the rest of the basin. South of New Ireland, another set of WNW-trending faults steps down to the oceanic crust of the Manus Basin. These faults are believed to be related to a transform fault along which New Britain slid southwest past New Ireland in the Pliocene-Pleistocene (Taylor, 1979).

The basin has an overall tilt down to the north, and the various sequences thin over the old outerarc high -- the Emirau-Feni Ridge -- in the north. Gulf profile NI20 (Figure 4), a typical profile across the central basin, shows thinning of the C-D and B-C sequences toward the ridge, and thinning of D-E and B-C sequences toward the old island-arc, which lay under or south of New Ireland. The maps of sediment fill above the C and D horizons (Figure 6) also indicate that a broad basin has long existed, maximum fill having been in the center of the present structural basin.

SOURCE-ROCK POTENTIAL

Because younger sequences were probably not

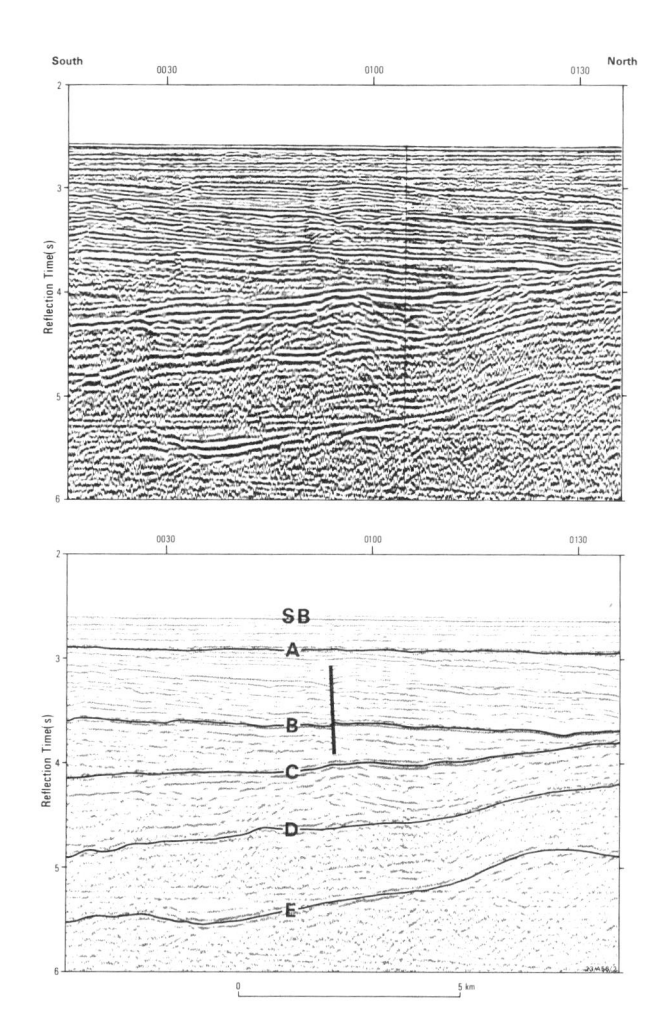

Figure 3. Part of multichannel Gulf seismic profile NI20, showing typical character and interpretation. Profile location shown on Figure 1. Sequence nomenclature given in Table 1. Note especially the complex character of sequence C-D, believed to consist of shelf carbonates, including reefs.

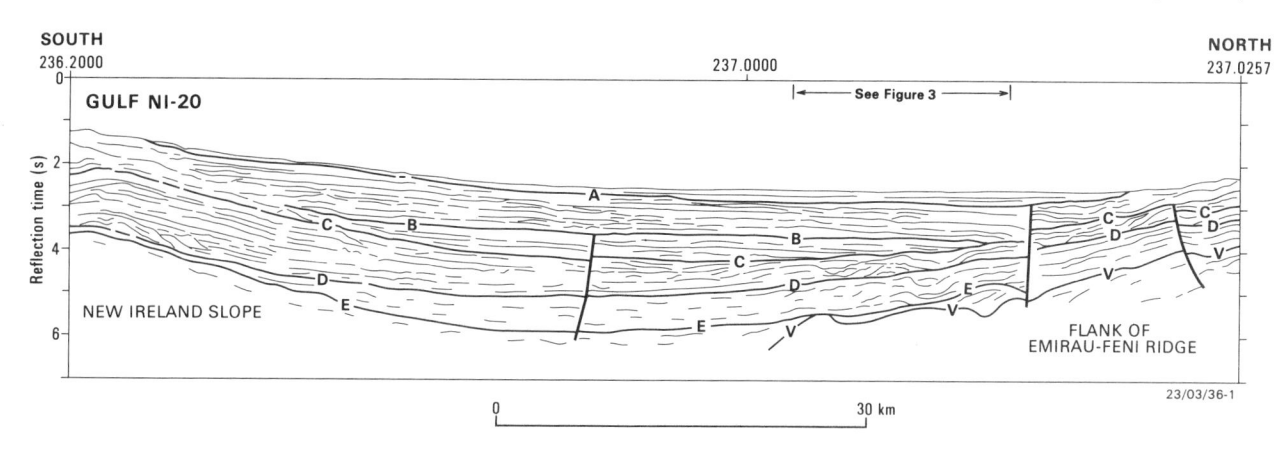

Figure 4. Line drawing of Gulf seismic profile NI20 off northwest New Ireland. Profile location shown on Figure 1. Sequence nomenclature given in Table 1. This typical New Ireland Basin profile shows overall thickening into the basin, and the prograding nature of the A-B and B-C sequences.

Table 1. Seismic sequences.

Sequence Velocity* Thickness	Character	Interpretation
SB-A 1800 m/s 200-300 m	Semi-transparent well-bedded sequences immediately beneath sea bed (SB). Conformably overlies high-frequency reflector (A), which is unconformable on underlying sequence in places.	Pleistocene to Recent hemipelagic oozes.
A-B 2000 m/s 500-1000 m	Well-bedded sequence with interbedding of strong reflectors and semi-transparent intervals. Progrades north-eastward. Conformably overlies shallowest continuous low-frequency reflector (B) that is conformable on underlying sequence.	Pliocene volcaniclastic turbidites resulting from volcanism, possibly triggered by subduction at the New Britain Trench just before and during the initial opening of the Manus Basin, interbedded with marls and chalks. Equivalent to the Rataman Formation on New Ireland.
B-C 2500 m/s 200-500 m	Well-bedded sequence like A-B. Progrades northeastward and is unconformable on regular underlying low-frequency reflector (C).	Late Miocene to earliest Pliocene chalks and marls, which accumulated as Globigerina oozes, interbedded with volcaniclastic turbidites. Equivalent to Punam Limestone on New Ireland.
C-D 3600 m/s 1000-2500 m	Mixed seismic character. Weak to strong parallel reflectors, dipping reflectors, and buildups. Unconformably overlies continuous low-frequency reflector (D), which is generally conformable on underlying sequence.	Early to late Miocene platform and upper slope limestones equivalent to thick Lelet Limestone of New Ireland. Potential petroleum source and reservoir rocks.
D-E 3900 m/s 1000 m	Well-bedded variably reflecting sequence, locally containing dipping reflectors, channels, and buildups. Unconformably overlies strong low-frequency reflector (E).	Early Miocene outer shelf and slope sediments, largely volcaniclastic, with some channels and possible carbonate banks. Equivalent to Lossuk River beds on New Ireland; potential petroleum source rocks.
E-V 5000 m/s 1000 m	Well-bedded variably reflecting, poorly defined sequence at limit of acoustic penetration. Unconformably overlies strong, diffracting irregular surface.	Probably Eocene to early Miocene volcanics and volcaniclastics equivalent to Jaulu Volcanics on islands. Diffracting surface beneath may either be oceanic basement or lie within Jaulu Volcanics.

* Velocity is average seismic interval velocity derived from the normal moveout velocities used for stacking, and from sonobuoy refraction velocities.

buried deeply enough to generate hydrocarbons (even offshore), the source-rock potential of the New Ireland Basin depends on hydrocarbon generation within the Miocene sequences, such as the Lossuk River beds, Lumis River volcanics and Tamiu siltstone, and the Miocene Lelet Limestone (Exon and Tiffin, 1984; Exon et al., 1986; Sandy, 1986). All these sequences are known to contain widespread

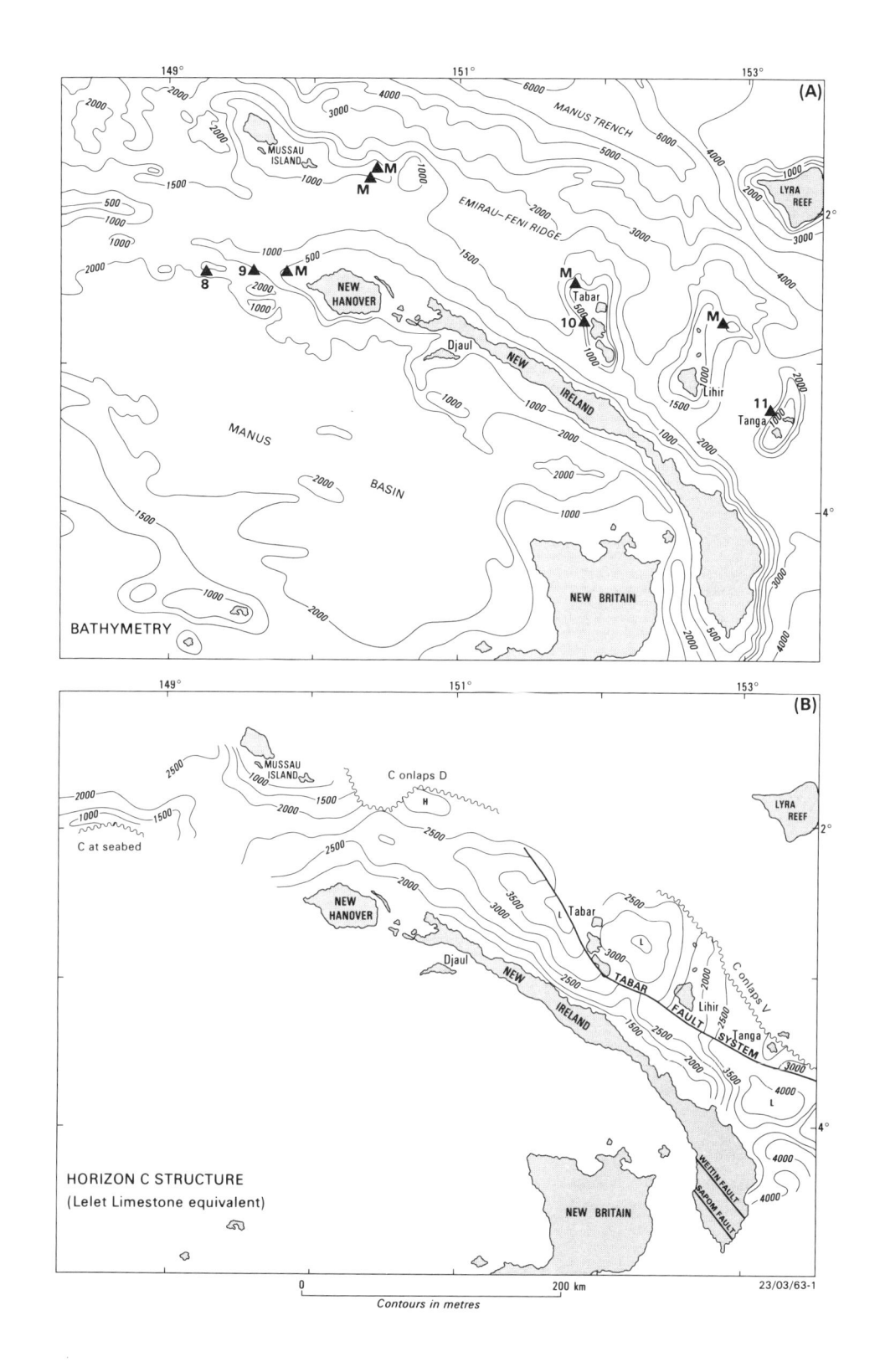

Figure 5. Maps of New Ireland Basin: A - Bathymetric map, showing 500 m contours. Triangles are dredge samples from *S.P. Lee* (prefix L) and *Machias* cruises (prefix M; Eade, 1979). B - Structure contour map on the top of the C horizon, believed to correspond to the top of the Miocene carbonate sequence. Contours, in meters below sea level, derived from *S.P. Lee* and Gulf profiles only, using interval velocities described by Marlow et al. (this volume).

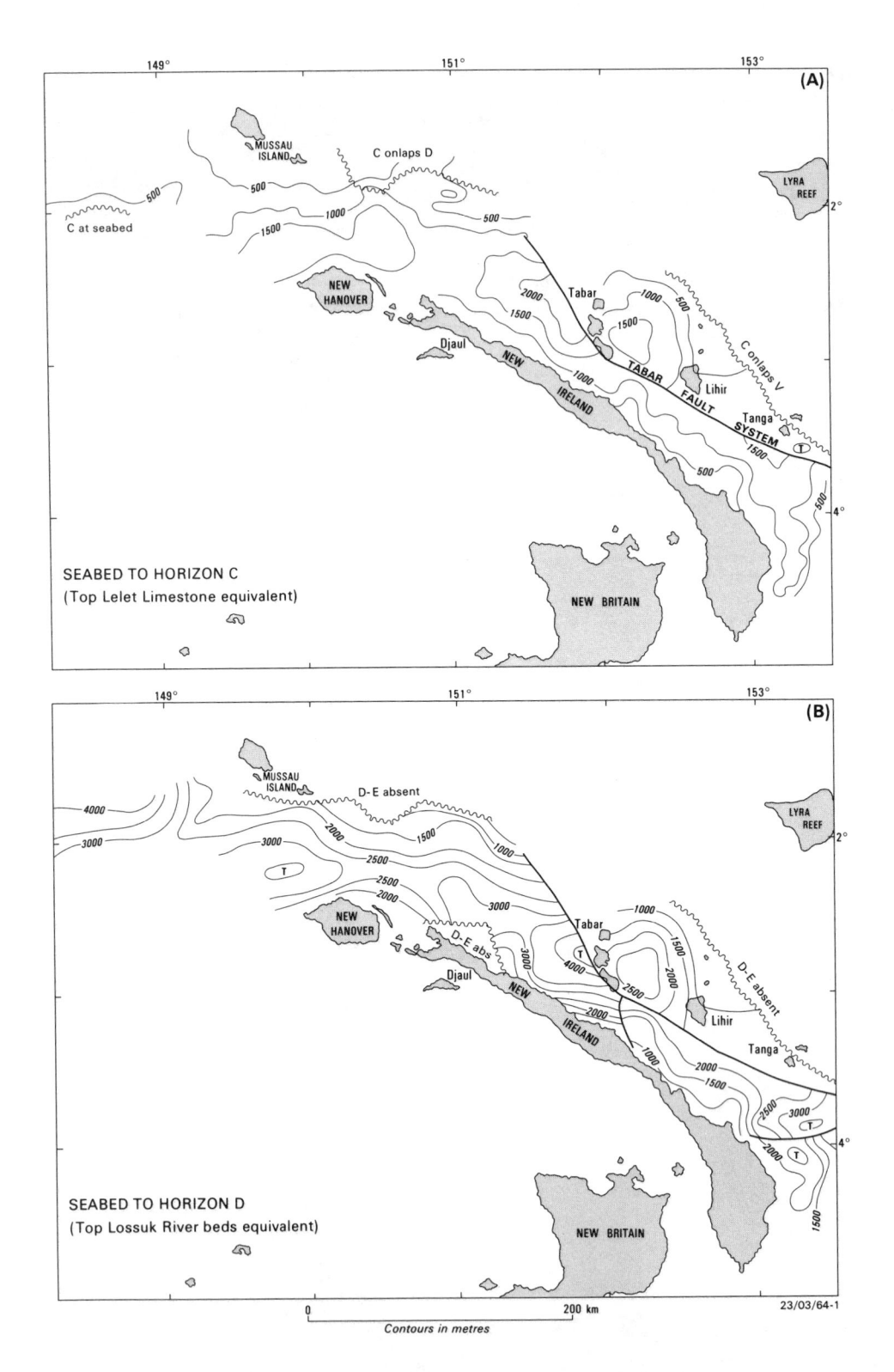

Figure 6. Isopach maps of New Ireland Basin, in meters: A - Thickness of sequence from seabed to C horizon, believed to represent the top of the Miocene carbonate sequence, the most likely petroleum reservoir in the basin. Contours derived from *S.P. Lee* and Gulf profiles only. B - Thickness of sequence from seabed to D horizon, believed to represent the top of a Miocene volcaniclastic sequence beneath Miocene carbonates. The most likely petroleum source rocks lie beneath this horizon. Contours derived from *S.P. Lee* and Gulf profiles only. Depocenters with thick sequences designated by T.

plant debris, algal material, and marine organisms.

In 1984, outcrop samples for source rock studies were collected from New Ireland by M. Glikson (Australian National University) and geologists from the GSPNG. Glikson (this volume) carried out studies in optical microscopy, reflectance, transmission electron microscopy, and elemental analysis on organic matter concentrates. The Australian Commonwealth Scientific and Industrial Research Organisation (CSIRO) Fossil Fuels Division (Rigby, 1986) carried out further analytical work, including total organic carbon (TOC), n-alkane distributions and carbon isotope analysis of coals and dispersed organic matter in the rocks, and stable carbon and oxygen isotope ratios of carbonates associated with the organic matter (Table 2). C.J. Boreham at BMR carried out total organic carbon and also Rock-Eval analyses of some samples (Table 3) which are plotted in a standard HI/Tmax cross-plot on Figure 7. The results of these various studies are summarized below by rock units. All these data were reviewed by Sandy (1986). Analyses of surface gases in six gravity cores from the offshore basin did not detect petroleum-related hydrocarbons (Kvenvolden, this volume).

Lossuk River Beds

Analyzed samples of the Lossuk River beds consisted of carbonaceous sandstone, siltstone, and shale. The organic matter includes material of probable bacterial origin (reflectance 0.1%), degraded terrestrial plant matter (reflectance 0.3-0.6%), and inertinite (reflectance more than 0.6%). The organic matter falls into 3 categories: (i) amorphous material, (ii) wood and semi-degraded wood with only minor amounts of pollen, and (iii) spores and fibrils. TOC concentrations were 0.2 to 4.8% (Tables 2 and 3). The samples are thermally immature, although the organic matter present has the potential to generate both oil and gas at increased

Table 2. Carbon and extract analyses from New Ireland outcrop samples (after Rigby, 1986).

Unit	Sample No.	*TOC %	Carbon as carbonate %	Benzene extract mg/gm TOC	Extract after hydrolysis pyrolysis mg/gm TOC	Lithology
Lossuk River beds	151	3.5	0.5	-	-	grey mudst.
	152	3.5	0.8	78	169	grey mudst.
	155	2.6	1.1	-	-	black mudst.
	158	4.8	4.8	-	-	calc. mudst.
Lumis River volcanics	160	-	6.4	-	-	black shale
	161	2.2	6.6	23.2	100	grey shale
	163	0.3	2.1	-	-	calc. mudst.
	165	3.3	7.2	-	73	grey shale
	167	5.3	0.05	22.6	98.1	black shale
Lelet Limestone	176	0.01	11.0	-	-	limestone
	177	0.01	10.7	-	-	limestone
	178	0.24	11.4	16.6	133	limestone
	181	0.01	12.4	-	-	limestone
	185	0.01	12.2	-	-	limestone
	207	0.02	12.5	-	-	limestone
	209	0.5	11.7	-	-	limestone
Matakan unit of Lelet Limestone	216	20.3	0.05	13.9	28.0	carb. mudst.
	217	59.3	0.06	36.9	101.5	coal
	220	62.8	0.07	42.1	135.0	coal

* Total organic carbon (TOC) was determined manometrically as CO_2, by combustion of the sediment in oxygen.

Table 3. Rock-Eval results from outcrop samples from New Ireland (C.J. Boreham, BMR).

No	Tmax °C	S1	S2	S3	S1+S2	S2/S3	Production Index S1/(S1+S2)	TOC %	Hydrogen Index	Oxygen Index (mg H&O/g TOC)	Lithology
\multicolumn{12}{c}{LOSSUK RIVER BEDS: NW NEW IRELAND}
152	432	0.22	1.30	4.76	1.52	0.27	0.14	2.74	47	174	grey mudst.
153	422	0.12	0.70	2.92	0.82	0.24	0.15	1.86	38	157	grey mudst.
154	431	0.21	1.19	4.28	1.40	0.28	0.15	2.51	47	171	black mudst.
155	426	0.18	0.93	3.40	1.11	0.27	0.16	2.19	42	155	black mudst.
157	307	0.04	0.01	1.20	0.05	0.01	0.80	0.21	5	571	shelly mudst.
\multicolumn{12}{c}{LUMIS RIVER VOLCANICS: NW NEW IRELAND}
159	407	0.19	0.52	2.71	0.71	0.19	0.27	0.90	58	301	black shale
162	403	0.13	0.25	1.90	0.38	0.13	0.34	1.08	23	176	N/A
163	423	0.03	0.9	0.85	0.12	0.11	0.25	0.30	30	283	calc. mudst.
165	422	0.03	0.28	0.82	0.31	0.34	0.10	0.68	41	121	grey shale
167	379	0.17	0.28	7.01	0.45	0.40	0.38	2.68	10	262	black shale
\multicolumn{12}{c}{LELET LIMESTONE: CENTRAL NEW IRELAND}
179	394	0.02	0.04	1.00	0.60	0.04	0.33	0.05	80	2000	limestone
180	390	0.01	0.05	0.55	0.06	0.09	0.17	0.07	71	786	limestone
183	277	0.00	0.00	0.20	0.00	0.00	N/A	0.40	0	500	limestone
\multicolumn{12}{c}{LELET LIMESTONE: MATAKAN UNIT}
216	378	1.28	8.59	9.59	9.97	0.91	0.13	10.92	80	88	carb.mudst.
217	397	14.59	60.92	46.94	75.51	1.30	0.19	43.60	140	108	coal
218	408	0.16	0.22	2.39	0.38	0.09	0.42	0.65	34	368	calc.siltst.
220	401	12.85	44.07	47.95	56.92	0.92	0.23	43.12	102	111	coal
221	414	0.16	0.50	0.32	0.66	1.58	0.24	0.97	52	33	sandstone

levels of maturity (Rigby, 1986).

Rock-Eval examination of five samples (Table 3 and Figure 7) showed that four samples contained significant amounts of pyrolyzable organic carbon. In these four, TOC was 1.86-2.74%; the volume of hydrocarbon present (S1) was low (0.04-0.22 mg/g); the volume of hydrocarbon that could be generated with further maturation (S2) was also low (0.70-1.30 mg/g); Tmax, the temperature at which maximum S2 generation occurred, was 422-432°C, indicating that the rocks are thermally immature; and the hydrogen index was low (38-47 mg hydrocarbon per gram TOC). All the samples plot as Type III in Figure 7, suggesting that the organic matter was predominantly derived from coals and woody matter (Teichmuller and Durand, 1983) and is prone to generate gas, although oxidation of organic matter can also produce kerogens that plot as Type III, even though the precursors were originally much richer in hydrogen. However, Rigby (1986) has demonstrated that all the Type III kerogens he has analyzed from New Ireland can produce waxy crude oils.

Lumis River Volcanics

Samples of the Lumis River volcanics analyzed are carbonaceous siltstone and shale. The organic matter consists of pollen and spores (much more abundant than in the Lossuk River beds), and wood and woody tissue; it has a vitrinite reflectance of 0.2-0.5%. TOC values are variable (Tables 2 and 3). Bimodal n-alkane distributions in benzene extracts suggest that the organic matter was derived from two sources: land plants, and algae or bacteria. Hydrous pyrolysis studies have demonstrated the potential of these sedimentary rocks to generate liquid hydrocarbons.

Rock-Eval analyses of five mudstone and shale samples (Table 3 and Figure 7) showed that TOC varied from 0.30 to 2.68%; S1 was low (0.03-0.19 mg/g); S2 was also low (0.25-0.9 mg/g); Tmax was 379-423°, indicating that the rocks are thermally immature; and the hydrogen index was low (10-58 mg hydrocarbon per gram TOC). All samples plot as Type III kerogens in Figure 7, suggesting either

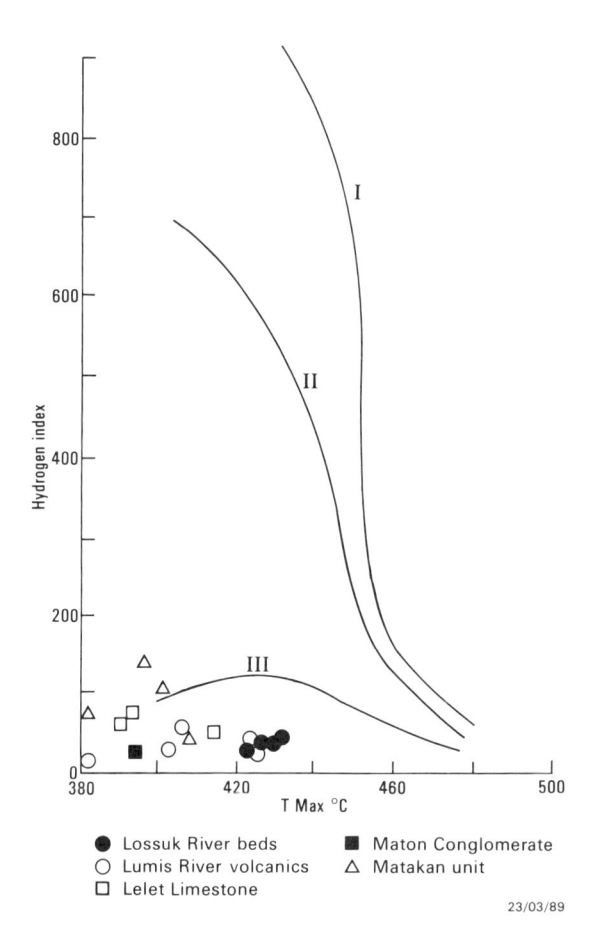

Figure 7. Standard van Krevelen plot of Rock-Eval data from New Ireland, showing kerogen typing and maturity.

that they are derived from coals and woody matter, or that significant post-depositional oxidation of the organic matter has occurred.

Lelet Limestone

Limestone samples contain sparse organic matter of probable algal origin. TOC values obtained were 0.6% or less (Tables 2 and 3), and source rock potential is generally poor. Rock-Eval analyses (Table 3) showed that the volume of hydrocarbon present was negligible. Consequently, the samples have no source-rock potential. Figure 7 shows that the material generally contains Type III kerogens, which may be derived from coals and woody matter, or may represent post-depositional oxidation of other organic material. However, the extract from at least one sample (178) suggests a major contribution from algal or bacterially derived organic matter.

Matakan Unit of Lelet Limestone

Coal and carbonaceous mudstone or siltstone samples, from this lagoonal sequence containing allochthonous cannel-coal and vitrinite-rich coal, gave vitrinite reflectances of 0.5% or less. The cannel-coal contains algae, spores and pollen, and woody and bacterial remains. Overall, vitric bands dominate the coals. Some coals contain abundant spores of plants from coastal mangrove swamps.

A carbonaceous mudstone and two coals gave TOC values of 20.3-62.8% (Table 2) and sulphur contents of 0.4-11.6% (Rigby, 1986). The samples are thermally immature. Rock-Eval analyses on five rocks are listed in Table 3. In the two coals, TOC was 43%; S1 was low (12.85 and 14.59 mg/g); S2 was also relatively low (44.07 and 60.92 mg/g); Tmax was 397 and 401°C, so the rocks are thermally immature. The hydrogen index was low, but higher than those in any other samples (102 and 140 mg hydrocarbons per gram TOC). The coals plot higher in Figure 7 than other samples, suggesting either that they may contain more Type II kerogens or that they are less oxidized.

Discussion

All samples were taken from outcrops or shallow auger holes, and clearly have been subjected to surface weathering -- a condition that complicates interpretation of the geochemical results. During weathering, hydrogen-rich organic components can be removed by oxidation. Reflectance studies and analysis of hydrocarbon extracts suggest that none of the samples have come within the oil-generation window.

Total organic carbon (TOC) values of about 2-5% in the carbonaceous sediment of the Lossuk River beds, and about 1-5% in the Lumis River volcanics (Tables 2 and 3), suggest that both units have some source-rock potential, as do the highly carbonaceous mudstones and coals in the Matakan unit of the Lelet Limestone. The remainder of the Lelet Limestone has low TOC and probably has negligible potential as a source rock. In all samples, low values of S1 (hydrocarbons present), S2 (producible hydrocarbons), and the hydrogen index (Table 3) suggest that the amount of hydrogen-rich material capable of producing oil is low. Values of Tmax are mostly around 400°C and do not exceed 432°C, which suggests that the sedimentary rocks are thermally immature. Most of the organic carbon appears to have been derived from woody plant matter, although pollen and spores are important in the Lumis River volcanics. The coals of the Matakan unit consist of two types: vitric and cannel coals. The cannel coals contain abundant algal and bacterial remains, in addition to spores, pollen, and woody remains, and consequently have good source-rock potential.

In general terms the high TOC values and dominance of woody material suggest that these rocks are capable of generating gas, with some exceptions. Rigby (1986) has shown that many of the samples are capable also of generating waxy crude oils. The source potential of samples from northwestern New Ireland is generally better than that of samples from the central and southern regions (Rigby, 1986). North and northwest of New Ireland maximum depths of burial have been, and are, much deeper than on New Ireland, which was uplifted in the Pliocene-Pleistocene (Exon and Tiffin, 1984). In basin axes, rocks in the C-D and older sequences (equivalents of the Lossuk River beds, Lumis River volcanics, Tamiu siltstone and Lelet Limestone) probably have been buried deeply enough to generate gas and oil.

MATURITY

As discussed above, few of the potential source rocks that are exposed on New Ireland approach the maturity necessary to generate hydrocarbons. However, New Ireland was uplifted in the Pliocene-Pleistocene, whereas the offshore basin contains 1500-2000 m of Pliocene-Pleistocene strata above the C horizon in places (Marlow et al., this volume; Figure 6), and so the older Miocene strata (C-E sequences) should be considerably more mature offshore than onshore.

The geothermal gradient offshore in the New Ireland Basin is unknown. However, the New Ireland, Bougainville, Cape Vogel and New Georgia Basins are tectonically similar, which suggests that gradients in all four basins may be similar and rather low overall. Low geothermal gradients are common in forearc basins, where isotherms in the underlying basement are depressed by the cooling effect of the descending slab of ocean floor within the subduction zone (Miyashiro, 1979). Studies by the GSPNG (Stewart, Francis, and Pederson, 1986) suggest that present-day gradients in the Bougainville Basin are very low (about 1.6°C/100m), but that paleogradients may have been higher. For example, a zeolitic facies is developed in uplifted basin strata (Blake and Miezitis, 1967), suggesting that parts of the basin were buried to depths where temperatures were 150-200°C.

The variability of maturation depths over limited distances in island-arc basins is illustrated by vitrinite and spore color-index determinations made on well samples from the Cape Vogel Basin (G. Francis, written comm.). The studies were carried out by Robertson Research Australia. Data from Nubiam No. 1 well indicate that Miocene strata are early mature at 2500 m and fully mature at 4200 m, whereas data from Goodenough No. 1 well, only 60 km away, indicate early maturity at 1550 m and full maturity at 3000 m. Nubiam No. 1 lies near the onlap margin of the forearc basin with the outerarc high, but Goodenough No. 1 lies closer to the original continental margin arc, and to the younger Woodlark spreading center.

Hobart and Weissel (1987) obtained heatflow and conductivity values from cores along a profile across the New Georgia Basin in the Solomon Islands. These values suggest extraordinarily low thermal gradients, probable temperatures ranging between 90°C and 110°C at 5000 m. Because some of the outcropping sedimentary rocks on New Ireland are almost mature, the thermal gradient was higher here than in New Georgia Basin. If the immediate sub-bottom temperature is 5°C, and if the temperature gradient were 2.5°C/100 m, the onset of hydrocarbon production (65°C) would lie at about 2400 m, and the onset of mature oil production (90°C) at about 3400 m.

From the information available, we consider any sedimentary source rocks buried 2500 m or more to have the potential to generate hydrocarbons, and those beneath 3500 m to generate oil. The two isopach maps (Figure 6), suggest that the lower part of the C-D sequence (our assumed Lelet Limestone equivalent) would be in the "oil window" only in the areas of thick sedimentation northwest of New Hanover, west of Tabar, and south of Tanga. The D-E sequence (our assumed equivalent of the Lossuk River beds, Tamiu siltstone and Lumis River volcanics) would be in the "oil window" over considerably more extensive areas.

Entry of strata into the hydrocarbon generation zone could nowhere have pre-dated the Pliocene. Thus, potential traps of virtually any age are worth prospecting, and the timing of the formation of structures is not critical. The seeming lack of continuous reservoir rocks suggests that long-range migration of hydrocarbons probably has not occurred.

POSSIBLE EXPLORATION TARGETS

Since the rock-types in the region are mostly volcaniclastic sediment and carbonates, potential reservoir-rock types are limited. Much of the volcaniclastic material is pyroclastic, and the alteration of glass to zeolites and clay minerals can be expected to have rapidly diminished any primary porosity. Epiclastic sandstones, especially if reworked by wave action or in distal turbidites, may be potential reservoir rocks. Clearly the best potential reservoir rocks are shelfal limestones, such as those of the Lelet Limestone. Petrographic examination of outcrop material from the Lelet Limestone has shown that considerable primary and secondary porosity exists, and Exon et al. (1986) report porosity of up to 25%. What porosity exists at depth remains a matter of conjecture.

Given the abundance of fine-grained volcaniclastic sediment and basinal limestone in the sequences, both onshore and offshore, the development of seals to petroleum migration would not appear to be a problem. A regional seal for the presumed shelf-limestone reservoir rocks would be provided by fine-grained parts of the Punam Limestone.

Because of the wide spacing of the multichannel seismic-reflection lines, which averages 30-40 km, no specific exploration targets can be identified. However some plays can be defined at this stage. They fall into the following categories:

Stratigraphic traps: carbonate banks, reefs, and fore-reef limestone, of Miocene age.

Structural traps: anticlines and fault traps.

Structural/stratigraphic traps: onlaps against older sequences.

Stratigraphic Traps

Because of the lack of evidence for reservoir rocks other than Miocene shelf carbonates, and the seismic interpretations, which suggest that some of these carbonates form reefs overlain by fine-grained seals, stratigraphic traps must be among the major exploration targets in the basin. The potential importance of such targets was first stressed by Exon and Tiffin (1984). Nayoan, Arpandi, and Siregar (1981) stated that about 10% of Indonesian production comes from Tertiary carbonate reservoirs -- mostly from Oligocene-Miocene reef complexes and other high-energy carbonate deposits within back-arc basins. Primary porosity dominates in major Indonesian fields, but secondary porosity is important sometimes; oil-producing intervals have porosities exceeding 20%. Most of the traps are purely stratigraphic, carbonate buildups being surrounded and capped by clastic material.

In the New Ireland region, Miocene shelf carbonate rocks are widespread and thick on New Ireland, and present on Manus, Mussau, and the Tabar Islands. All these areas appear to have been uplifted in the Pliocene-Pleistocene, so Miocene shelf carbonates (C-D sequence) may have been deposited widely on a shelf that extended over much of the present New Ireland Basin. The typical character of the C-D sequence is well illustrated in Figure 3, where buildups appear to be separated by basinal strata and chaotic zones. The sequence is widespread, retains its character, and varies little in thickness, all of which suggests that it is of sedimentary rather than igneous origin. Alternatively, the section may consist largely of volcaniclastic sediment deposited by turbidity and other gravity flows, but we favor the former interpretation.

The buildups in the C-D sequence do not have an associated magnetic anomaly -- a condition that supports their interpretation as sedimentary features. Interval velocities calculated from refraction and stacking velocities are around 3600 m/s (Exon and Tiffin, 1984; Marlow et al., this volume). These velocities are considerably higher than those in the overlying B-C sequence (about 2500 m/s), which we believe to consist of chalk, marl, and turbidites.

Probable reefal buildups can be identified on most seismic lines between Tabar and Mussau, but similar buildups are absent elsewhere, indicating facies changes on the margins of the basin, and constraining the search for carbonate stratigraphic traps. Hydrocarbons could enter the buildups and associated fore-reef deposits either from nearby strata within the C-D sequence, such as the carbonaceous Matakan unit, or from carbonaceous mudstones and siltstones in the D-E sequence. In the major depocenters, west of the Tabar Islands and north of New Hanover (Figure 6), hydrocarbons also could be supplied from immediately overlying sediment.

Structural Traps

Anticlines are not widespread in the New Ireland Basin, but fault blocks are more common. Both structural types are limited mostly to the basin margins -- along the Emirau-Feni Ridge, and along the

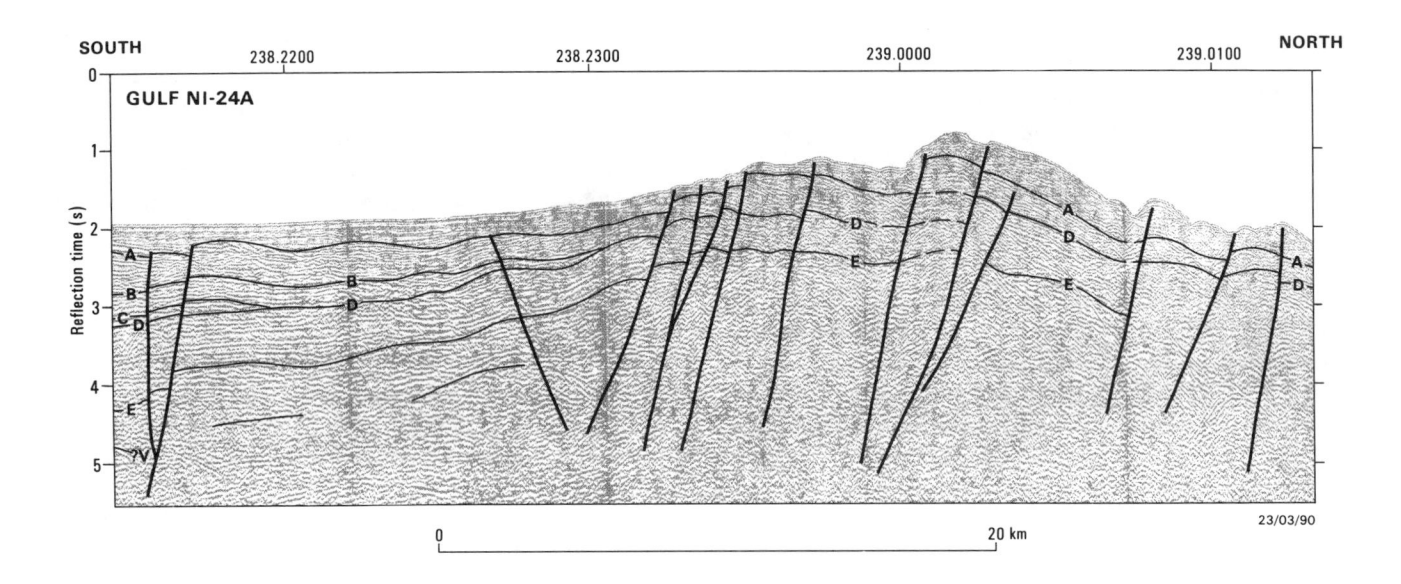

Figure 8. Part of Gulf multichannel seismic profile NI24A across Emirau-Feni Ridge, showing thinning of all sequences onto it, pinching out of BC and CD sequences, and large south-dipping normal faults. Location shown on Figure 1.

New Ireland-New Hanover high. Figure 8 is a typical seismic profile crossing the Emirau-Feni Ridge. It shows a thick basinal section to the south, an overall anticlinal form to the ridge, and numerous south-dipping normal faults. The B-C and C-D sequences are apparently missing from beneath the ridge, which must have been high in the Miocene.

The Tabar-to-Feni islands are horsts (Exon and Tiffin, 1984), sometimes flanked by anticlinal folds. One anticline is shown in Figure 9A -- an illustration of a seismic profile that crosses southwest of Tanga (*S.P. Lee* profile 401). A basement ridge is flanked by depocenters, and sequences A-B, B-C and C-D thin over it. At the center of the ridge (around CDP 3000) is a flat, high-amplitude "bright spot" (Marlow and Exon, 1985). The bright spot is better displayed in a true-amplitude section (Figure 9B), which shows it to be about 2 km wide on this line. Table 4 summarizes velocity and thickness information for the sequences above the bright spot; the two-way time from the seabed to the bright spot of 1.2 s represents depth of burial of about 1700 m. Water depth is about 2500 m, but it is possible that the bright spot extends southward into shallower water. This bright spot, which cuts across the sedimentary structure, may well represent a hydrocarbon accumulation (a gas/oil or gas/water contact).

New Ireland and New Hanover are cut by a series of major faults that extend offshore, especially on the southwestern side of the islands where the strata drop down to the Manus Basin (Exon et al., 1986, Figure 6). Many of the faults are simple normal faults, but some of those to the southwest also involve northwest-trending transcurrent movements, like those on the Weitin and Sapom Faults. Petroleum potential on the islands is limited, as there is little or no overburden on the Lelet Limestone, but there may be some petroleum potential on the flanks of the islands.

Table 4. Depth to the top of the flat bright spot at CDP 3000 along *S.P. Lee* seismic profile 401, southwest of Tanga.

	Two-way travel time (sec)	Interval velocity (m/sec)	Thickness (m)	Sum of total thickness
Seabed - A	0.16	1800	144	144
A - C*	0.35	2000	350	494
C - D	0.30	3600	540	1034
D - Bright spot	0.35	3900	683	1717

* Note: Location of CDP 3000 on Line 401 is 3.719°S, 152.945°E. Water depth is 2500 m above the bright spot. Section B-C pinches out on the flank of the anticline adjacent to the bright spot.

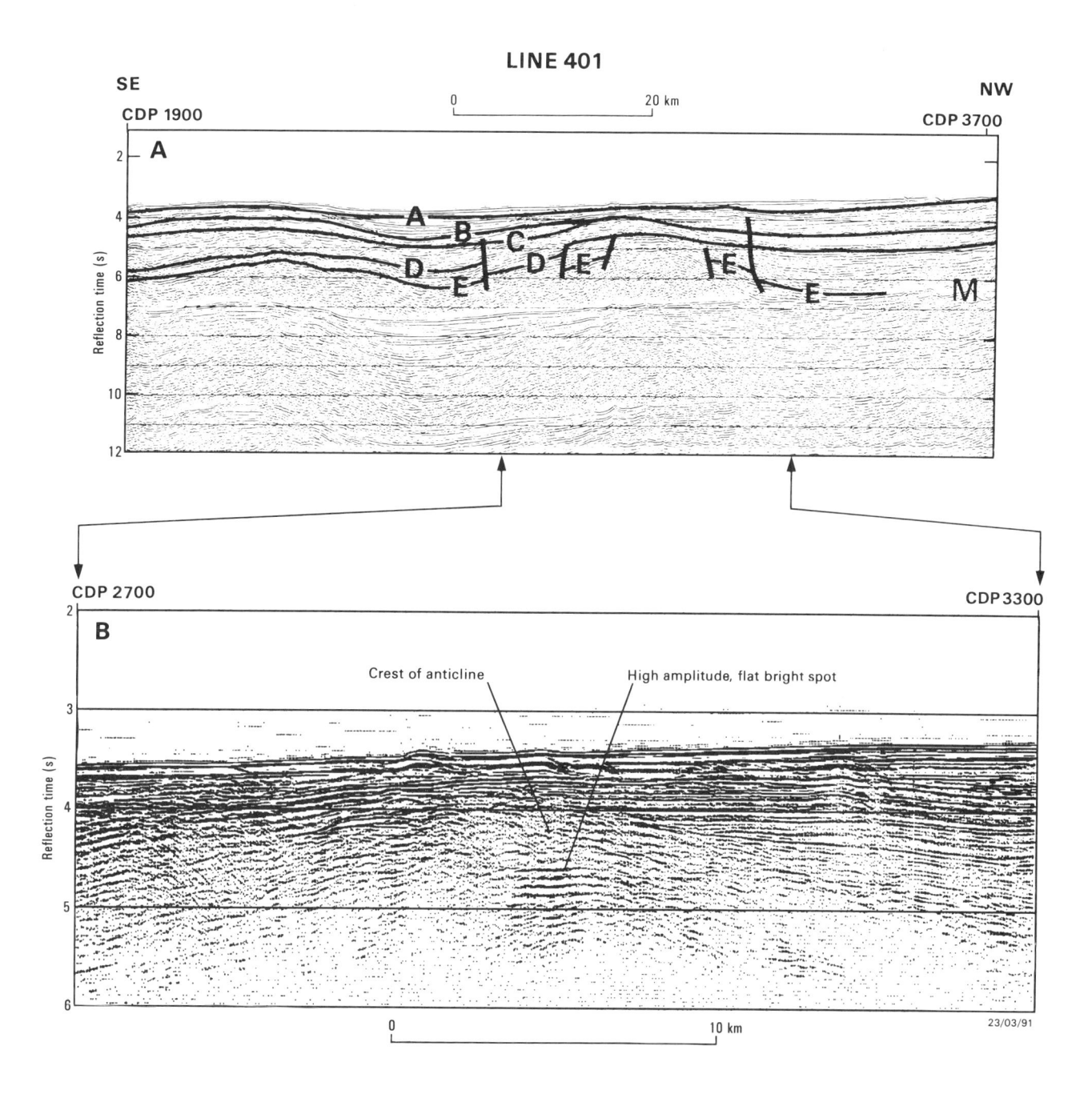

Figure 9. *S.P. Lee* multichannel seismic profile 401 running parallel to basin axis south of Tanga island. Location shown on Figure 1. A - General interpretation of profile showing thinning across anticlinal feature. B - Flat, high-amplitude bright spot in anticlinal area. This true amplitude display indicates how strong the flat-lying reflector is, and its cross-cutting nature. It may represent a gas/water or oil/water interface.

Structural/Stratigraphic Traps

Most seismic lines across the basin show onlap against the highs (e.g. Figures 3, 4, 8 and 9). In general, seismic sequences C-D and A-B (Lelet Limestone equivalent and Rataman Formation equivalent) are the most persistent, but they too thin rapidly or pinch out toward the highs on some lines. Both the C-D and D-E sequences commonly pinch out against the Emirau-Feni Ridge (Figures 5 and 6). Some sequences show depositional downlap, for example sequence B-C in Figure 4, but most such downlaps do not provide potential traps. Overall, there are numerous possible onlap traps developed on the margins of the basin, but they are far from being accurately defined.

DISCUSSION AND CONCLUSIONS

The New Ireland Basin contains possible Miocene source rocks and reservoir rocks, and vitrinite reflectance and geochemical data indicate that submature sedimentary rocks are present on land. Were these same rocks deeply buried offshore, where more than 5 km of sedimentary section exist, they would be capable of generating hydrocarbons. The flat bright spot on *S.P. Lee* profile 401 suggests that hydrocarbons have been generated.

Our interpretation of the five offshore seismic sequences suggests that a 1000m thick Miocene clastic sequence (D-E) underlies a 1000-1500 m thick Miocene carbonate shelf sequence (C-D). This sequence is overlain by three bathyal carbonate and turbidite sequences, totalling 1000-2000 m thick, of late Miocene to Recent age.

The structure contour map of the C horizon (top of the interpreted carbonate reservoir sequence) shows highs along New Hanover and New Ireland, and along the Emirau-Feni Ridge (Figure 5B). The major Tabar fault system appears to separate the Tabar-Tanga group of islands from structurally low areas to the southwest, where the C horizon lies as much as 4000 m below sea level.

The isopach map of the section overlying the C horizon shows the influence of these structural features (Figure 6A). Thickness varies from zero on New Ireland and 500 m along the Emirau-Feni Ridge, to more than 1500 m in depocenters north of New Hanover, southwest of Tabar and Tanga, and between Tabar and Lihir. The isopach map of the section overlying the D horizon (top of the interpreted hydrocarbon source rocks) shows as much as 3500 m of overburden northwest of New Hanover, 4500 m west of Tabar, 3500 m south of Tanga, and 2500 m between Tabar and Lihir (Figure 6B). If thermal gradients are in the normal range of 25-30°C/1000 m, both C-D and D-E sequences are buried deeply enough to generate hydrocarbons.

The general structural configuration of the New Ireland Basin suggests that any hydrocarbons generated in the basin could migrate a limited distance up-dip to the southwest or northeast, where there are pinchouts in most sequences as well as normal faults. However, the prime exploration targets are stratigraphic traps in the C-D sequence, largely as buildups that we believe to be Miocene reefs, and associated fore-reef debris-flow deposits.

A major problem with petroleum exploration in the New Ireland Basin is the water depth. Only about one-third of the basin is exposed or is less than 1000 m deep. Probably the areas with the greatest petroleum potential are those where buildups are widespread -- in the areas bounded by the Emirau-Feni Ridge and New Hanover-New Ireland, between Mussau and Tabar. Here, shallow water surrounds the islands of Mussau, Emirau, and Tench, and lies along the coasts of New Hanover and New Ireland. Shallow-water areas farther east are also possibilities, especially along the New Ireland coast, where the Miocene Lelet Limestone is present. Prospects around the Tabar-to-Feni group are difficult to estimate. Quaternary volcanics on the islands, and the apparently abundant volcanic material in nearby seismic sections, may tend to downgrade the prospects, despite the possible advantage of higher local heat flow.

To conclude, positive features of the New Ireland Basin for exploration include thick sections, adequate structures, and probable source and reservoir rocks. Negative features are limited shallow-water areas and the lack of offshore well data.

ACKNOWLEDGMENTS

This paper has been reviewed by G. Francis (GSPNG), G. O'Brien and J. Colwell (BMR), and D. Rigby (CSIRO), and extensively modified as a result of their comments. Technical editing was carried out by Catherine Campbell (United States Geological Survey). The figures were drawn by Terry Kimber and Lana Murray at BMR. We are most grateful for all these contributions. Our paper draws freely on other papers in this volume. Exon publishes with the permission of the Director, Bureau of Mineral Resources, Canberra.

REFERENCES

Blake, D.H., and Miezitis, Y., 1967, Geology of Bougainville and Buka Islands, New Guinea: Australian Bureau of Mineral Resources Bulletin 93 (PNG1), 56p.

Brown, C.M., 1982, Kavieng, Papua New Guinea - 1:250,000 Geological Series: Geological Survey of Papua New Guinea, Explanatory Notes SA/56-9, 26 p.

Eade, J.V., 1979, Papua New Guinea offshore survey, Cruise 1-79(1): CCOP/SOPAC Cruise Report 24 (unpublished), 10 p.

Exon, N.F., and Tiffin, D.L., 1984, Geology and petroleum prospects of offshore New Ireland Basin in northern Papua New Guinea, *in* Watson, S.T., ed., Transactions of the Third Circum-Pacific Energy and Mineral Resources Conference, p. 623-630.

Exon, N.F., Stewart, W.D., Sandy, M.J., and Tiffin, D.L., 1986, Geology and offshore petroleum prospects of the eastern New Ireland Basin, northeastern Papua New Guinea: Bureau of Mineral Resources Journal of Australian Geology and Geophysics, v. 10, p. 39-51.

Hobart, M., and Weissel, J., 1987, Geothermal surveys in the Solomon Islands-Woodlark Basin Region, *in* Taylor, B., and Exon, N.F., eds., Marine geology, geophysics, and geochemistry of the Woodlark Basin - Solomon Islands: Circum-Pacific Council for Energy and Mineral Resources, Earth Science Series, v. 7, p. 49-66.

Marlow, M.S., and Exon, N.F., 1985, Basin development and regional tectonics of the New Ireland and Manus regions: Australian Bureau of Mineral Resources Record 1985/32, p. 11-18.

Miyashiro, A., 1979, Metamorphism and plate convergence, *in* Strangway, D.W., ed., The continental crust and mineral deposits: Geological Association of Canada Special Paper 20, p. 591-605.

Nayoan, G.A.S., Arpandi, and Siregar, M., 1981, Tertiary carbonate reservoirs in Indonesia, *in* Halbouty, M.T., ed., Energy Resources of the Pacific Region: American Association of Petroleum Geologists Studies in Geology 12, p. 133-145.

Rigby, D., 1986, The source and isotopic composition of organic and carbonate carbon in rock outcrop samples from New Ireland: CSIRO Institute of Energy and Earth Resources, Restricted Investigation Report 1648R, 43 p.

Sandy, M.J., 1986, Reservoir and source rock potential of New Ireland outcrop samples: Geological Survey of Papua New Guinea Report 86/19, 21 p.

Stewart, W.D., and Sandy, M.J., 1986, Cenozoic stratigraphy and structure of New Ireland and Djaul Island, New Ireland Province, Papua New Guinea: Geological Survey of Papua New Guinea Report 86/12, 96 p.

Stewart, W.D., Francis, G., and Pederson, S.L., 1986, Hydrocarbon potential of the Bougainville and southeastern New Ireland Basins, Papua New Guinea: Geological Survey of Papua New Guinea Report 86/11, 13 p.

Taylor, B., 1979, Bismarck Sea: evolution of a back-arc basin: Geology, v. 7, p. 171-174.

Teichmuller, M., and Durand, B., 1983, Fluorescence microscopical rank studies on liptinites and vitrinites in peats and coals, and comparison with the results of the Rock Eval pyrolysis: International Journal of Coal Geology, v. 2, p. 197-230.

Wallace, D.A., Chappell, B.W., Arculus, R.J., Johnson, R.W., Perfit, M.R., Crick, I.H., Taylor, G.A.M., and Taylor, S.R., 1983, Cainozoic volcanism in the Tabar, Lihir, Tanga and Feni islands, Papua New Guinea: geology, whole-rock analyses, and rock-forming mineral composition: Bureau of Mineral Resources Australia, Report 243; BMR Microform MF 197, 27 p.

Marlow, M.S., Dadisman, S.V., and Exon, N.F., editors, 1988, Geology and offshore resources of Pacific island arcs—New Ireland and Manus region, Papua New Guinea, Circum-Pacific Council for Energy and Mineral Resources Earth Science Series, v. 9: Houston, Texas, Circum-Pacific Council for Energy and Mineral Resources.

MULTICHANNEL SEISMIC-REFLECTION DATA COLLECTED AT THE INTERSECTION OF THE MUSSAU AND MANUS TRENCHES, PAPUA NEW GUINEA

H.F. Ryan, M.S. Marlow
U.S. Geological Survey, Menlo Park, California, 94025, USA

ABSTRACT

Interpretations of recently acquired multichannel seismic-reflection data near the intersecting Mussau and Manus Trenches provide constraints regarding the recent tectonic history of the northernmost Papua New Guinea region, particularly with respect to the southern and eastern boundaries of the Caroline plate. Our data confirm that east-directed convergence is active along the Mussau Trench, which forms the southeastern boundary of the Caroline plate. Activity along the Manus Trench is equivocal. East of the Mussau Trench, the Manus trench-fill is deformed. This deformation is probably related to the interaction of the Pacific plate with the Manus Trench. West of the Mussau Trench, data indicative of activity along the Manus Trench is ambiguous. The western part of the Manus Trench may be a relict feature and therefore, could not form the southern boundary of the Caroline plate. Our data, however, do not preclude the possibility of strike-slip motion along the Manus Trench.

INTRODUCTION

Papua New Guinea lies within one of the most tectonically complex regions in the world--a region with as many as 10 possible plate boundaries (Johnson, 1979). Two distinctive bathymetric features of the tectonic framework are the Mussau and Manus Trenches that intersect nearly at right angles about 650 km north of the island of New Guinea (Figure 1). The Manus Trench stretches arcuately from the New Guinea Trench to the Lyra Trough; the linear Mussau Trench extends from the Manus Trench to almost 3°N latitude and trends about due north. Both the Mussau and Manus Trenches have axial depths greater than 6,000 m and outer swells of the seaward trench wall that are characteristic of subduction zones. However, uncharacteristic of convergent margins, these trenches lack significant seismicity and arc-type volcanism (Johnson and Molnar, 1972).

Based on plate motions, both the Mussau and Manus Trenches have been considered to be active plate boundaries (Weissel and Anderson, 1978). North of Manus Trench and west of Mussau Trench lies the postulated Caroline plate. The Caroline plate is believed to underthrust the Pacific plate along the southern part of the Mussau Trench (Hegarty, Weissel, and Hayes, 1983). The Manus Trench was a well-documented convergent plate boundary, active from the Eocene to the early Miocene (Exon and Tiffin, 1984). However, current activity along the trench is equivocal. If this trench is still active, a small plate (the North Bismarck plate) must exist in the northern part of the Bismarck Sea wedged between the Manus Trench on the north and the ridge-transform fault zone that bisects the Manus Basin on the south (Figure 1; Johnson and Molnar, 1972; Curtis, 1973; Hedervari and Papp, 1977; Johnson, 1979). However, if the Manus Trench is inactive, the southern boundary of the Caroline plate may be located at the Manus Basin fault zone, which forms the northern margin of the South Bismarck plate,

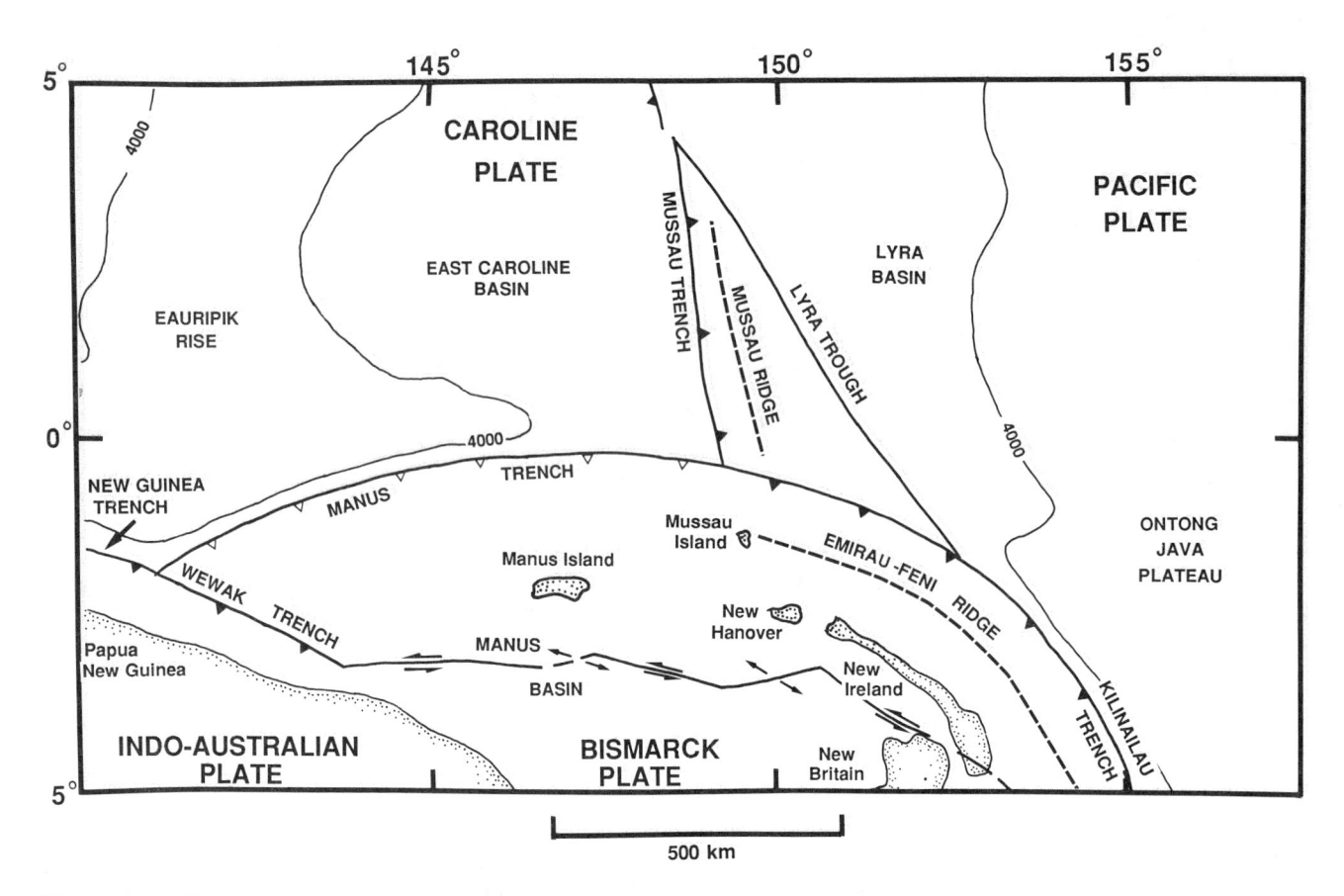

Figure 1. Map showing location of plate boundaries near the intersection of the Mussau and Manus Trenches (modified from Hamilton, 1979). Open triangles represent an inactive trench; closed triangles represent an active trench. The 4-km bathymetric contour delineates the Eauripik Rise and Ontong Java Plateau. The hypothesised North Bismarck plate would lie between the Manus Trench and the Bismarck plate.

although the eastern boundary is poorly defined (Figure 1).

Several interrelated tectonic events are involved in the formation and reorganization of the minor plates in the region of Papua New Guinea during the late Cenozoic. 1.) At 8-10 Ma, the convergence rates between the Pacific and Indo-Australian plates increased from about 30 to 96 mm/yr (Molnar et al., 1975). 2.) At 3.5 Ma, the Manus Basin opened up, displacing the Bismarck Sea southwest with respect to the Pacific plate (Taylor, 1979). 3.) During late Quaternary time, the New Britain island arc collided with mainland Papua New Guinea (Johnson, 1979). The evolution of the Mussau and Manus Trenches into active plate boundaries is probably related to these events. However, the lack of seismicity and volcanism along the trenches implies that, if active, these features are young. The purpose of this report is to determine whether the Mussau and Manus Trenches are active plate boundaries and to document the structural style of the trench-trench intersection.

TECTONIC HISTORY

The Cenozoic tectonic history of Papua New Guinea has been reviewed by Johnson (1979) and Kroenke (1984). The tectonic setting of the Mussau Trench based on analysis of single-channel seismic-reflection, gravity and magnetic data is outlined below (Kogan, 1976; Erlandson et al., 1976; Weissel and Anderson, 1978; Hegarty, Weissel, and Hayes, 1983).

The Mussau Trench is bordered on the west by the East Caroline Basin. Samples of basalt drilled at Deep Sea Drilling Project (DSDP) sites 62 and 63, and magnetic anomaly correlations indicate that the East Caroline Basin is floored by lower to middle-Oligocene oceanic crust (Winterer et al., 1971; Weissel and Anderson, 1978). This crust is characterized by a rectilinear pattern of northeast- and northwest-trending ridges and valleys that are covered by thick Neogene calcareous strata (Erlandson et al., 1976). These strata extend to the eastern margin of the basin where they dip undisturbed into the Mussau Trench (Erlandson et al., 1976). About 25 km east of, and parallel to, the Mussau Trench is the Mussau Ridge, a steep-sided bathymetric feature with little sediment cover (Figure 2). Dredge samples from the lower ridge slope consist primarily of altered oceanic-type basalt (Udintsev et al., 1974).

According to Hegarty, Weissel, and Hayes (1983), the oceanic crust on opposite sides of the Mussau Trench is the same age, as indicated by the interpretation of magnetic anomalies. They suggest that the Mussau Trench was an intraoceanic fracture zone within the East Caroline Basin that developed into an active margin when the Caroline plate formed; part of the Oligocene oceanic crust of the Caroline Basin remained attached to the Pacific plate. Using a simple flexural-beam gravity model, Hegarty, Weissel, and Hayes (1983) estimate the amount of crustal shortening between the Caroline and Pacific plates to be about 7 km at 2°N latitude since about 1 Ma. The Mussau Trench is an example of a new subduction zone that is forming along an old oceanic fracture zone.

The Manus Trench is older than the Mussau Trench and was the site of active subduction from Eocene until earliest Miocene time (Exon and Tiffin, 1984; Kroenke, 1984). This paleo-subduction zone had all of the elements typical of modern convergent margins: a volcanic island arc (e.g. New Ireland, New Hanover, and Manus Islands), a forearc basin (New Ireland Basin), an outer-arc high (Emirau-Feni Ridge), and a trench (Manus Trench). The collision of the Eauripik Ridge and Ontong Java Plateau with the Manus Trench may have terminated subduction along this trench in earliest Miocene time (de Broin, Aubertin, and Ravenne, 1977; Kroenke, 1984). The remainder of the Miocene was a time of relative tectonic quiescence in the region. During early Pliocene time, the opening of Manus Basin resulted in the displacement of New Britain to the southeast with respect to New Ireland along transcurrent faults, and in the associated deformation of the New Ireland Basin (Exon and Tiffin, 1984).

Although the Manus Trench retains the geomorphic features of the paleo-subduction zone, little evidence exists for subduction along the trench since the earliest Miocene. However, the recent uplift of limestone terraces on Mussau Island indicates active deformation south of the Manus Trench (Exon et al., 1986). Pliocene and younger volcanism has occurred along the Tabar to Feni Islands, but the petrogenesis of these rocks is enigmatic (Johnson et al., this volume).

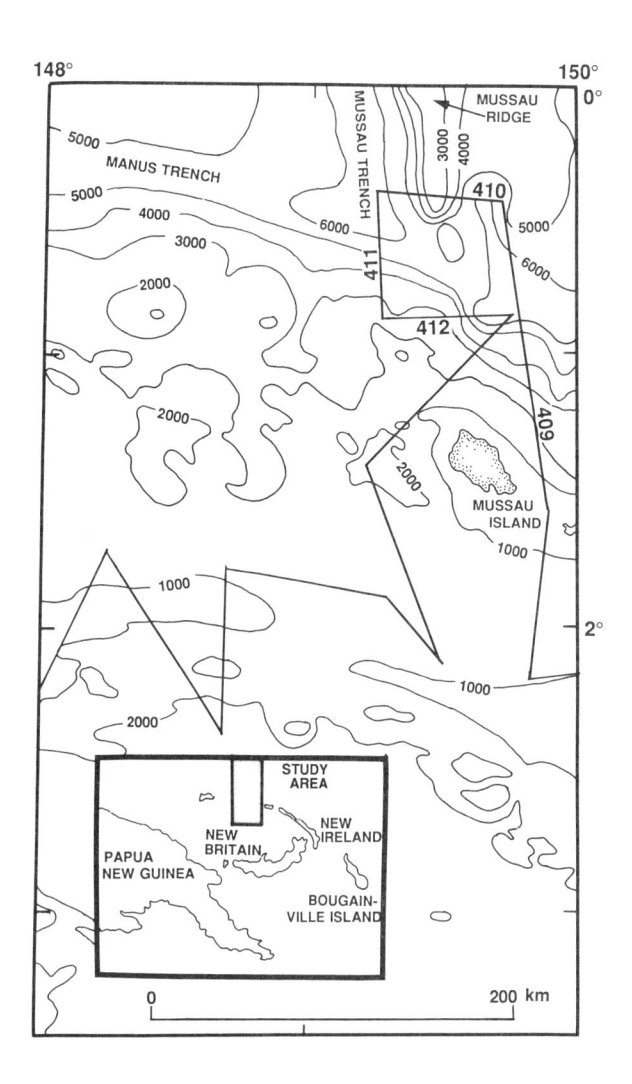

Figure 2. Location of 1984 *Lee* multichannel seismic-reflection tracklines near the Mussau and Manus Trenches (heavy lines) and 1,000-m bathymetric contours. Inset map shows the location of the study area.

DATA ACQUISITION AND PROCESSING

In 1984, the U.S. Geological Survey ship R/V *S.P. Lee* collected about 250 km of 24-fold seismic-reflection data near the Mussau and Manus Trench intersection (Figure 2). The multichannel survey was conducted with a five-air-gun, 2,000-psi tuned array that totaled 20 liters. Data were recorded using a 2,400-m, 24-channel streamer and high-density digital recorder. A shot spacing of 50 m was controlled by satellite navigation. Gravity, magnetic, and high resolution seismic data were also collected.

The multichannel seismic-reflection data were processed at both the U.S. Geological Survey, and at the Australian Bureau of Mineral Resources (BMR) in Canberra. Preliminary processing included demultiplexing, trace editing, velocity analysis, stacking, deconvolution, and band-pass filtering. Selected profiles were migrated to remove diffractions and restore dips.

INTERPRETATION

Mussau Trench and Ridge

Multichannel seismic-reflection data collected perpendicular to the southernmost Mussau Trench (line 410; Figure 2) indicate convergence between the Caroline and Pacific plates, which has been documented to the north (e.g. Erlandson et al., 1976; Hegarty, Weissel, and Hayes, 1983), continues as far south as the Manus Trench. A well-bedded sediment package below a zone of chaotic reflectors is observed beneath the Mussau Trench and can be traced at least 3 km eastward below the inner trench wall (Figure 3). Single-channel seismic-reflection data recorded across a portion of the trench between lines 410 and 411 show at least 300 ms (two-way travel time) of flat-lying trench fill. These strata may have been uplifted at the inner trench wall, although the evidence is inconclusive (Figure 4).

We believe the thick, well-bedded sediment package is East Caroline Basin strata that are subducting beneath the lower trench slope of the Mussau Trench. Although profile 410 was not continued far enough to correlate the underthrusting strata directly with the basinal strata to the west, the underthrusting strata are similar in thickness and

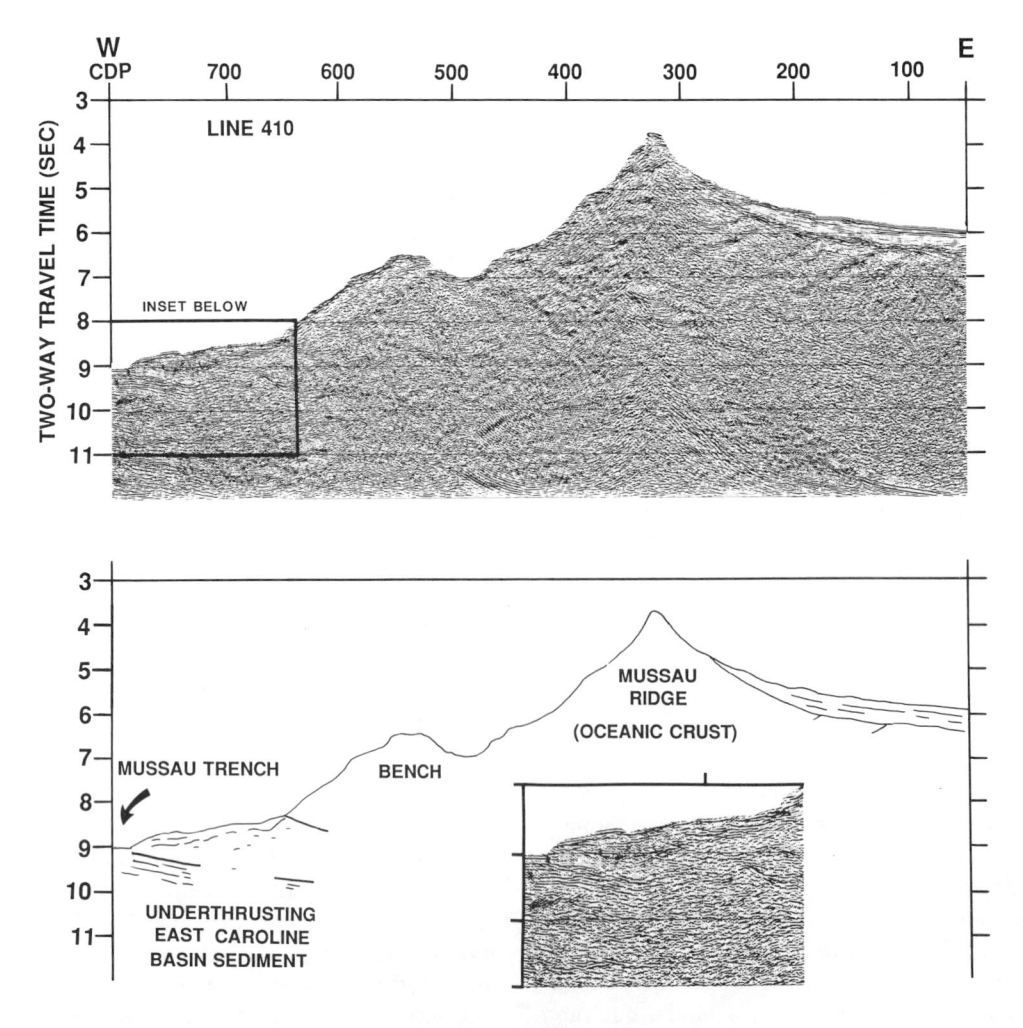

Figure 3. Multichannel line 410 collected perpendicular to the Mussau Trench and Ridge. Data were stacked, migrated, and deconvolved. CDP (Common Depth Point) spacing is 50 m. A coherent zone of reflectors dips westward beneath the inner trench slope. Inset shows an enlarged view of Mussau Trench.

character to the well-indurated Neogene sediment of the East Caroline Basin observed farther to the north (Kroenke et al., 1971; Erlandson et al., 1976). The underthrusting sediment along profile 410 is subducted beneath a decollement. We are not sure whether the chaotic reflectors above the decollement are offscraped and deformed trench-fill or slumped deposits from the adjacent Mussau Ridge. The volume of deformed material is small because of the slow-rate of trench sedimentation, the selective subduction of much of the basinal strata, the low convergence rate, and the short time of convergence.

A major topographic feature that parallels the Mussau Trench is the Mussau Ridge (Figure 2)-- a ridge that was initially considered to be an incipient volcanic arc (Hamilton, 1979). However, Hegarty, Weissel, and Hayes (1983) modeled the gravity anomaly across the ridge as an overthrust slab of oceanic crust formed initially at a fracture zone within the Caroline Basin oceanic crust. At the western base of the ridge, a bench is present (Figure 3). The bench has been interpreted as an accretionary complex (Erlandson et al., 1976) and as downfaulted Pacific crust (Hegarty, Weissel, and Hayes, 1983). The rocks beneath the bench are acoustically similar to the ridge, suggesting that they do not constitute an accretionary complex.

Manus Trench

The Manus Trench has been thought to form part of the southern margin of the Caroline plate west of the Mussau Trench (Weissel and Anderson, 1978). East of the Mussau Trench, however, a trapped piece of the former Caroline plate that is behaving as part of the Pacific plate abuts the Manus Trench. The entire length of the Manus Trench has been thought to form the northern margin of the North Bismarck plate (Johnson and Molnar, 1972). The activity of the Manus Trench both east and west of the Mussau Trench has important implications for locating the southern Caroline plate boundary.

Line 411 (Figure 5) was collected perpendicular to the Manus Trench in the tectonically complex zone at the intersection of the Mussau and Manus Trenches (Figure 2). Because the profile crossed the trench intersection, structures are not easily resolved. Less than 200 m of flat-lying strata exist in the trench and show little evidence for deformation along the inner trench wall of Manus Trench (Figure 5). Resolvable reflection horizons beneath the lower trench slope dip to the north rather than to the south in the predicted direction of underthrusting

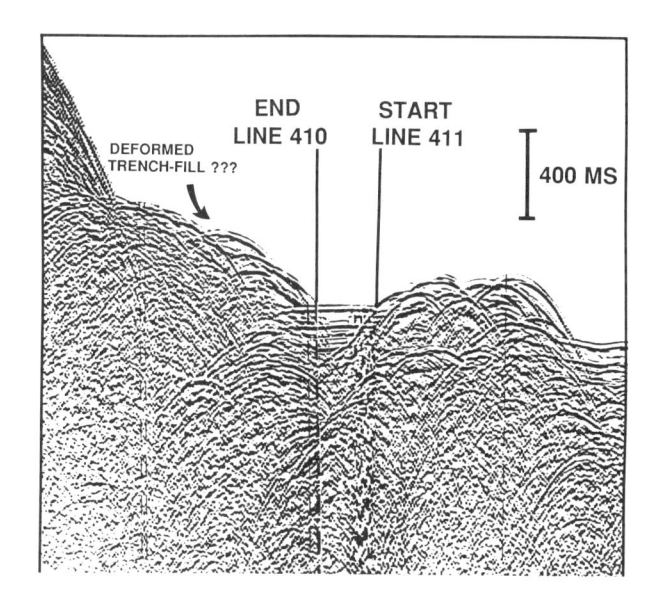

Figure 4. Single-channel monitor record between lines 410 and 411. The trench fill appears to be deformed at the end of line 410, although the deformed sediment could be the result of slumping.

(Figure 5). North of Manus Island, two single-channel seismic-reflection profiles across the Manus Trench show that the sediment in the trench is undeformed (Figure 6; D. Tiffin, written comm., 1985).

The lack of deformed trench sediment suggests that significant convergence has not occurred along the Manus Trench west of the Mussau Ridge in the geologically recent past. We note, however, that the slow convergence rate and the obliquity of convergence (> 65° from the trend of the trench axis)

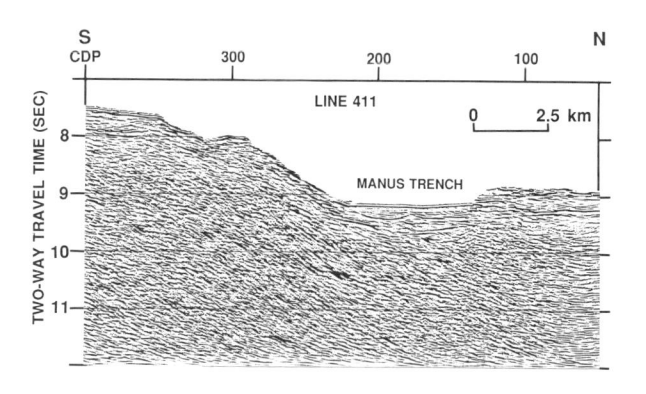

Figure 5. Multichannel line 411 collected perpendicular to the Manus Trench at the intersection with the Mussau Trench. Data were dip-filtered, deconvolved, stacked, and migrated. CDP spacing is 50 m. The trench sediment is undeformed at the inner trench slope.

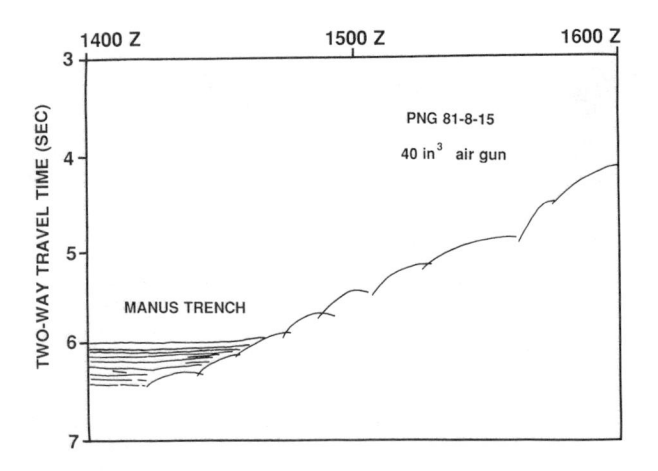

Figure 6. Interpretation of single-channel profile across the Manus Trench near Manus Island. Trench sediment onlaps the inner trench slope.

predicted for the Manus Trench (Weissel and Anderson, 1978) may account for the lack of evidence for compressional deformation and permits the possibility of a component of active transcurrent faulting here.

Line 409 was collected perpendicular to the Manus Trench about 45 km east of the Mussau Trench (Figure 2). The trench sediment is deformed into broad open folds against the inner trench slope (Figure 7); folded trench strata are also observed on a seismic-reflection profile collected immediately west of line 409 (de Broin, Aubertin, and Ravenne, 1977). Seaward from the folded strata on the inner trench wall, south-dipping thrust faults deform the trench fill. Irregular sub-trench-fill reflectors are undeformed and dip gently southward (Figure 7); these reflectors can be directly correlated with basinal strata to the north.

We interpret the folded strata beneath the inner trench wall along line 409 to be an accretionary complex (Figure 7). The deformation of trench strata indicates that compression occurs between the Pacific plate and the region to the immediate south. We infer that the undeformed basement sequence below the folded and thrust-faulted trench fill is pelagic basinal strata deposited on irregular oceanic crust that preferentially subducted below a decollement (Figure 7). This style of deformation is similar to that along the Mussau Trench (Figure 3), and to that along other margins where the selective subduction of basinal pelagic sediment occurs (e.g., the Nankai Trough, Leggett, Aoki, and Toba, 1985, Figure 4). The deformation at the Manus Trench along line 409 suggests that the eastern part of the trench is an active feature. A similar style of deformation occurs along the Kilinailau Trench to the south (T. Bruns, oral comm., 1987).

Line 409 crosses the trench-slope break associated with the Manus Trench (the Emirau-Feni Ridge) near Mussau Island. Our data suggest that the ridge is underlain by igneous rock. The lowermost trench slope is poorly reflective; however, farther up the slope (CDP 250-1050), high-amplitude reflection horizons with high interval velocities (5-6 km/s) are present (Figure 8). These horizons are associated with high-frequency magnetic anomalies. Seaward of the trench-slope break is a slope basin filled with about 1 km of sediment (Figure 8). The basin is bounded to the south by a pronounced scarp, and strata within the basin are gently folded. The steepness of the scarp may indicate that deformation is recent. Igneous intrusive bodies within 55 km of the trench axis is anomalous and probably not related to subduction along the Manus Trench (Exon and Tiffin, 1984; Johnson, 1979).

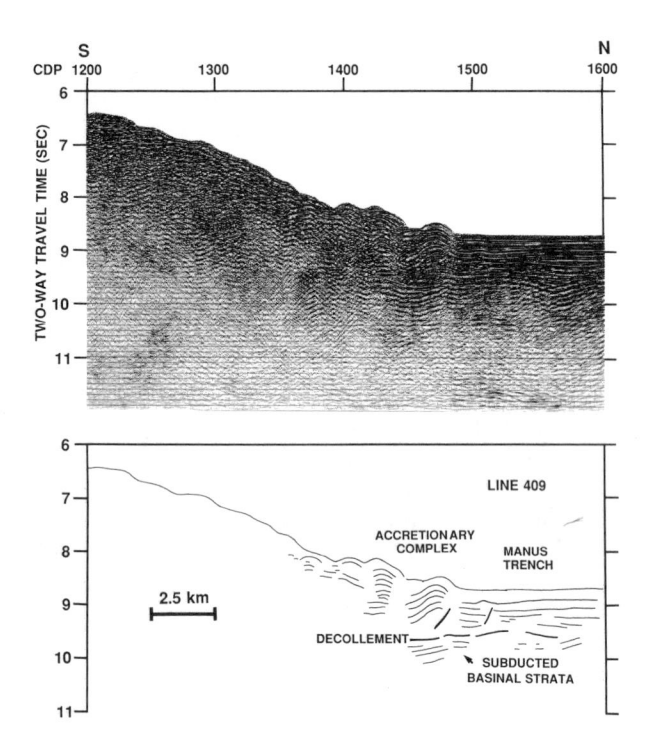

Figure 7. Multichannel line 409, CDP 1200-1600, collected across the Manus Trench east of the Mussau Ridge. Data were dip-filtered, stacked, and migrated; the data are ungained. CDP spacing is 50 m. Note the well-developed folds south of the trench.

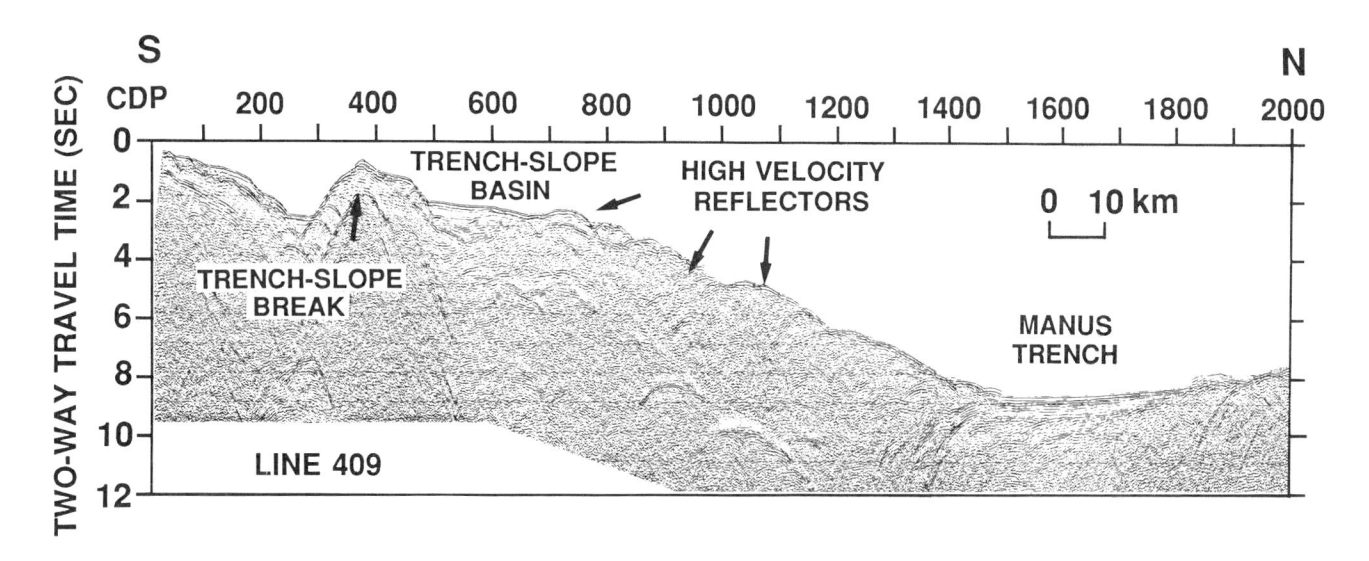

Figure 8. Multichannel line 409, CDP 50-2000 from the Emirau-Feni Ridge to the Manus Trench. Data were filtered and stacked. CDP spacing is 50 m. Region of high amplitude reflectors corresponds to the zone of high amplitude magnetic anomalies and high interval velocities. The steep scarp (at CDP 475) forms the southern edge of the trench-slope basin.

CONCLUSIONS

Multichannel seismic-reflection data show that active convergence occurs along the southern part of the Mussau Trench near its intersection with the Manus Trench (about 1°S). The minor amount of deformation at the Mussau Trench, combined with the lack of seismicity and volcanism along this proposed plate boundary, suggests that convergence has been active for only a short duration and at slow convergence rates.

The southern margin of the Caroline plate remains to be resolved. If the Manus Trench west of the Mussau Trench forms the southern Caroline plate boundary, then there must be a small plate (the North Bismarck plate) between the South Bismarck plate and the Caroline plate (Figure 1). If not, then the northern margin of the South Bismarck plate is the southern Caroline plate margin. Although the Manus Trench is a deep bathymetric feature with a pronounced outer-arc swell, unequivocal evidence for deformation along the inner trench wall is not apparent in seismic-reflection data. Moreover, the seismological and petrological data currently available along the western part of the Manus Trench provide no compelling evidence that the Manus Trench is an active convergent plate boundary. If the trench is active, the amount of recent underthrusting along the trench is negligible.

East of the Mussau Ridge, the Manus Trench appears to be active. Seismically imaged basinal strata are underthrusting the Manus Trench beneath a decollement, and the overlying trench sediment is deformed into broad open folds. The continuation of this deformation to the south along the Kilinailau Trench supports the existence of a plate boundary. Additional data are needed, particularly along the western segment of the Manus Trench, to determine whether a North Bismarck plate exists.

ACKNOWLEDGMENTS

We would like to thank Captain Al and the crew of the R/V *S.P. Lee* for their support during data acquisition and Frank Brassil and David Falvey for their hospitality while in Canberra. Patrick McClellan, Terry Bruns, and Michael Fisher provided thoughtful reviews of the manuscript.

REFERENCES

Curtis, J.W., 1973, The spatial seismicity of the Papua New Guinea and the Solomon Islands: Journal of the Geological Society of Australia, v. 20, p. 1-20.

de Broin, C.E., Aubertin, F., and Ravenne, C., 1977, Structure and history of the Solomon-New Ireland region: International Symposium Geodynamics South-West Pacific, Noumea (New Caledonia), 1976, Paris, Editions Technip., p. 37-50.

Erlandson, D.L., Orwig, T.L., Kiilsgaard, G., Mussells, J.H., and Kroenke, L.W., 1976, Tectonic interpretations of the East Caroline and Lyra Basins from reflection-profiling investigations:

Geological Society of America Bulletin, v. 87, p. 453-462.

Exon, N.F., Stewart, W.D., Sandy, M.J., and Tiffin, D.L., 1986, Geology and offshore petroleum prospects of the eastern New Ireland Basin, northeastern Papua New Guinea: Bureau of Mineral Resources Journal of Australian Geology and Geophysics, v. 10, p. 39-51.

Exon, N.F. and Tiffin, D.L., 1984, Geology of offshore New Ireland basin in northern Papua New Guinea, an its petroleum prospects, *in* Watson, S.T., ed., Transactions of the Third Circum-Pacific Energy and Mineral Resources Conference, Honolulu, 1982, p. 623-630.

Hamilton, W., 1979, Tectonics of the Indonesian Region: U.S. Geological Survey Professional Paper 1078, 338 p.

Hedervari, P. and Papp, Z., 1977, Seismicity maps of the New Guinea-Solomon Islands region: Tectonophysics, v. 42, p. 261-281.

Hegarty, K.A., Weissel, J.K., and Hayes, D.E., 1983, Convergence of the Caroline and Pacific plates--Collision and subduction, *in* Hayes, D.E., ed., The Tectonics and Geologic Evolution of Southeast Asian Seas and Islands: Part 2, Geophysical Monograph Series, American Geophysical Union, Washington D.C., v. 27, p. 326-348.

Johnson, R.W., 1979, Geotectonics and volcanism in Papua New Guinea: a review of the late Cainozoic: Bureau of Mineral Resources Journal of Australian Geology and Geophysics, v. 4, p. 181-207.

Johnson, T. and Molnar, P., 1972, Focal mechanisms and plate tectonics of the southwest Pacific: Journal of Geophysical Research, v. 77, p. 5000-5032.

Kogan, M.G., 1976, Gravity anomalies and main tectonic units of the southwest Pacific: Journal of Geophysical Research, v. 81, p. 5240-5248.

Kroenke, L., Moberly, R., Winterer, E.L., and Heath, G.R., 1971, Lithologic interpretation of continuous reflection profiling, Deep Sea Drilling Project, Leg 7, *in* Winterer, E.L., et al., 1971, Initial Reports of the Deep Sea Drilling Project, Volume VII: U.S. Government Printing Office, Washington, D.C., p. 1161-1226.

Kroenke, L.W., 1984, Cenozoic tectonic development of the southwest Pacific: UN ESCAP, CCOP/SOPAC Technical Bulletin, no. 6, 122 p.

Leggett, J., Aoki, Y., and Toba, T., 1985, Transition from frontal accretion to underplating in a part of the Nankai Trough accretionary complex off Shikoku (SW Japan) and extensional features on the lower trench slope: Marine and Petroleum Geology, v. 2, p. 131-141.

Molnar, P., Atwater, T., Mammerickx, J. and Smith, S.M., 1975, Magnetic anomalies, bathymetry and tectonic evolution of the South Pacific since the late Cretaceous: Geophysical Journal of the Royal Astronomical Society, v. 40, p. 383-420.

Taylor, B., 1979, Bismarck Sea: evolution of a back-arc basin: Geology, v. 7, p. 171-174.

Udintsev, G.B., Dimitriyev, L.V., Sharas'kin, A.Ya., Agapova, G.V., Zenkovich, N.L., Berenev, A.F., Kurentsova, N.A., and Suzyumov, A. Ye., 1974, New data on trench-faults in the southwestern Pacific: Geotectonics, 1974, no. 2, p. 65-69.

Weissel, J.K. and Anderson, R.N., 1978, Is there a Caroline plate?: Earth and Planetary Science Letters, v. 41, p. 143-158.

Winterer, E.L., et al., 1971, Initial reports of the Deep Sea Drilling Project, Vol. VII: Washington, D.C., U.S. Government Printing Office, 1757 p.

Marlow, M.S., Dadisman, S.V., and Exon, N.F., editors, 1988, Geology and offshore resources of Pacific island arcs—New Ireland and Manus region, Papua New Guinea, Circum-Pacific Council for Energy and Mineral Resources Earth Science Series, v. 9: Houston, Texas, Circum-Pacific Council for Energy and Mineral Resources.

GEOPHYSICAL STUDY OF A MAGMA CHAMBER NEAR MUSSAU ISLAND, PAPUA NEW GUINEA

S.V. Dadisman, M.S. Marlow
U.S. Geological Survey, Menlo Park, California 94025, USA

ABSTRACT

Analysis of 24-channel seismic-reflection data collected near Mussau Island, Papua New Guinea, shows a high-amplitude, negative-polarity reflection that we believe is from the top of a magma chamber. The reflecting horizon lies at a depth of about 4.4 s subbottom and can be traced laterally for 2.6 km. On shot gathers, the reflection demonstrates normal moveout appropriate for an in-plane event. The frequency spectrum of the reflection shows a decrease in high-frequency content when compared to the sea floor reflection, as would be expected for a deep subsurface event. The polarity of the reflection event is negative, suggesting that the reflection horizon is the top of a low-velocity zone. Magnetic data indicate that the ridge containing the reflecting horizon is magnetic, and the geology of Mussau Island suggests that the ridge is volcanic in origin. We speculate that the high-amplitude reflection is from the top of a magma chamber some 7-11 km deep.

INTRODUCTION

Multichannel seismic-reflection data have been used to document axial magma chambers within active spreading centers (e.g. Herron et al., 1978; Hale, Morton, and Sleep, 1982; Morton and Sleep, 1985; and Detrick et al., 1987). In this paper we introduce multichannel seismic-reflection data, collected in 1984 by the Australia-New Zealand-United States Tripartite for the United Nations Committee for Coordination of Joint Prospecting for Mineral Resources in the South Pacific Offshore Areas (CCOP/SOPAC), that may image a reflection from the top of a magma chamber within an island arc.

This reflection horizon occurs 90 km south of the intersection of Manus and Mussau Trenches and 15 km west of Mussau Island, northeast of mainland Papua New Guinea (Figure 1). The horizon is laterally continuous for 2.6 km and lies at an estimated depth of 7 to 11 km. This study describes the geophysical characteristics of the reflection in order to determine: 1) if the event is a deep feature that lies within the plane of the reflection profile, and 2) if the horizon is the top of a low-velocity zone. We then consider whether or not the low-velocity zone delineates a magma body or chamber. However, the tectonic origin of the magma chamber is uncertain.

REGIONAL SETTING

The regional tectonic setting of the area around Mussau Island is discussed in detail by Ryan and Marlow (this volume). Eastward-directed subduction of the Caroline plate beneath the Pacific plate appears to be incipient along Mussau Trench (Figure 1). Convergence along Manus Trench appears to be reactivated only to the east of the intersection with Mussau Trench. However, an active Benioff zone related to subduction is absent beneath Mussau Island and the Tabar-to-Feni island chain that lies farther to the southeast (Johnson et al., this volume). The lack of seismicity suggests that subduction along

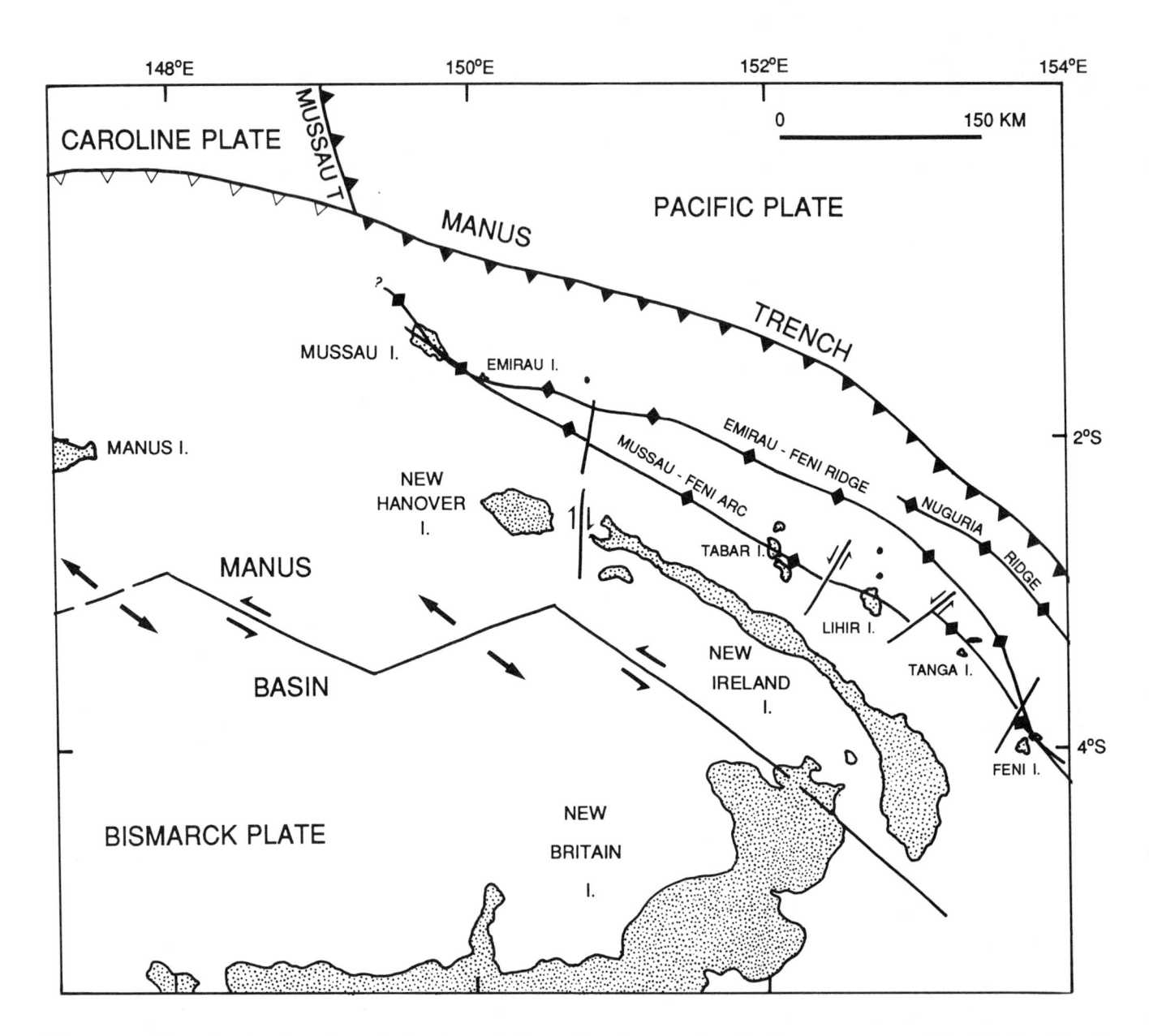

Figure 1. Map showing location of plate boundaries and major geomorphic features surrounding Mussau Island (modified after Taylor, 1979; de Broin, Aubertin, and Ravenne, 1977; Hegarty, Weissel, and Hayes, 1983; Exon et al., 1986) Open triangles represent an inactive trench; closed triangles represent an active trench. Diamonds represent ridges or island arcs. Solid lines denote transform faults.

Manus Trench is intermittent or just has begun and therefore is not associated with a Quaternary island arc (Johnson, 1979).

Little has been published about the geology of Mussau Island, hence little is known of the geologic history of the island. White and Warin (1964) published a generalized geologic map of Mussau based on one field traverse and an uncontrolled photo-mosaic. The central hills of the island are reported to be underlain by volcanic rocks (basalts and tuffs were noted), and are surrounded by a coral terrace 3 to 8 km wide. The oldest rocks on the island are described by Exon et al. (1986) and Stewart and Sandy (this volume) as typical island-arc volcanics, similar to the Jaulu Volcanics of middle Eocene to early Miocene age exposed on the nearby islands of New Ireland, New Hanover, and Tabar. Exon et al. (1986) believe that Mussau was part of a mid-Tertiary volcanic arc associated with Manus Trench. However, Mussau Island is located only 55 km south of Manus Trench in the Manus forearc region (Figure 1)—a distance that is about 100-150 km less than the rest of the islands in the mid-Tertiary arc and is anomalously close for subduction related arc volcanism in general (Dickinson and Seely, 1979). Mussau may have been transported trenchward by strike-slip faulting since the mid-Tertiary, or the island may have formed in place in the forearc. Mussau is now part of the Emirau-Feni Ridge, which is believed to be a mid-Tertiary outer arc high (Figure 1; Exon et al., 1986; Ryan and Marlow, this volume).

Further reconnaissance mapping on Mussau Island has revealed early Miocene shelf limestone, conglomerates containing pebbles of early Miocene planktonic foraminiferal limestone, and Pliocene planktonic foraminiferal limestone exposed in a series of recently uplifted terraces (Exon et al., 1986). Fresh, presumably young, gabbros also are present on Mussau (C.M. Brown, oral comm.), indicating that a second phase of igneous activity may have occurred on the island during the late Cenozoic.

Mussau Island is aligned southeastward with the Tabar-to-Feni islands, and together they form the Mussau-Feni island arc (Figure 1). Most of the volcanism in the Tabar-to-Feni chain occurred during the Pliocene and Pleistocene; however, mid-Miocene volcanic rocks are exposed on Tabar, and some volcanic activity along the chain may have occurred in the Holocene (Wallace et al., 1983; Johnson et al., this volume). At present, all four island groups in the Tabar-to-Feni chain exhibit thermal activity that probably results from the cooling of magma bodies (Wallace et al., 1983; Johnson et al., this volume). Although the tectonic origin of the late Cenozoic volcanism in the Tabar-to-Feni chain is enigmatic, Johnson et al. (this volume) are convinced that the rocks have a definite island arc-type geochemical signature and are somehow related to subduction processes. Taylor (1979) showed—by plate reconstructions of the region before the opening of Manus Basin—that Mussau and the Tabar-to-Feni chain were underlain by the northeast-dipping slab associated with the New Britain Trench. The Mussau-Feni arc volcanism may have started in response to subduction along the New Britain Trench and continued for sometime after the opening of Manus basin in the Pliocene. Alternatively, Johnson (1979) suggests that faulting associated with the opening of the Manus Basin may be responsible for the Late Cenozoic volcanism in the Tabar-to-Feni chain.

DATA COLLECTION AND PROCESSING

Multichannel seismic-reflection data were col-

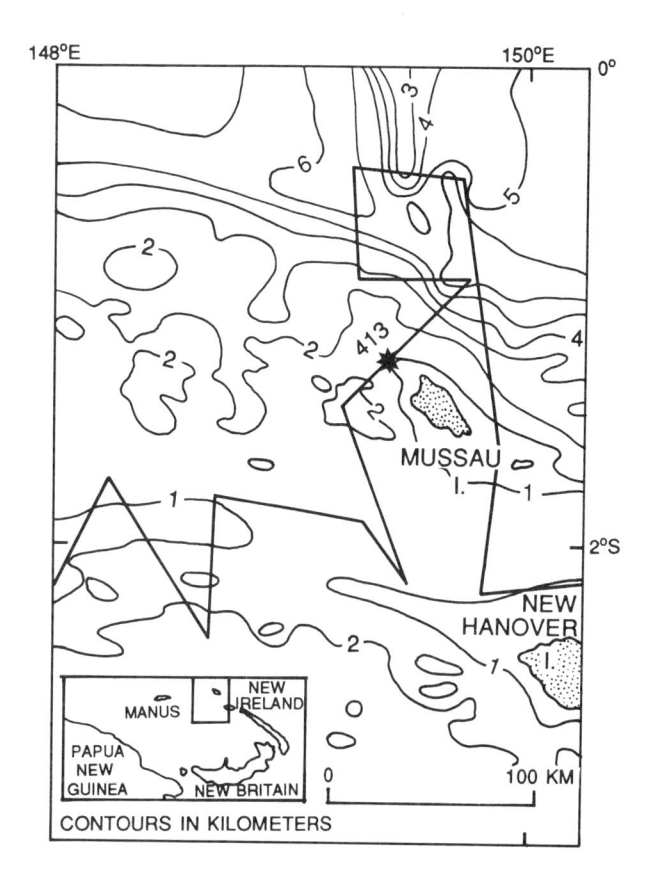

Figure 2. Trackline map showing the location of multichannel seismic-reflection data collected near Mussau Island. Star marks location of the deep reflector. Inset shows the location of study area.

lected aboard the U.S. Geological Survey research vessel *S.P. Lee*, using a 2,400-m, 24-channel hydrophone streamer, and were recorded by a Globe Universal Sciences (GUS)[1] model 4200 high-density digital recorder. The sound source employed a five-air-gun array that totaled 21 liters and operated at 2,000 psi. The 50-m shot spacing was controlled by satellite navigation. Magnetic, gravity, bathymetric, and high-resolution single-channel seismic-reflection data were collected concurrently.

Multichannel reflection data were processed using Digicon's Interactive Seismic Computer (DISCO) software. A preliminary stack of line 413 revealed a high-amplitude reflection about 4.4 s below the sea floor (Figures 2 and 3). Part of the line was reprocessed to enhance this reflection. The

---

[1] Any use of trade names is for purpose of identification only and does not imply endorsement by the U.S. Geological Survey, the New Zealand Geological Survey, or the Bureau of Mineral Resources, Australia.

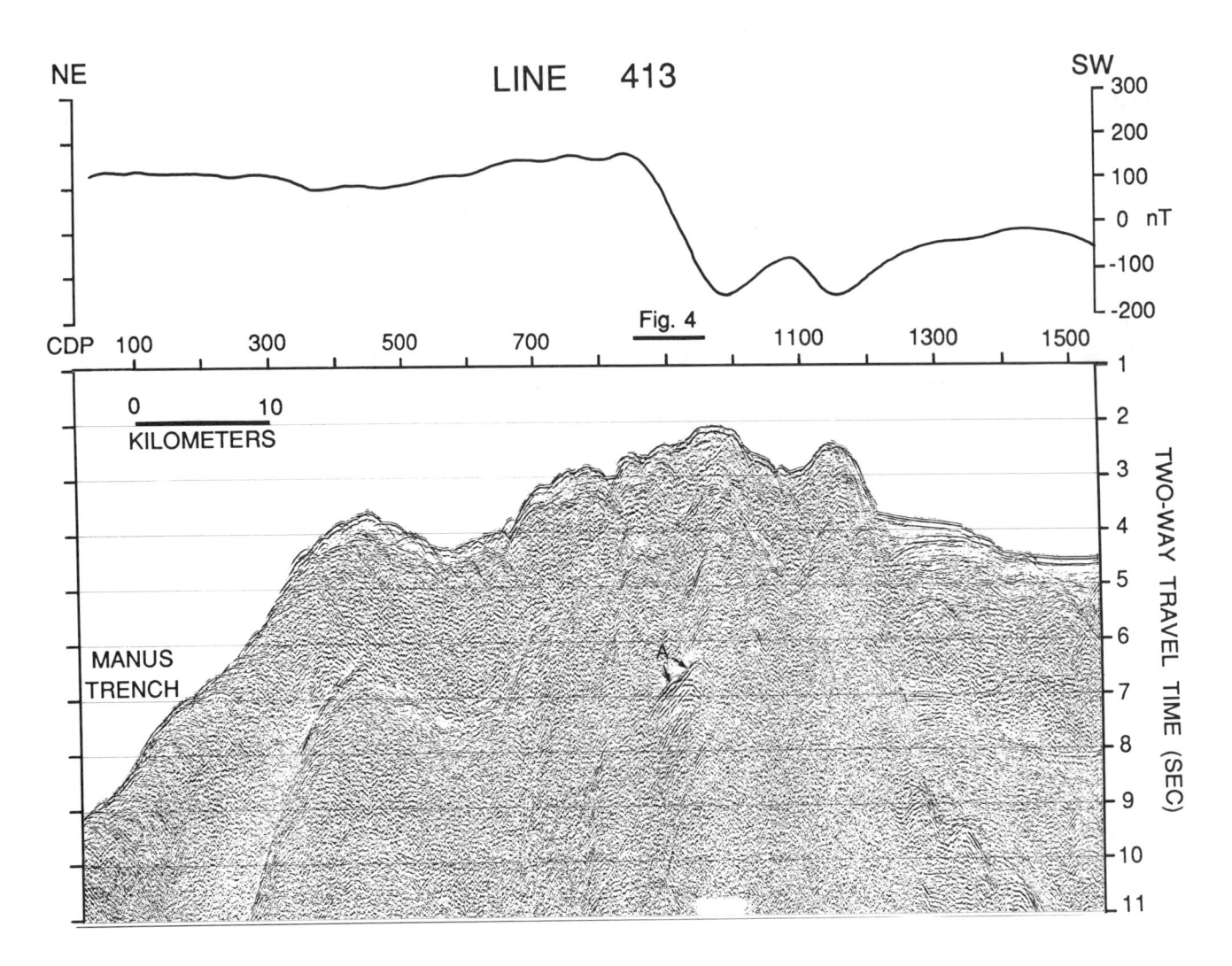

Figure 3. Seismic-reflection and magnetic profiles collected along line 413 (see Figure 2 for location). The deep reflection is marked A and is located on the seismic profile near CDP 900 at 5.5 s. A 300-gamma anomaly is centered at CDP 925. Bar shows location of Figure 4. Positive polarity is plotted in black.

reprocessing included pre-stack migration, deconvolution, normal moveout (NMO) correction, common depth point (CDP) stacking, and bandpass filtering. Shot gathers and frequency spectra of the reflection were also examined for inconsistencies, such as abnormal moveout and large high-frequency content.

The clearest image of the reflection was produced by first pre-stack migrating the data at a velocity of 1500 m/s to remove diffraction tails associated with the irregularly dipping surface of the horizon. The pre-stack migration was done by using a finite-difference algorithm that unfortunately produces aliased events directly above a steeply dipping reflection (Digicon Inc., written comm.; Figures 4 and 5). The dip-aliased events mimic the dip of the reflection; they were not removed to avoid artificially biasing the data.

After pre-stack migration, predictive deconvolution was applied to the data to remove the bubble pulse reverberations that follow primary reflections. Deconvolution not only helps us delimit individual reflections, but also allows us to determine reflection polarity. Polarity is significant in this study because a reflection from a high to low velocity boundary will cause the waveform to reverse from positive to negative polarity. A change downward from high to low velocity across a boundary can result from the presence of a less dense solid, a gas, or a liquid beneath the boundary.

The data were stacked by CDP, using a velocity of 2,500 m/s, which corrected for normal moveout. The stacked sections shown in Figures 3 and 4 were frequency filtered using a 5-8-30-35 Hz bandpass filter.

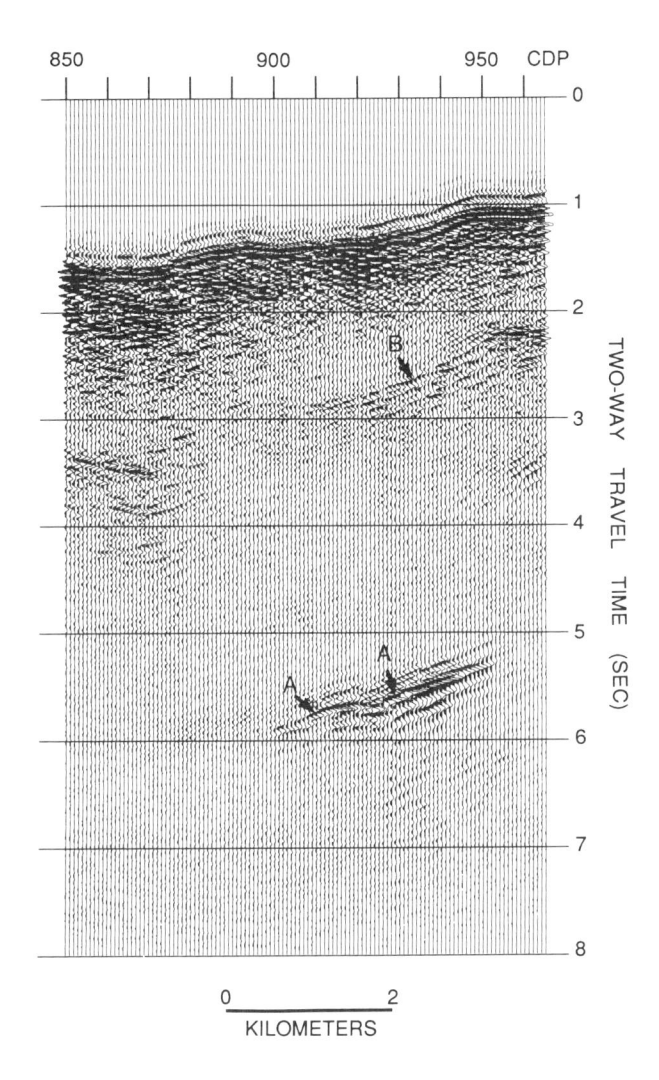

Figure 4. Ungained section of line 413 containing the deep reflection A (see text for processing sequence). Negative polarity is plotted in black. The low-amplitude events, 50-200 ms above the deep reflection, are the dip-aliased events produced by the prestack migration program (see Figure 5). Reflection B is the first water-bottom multiple.

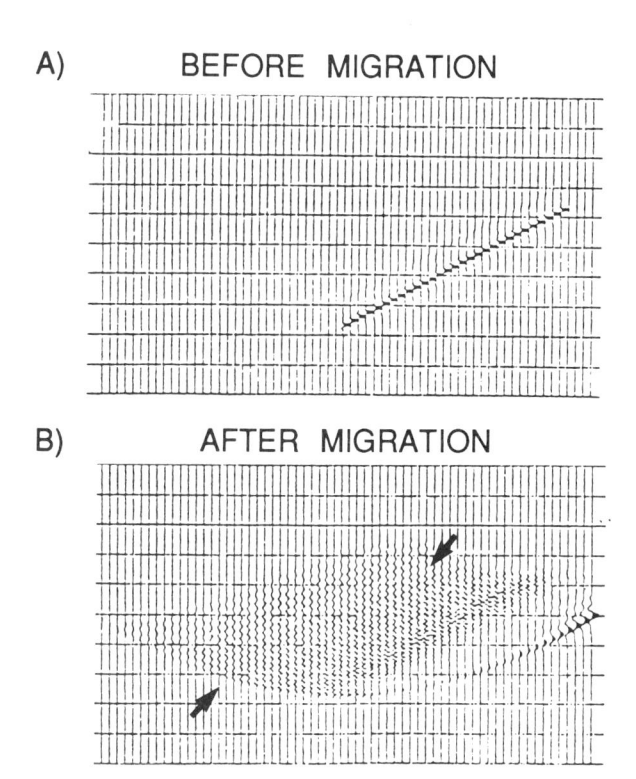

Figure 5. Example of dip aliasing produced by migration of steeply dipping surfaces using a finite-difference algorithm (Digicon, written comm.).

DISCUSSION

Interpretation of Processed Results

Figure 4 shows the final processed section of part of line 413. The reflection marked A is the deep event we want to analyze to determine whether it is from the deep subsurface. The events some 50 to 200 ms directly above reflection A are the dip-aliased events produced by the prestack migration program The reflections directly below event A are reverberations not totally removed by predictive deconvolution. The first water-bottom multiple is marked by event B.

The most conspicuous trait of reflection A is the amplitude (Figure 4). The amplitude of A is greater than that of the first water-bottom multiple B and approaches that of the sea-floor reflection (Figure 6). Large reflection amplitudes generally result from large changes in acoustic impedance (the product of rock velocity and density). If the reflection is from a deep subsurface feature, then the large amplitude of A represents a large increase or decrease in velocity or density.

The polarity of a reflection is significant because reflections from a boundary underlain by a low-velocity medium has a negative reflection coefficient and will reverse the polarity of the waveform from that reflected from the sea floor. Because the polarity of event A was difficult to determine, we used several approaches, all of which involved deconvolving these data. Deconvolution removes the reverberation produced by the bubble pulse and therefore is essential to resolve reflection polarity. These reverberations are both positive and negative in polarity and are superimposed on the primary reflections. To ascertain the polarity, the final

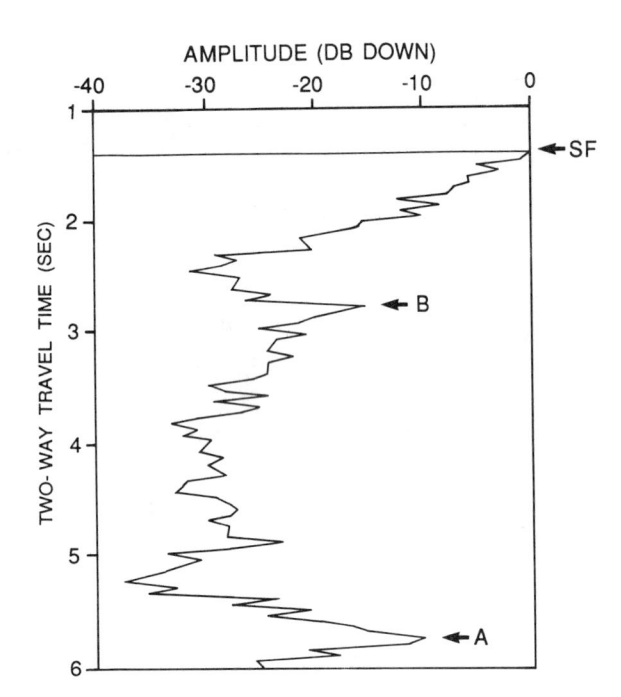

Figure 6. Trace amplitude plot for CDP 921. Amplitudes are all relative to the sea floor (SF). Note that the amplitude of the deep reflection (A) is much stronger than the first water-bottom multiple (B).

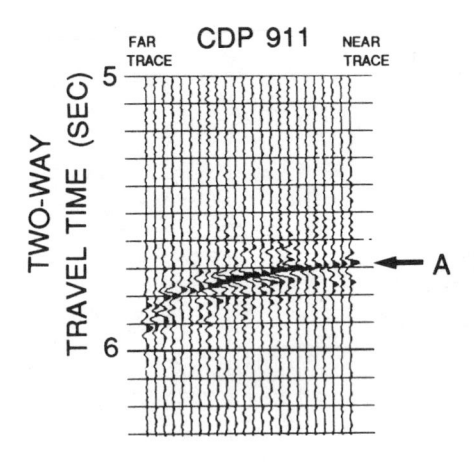

Figure 7. Deconvolved gather for CDP 911. Negative polarity is plotted in black. Note the first high-amplitude reflection has negative polarity (arrow A)(Tanner and Sheriff, 1977).

processed section shown in Figure 4 was examined. The first high-amplitude event recognized is negative in polarity (event A). Unfortunately, the dip-aliased events produced by the prestack migration program make the isolation of event A difficult and may have adversely affected its amplitude and polarity. Instead, deconvolved CDP-gathers were examined for polarity reversal, as shown in Figure 7. Because the data here were not migrated, diffraction tails obscure many of the gathers. However, Figure 7 shows that the first high-amplitude event A is indeed negative in polarity. Finally, individual, deconvolved, stacked, CDP gathers were examined to see if the shape of the wavelet at the sea floor was also the same, but reversed at the deep reflection and at the first water-bottom multiple. We were able to find some traces that showed the reversal of the wavelet at both the multiple and at event A (Figure 8). Thus, both stacked and unstacked, migrated and unmigrated data show event A to have negative polarity.

Shot gathers and frequency content of event A were examined for abnormal moveout and for large high-frequency content that might prove that reflection A is an out-of-plane event, such as "sideswipe" from a target to the side of the trackline. Figure 9 shows a deconvolved shot gather from shot 915. The moveout is appropriate for an in-plane reflection, implying that event A is a subsurface feature (Larner et al., 1983). Figure 10 shows plots of the frequency spectrum of the sea-floor reflection and the deep event A. High frequencies are attenuated by extended travel time through rock much faster than are low frequencies. Figure 10 reveals that the high-frequency content of event A is reduced relative to the sea floor reflection, as would be expected for a deep subsurface event, also suggesting that event A is a deep feature located within the plane.

Calculation of accurate interval velocities from the data were not possible, because the data were collected over rough bathymetry, and because the reflector is also an irregular surface. Depth estimates were obtained by assigning minimum and maximum average interval velocities (3000 m/s and 5000 m/s respectively) for the material above the reflector. These interval velocities are rough approximations based on estimated velocities of the onshore and offshore rocks. An average interval velocity of 3000 m/s above the reflection horizon would place the reflector at a depth of 7 km. Likewise, a velocity of 5000 m/s above the horizon would yield a depth of 11 km.

Geologic Implications

The collection and subsequent processing of marine multichannel data from areas dominated by rough bathymetry can produce artifacts that might

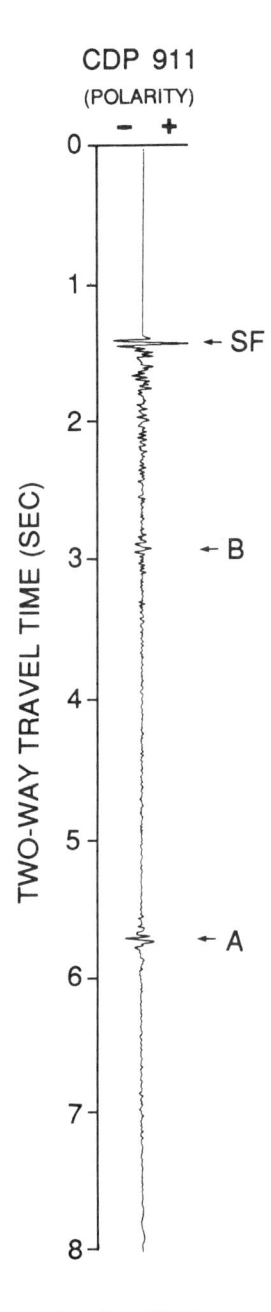

Figure 8. Trace plot for CDP 911. The trace has been deconvolved before stacking. Compare the signature of the waveform at the sea floor (SF) to that found at the deep reflection (A). The polarity at the sea floor is positive the polarity of the deep reflection is negative. Next compare the polarity of the deep

be mistaken for geology. Nevertheless, we believe that reflection event A as shown on Figure 4 is a deep subsurface feature and not a processing artifact, multiple reflection, or sidescatter. First, the reflection is clearly visible on an ungained shot gather (Figure 9), and therefore is not a processing artifact. Second, additive, long-path multiples might

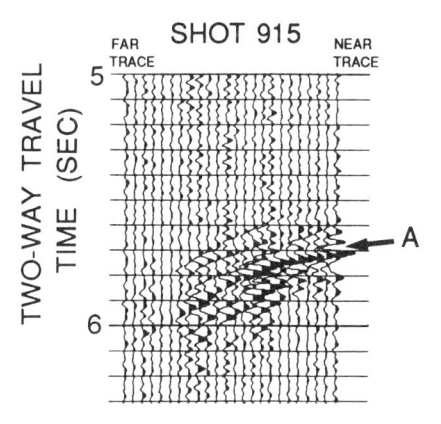

Figure 9. Shot gather for CDP 915. The reflection event at 5.7 s shows hyperbolic moveout (normal moveout) characteristic of an in-plane reflection. Positive polarity is plotted in black.

also produce all the geophysical characteristics of reflection A. However, continuous reflections are absent above the deep reflector, negating multiples as the source of reflection A (Figure 4). Third, the high-frequency content of the deep reflector is reduced, as would be expected for a deep subsurface event (Figure 10). Also, the shot gathers demonstrate appropriate moveout for event A, which suggests that the deep reflector is located within the plane of the reflection profile (Figure 9; Larner et al., 1983).

Reflection A has negative polarity, indicating that the reflector is the top of a low-velocity zone. This zone could have a low velocity if either sediment, gas, or liquid magma are present. The estimated depth of the zone, some 7-11 km, precludes low-density sediment. We favor a magma interpretation based on the regional considerations discussed below. However, gas or a gas-charged magma may be present.

Mussau Island is part of the Mussau-Feni island arc which includes the Tabar-to-Feni islands to the southeast (Figure 1). Although reconnaissance mapping and extensive geochemical and petrological work have been done on the Tabar-to-Feni islands, their tectonic history remains an enigma (see Johnson, 1979; Wallace et al., 1983; Johnson et al., this volume). These islands are aligned with Mussau and share similar geologic histories. Like Mussau, the Tabar-to-Feni islands have undergone two periods of igneous activity. The first occurred in the mid Tertiary and formed both Mussau and Tabar (Exon et al., 1986; Stewart and Sandy, this volume). The second occurred in the late Cenozoic and resulted in alkalic

A) SEA-FLOOR REFLECTION AMPLITUDE SPECTRUM

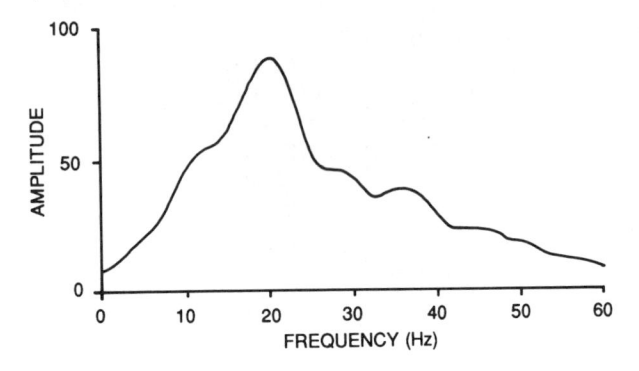

B) REFLECTION A AMPLITUDE SPECTRUM

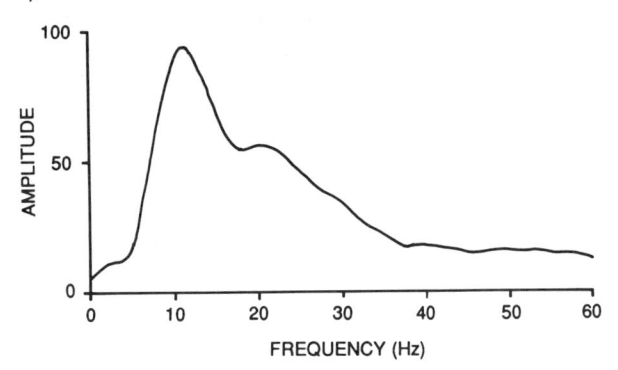

Figure 10. A) Plot of the frequency content versus relative amplitude for summed CDPs 915-925 at the sea floor (1.3-1.5 s). Most of the sea floor data are in the 20 Hz range. B) Plot of the frequency content versus relative amplitude for summed CDPs 915-925 at the deep reflection A (5.6-5.8 s). Most of the data for the deep reflection are in the 10 Hz range.

volcanism in the Tabar-to-Feni islands and intrusive gabbros on Mussau. All four island groups of the Tabar-to-Feni chain currently exhibit thermal activity that may be caused by cooling magma bodies (Wallace et al., 1983; Johnson et al., this volume). Raised limestone terraces exposed on Mussau and on the Tabar-to-Feni islands suggest recent tectonic activity along the Mussau-Feni arc. The young intrusives on Mussau, and the igneous and thermal activity in the Tabar-to-Feni islands, together with the uplift of the entire Mussau-Feni arc, support the idea of an active or at least warm magma chamber as the cause of reflection event A near Mussau Island (Figures 1 and 4). Finally, the estimated depth of 7-11 km for the low-velocity zone is also consistent with depths proposed by Gill (1981) for shallow magma chambers within volcanic island arcs.

CONCLUSIONS

The high-amplitude reflection event A recorded near Mussau Island is from an in-plane horizon above a low-velocity zone located beneath the volcanic Mussau-Feni island arc (Figures 1 and 4). Onshore and regional geology, and the interpretation of seismic-reflection and magnetic data, all suggest that the low-velocity zone is the top of a magma chamber some 7-11 km beneath the ridge crest.

The tectonic origin of the proposed magma chamber is uncertain. We suspect that the chamber is related to late Cenozoic igneous activity on Mussau Island and igneous and thermal activity of the same age in the adjacent Tabar-to-Feni islands. Late Cenozoic igneous activity all along the Mussau-Feni arc may be a thermal remnant from a now-abandoned subduction zone beneath the arc (see Johnson et al., this volume).

ACKNOWLEDGMENTS

We thank Captain Allan McClennaghan and the crew of the *S.P. Lee* for their assistance and enthusiasm in the collection of the data used in this study. Discussions with Holly Ryan (USGS) provided focus and aided our interpretation of the data. The manuscript was critically reviewed and strengthened by Michael Fisher and Jack Vedder (both at the USGS).

REFERENCES

de Broin, C.E., Aubertin, F., and Ravenne, C., 1977, Structure and history of the Solomon-New Ireland region: International Symposium Geodynamics South-West Pacific, Noumea (New Caledonia), 1976, Paris, Editions Technip, p. 37-50.

Detrick, R.S., Buhl, P., Mutter, J., Vera, E., Orcutt, J., Madsen, J., and Brocher, T., 1987, Multichannel seismic imaging of the crustal magma chamber along the East Pacific Rise: Nature, v. 326, p. 35-41.

Dickinson, W.R., and Seely, D.R., 1979, Structure and stratigraphy of forearc regions: American Association of Petroleum Geologists Bulletin, v. 63, no. 1, p. 1-31.

Exon, N.F., Stewart, W.D., Sandy, M.J., and Tiffin, D.L., 1986, Geology and offshore petroleum prospects of the eastern New Ireland Basin, northeastern Papua New Guinea: Bureau of Mineral Resources Journal of Australian Geology and Geophysics, v. 10, p. 39-51.

Gill, J.B., 1981, Orogenic andesites and plate tectonics: New York, Springer-Verlag, 390 p.

Hale, L.D., Morton, C.J., and Sleep, N.H., 1982, Reinterpretation of seismic reflection data over the East Pacific Rise: Journal of Geophysical Research, v. 87, no. B9, p. 7707-7717.

Hegarty, K.A., Weissel, J.K., and Hayes, D.E., 1983, Convergence of the Caroline and Pacific plates--Collision and subduction, *in* Hayes, D.E., ed., The tectonics and geologic evolution of southeast Asian seas and islands: Part 2, Geophysical Monograph Series, American Geophysical Union, Washington D.C., v. 27, p. 326-348.

Herron, T.J., Ludwig, W.J., Stoffa, P.L., Kan, T.K., and Buhl, P., 1978, Structure of the East Pacific Rise crest from multichannel seismic reflection data: Journal of Geophysical Research, v. 83, no. B2, p. 798-804.

Johnson, R.W., 1979, Geotectonics and volcanism in Papua New Guinea; a review of the late Cainozoic: Bureau of Mineral Resources Journal of Australian Geology and Geophysics, v. 4, p. 181-207.

Larner, K., Chambers, R., Yang, M., Lynn, W., and Wai, W., 1983, Coherent noise in marine seismic data: Geophysics, v. 48, no. 7, p. 854-886.

Morton, J.L., and Sleep, N.H., 1985, Seismic reflections from a Lau Basin magma chamber; *in*, Scholl, D.W., and Vallier, T.L., (comp. and eds.), Geology and offshore resources of Pacific island arcs -- Tonga region, Circum-Pacific Council for Energy and Mineral Resources Earth Science Series, v. 2: Houston, Texas, Circum-Pacific Council for Energy and Mineral Resources, p. 441-453.

Tanner, M.T., and Sheriff, R.E., 1977, Application of amplitude, frequency, and other attributes to stratigraphic and hydrocarbon determination: American Association of Petroleum Geologists, Memoir 26, p. 301-327.

Taylor, B., 1979, Bismarck Sea: evolution of a back-arc basin: Geology, v. 7, p. 171-174.

Wallace, D.A., Johnson, R.W., Chappell, B.W., Arculus, R.J., Perfit, M.R., and Crick, I.H., 1983, Cainozoic volcanism of the Tabar, Lihir, Tanga, and Feni Islands, Papua New Guinea: geology, whole-rock analyses, and rock-forming mineral compositions: Bureau of Mineral Resources, Australia, Report 243, 62 p.

White, W.C., and Warin, O.N., 1964, A survey of phosphate deposits in the south-west Pacific and Australian waters: Bureau of Mineral Resources of Australia Bulletin, Geology and Geophysics, no. 69, 173 p.

Marlow, M.S., Dadisman, S.V., and Exon, N.F., editors, 1988, Geology and offshore resources of Pacific island arcs—New Ireland and Manus region, Papua New Guinea, Circum-Pacific Council for Energy and Mineral Resources Earth Science Series, v. 9: Houston, Texas, Circum-Pacific Council for Energy and Mineral Resources.

WIDESPREAD LAVA FLOWS AND SEDIMENT DEFORMATION IN A FOREARC SETTING NORTH OF MANUS ISLAND, NORTHERN PAPUA NEW GUINEA

M. S. Marlow
U.S. Geological Survey, Menlo Park, California 94025, USA

N. F. Exon
Bureau of Mineral Resources, Geology and Geophysics, Canberra, A.C.T. 2601, Australia

D. L. Tiffin
U.N. ESCAP, CCOP/SOPAC, c/o Mineral Resources Department, Suva, Fiji

ABSTRACT

The Manus forearc is 150 km wide and extends north from the Manus Island part of the West Melanesian arc to the Manus Trench north of the Bismarck Sea. Geophysical profiles collected over the forearc have revealed extensive areas of magnetic, highly reflective, and flat-lying layers interpreted as outpourings of lava. The layers or flows are buried by 100-400 m of hemipelagic(?) sediment over an area of more than 8,000 km^2 of the forearc. The flows occupy the broad, flat regions of the forearc; their thickness, age, and source are unknown. They may be related to the West Melanesian arc to the south, or to subduction-related volcanism from convergence at the Manus Trench to the north. Alternatively, the flows may be related to back-arc spreading that occurred north of the arc in either middle Miocene or middle Pliocene time.

Gravity, seismic-refraction, and seismic-reflection data indicate that the forearc is underlain around the flows by a thick mass of deformed sedimentary strata, part of which may have been accreted to the forearc by southward-dipping thrust faults. Earthquake studies confirm the existence of thrust events associated with southward-dipping faults in the Manus forearc. The style of accretionary deformation in this forearc is perhaps similar to that of the more extensively studied convergence zones in the Pacific and Atlantic regions, although the ages of the strata and the deformational periods in the Manus forearc are unknown. Deformation may be related to the opening of Manus Basin about 3.5 Ma and the resultant splitting and northward migration of the Manus Island segment of the arc away from New Britain.

INTRODUCTION

Previous Work

Geophysical profiles that cross the Manus forearc and Manus segment of the West Melanesian volcanic arc are, in part, tracks of vessels transiting through the area (see Murauchi and Asanuma, 1977; and de Broin et al., 1977). Marine geophysical surveys of the Bismarck Sea are reported in some detail by Connelly (1974, 1976), Willcox (1977), Taylor (1979), Exon et al. (1986), and Taylor and Exon (1987). Several single-channel seismic-reflection lines were run in 1981 by CCOP/SOPAC across the forearc as reported by Exon and Tiffin (1984). Connelly (1976) was the first to note that parts of the forearc region are characterized by magnetic anomalies indicative of igneous rocks beneath sedimentary layers in the forearc. Francis (this volume) discusses the geology of Manus Island.

1984 Lee Cruise

As part of the tripartite (Australia, New Zealand, and the USA) CCOP/SOPAC cruise of the R/V *S.P. Lee*, multichannel seismic-reflection profiles were collected in 1984 across the West Melanesian arc and Manus forearc (Figure 1; Marlow and Exon, 1985). Bathymetric, magnetic, gravity, and separate single-channel seismic-reflection data were also collected, and a few dredge and core stations were occupied (see Childs and Marlow, this volume; Carlson, this volume; Johnson et al., this volume). The multichannel seismic-reflection data were processed at the Bureau of Mineral Resources (BMR) in Canberra, Australia, and at the U.S. Geological Survey in Menlo Park, California. For details of the processing, see Ryan and Marlow (this volume).

Plate Motions and Models in Papua New Guinea Region

At least six and possibly as many as ten plate boundaries transect the region of Papua New Guinea, making this region one of the most complex tectonic areas of the world (Johnson, 1979). The postulated plate boundaries that outline the West Melanesian arc define parts of the Caroline, Bismarck, Indo-Australia, and Pacific plates (Figure 1). A fifth plate, the North Bismarck plate, was postulated by Johnson and Molnar (1972) and is delimited on the north by the Manus Trench, and on the south by the active spreading ridges and transform faults defining the northern boundary of the Bismarck plate. The existence of the North Bismarck plate and its northern boundary is conjectural and is discussed in detail by Ryan and Marlow (this volume). The Manus forearc is delimited geomorphically to the north by the Manus Trench, to the south by the West Melanesian arc that includes the Admiralty Islands of Manus and Rambutyo, to the west by mainland Papua New Guinea, and to the east by Mussau Island (Figure 1).

Paleomagnetic data, recently collected from the islands in the Papua New Guinea region and the northern Solomon Islands, have allowed a tentative reconstruction of Cenozoic plate motions in the region (Figure 2; Falvey and Pritchard, 1984). Falvey and Pritchard's model postulates that the islands of New Ireland, Manus, and New Britain were part of an early island arc system that evolved during the Eocene and Oligocene and included the Solomon Islands. Subduction of the Pacific plate was directed to the south beneath this island arc. By 30 Ma, back-arc spreading in the ancestral Solomon Sea had rotated the island arc into a nearly linear, northeast-facing structure, with subduction of the Pacific plate directed to the southwest. Arc reversal, affecting the New Britain-Manus Islands portion of the arc, occurred about 15 Ma or later, with the Solomon Sea then being subducted to the north beneath New Britain and Manus. Back-arc spreading began in the late Oligocene in the region north of New Britain and Manus. By 10 Ma the rest of the Solomon Island arc had reversed polarity, causing the oceanic crust of the Solomon Sea to subduct to the north and northeast beneath New Britain, Manus, New Ireland, and the northern Solomon Islands. About 3.5 Ma, back-arc spreading began north of New Britain, opening the Manus Basin and translating New Britain to the southeast past New Ireland and into collision with mainland Papua New Guinea, but leaving Manus nearly stationary. The opening of Manus Basin is continuing to the present as evidenced by magnetic anomalies and shallow seismicity along the transform faults in the basin (Taylor et al., 1986; Eguchi et al., 1987).

In a different model, Kroenke (1984) postulated that the West Melanesian-North Solomon Island arcs formed in latest Eocene or earliest Oligocene and were active through most of the Oligocene, perhaps into the earliest Miocene. In his model, the cessation of arc volcanism in the early Miocene ($\sim$22 Ma) is related to the arrival into the subduction zone of two large oceanic plateaus: the Eauripik Ridge north of the Manus Trench (see Figure 3.2 of Kroenke, 1984) and the Ontong Java Plateau north of the Kilinailau-North Solomon Trenches (Figure 1). Following the cessation of volcanism, a short-lived rifting event of sea-floor spreading is postulated by Kroenke (1984) for the area between Manus Island and Manus Trench during the middle Miocene, which caused the translation of Manus Island to its present position relative to the Manus Trench. As discussed below, this rifting event may be responsible for the widespread lava flows in the Manus forearc.

GEOPHYSICAL PROFILES AND INTERPRETATION

Four widely-spaced geophysical profiles cross

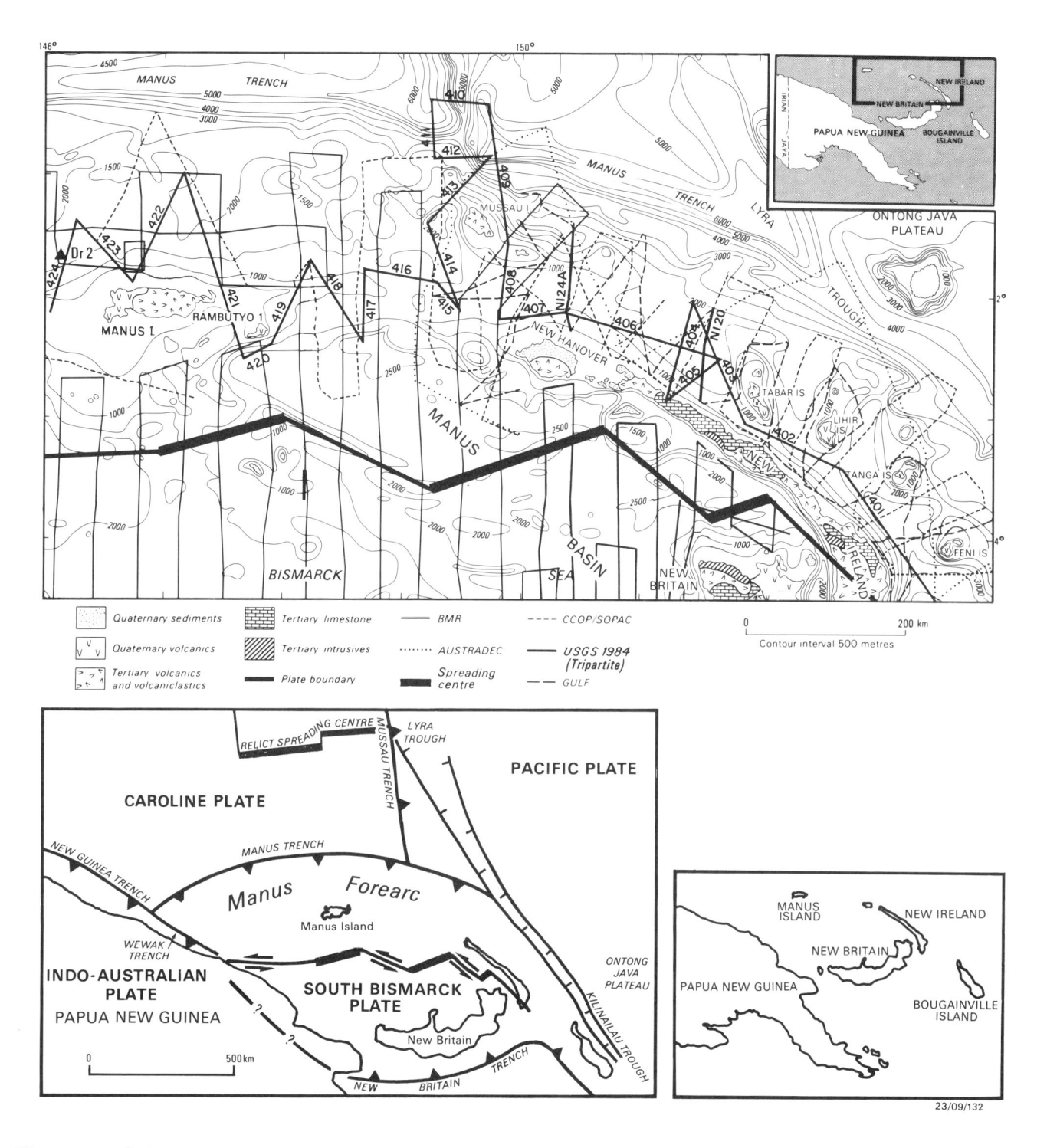

Figure 1. Index maps of Manus region of Papua New Guinea showing the Manus forearc study area and plate outlines. Top map shows bathymetry contours in meters (after Taylor, 1979) and ship tracks of geophysical lines. R = Rambutyo Island.

the forearc region north of Manus Island (Figure 1). Data collected along these lines include multichannel seismic-reflection data and illustrate salient structural features of the Manus forearc (Figures 3-7). An additional single-channel seismic-reflection profile is shown to illustrate high-resolution acoustic details not so apparent on the multichannel lines (Figure 8).

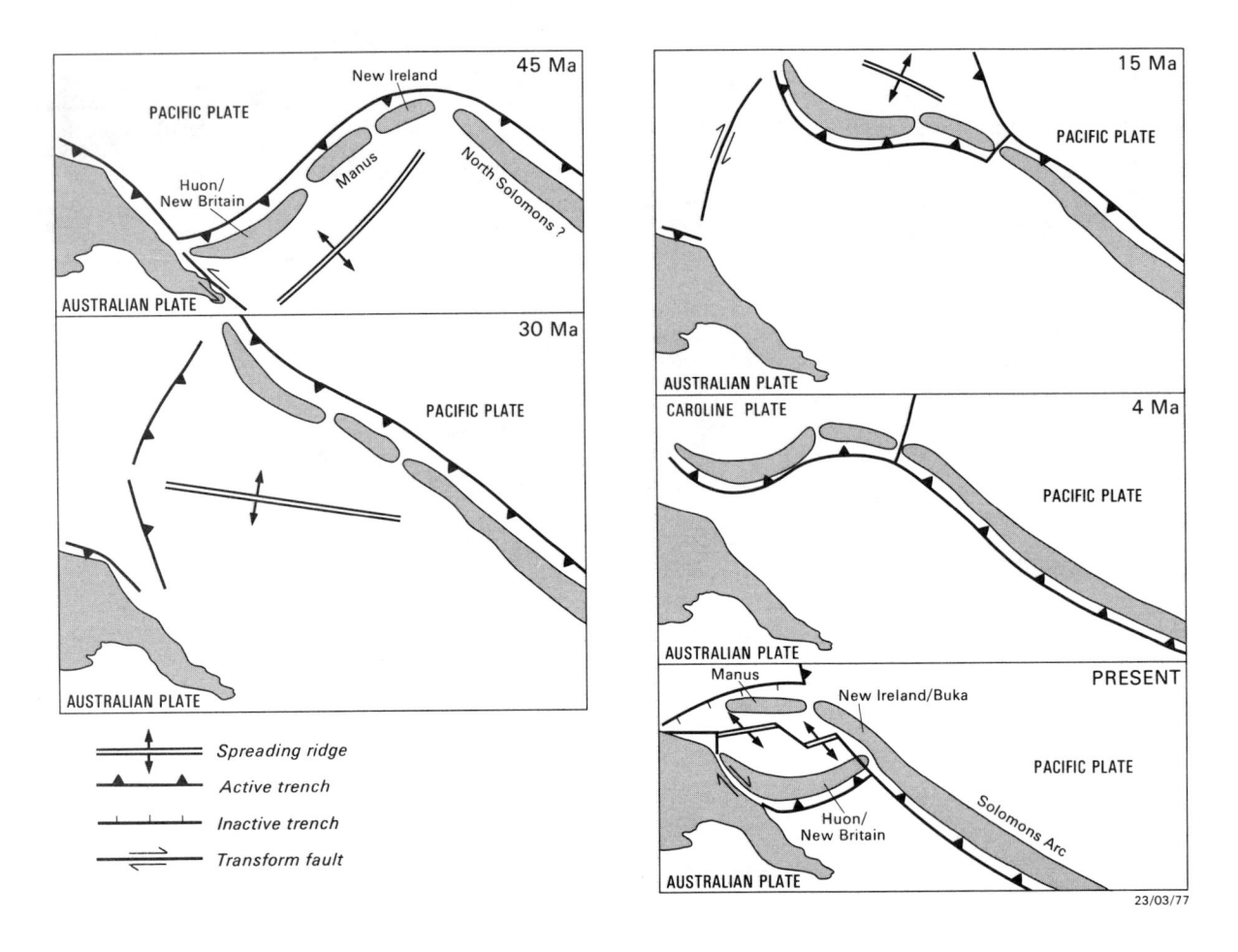

Figure 2. Plate tectonic reconstruction of the New Guinea region from 45 Ma to the present (from Falvey and Pritchard, 1984). An ancient island arc formed east or southeast of mainland Papua New Guinea in the early Eocene, facing northwest by 45 Ma. By 30 Ma the arc faced northeast and began undergoing reversal in polarity along the northern portion. Total polarity reversal of the arc was complete by 4 Ma, and the Manus Basin began opening 3.5 Ma, with the arcs evolving into their present configuration. A backarc spreading ridge is present north of New Britain and Manus at 15 Ma.

Line 423

Line 423 extends from near the western end of Manus Island to the northwest part-way across the forearc (Figures 1 and 3). This line shows progressively increasing deformation to the north of the sediment ramp flanking the northern side of the West Melanesian arc. The deformed section, more than 3 s (3 km) thick, is associated with a gravity low of +15 mgal midway up the slope near CDP 450, where the strata are highly deformed by broad folding and arching.

The magnetic and seismic-reflection data show layers interpreted as lava flows that extend from near CDP 700 to the north end of the line. The reflectors, which are labeled lava flows in Figure 3, are associated with magnetic anomalies with wavelengths of 5 km and amplitudes of 100 nT. To test whether the reflective and magnetic layers could be lava flows near the sea floor, we modeled the magnetic anomalies using thin bodies shown in Figure 4. The anomalies can be modeled using shallow bodies near the sea floor a few hundred meters thick with magnetizations both in the direction of the earth's present field and reversed in direction. The normal and reversed directions suggest that the bodies may be composite flows spanning several million years and different eruption cycles.

We were able to resolve dipping reflectors, presumably within the deformed sediment apron, beneath two knolls that stand above the surrounding flows near CDP's 860 and 1040. Also, unlike the flat-lying flows observed on other lines (see below), the lava flows dip gently to the north from CDP 660

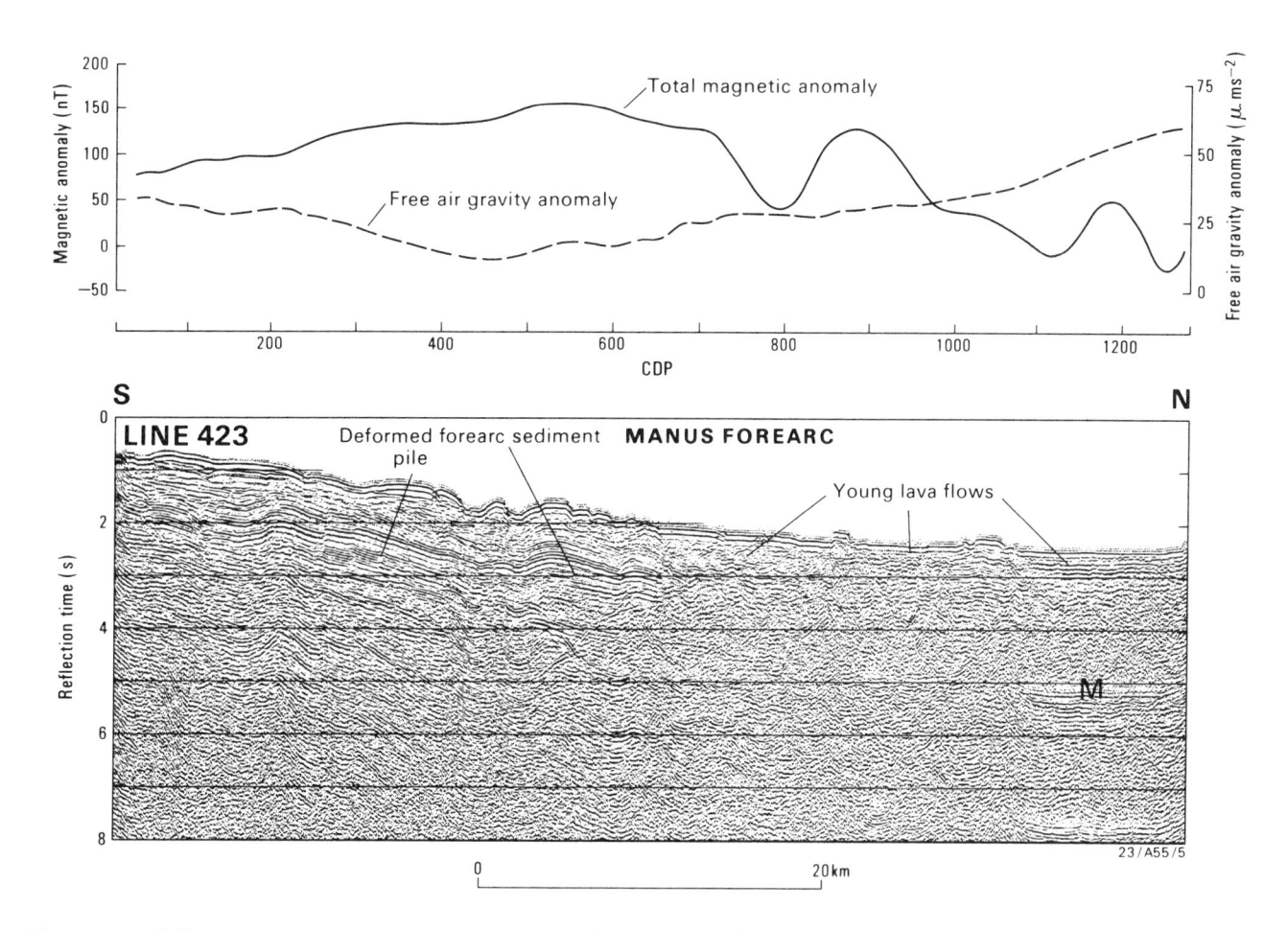

Figure 3. Migrated 24-channel seismic-reflection (time section), magnetic, and gravity profiles along Line 423 across the Manus forearc. See Figure 1 for location. Reflection time is two-way time in seconds. CDP refers to Common Depth Points spaced at 50 m along the line. "M" at 5 s near the northern end of the line refers to the strong multiple.

to CDP 1140, north of which they are flat lying (note the strong multiple in Figure 3).

Line 421

Line 421 trends northwest-southeast across the West Melanesian arc and the Manus forearc, but in Figure 5 we have shown only the part of the line that crosses the central part of the forearc in water depths of 1500 to 2200 m. At the northern end and along the southern half of the line, chaotic subbottom reflectors underlie a hummocky, rolling sea floor. Within this chaotic section, numerous folded reflectors and layers are offset by thrust faults dipping to the south. The sea floor in this area is offset by faulting, implying young tectonic activity, a supposition supported by earthquakes in the forearc (see McCue, this volume). We were able to resolve deformed layers as deep as 3 s (3 km) beneath the sea floor; because no reflection from basement is evident from beneath the sediment pile, we suspect that the pile is thicker than 3 km. We interpret this section as sediment that was first deposited on a subducting plate near the Manus Trench and then scraped off during convergence along the trench. An alternative interpretation is that the deformed sediment may be part of a sediment apron that was derived from the West Melanesian arc to the south and that the sediment body has been deformed in place. The thrust faults may be imbricate thrusts that are still active.

Flat-lying, highly reflective layers, reverberant enough to create a strong multiple, underlie the sea floor in the central part of Line 421 (between CDP's 2600 and 3100, Figure 5). Two magnetic anomalies with amplitudes of 200 nT occur over the reflective beds. These reflectors lie along the bathymetrically low portions of the profile, between bathymetric highs or knolls. Because the reflecting layers are flat

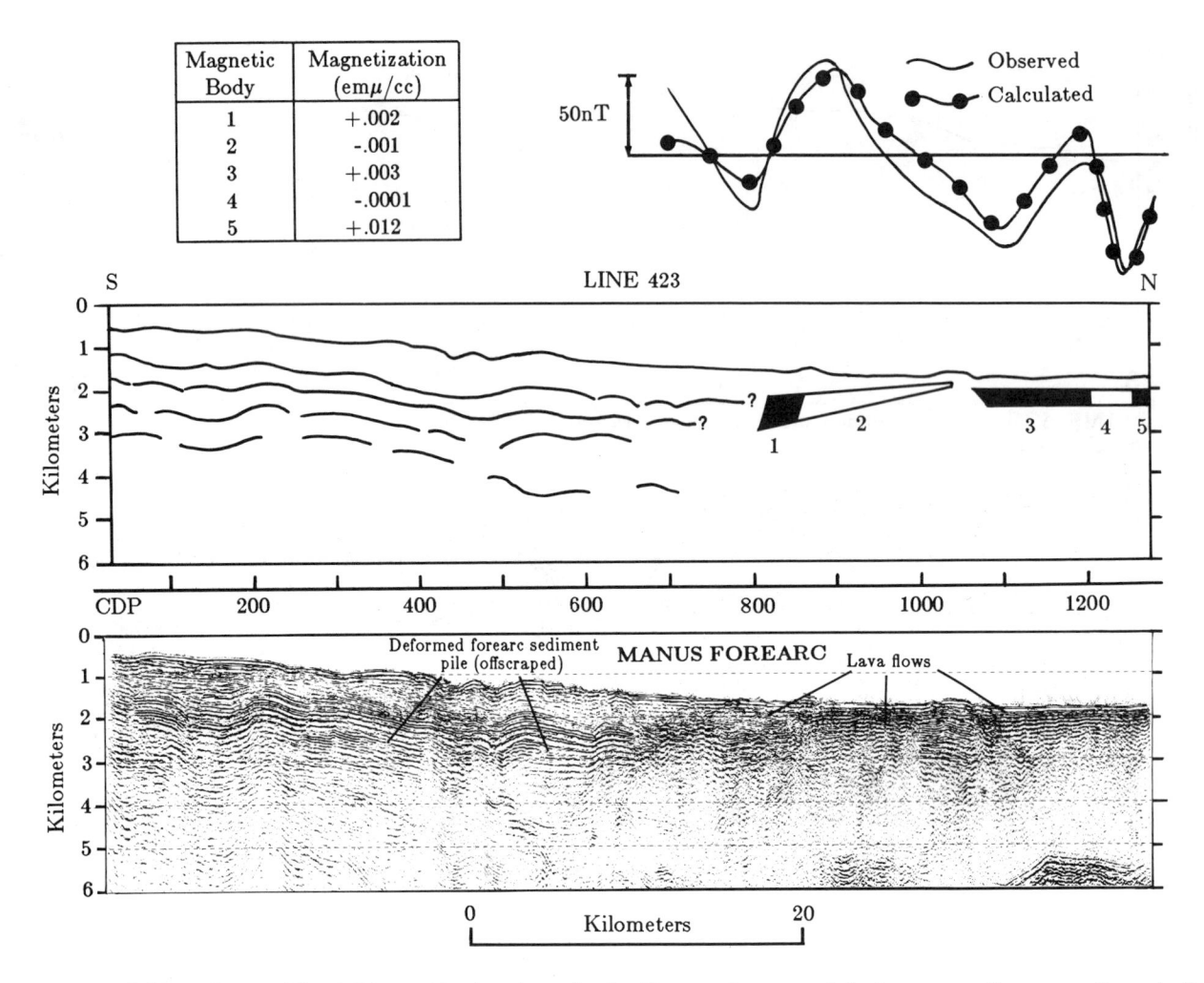

Magnetic Body	Magnetization (emµ/cc)
1	+.002
2	-.001
3	+.003
4	-.0001
5	+.012

Figure 4. Magnetic model of Line 423 showing the bodies used to model the magnetic anomalies at the northern end of the line. Seismic section is true amplitude and depth-corrected. A long wavelength regional magnetic anomaly was removed from the observed profile shown in Figure 3 before modeling. The bodies are near the sea floor and are only a few hundred meters thick and magnetization is is in the direction of the present-day Earth's field and reversed (shaded bodies). Values of magnetization are listed in table (+ = normal, - = reversed).

lying, highly reflective, and magnetic and occupy bathymetric lows in the forearc, we suspect that they are lava flows. A thin pelagic or hemipelagic layer may overlie the lava flows, but the outgoing signal of the multichannel seismic-reflection system has a long bubble pulse of 200 to 300 ms, making the identification of thin sediment layers on the sea floor difficult (see Mann, 1985 for a discussion of seismic-reflection data collected on the R/V *S.P. Lee*; see also CCOP/SOPAC Line 15 below).

Line 422

Line 422 extends to the southwest across the forearc toward Manus Island (Figures 1 and 6). A stratified sediment apron at least 3 s (3 km) thick and composed, in part, of undeformed sediment, dips gently northward beneath the slope north of the island. Although we did not detect a basement reflector beneath the apron, gravity data suggest that the maximum sediment thickness along Line 422 is near CDP 1300 (Figure 6). Near the center of the profile, the sedimentary apron is deformed into broad folds and offset by faults that in places extend to the sea floor. This deformed section is interpreted as part of the apron related to the slope off Manus Island and not to any offscraped section derived from the Manus Trench.

Magnetic data and seismic-reflection data support the interpretation that lava flows are found along Line 422 between CDP's 200 and 400 (Figure

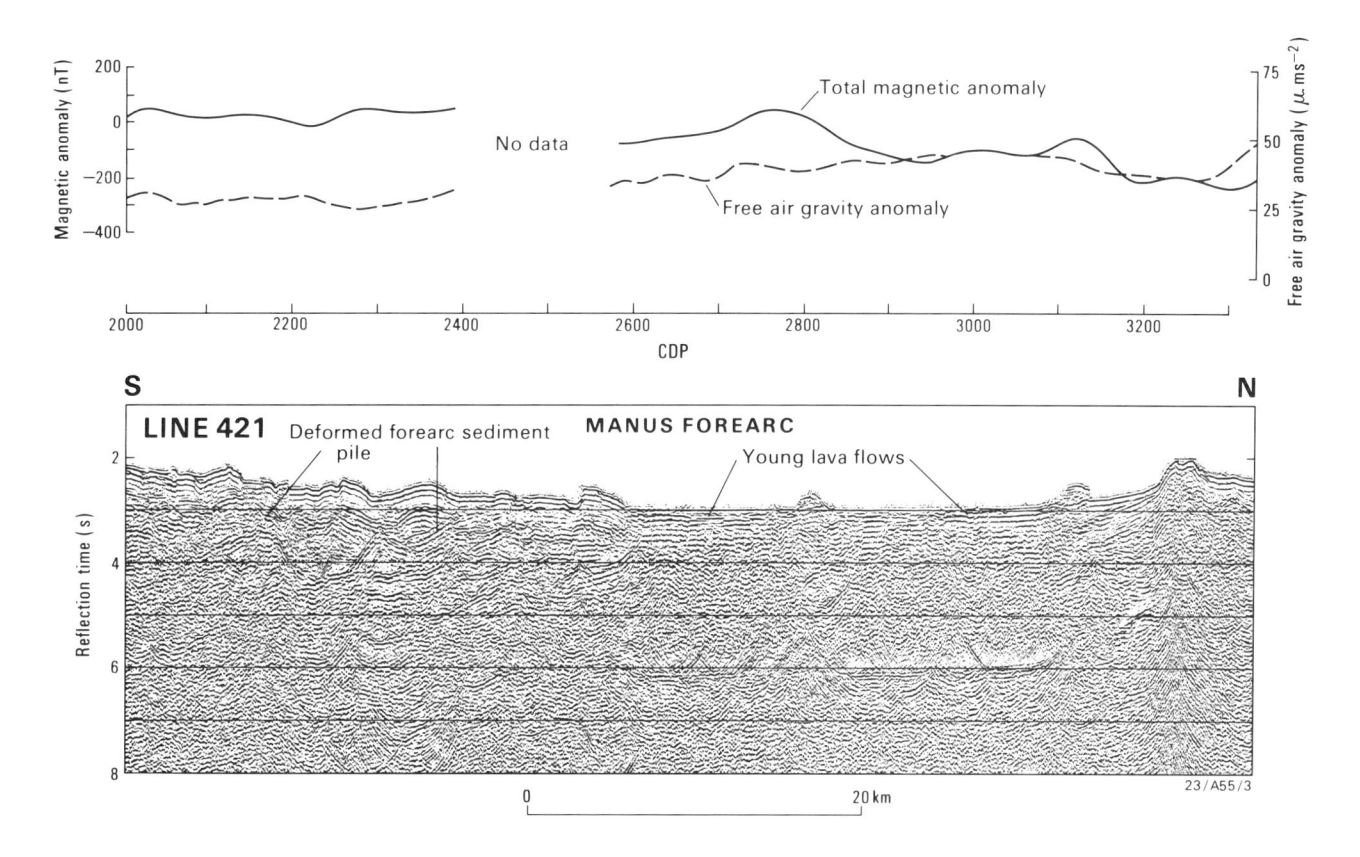

Figure 5. Migrated 24-channel seismic-reflection, magnetic, and gravity profiles along Line 421 across the Manus forearc. See Figure 1 for location. Reflection time is two-way time in seconds. CDP refers to Common Depth Points spaced at 50 m along the line. This is part of Line 421 from CDP 2000 to CDP 3412 (end). A strong multiple is present near 6 s at the center of the profile.

6). Unlike the flows found on Line 421, however, those on Line 422 are buried by about 400-ms (400-m) of sediment.

Line 424

The westernmost profile crosses the West Melanesian arc 50 km west of Manus Island (Figures 1 and 7). Along the southern end of the line, south-dipping reflectors are offset near CDP 1400 by a high-angle normal fault, which probably resulted from relative uplift of the arc. The tilted beds are interpreted as volcaniclastic sediment derived from the arc. The northern flank of the arc is characterized by highly reflective and magnetic layers interpreted as lava flows that are probably related to volcanism along the arc. An outcrop of these flows along the terrace near CDP 1000 was sampled by dredging and a sample of ocean-island ferrobasalt was recovered (see Johnson et al., this volume). Another reflective and magnetic unit occurs at the northern end of the line (Figure 7). This unit is on strike with similar units on Line 423 (Figure 3) and is interpreted as part of the extensive lava flows in the Manus forearc.

Chaotic reflectors near CDP's 400 to 600 on Line 424 are interpreted as representing deformed sediment (Figure 7). The deformed section is characterized by a gravity low and is similar to the deformed zones acoustically resolved on the other forearc crossings.

The sea floor on Line 424 is underlain by a thin, 100-ms (100-m), layer interpreted as a pelagic or hemipelagic unit, except across the normal fault along the West Melanesian arc near CDP 1400 (Figure 7). Here, the pelagic unit is discontinuous and the sea floor is offset, implying young tectonic activity within the forearc.

Line 420

We have included one short profile, Line 420,

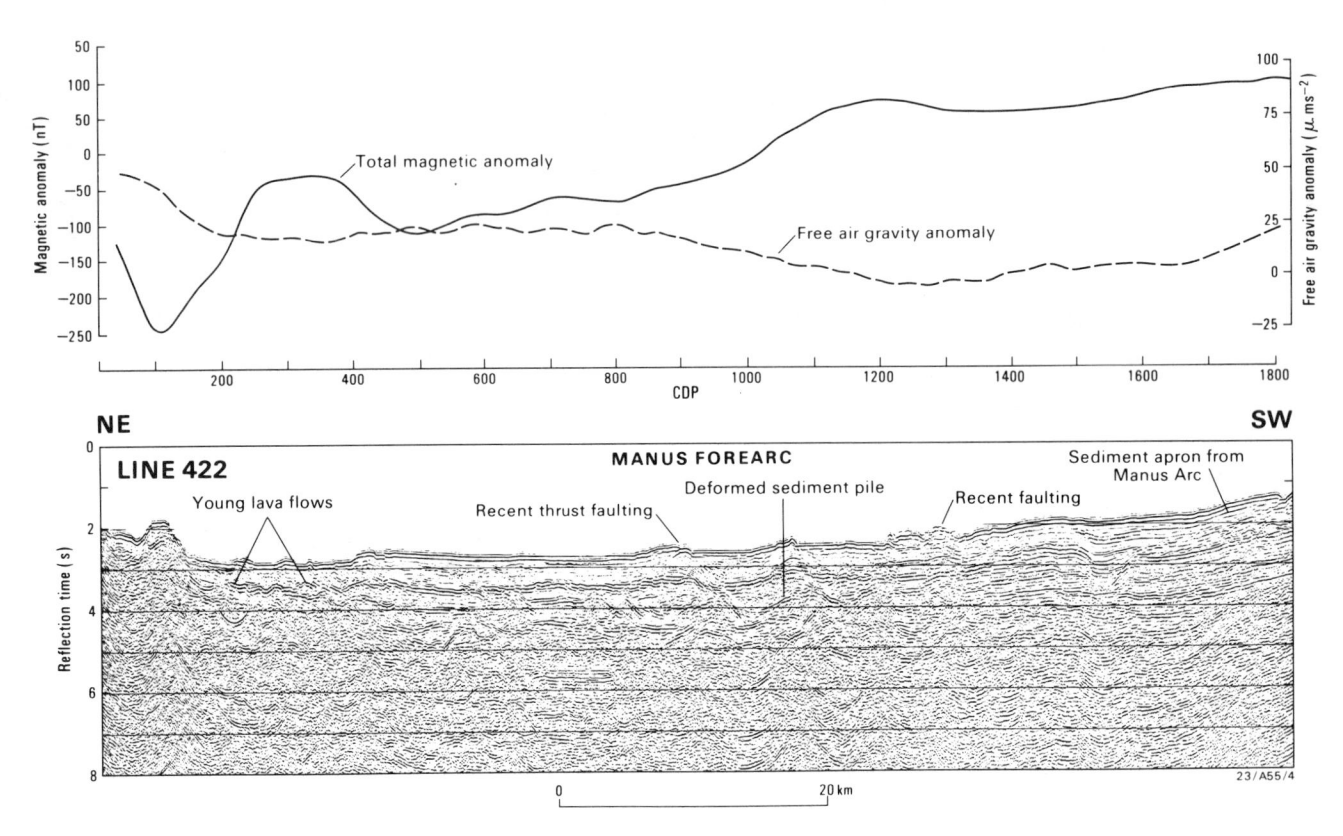

Figure 6. Migrated 24-channel seismic-reflection, magnetic, and gravity profiles along Line 422 across the Manus forearc. See Figure 1 for location. Reflection time is two-way time in seconds. CDP refers to Common Depth Points spaced at 50 m along the line.

on the northern edge of Manus Basin (Figures 1 and 8) to contrast with the lines showing lava flows on the north side of the West Melanesian arc near Manus Island. Manus Basin is an active back-arc spreading basin south of the West Melanesian arc that began to open at 3.5 Ma (Taylor, 1979). The basin is floored by young volcanic outpourings and striped by magnetic anomalies. The reflection and magnetic characteristics of the flows on Line 420 (Figure 8) are similar to those flows on Lines 421 to 424 (Figures 3 to 7), indicating that both back-arc and the forearc are underlain by volcanic flows.

CCOP/SOPAC Line 15

A high-resolution seismic-reflection profile across the far western Manus forearc shows the acoustic details of the forearc lava flows and overlying sediment cover (Figure 9). A strongly reflective unit is present beneath a thin, 100-ms (100-m) surface layer. The strong reflector is presumably the top of the forearc flows, and the thin overlying layer is a presumed pelagic or hemipelagic mantling layer, deposited after the volcanic activity ceased. Resolution of this thin upper layer on this and other CCOP/SOPAC single-channel lines forms the basis for interpreting pelagic layers as present on the multichannel seismic-reflection profiles in the forearc, where the length of the bubble pulse in the multichannel data often makes the interpretation of the first 100 ms beneath the sea floor difficult.

DISCUSSION

Volcanism in the Forearc

Summary discussions of volcanism and related igneous activity in modern and ancient forearcs are found in Gill (1981), Kobayashi (1983b), and Jakes and Miyake (1984). The forearc is that region between the magmatic island arc and the associated, coeval trench (see Dickinson and Seely, 1979 for further definitions). Igneous rocks present in modern forearcs may have formed elsewhere, and then been tectonically transported and emplaced in the forearc. Or, sites of igneous activity along an ancient vol-

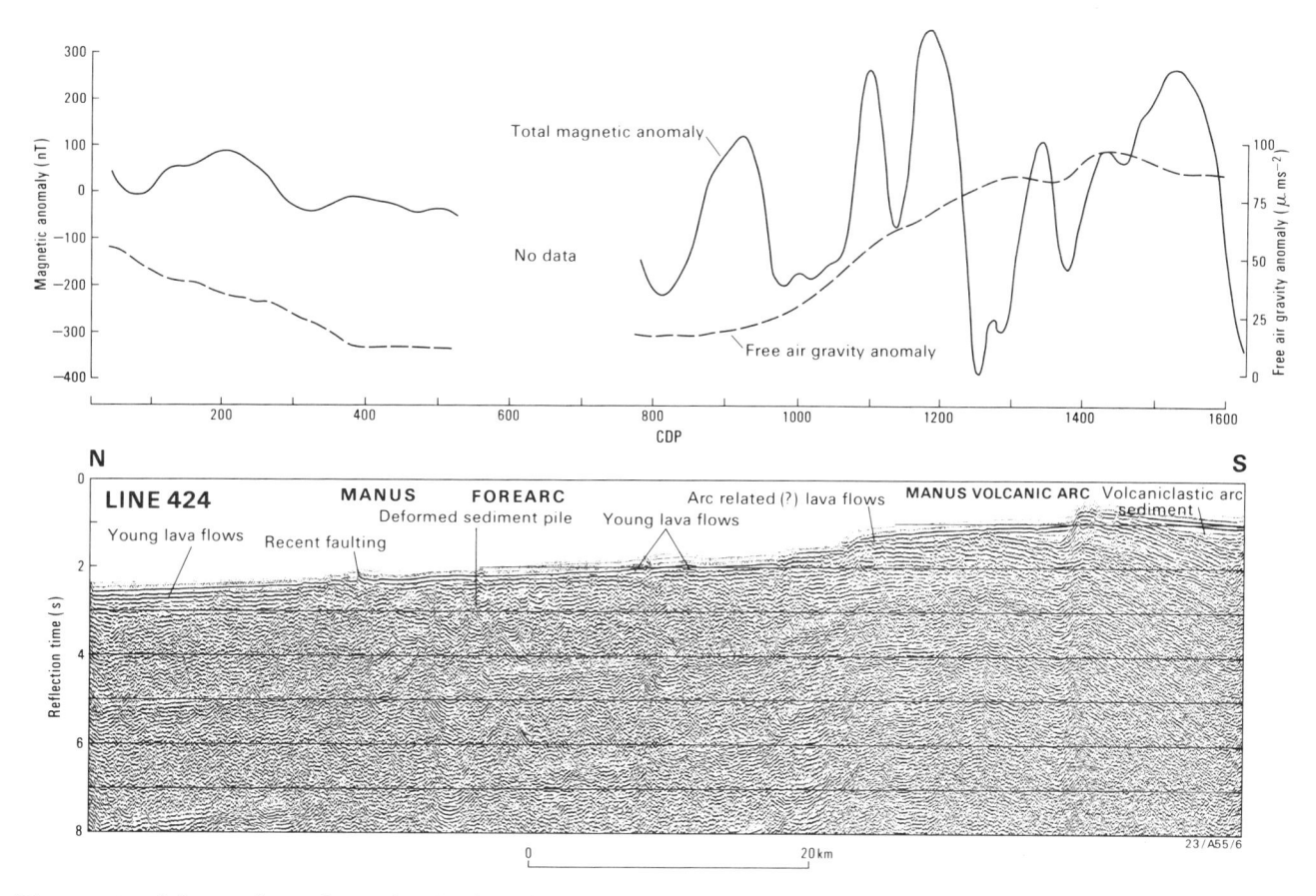

Figure 7. Migrated 24-channel seismic-reflection, magnetic, and gravity profiles along Line 424 across the Manus forearc. Terrace on the sea floor near CDP 1000 was dredged and ferrobasalt recovered (see Johnson et al., this volume). See Figure 1 for location. Reflection time is two-way time in seconds. CDP refers to Common Depth Points spaced at 50 m along the line.

canic arc may in time migrate away from a trench, leaving behind a remnant volcanic arc within an area that later evolves into a forearc. Whether igneous activity occurs *in situ* during forearc formation is debatable, and resolution of this problem awaits the collection of more data from ancestral and modern forearcs.

Table 1 summarizes the known examples of volcanic and plutonic rocks and suspected occurrences of volcanism and plutonism in modern and ancient forearcs. The only known occurrence of modern volcanism in a forearc is in the New Georgia Group of the Solomon Islands, where eruptions occur within 30 km, as well as seaward, of the coeval trench (Table 1). Here, volcanism in the forearc is attributed to subduction of the Woodlark spreading center beneath the Solomon Islands.

Kobayashi (1983a) suggests that forearc volcanism probably is a short-lived phenomenon related to one of three possible causes:

(1) Initial subduction. During the initiation of subduction, the asthenosphere under the forearc is not cooled by the subducting oceanic lithosphere, but instead, the lithosphere brings water and carbon dioxide into the hot asthenosphere. If the forearc region is extensionally deformed during the onset of subduction, then magnesian andesite can form and be injected into the overlying lithosphere. A possible example of such volcanism is the formation of boninite in the Bonin Islands (Table 1). Jakes and Miyake (1984) suggest that magma generated by initial subduction will pool beneath the less dense, overlying sediment of the forearc, forming an ophiolite. Because the magma does not reach the surface, exposures of forearc volcanic rocks are rare.

(2) Subduction of an active spreading ridge. According to Marshak and Karig (1977), igneous activity in forearcs may be caused by a short-lived event, such as the subduction of an actively spreading ridge or the passage of a triple junction beneath the forearc. When an active spreading ridge and adjacent young oceanic crust are subducted beneath the forearc, the crust under the forearc is heated by the downgoing slab; this may result in forearc vol-

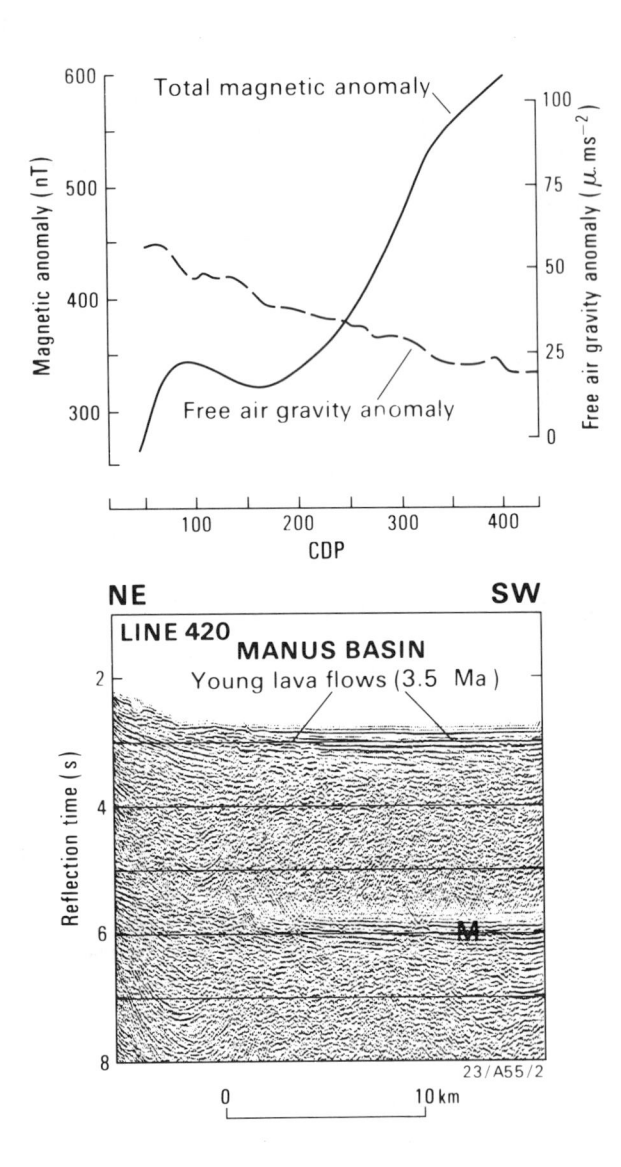

Figure 8. Migrated 24-channel seismic-reflection, magnetic, and gravity profiles along Line 420 across the Manus Basin. See Figure 1 for location. Reflection time is two-way time in seconds. CDP refers to Common Depth Points spaced at 50 m along the line. "M" refers to multiple reflection.

canism. Such volcanism may have occurred in southwest Japan as a result of the subduction of young Shikoku Basin crust (Table 1; Kobayashi, 1983a). Gill (1981) also proposes subduction of an actively spreading ridge as the cause of what he considers to be rare magmatism in the forearc.

(3) Slab disruption. If a continuous slab of lithosphere that has been underthrusting for a long time is fractured beneath the forearc, then hot asthenospheric material that invades the break may rise, creating volcanism within the forearc.

Volcanism in the Manus forearc differs significantly from magmatic activity in the other forearcs listed in Table 1. The volcanic flows north of Manus Island cover at least 8,000 km^2 of the forearc (Figure 10), an area much larger than any known occurrence of volcanic rocks in other forearcs. Furthermore, the Manus forearc flows are flat-lying to gently dipping and cover large areas, unlike the spotty or isolated occurrences of magmatic rocks in other forearcs. Connelly (1976) first noted that part of the forearc just west of Manus Island is underlain by igneous crust that can be modeled magnetically using blocks 1-2 km thick beneath a sedimentary cover several hundred meters thick.

Of the three causes of forearc volcanism discussed above, that of initial subduction is unlikely to have caused the Manus forearc volcanism. The Eocene initiation of subduction that built the Oligocene-early Miocene West Melanesian arc would not produce the near surface characteristics of the flows. The lack of a well-defined Benioff zone beneath the forearc (Denham, 1969; Johnson and Molnar, 1972; Curtis, 1973; Johnson, 1979) suggests that recent reactivation of the Manus Trench to form the southern boundary of the Caroline plate (Weissel and Anderson, 1978) has not introduced sufficient material beneath the forearc to initiate subduction-related volcanism.

The second possible cause of forearc volcanism discussed by Kobayashi (1983a), that of volcanism induced by subduction of an active spreading ridge, is also unlikely to apply to the Manus forearc. The relict spreading center of the Caroline Basin in the Falvey and Pritchard (1984) model remains today north of the Manus Trench and has not been subducted.

The last mechanism for forearc volcanism proposed by Kobayashi (1983a), that of slab disruption, may explain the volcanic flows in the Manus forearc. When subduction from the north beneath the Manus Island forearc ceased (at 15 Ma according to Falvey and Pritchard, 1984), the subducting slab may have broken up beneath the forearc. Such fracturing, according to Kobayashi (1983a), could lead to eruptive magmatism in the forearc. If the inferred timing of such a break up is correct (at 15 Ma), then the age of the flows north of Manus should be Miocene.

Another possible mechanism that could have caused the volcanism surrounding Manus Island is that of back-arc spreading, when subduction was to the north along a trench south of Manus Island. Kroenke (1984) suggests that a middle Miocene episode of sea-floor spreading occurred between Manus Island and Manus Trench. Using sedimentation rates from DSDP Site 63 (7 m/m.y. for 9.5-0

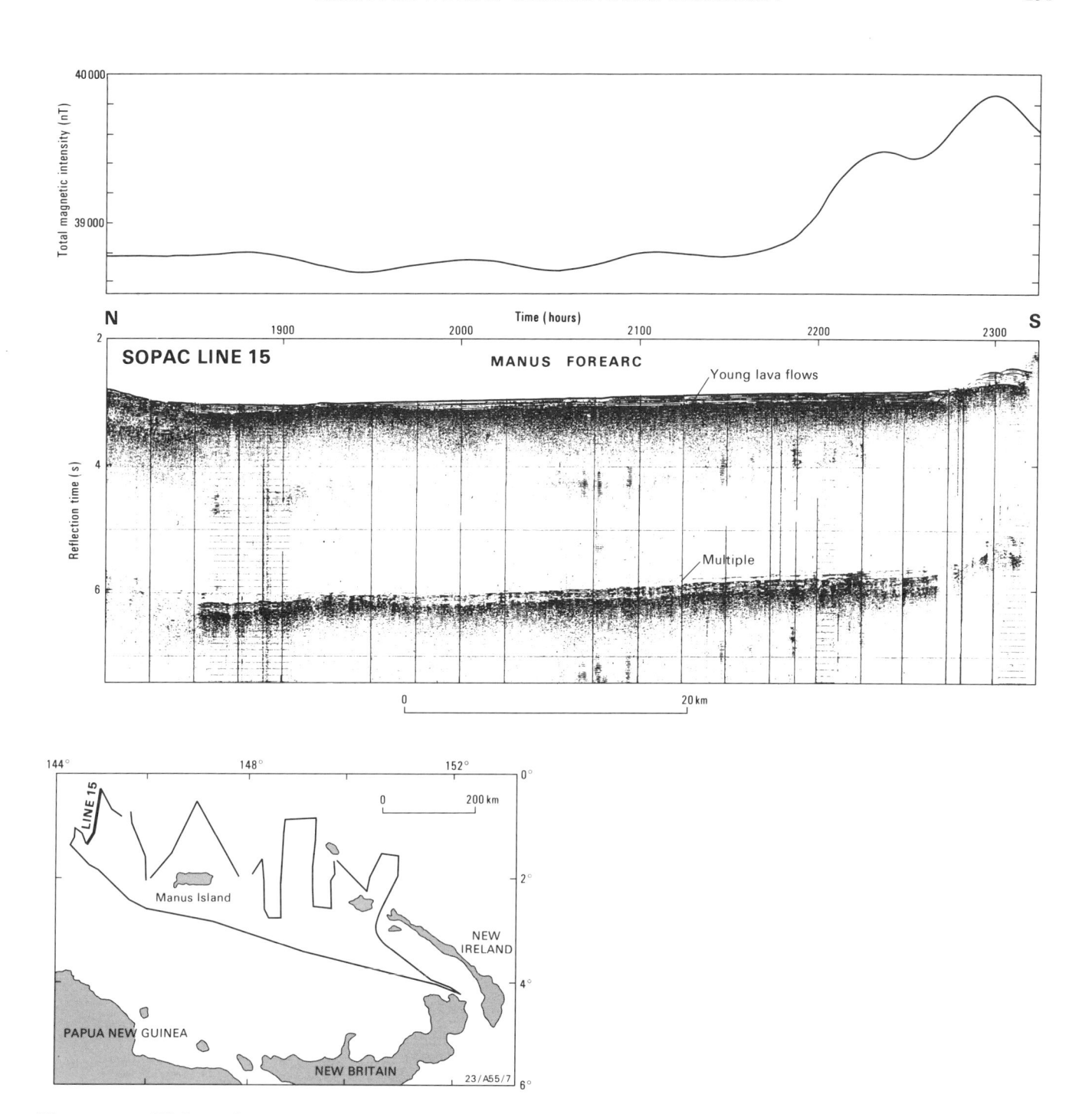

Figure 9. High-resolution seismic-reflection profile, CCOP/SOPAC Line 15 across the western Manus forearc, and location map. This profile, 200 km from Manus Island, is a single-channel line shot with a 655 cm³ (40 cu. in) air gun. Reflection time is two-way time in seconds.

Ma, 28 m/m.y. for 11-9.5 Ma, and 14 m/m.y. for 15-11 Ma; Winterer et al., 1971) in the nearby Caroline Basin north of the Bismarck Sea, we calculate that a minimum of 165 m of pelagic sediment should have accumulated in the Manus forearc during the last 15 million years. Thus, the 100 to 400 m of sediment observed burying the lava flows in the forearc (Figures 3-7) are consistent with a general age of Miocene for the flows based on these nearby pelagic sedimentation rates in the Caroline Basin. However, the thick sediment apron that flanks the West Melanesian arc (Figures 3 and 6) suggests that turbidite flows and hemipelagic deposition also contribute to sedimentation in the Manus forearc and that

Table 1. Modern and ancient volcanic arcs with volcanic flow rocks, active volcanism, or intrusive activity in the forearc region.

Region	Rock Type	Comments	References
Southern Alaska	Pillow basalt, sheeted dolerite, and gabbro	Paleocene to Eocene(?) mafic sequence of igneous rocks on Knight Island; Cretaceous pillow basalts, sheeted dikes, and gabbro intrusions on the Resurrection Peninsula. May have been related to a leaky transform fault in the forearc.	Tysdal et al. (1977); Miyake (1985)
Eastern Aleutian Arc	Granodiorite, granite, basalt, and andesite;	Dates range from 62 to 40 Ma.. Attributed to passage of ridge-trench-trench triple junction (Kula-Farallon Ridge).	Marshak and Karig (1977); Hill et al. (1981); Moore et al. (1983); Hudson et al. (1977)
Eastern Aleutian Arc	Basalt	K/Ar age of 22.5 Ma. on a basalt sample from Unimak Seamount along with early and middle Miocene sediment recovered in a dredge haul 50 to 60 km from the trench. Attributed to arc volcanism followed by truncation of the arc margin.	Bruns et al. (1987a,b)
Japan (Oyashio paleoland)	Dacite conglomerate	K/Ar age of 23.4 ±5 Ma. DSDP Site 439. Below Oligocene sandstone; suggested tectonic erosion of ancient forearc that once lay seaward of the drill site.	von Huene et al. (1980); Kobayashi (1983b)
Southwest Japan	Siliceous igneous (extrusive and intrusive)	Age 17 to 13 Ma. Belt less than 50 km from coeval trench. Attributed to passage of Japan-Ryukyu-Izu-Bonin Trench triple junction and associated passage of Izu-Bonin volcanic arc along the accretionary Japanese forearc (Marshak and Karig, 1977). Or volcanism may be related to subduction of young crust of the Shikoku Basin (Kobayashi, 1983a).	Marshak and Karig (1977); Shibata (1978); Kobayashi (1983a); Miyake (1985)
Bonin Islands	Boninite	K/Ar ages of 40-44 Ma. Islands are 100 km or less from Bonin Trench. Presently active volcanic island arc is 250 km from associated trench.	Kuroda and Shiraki (1975); Shiraki et al. (1978, 1980); Cameron et al. (1979); Kobayashi (1983a,b); Dobson (1987)
Mariana Arc	Boninite, arc tholeiitic basalt and differentiates	Eocene age. DSDP Site 458. Island arc affinities are a product of subduction-related volcanism, suggesting subduction erosion of the forearc.	Hussong et al. (1982); Bloomer (1983)
Mariana Trench	Gabbro, ultramafic rocks, boninite, altered basalt, andesite and dacite, and alkalic basalt	Landward slope of trench. Probably a late Eocene arc complex exposed by subduction erosion. May include some offscraped seamount rocks.	Bloomer (1983); Bloomer and Hawkins (1983)
Yap-Mariana Trenches	Arc tholeiitic basalt, alkalic basalt	K/Ar ages of 7.8 and 10.8 Ma; dredged from forearc close to trenches.	Beccaluva et al. (1980, 1986); Crawford et al. (1981, 1986)
Western Sumatra	Scattered intrusive and extrusive rocks	Probable Oligocene age. Within 100 km of trench and seaward of associated volcanic arc axis positions throughout Cenozoic time. Attributed to northwestward migration of ridge-trench-trench triple junction along Sunda Trench.	Marshak and Karig (1977)

sedimentation rates in the forearc are higher than in the abyssal Caroline Basin to the north. Thus, the 100 to 400 m of sediment mantling the flows in the forearc imply a maximum age of Miocene for the flows and they could be considerably younger.

Volcanism Related to the West Melanesian Arc

The suspected volcanic flows in the Manus forearc, as discussed above, may be related to magmatism in the forearc itself. Alternatively, the flows

Table 1. Continued.

Region	Rock Type	Comments	References
Eastern Papua New Guinea	Basalt, nephelinite, basanite, tephrite, trachybasalt, and trachyandesite	Tabar to Feni Islands off the east coast of New Ireland; pre-middle-Miocene but mostly Quaternary.	Johnson et al. (1976); Johnson (1979)
Solomon Islands	High K_2O/TiO_2 basalt and basaltic andesite	Volcanism in the New Georgia Group within 30 km of associated trench. Attributed to subduction of the Woodlark spreading center beneath the Solomon Islands.	Stanton and Bell (1969); Weissel et al. (1982); Taylor and Exon (1982, 1987); Dunkley (1983); Ramsay et al. (1984); Perfit and Langmuir (1984)
Southern Chile	Granodiorite plutons and porphyritic stocks and sills	Age 4.0 to 3.5 Ma. 150 km seaward of the Quaternary volcanic arc associated with the Peru-Chile Trench. Attributed to subduction of the Chile Rise (Chile margin triple junction).	Forsythe and Nelson (1985); Forsythe et al. (1986)
Nicoya Peninsula	Tholeiitic basalt	Late Cretaceous. Oceanic plateau, intraplate oceanic volcanism, or primitive arc volcanism.	Lundberg (1983)
Southern California	Basalt, pyroxene andesite, and dacite	Age 25 to 15 Ma. Rocks occur in the northwestern Los Angeles Basin and the continental borderland. Erupted within 0 to 100 km of contemporaneous plate boundary (ridge-trench-transform triple junction) and may have been emplaced in an extensional environment as a result of ridge subduction.	Hurst (1982); Miyake (1985); See also Gill (1981) for summary and references
Western California	Gabbro	Gabbro intrusions in Franciscan assemblage. Gabbro intruded into accretionary prism while convergence and subduction active as evidenced by blueschist-facies metamorphism. Mantle origin for magma. May be related to bending of plate or intersection of a fracture zone with margin.	Echeverria (1980); Johnson and O'Neil (1984)
Western Washington	Tholeiitic basalt	Crescent volcanic rocks beneath the Olympic Peninsula of early and middle Eocene age and as much as 15 km thick; apparently extruded on the Juan de Fuca plate and western margin of the North America plate. May be offscraped seamounts of an oceanic island chain (Snavely et al., 1968). Later volcanism (45 to 32 Ma) in forearc consisted of basaltic pillow lavas and tuff - from the ancestral Yellowstone hotspot? (Duncan and Kulm, 1987).	Snavely et al. (1968); Cady (1975); Miyake (1985); Duncan and Kulm (1987)
Troodos, Cyprus	Andesite-dacite-rhyolite assemblage and basalt-basaltic andesite assemblage	Overlain by Cretaceous sediment. Chemistry of lava indicates origin similar to forearc rocks of the Mariana and Bonin arcs.	Robinson et al. (1983); Miyake (1985)

could conceivably be related to volcanism in the West Melanesian arc to the south. Magmatism in this arc, including Manus Island, has been continuous from the Eocene to the present (Jaques, 1980; Johnson, 1979; see Francis, this volume).

Line 424 (Figure 7) reveals a continuous series of gently tilted, magnetic, and highly reflective shallow reflectors interpreted as volcanic flows dipping north as a series of terraces stepping down into the forearc. Conceivably, these flows may be cogenetic and their source could be from within the West Melanesian arc (Admiralty Islands). Because of the wide spacing of geophysical lines across the Manus forearc, all the flows interpreted in Figure 10 could be derived from the West Melanesian arc to the south of the forearc. Dredging of the terrace near CDP 1000 on Line 424 recovered a ferrobasalt of ocean-island affinity some 50 km west of Manus Island; the dredge sample is similar to the ferrobasalts exposed in Southwest Bay on Manus Island

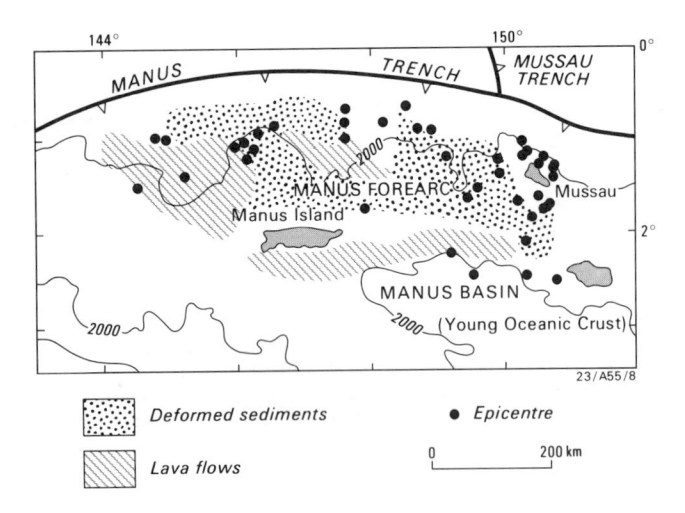

Figure 10. Lava flows, deformed sediment masses, and earthquake epicenters (see McCue, this volume) in the volcanic arc and forearc region around Manus Island. The lava flows cover more than 8,000 km² of the forearc. Bathymetric contours in meters.

(Johnson et al., this volume), and the flows forming the dredged terrace are geophysically similar to flows interpreted farther north in the forearc. However, if all the flows originated solely in the West Melanesian arc, then there must be multiple sources in order to have spread out over 8,000 km² of the forearc. Flows well north of the dredge location must be sampled to test whether they still have chemical affinities to the volcanic rocks exposed on Manus Island.

Sediment Deformation in the Forearc

In addition to the flat reflectors interpreted as volcanic flows, dipping and disrupted reflectors occur beneath the Manus forearc (Figures 3-7). The reflectors are broken by southward-dipping thrust faults. Earthquake studies of two events in the forearc also suggest thrusting mechanisms along southward-dipping faults within the forearc (McCue, this volume). In addition, scattered shallow earthquakes occur in the areas of disrupted reflectors, but only rarely in the areas of suspected lava flows (Figure 10).

Only shallow earthquakes occur beneath the Manus forearc, where no Benioff zone exists (Denham, 1969; Johnson and Molnar, 1972; Curtis, 1973; Johnson, 1979), and the sediment fill in the Manus Trench to the north of the forearc shows no evidence of recent convergence (Ryan and Marlow, this volume). Thus, plate convergence is not active along the Manus forearc in the sense of the classical convergence zones around the Pacific Ocean. However, shallow earthquakes and faults that offset the sea floor show young tectonic activity in the forearc.

The young deformation of sediment in the Manus forearc may be related to the opening of Manus Basin south of the West Melanesian arc. According to Taylor (1979), Manus Basin began to open about 3.5 Ma along back-arc spreading centers that trend roughly east-west across the basin. Thus, the new plate north of the spreading centers and south of the West Melanesian arc is growing and moving north. We speculate that such motion may be pushing the West Melanesian arc into the Manus forearc, resulting in faulting and earthquakes in the forearc. Such deformation may be occurring along old zones of weakness formed by earlier periods of plate convergence along the Manus Trench (see Ryan and Marlow, this volume).

SUMMARY

Geophysical data, including seismic-reflection and magnetic profiles, show that more than 8,000 km² of the Manus forearc is underlain by flat-lying to gently dipping lava flows. These flows occupy the broad, low-lying regions of the forearc, and their age and thickness are unknown. The flows are overlain by 100 to 400 m of hemipelagic(?) sediment. Sedimentation rates in the Caroline Basin to the north suggest a possible Miocene or Pliocene age for the base of the sediment cover. However, because the flows and their sediment cover have not been sampled, this age is conjectural. Several hypotheses are postulated in the literature for the generation of magma in forearcs. The most likely mechanism for the Manus forearc volcanism is that of back-arc spreading, when subduction was directed to the north from a trench south of Manus Island.

Large areas of the Manus forearc are underlain by sediment masses that are deformed by folding and thrust faulting. Evidence for this tectonic activity, which ruptures the sea floor in many places, comes from seismic-reflection and gravity profiles as well as earthquake solutions for epicenters in the forearc. No well-defined Benioff zone dips beneath the arc and forearc, which are characterized by scattered, shallow earthquakes. The sediment masses may have accumulated during earlier periods of convergence between the West Melanesian arc (North Bismarck plate?) and the Caroline plate to the north. Alternatively, the sediment may have been deposited in a forearc basin and deformed in place. Recent tectonic activity is now probably related to northward-directed stress of the Manus Island segment of the West Melanesian arc and northern Manus Basin against the Manus forearc as a result of the opening of Manus Basin to the south.

The Manus region of northern Papua New Guinea has been surveyed by widely spaced geophysical lines, and only a few of these lines included multichannel seismic-reflection data (Figure 1). Offshore samples and refraction data for the area are also sparse, and no offshore wells have been drilled. The enigmatic problem of volcanism in the Manus forearc and the relation of that volcanism to the West Melanesian arc and to the Manus Basin can only be addressed by systematic multichannel seismic-reflection surveys and by drilling.

ACKNOWLEDGMENTS

We thank David Falvey of the Australian Bureau of Mineral Resources and the American Association of Petroleum Geologists for permission to publish Figure 2. We also thank Michael Fisher and Tracy Vallier for their thoughtful reviews.

REFERENCES

Beccaluva, L., Maciotta, G., Savelli, C., Serri, G., and Zeda, O., 1980, Geochemistry and K/Ar ages of volcanics dredged in the Philippine Sea (Mariana, Yap, Palau Trenches and Parece Vela Basin), in Hayes, D.E., ed., The tectonic and geologic evolution of Southeast Asian seas and islands: American Geophysical Union, Monograph Series, v. 23: Washington, D.C., p. 247-268.

Beccaluva, L., Serri, G., and Dostal, J., 1986, Geochemistry of volcanic rocks from the Mariana, Yap and Palau Trenches bearing on the tectono-magmatic evolution of the Mariana Trench-arc-backarc system, in Wezel, F.-C., ed., The origin of arcs: Developments in Geotectonics 21, Amsterdam, Elsevier, p. 481-508.

Bloomer, S.H., 1983, Distribution and origin of igneous rocks from the landward slopes of the Mariana Trench: Implications for its structure and evolution: Journal of Geophysical Research, v. 88, p. 7411-7428.

Bloomer, S.H., and Hawkins, J.W., 1983, Gabbroic and ultramafic rocks from the Mariana Trench: An island arc ophiolite, in Hayes, D.E., ed., The tectonic and geologic evolution of Southeast Asian seas and islands Part 2: American Geophysical Union Monograph Series, v. 27, Washington, D.C., p. 294-317.

Bruns, T.R., Vallier, T.L., Pickthorn, L.B., and von Huene, R., 1987a, Early Miocene calc-alkaline basalt dredged from the Shumagin margin, Alaska, in Hamilton, T.D., and Galloway, J.P., eds., Accomplishments in Alaska in 1986: U.S. Geological Survey Circular 998, p. 143-146.

Bruns, T.R., von Huene, R., Cullotta, R.C., Lewis, S.D., and Ladd, J.W., 1987b, Geology and petroleum potential of the Shumagin margin, Alaska, in Scholl, D.W., Grantz, A., and Vedder, J., eds., Geology and resource potential of the continental margin of western North America and adjacent ocean basins—Beaufort Sea to Baja California: Circum-Pacific Council for Energy and Mineral Resources, Earth Science Series, v. 6, Houston, Texas, p. 157-189.

Cady, W.M., 1975, Tectonic setting of the Tertiary volcanic rocks of the Olympic Peninsula, Washington: U.S. Geological Survey Journal of Research, v. 3, p. 573-582.

Cameron, W.E., Nisbet, E.G., and Dietrich, V.J., 1979, Boninites, komatites, and ophiolitic basalts: Nature, v. 280, p. 550-553.

Connelly, J.B., 1974, A structural interpretation of magnetometer and seismic profiler records in the Bismarck Sea, Melanesian Archipelago: Journal of the Geological Society of Australia, v. 21, p. 459-469.

Connelly, J.B., 1976, Tectonic development of the Bismarck Sea based on gravity and magnetic modeling: Royal Astronomical Society Geophysical Journal, v. 46, p. 23-40.

Crawford, A.J., Beccaluva, L., and Serri, G., 1981, Tectono-magmatic evolution of the West Philippine-Mariana region and the origin of boninites: Earth and Planetary Science Letters, v. 54, p. 346-356.

Crawford, A.J., Beccaluva, L., Serri, G., and Dostal, J., 1986, The petrology, geochemistry and tectonic implications of volcanics dredged from the intersection of the Yap and Mariana trenches: Earth and Planetary Science Letters, v. 80, p. 265-280.

Curtis, J.W., 1973, The spatial seismicity of Papua New Guinea and the Solomon Islands: Journal of the Geological Society of Australia, v. 20, p. 1-20.

de Broin, C.E., Aubertin, F., and Ravenne, C., 1977, Structure and history of the Solomon-New Ireland Region: International Symposium on the Geodynamics in South-West Pacific, Noumea, New Caledonia, 1976: Paris, Edition Technip, p. 37-50.

Denham, D., 1969, Distribution of earthquakes in the New Guinea-Solomon Islands Region: Journal of Geophysical Research, v. 74, p. 4290-4299.

Dickinson, W.R., and Seely, D.R., 1979, Structure and stratigraphy of forearc regions: American Association of Petroleum Geologists Bulletin, v. 63, p. 2-31.

Dobson, P. F., 1986, The petrogenesis of boninite; a field, petrologic, and geochemical study of the volcanic rocks of Chichi-Jima, Bonin Islands, Japan: Stanford, Stanford University Ph.D. Dissertation, 178 p.

Duncan, R.A., and Kulm, L.D., 1988, Plate tectonic evolution of the Cascades arc-subduction complex, in Plafker, G. and Jones, D., eds., Geology of Alaska: Decade of North American Geology, Geological Society of America, v. 2 [in press].

Dunkley, P.N., 1983, Volcanism and the evolution of the ensimatic Solomon Islands arc, in Shimozuru D., and Yokoyama, I., eds., Arc Volcanism: Physics and and Tectonics: Tokyo, Terra Scientific Publishing Company, p. 225-241.

Echeverria, L.M., 1980, Oceanic basaltic magmas in accretionary prisms: The Franciscan intrusive gabbros: American Journal of Science, v. 280, p. 697-724.

Eguchi, T., Fujinawa, Y., Ukawa, M., and Bibot, L., 1987, Microearthquakes along the back-arc spreading system in the Eastern Bismarck Sea: Geo-Marine Letters, v. 6, p. 235-240.

Exon, N.F., and Tiffin, D.L., 1984, Geology and petroleum prospects of offshore New Ireland Basin in northern Papua New Guinea, in Watson, S.T., ed., Transactions of the Third Circum-Pacific Energy and Mineral Resources Conference, Honolulu, Hawaii, 1982, p. 623-630.

Exon, N.F., Stewart, W.D., Sandy, M.J., and Tiffin, D.L., 1986, Geology and offshore petroleum prospects of the eastern New Ireland Basin, northeastern Papua New Guinea: Bureau of

Mineral Resources Journal of Australian Geology and Geophysics, v. 10, p. 39-51.

Falvey, D.A., and Pritchard, T., 1984, Preliminary paleomagnetic results from northern Papua New Guinea: Evidence for large microplate rotations, in Watson, S.T., ed., Transactions of the Third Circum-Pacific Energy and Mineral Resources Conference, Honolulu, Hawaii, 1982, p. 593-599.

Forsythe, R., and Nelson, E., 1985, Geological manifestations of ridge collision: Evidence from the Golfo de Penas-Taitao basin, southern Chile: Tectonics, v. 4, p. 477-495.

Forsythe, R.D., Nelson, E.P., Carr, M.J., Kaeding, M.E., Herve, M., Mpodozis, C., Soffia, J.M., and Harambour, S., 1986, Pliocene near-trench magmatism in southern Chile: A possible manifestation of ridge collision: Geology, v. 14, p. 23-27.

Gill, J. B., 1981, Orogenic Andesites and Plate Tectonics: Berlin, Springer-Verlag, 390 p.

Hill, M., Morris, J., and Whelan, J., 1981, Hybrid granodiorites intruding the accretionary prism, Kodiak, Shumagin, and Sanak Islands, southwest Alaska: Journal of Geophysical Research, v. 86, p. 10569-10590.

Hudson, T., Plafker, G., and Lanphere, M. A., 1977, Intrusive rocks of the Yakutat-St. Elias area, south-central Alaska: U.S. Geological Survey Journal of Research, v. 5, p. 155-172.

Hurst, R.W., 1982, Petrogenesis of the Conejo volcanic suite, southern California: Evidence for mid-ocean ridge-continental margin interactions: Geology, v. 10, p. 267-272.

Hussong, D.M., Uyeda, S., et al., 1982, Initial Reports of Deep Sea Drilling Project, v. 60: Washington, D.C., U.S. Government Printing Office, 929 p.

Jakes, P., and Miyake, Y., 1984, Magma in forearcs: Implication for ophiolite generation: Tectonophysics, v. 106, p. 349-358.

Jaques, A.L., 1980, Admiralty Islands Papua New Guinea, Geological Series-Explanatory Notes: Geological Survey of Papua New Guinea, 25 p., 1 sheet, scale 1:250,000.

Johnson, C.M., and O'Neil, J.R., 1984, Triple junction magmatism: A geochemical study of Neogene volcanic rocks in western California: Earth and Planetary Science Letters, v. 71, p. 241-262.

Johnson, R.W., Wallace, D.A., and Ellis, D.J., 1976, Feldspathoid-bearing potassic rocks and associated types from volcanic islands off the coast of New Ireland, Papua New: A preliminary account of geology and petrology, in Johnson, R.W., ed., Volcanism in Australasia: New York, Elsevier, p. 297-316.

Johnson, R.W., 1979, Geotectonics and volcanism in Papua New Guinea: A review of the late Cainozoic: Bureau of Mineral Resources Journal of Australian Geology and Geophysics, v. 4, p. 181-207.

Johnson, T., and Molnar, P., 1972, Focal mechanisms and plate tectonics of the southwest Pacific: Journal of Geophysical Research, v. 77, p. 5000-5032.

Kobayashi, K., 1983a, Cycles of subduction and Cenozoic arc activity in the northwestern Pacific margin, in Hilde, T.W.C., and Uyeda, S., eds., Geodynamics of the Western Pacific-Indonesian Region: American Geophysical Union, Geodynamics Series, v. 11, Washington, D.C., p. 287-301.

Kobayashi, K., 1983b, Fore-arc volcanism and cycles of subduction, in Shimozuru, D., and Yokoyama, I., eds., Arc Volcanism: Physics and Tectonics: Tokyo, Terra Scientific Publishing Company, p. 153-163.

Kroenke, L.W., 1984, Cenozoic tectonic development of the southwest Pacific: United Nations Economic and Social Commission of Asia and the Pacific CCOP/SOPAC Technical Bulletin, no. 6, 122 p.

Kuroda, N., and Shiraki, K., 1975, Boninite and related rocks of Chichijima, Bonin Islands, Japan: Reports of Faculty of Science, Shizuoka University, v. 10, p. 145-155.

Lundberg, N., 1983, Development of forearcs of intraoceanic subduction zones: Tectonics, v. 2, p. 51-61.

Mann, D.M., 1985, Multichannel profiles collected in 1982 in the Tonga region of the South Pacific, in Scholl, D.W., and Vallier, T.L., compilers and editors, Geology and offshore resources of Pacific island arcs-Tonga region: Circum-Pacific Council for Energy and Mineral Resources Earth Science Series, v. 2, Houston, Texas, p. 49-53.

Marlow, M.S. and Exon, N.F., 1985, Basin development and regional tectonics of the New Ireland and Manus regions: Bureau of Mineral Resources, Australia Record 1985/32, p. 11-18.

Marshak, R.S., and Karig, D.E., 1977, Triple junctions as a cause for anomalously near-trench igneous activity between the trench and volcanic arc: Geology, v. 5, p. 233-236.

Miyake, Y., 1985, MORB-like tholeiites formed within the Miocene forearc basin, southwest Japan: Lithos, v. 18, p. 23-34.

Moore, J.C., Byrne, T., Plumley, P.W., Reid, M., Gibbons, H., and Coe, R.S., 1983, Paleogene evolution of the Kodiak Islands, Alaska: Consequences of ridge-trench interaction in a more southerly latitude: Tectonics, v. 2, p. 265-293.

Murauchi, S., and Asanuma, T., 1977, Seismic reflection profiles in the western Pacific, 1965-74: Contribution from Geodynamics Project of Japan 77-1: Tokyo, University of Tokyo Press, 232 p.

Perfit, M.R., and Langmuir, C.H., 1984, Geochemical effects of ridge subduction in the Solomon Islands-Woodlark Basin region: Geological Society of America Abstracts with Programs, v. 16, p. 622.

Ramsay, W.R.H., Crawford, A.J., and Foden, J.D., 1984, Field setting, mineralogy, chemistry, and genesis of arc picrites, New Georgia, Solomon Islands: Contributions to Mineralogy and Petrology, v. 88, p. 386-402.

Robinson, P.T., Melson, W.G., O'Hearn, T., and Schmincke, H., 1983, Volcanic glass compositions of the Troodos ophiolite, Cyprus: Geology, v. 11, p. 400-404.

Shibata, K., 1978, Contemporaneity of Tertiary granites in the outer zone of southwest Japan: Journal of the Geological Survey of Japan, v. 29, p. 551-554.

Shiraki, K., Kuroda, N., Maruyama, S., and Urano, H., 1978, Evolution of the Tertiary volcanic rocks in the Izu-Mariana arc: Bulletin Volcanologique, v. 41, p. 548-562.

Shiraki, K., Urano, H., and Maruyama, S., 1980, Clinoenstatite in boninites from the Bonin Islands, Japan: Nature, v. 285, p. 31-32.

Snavely, P.D., Jr., MacLeod, N.S., and Wagner, H.C., 1968, Tholeiitic and alkalic basalts of the Eocene Siletz River volcanics, Oregon Coast Range: American Journal of Science, v. 266, p. 454-481.

Stanton, R.L., and Bell, J.D., 1969, Volcanic and associated rocks of the New Georgia Group, British Solomon Islands Protectorate: Overseas Geology and Mineral Resources, v. 10, p. 113-145.

Taylor, B., 1979, Bismarck Sea: Evolution of a back-arc basin: Geology, v. 7, p. 171-174.

Taylor, B., Crook, K., and Sinton, J., 1986, Fast spreading and sufide deposition in the Manus backarc basin: American Association of Petroleum Geologists Bulletin, v. 70, p. 936-937.

Taylor, B., and Exon, N., 1982, Forearc volcanism and subduction without a trench: Peculiarities of ridge subduction in the Woodlark-Solomons Region: EOS Transactions of the American Geophysical Union, v. 63, p. 1120-1121.

Taylor, B., and Exon, N., 1987, Marine Geology, Geophysics, and Geochemistry of the Woodlark Basin - Solomon Islands: Circum-Pacific Council for Energy and Mineral Resources Earth

Science Series, v. 7, Houston, Texas, 853 p.

Tysdal, R.G., Case, J.E., Winkler, G.R., and Clark, S.H.B., 1977, Sheeted dikes, gabbro, and pillow basalt in flysch of coastal southern Alaska: Geology, v. 5, p. 377-383.

von Huene, R., Langseth, M., Nasu, N., and Okada, H., 1980, Summary, Japan Trench transect *in*, Scientific Party, Initial Reports of Deep Sea Drilling Project, v. 56-57, Part 1: Washington, D.C., U.S. Government Printing Office, p. 473-488.

Weissel, J. K. and Anderson, R. N., 1978, Is there a Caroline plate?: Earth and Planetary Science Letters, v. 41, p. 143-158.

Weissel, J.K., Taylor, B., and Karner, G.D., 1982, The opening of the Woodlark basin, subduction of the Woodlark spreading system, and the evolution of northern Melanesia since mid-Pliocene time: Tectonophysics, v. 87, p. 253-277.

Willcox, J.B., 1977, Some gravity models of the continental margin in the Australian region: Australian Society of Exploration Geophysicists Bulletin, v. 8, p. 118-124.

Winterer, E.L., et al., 1971, Initial Reports of the Deep Sea Drilling Project, Volume VII, Part 1: Washington, D.C., U.S. Government Printing Office, 841 p.

Part 3 – Summary

Marlow, M.S., Dadisman, S.V., and Exon, N.F., editors, 1988, Geology and offshore resources of Pacific island arcs—New Ireland and Manus region, Papua New Guinea, Circum-Pacific Council for Energy and Mineral Resources Earth Science Series, v. 9: Houston, Texas, Circum-Pacific Council for Energy and Mineral Resources.

GEOLOGY AND OFFSHORE RESOURCE POTENTIAL OF THE NEW IRELAND - MANUS REGION—A SYNTHESIS

N.F. Exon
Bureau of Mineral Resources, Geology and Geophysics, Canberra, A.C.T., 2601, Australia

M.S. Marlow
U.S. Geological Survey, Menlo Park, California 94025, USA

ABSTRACT

The equatorial New Ireland-Manus region forms a convex-northward arc about 900 km long and 200 km wide, extending from New Ireland in the east, almost to the Papua-New Guinea mainland in the west. The region is flanked by Eocene to Oligocene trenches in the north (part of which may be newly reactivated), and a Pliocene to Holocene back-arc basin in the south. Between the trenches and the back-arc basin is an area with a well-developed sedimentary basin in the east—the New Ireland Basin— and a more complex, poorly understood depositional area in the west.

The area between the trenches and the back-arc basin consists of an Eocene to Oligocene arc in the south, and a complementary forearc basin in the north, which contains as much as 6000 m of Miocene and younger sedimentary and volcaniclastic rocks. The region was tilted down to the north and up to the south in the Pliocene, so that most island areas are in the south. Other islands occur along an outer-arc high (the Emirau-Feni Ridge) in the north; in the east are the Pliocene to Pleistocene volcanic islands of the Tabar-Feni group.

After a geoscience cruise of the R/V *S.P. Lee*, geologists and geophysicists from Australia, the USA, Papua New Guinea and CCOP/SOPAC (Suva), prepared this report covering onshore and offshore geology and petroleum prospects in the New Ireland-Manus region. The region lies in an area of microplates between the Pacific and Australia-India plates, and is tectonically complex. Although undrilled, the region appears to have some petroleum prospects, because deeply buried potential source and reservoir rocks are present as well as widespread and thick, fine-grained seal rocks. Both stratigraphic and structural traps would have developed much of their present configuration by the Pliocene, when potential source rocks would have entered the hydrocarbon generation zone.

INTRODUCTION

The New Ireland-Manus region (Figure 1) is underlain, in part, by a sequence of flows and volcaniclastic rocks that began to accumulate in the Eocene, forming the western end of the West Melanesian Arc. In the Eocene and Oligocene, New Britain, Manus Island, New Hanover and New Ireland were all part of a north-facing volcanic arc bounded by the Manus Trench (Kroenke, 1984). There followed a long period of tectonic quiescence in the Miocene, during which time limestone was deposited in places on the arc (Stewart and Sandy, this volume; Francis, this volume). After early Pliocene volcanism on a southwest-facing arc, that was related to the New Britain Trench, the back-arc

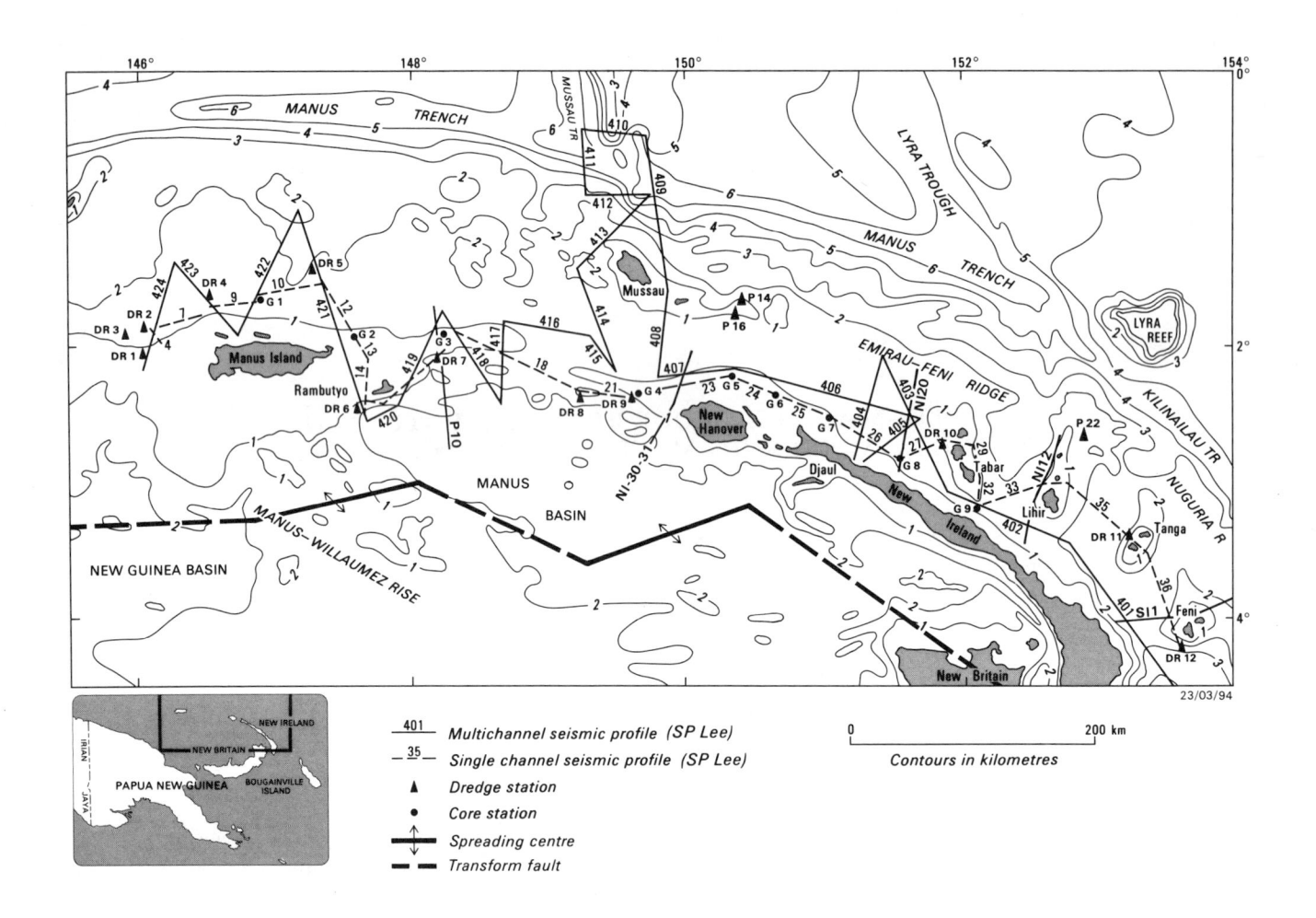

Figure 1. Bathymetric map of the New Ireland-Manus region, showing *S.P. Lee* geophysical tracks and sampling stations, and other key reflection seismic profiles illustrated in Figures 4 and 5 (P10 = CCOP/SOPAC profile PN81-8-10; NI and SI = Gulf profiles). South Bismarck plate lies south of spreading centers and transform faults.

Manus Basin started to form along spreading centers linked by transform faults (Taylor, 1979). This spreading was related to left lateral movement between New Britain and New Ireland, and probably resulted in the uplift of the southern side of the New Ireland-Manus region. Throughout the Miocene, Pliocene, and Pleistocene, a thick pile of volcaniclastic and carbonate sediments was deposited on the flows and volcaniclastics of the arc basement, between the main volcanic buildup of the West Melanesian Arc to the south and the forearc high (Emirau-Feni Ridge) to the north (Marlow, Exon, and Tiffin; Marlow et al., both this volume). Stratigraphic columns from Manus, New Hanover, and New Ireland illustrate the complex geological history of the region (Figure 2).

This paper summarizes geological information for the entire region between the Manus Trench and the Manus Basin—an arcuate area about 900 km long and 200 km wide. Water depths range from 0 to 6400 m, generally deepening northward from Manus Island, New Hanover, and New Ireland (Figure 1). East of Mussau lies the structurally simple New Ireland Basin, containing more than 6000 m of sedimentary fill in places (Marlow et al., this volume). The area west of Mussau is more complex and poorly understood (Marlow, Exon, and Tiffin, this volume). We believe sedimentary rocks deposited in the New Ireland-Manus region have some petroleum potential. However, as no exploration wells have been drilled, any assessment of this potential must be based on understanding the region's geological history.

Regional geological and geophysical data available before the 1984 cruise of R/V *S.P. Lee* were synthesised by Hohnen (1978), Johnson (1979),

SYNTHESIS

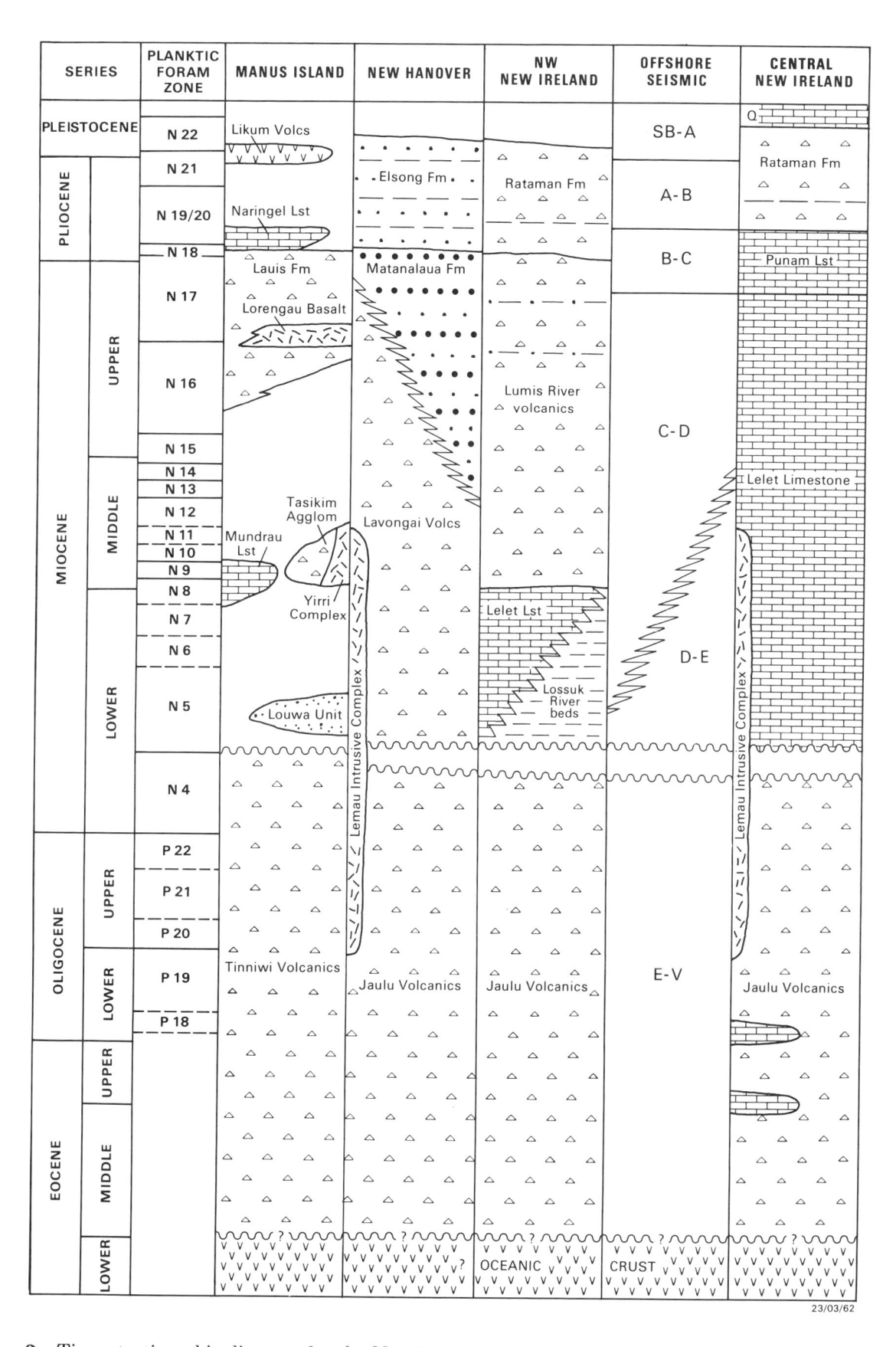

Figure 2. Time stratigraphic diagram for the New Ireland-Manus region.

Jaques (1980), Brown (1982), Exon and Tiffin (1984), Kroenke (1984), and Exon et al. (1986). The present paper deals largely with the two subregions separately—the New Ireland Basin and the Manus regions—because of the considerable differences both in framework geology and in resource potential.

GEOMORPHOLOGY AND REGIONAL TECTONICS

The arcuate New Ireland-Manus region (Figures 1 and 3) is bounded to the north by the Manus and Kilinailau Trenches, and to the south by the Manus Basin, the New Guinea Basin, and the intervening Manus-Willaumez Rise. The geomorphology and regional tectonics of the region were described by Taylor (1975, 1979) and Johnson (1979). Kroenke, Jouannic, and Woodward (1983) produced a regional bathymetric map of the area. Detailed bathymetric maps of the region were produced by Exon and Tiffin, based on CCOP/SOPAC geophysical surveys made in 1981, and the geomorphology was briefly described by Exon and Tiffin (1984) and Exon et al. (1986). A new bathymetric map of the region, incorporating the 1984 *S.P. Lee* data, is described by Chase (this volume).

Reviews of the regional tectonics of the New Ireland-Manus region include those of de Broin, Aubertin, and Ravenne (1977), Johnson (1979), Taylor (1979), Exon and Tiffin (1984), Falvey and Pritchard (1984), and Kroenke (1984). The region lies in an area of microplates between the Pacific plate, that is moving west-northwest at 10.7 cm/yr, and the Australia-India plate, that is moving north-northeast at 7 cm/yr. The New Ireland-Manus region is separated from the South Bismarck plate by spreading centers and transform faults in the Bismarck Sea. The New Ireland-Manus region may form a North Bismarck microplate bounded by the trenches to the north (McCue, this volume), part of the Pacific plate, (Taylor, 1979), or part of the Caroline plate.

Manus-Kilinailau Trench System

The Manus Trench (Figure 1) extends westward in a simple arc 1500 km long, from northeast of Tabar Island to near Wuvulu Island off westernmost Papua New Guinea. The trench is probably the surface expression of a southward-plunging subduction zone that was active in the Eocene and Oligocene. Today the trench is inactive, or virtually so, over much of its extent (Ryan and Marlow, this volume; McCue, this volume). Its eastern extension is the Kilinailau Trench. We will describe the Manus-Kilinailau Trench system from east to west.

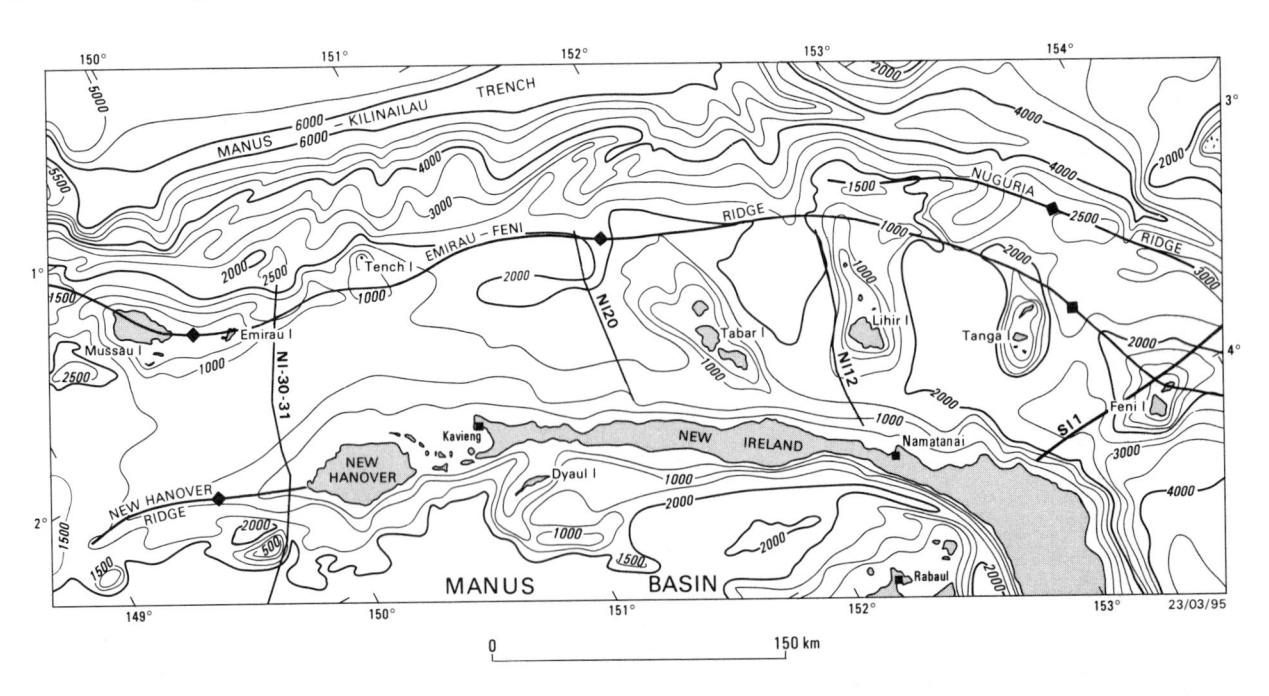

Figure 3. Bathymetric map of the New Ireland-Manus region, showing key Gulf seismic-reflection profiles illustrated in Figures 4 and 5. Major structural highs are Emirau-Feni Ridge (EFR), Nuguria Ridge (NR) and New Hanover (NHR)-New Ireland high.

The Kilinailau Trench is a feature variously interpreted as being a sinistral transcurrent fault zone (de Broin, Aubertin, and Ravenne, 1977), an inactive trench (Kroenke, 1972), or a subduction zone (Hamilton, 1979). Its relatively shallow axial depth of less than 4600 m appears to be related to the collision of the Ontong Java Plateau with the New Ireland region in the early Miocene (Kroenke, 1984). The Ontong Java Plateau extends westward as far as the Nuguria and Sable Islands, which lie across the Kilinailau Trench opposite Feni Island. Near the plateau, axial water depths in the trench are shallowest. Lyra Reef (Figure 1) caps a large seamount, which apparently collided with the New Ireland region and, like the Ontong Java Plateau, clearly constricts the Kilinailau Trench.

The Lyra Trough (Figure 1), a northwest-trending feature more than 5500 m deep in places, may be a graben related to marginal basin formation (Erlandson et al., 1976), an inactive trench, or a strike-slip fault (Hamilton, 1979). We believe that the trough is probably a fossil feature, related to the history of the Pacific plate, rather than a feature of importance to the New Ireland region.

The eastern Manus Trench, between the north-trending Mussau Trench and Lyra Reef (Figure 1), is as much as 6400 m deep and asymmetrical, the southern slope being steeper. East of the Mussau Trench, along *S.P. Lee* profile 409 (Figure 1), the sediment fill in the Manus Trench is deformed against its southern flank, indicating compression (Ryan and Marlow, this volume). The deformation may be related to under-thrusting of the Pacific plate or perhaps to collision of the nearby Mussau Ridge with the Manus Trench (Ryan and Marlow, this volume).

The Mussau Trench and adjacent Mussau Ridge separate the Caroline plate to the west from the Pacific plate to the east (Figure 1). The Trench is more than 6000 m deep, and the Ridge rises to 2800 m in depth 200 km north of Mussau Island. The Caroline plate underthrusts the Pacific plate along the southern part of the Mussau Trench (Hegarty, Weissel, and Hayes, 1983). *S.P. Lee* seismic profiles 409 to 413 box the Manus-Mussau Trench triple junction (Figure 1), and provide constraints on the junction's history (Ryan and Marlow, this volume). The profiles suggest that convergence is recent and slow along the southern Mussau Trench.

The western Manus Trench extends westward 900 km from its intersection with the Mussau Trench. It shallows steadily westward until it disappears near Wuvulu Island at 3500 m (Kroenke, Jouannic and Woodward, 1983). Although Weissel and Anderson (1978) and Hamilton (1979) claimed that the Manus Trench south of the Caroline plate is an active subduction zone, neither earthquake records (McCue, this volume) nor seismic-reflection profiles (Ryan and Marlow, this volume; Marlow, Exon, and Tiffin, this volume) provide any evidence of activity. We consider, as did Exon and Tiffin (1984), that subduction along the Trench probably ended in the earliest Miocene, an age derived from study of the New Ireland-Manus volcanic arc.

Outer-Arc Highs and Trench Slope

According to Karig and Sharman (1975), an outer-arc structural high is a prominent feature of most island arcs and may consist either of a melange zone or of a basement high, depending on the type of the subduction. De Broin, Aubertin, and Ravenne (1977) mapped what they called the "North Eastern Ridge" in the West Melanesian forearc and showed it extending from east of Guadalcanal in the Solomon Islands to Mussau in the New Ireland Basin. The western part of this feature was renamed the "Emirau-Feni Ridge" by Exon et al. (1986), and its location is shown in Figure 3.

The Emirau-Feni Ridge includes Mussau, Emirau, Feni, and Green Islands, but is north of the Tabar, Lihir, and Tanga groups. West of Lihir Island, the ridge marks the crest of the slope above the Manus Trench. Farther east another high, the Nuguria Ridge, separates the Emirau-Feni Ridge from the Kilinailau Trench. Water depths along the Emirau-Feni Ridge exceed 2000 m only where it is cut by canyons. Whether the ridge persists west of Mussau is unclear, because of tectonic complexity and the wide line spacing.

De Broin, Aubertin, and Ravenne (1977) suggested that the Emirau-Feni Ridge is cut by transcurrent faults normal to the trench, and that the volcanic islands of Tabar, Lihir, Tanga, and Feni stand on these faults. Although the islands are indeed perched on north-to-northeast-trending horst blocks (Exon and Tiffin, 1984), and normal faults with similar strike directions cut New Ireland (Stewart and Sandy, this volume), the seismic-reflection profiles are too widely spaced to map the faults accurately and determine whether there has been transcurrent movement. On many seismic profiles, layered sequences in the New Ireland Basin thin northward against the Emirau-Feni Ridge (Figures 4 and 5). The layered sequences also generally thin along the ridge from west to east. We believe that the base-

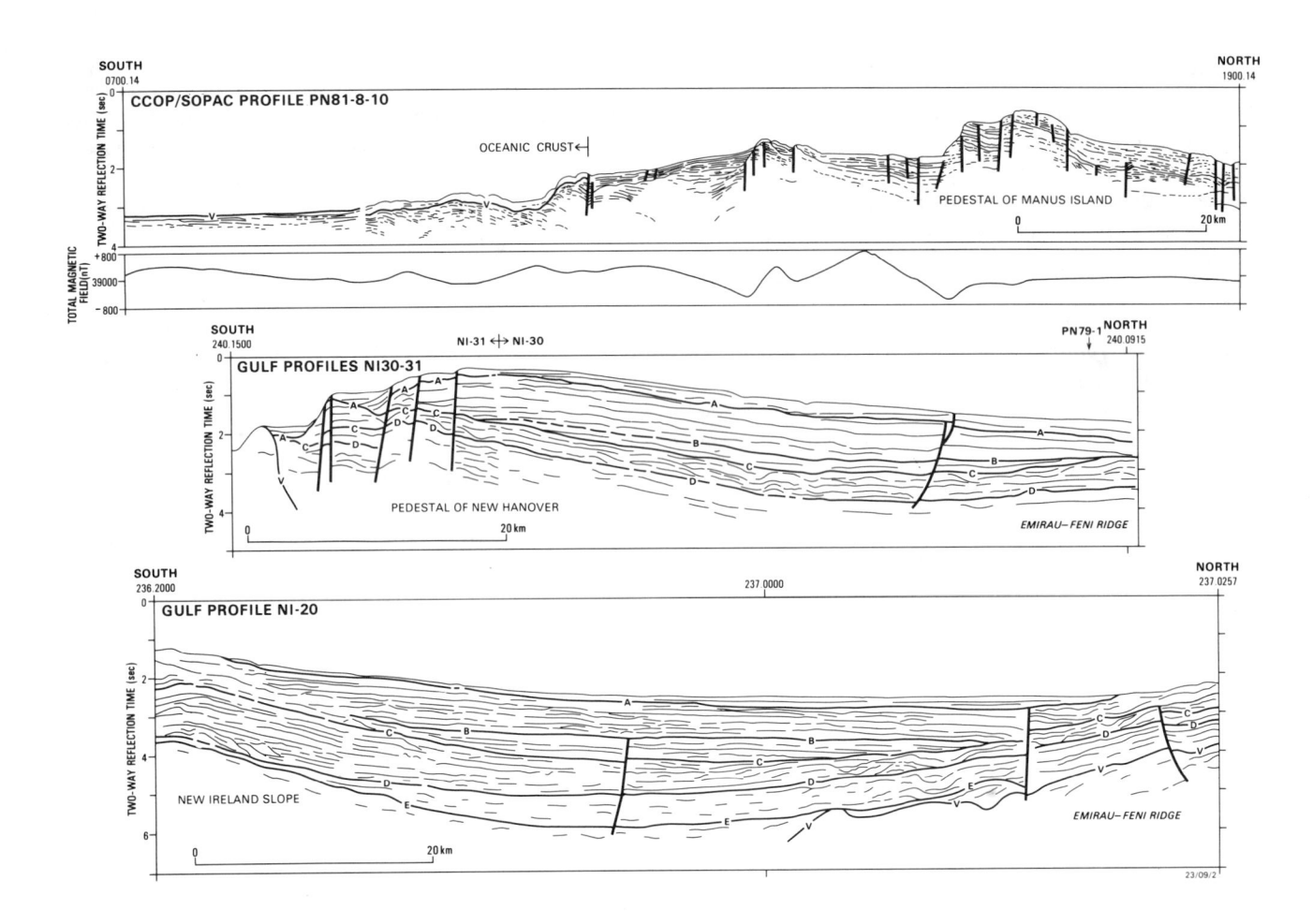

Figure 4. Line drawings of Gulf seismic-reflection profiles NI 20 and NI 30-31 and CCOP/SOPAC seismic-reflection profile PN81-8-10 in the New Ireland-Manus region. Locations shown on Figures 1 and 3. After Exon and Tiffin, 1984.

ment rocks of the ridge are probably oceanic basement, but they have not been sampled.

Exon and Tiffin (1984) first showed that a continuous, northwest-trending, bathymetric high—the 400-km-long Nuguria Ridge—lies beyond the Emirau-Feni Ridge, northeast of Lihir, Tanga, Feni, and Green Islands (Figure 3). The ridge is abruptly terminated to the west by a canyon that runs into the Lyra Trough, and to the east by a bathymetric deep north of Buka Island (Vedder, Bruns, and Cooper, in press). The crest of the ridge is generally 1500-2500 m deep. Nuguria Ridge probably consists of volcanic basement rocks overlain by a little sediment (Figure 5; Bruns, Vedder, and Culotta, in press). We suggest that the ridge may be a sliver of island-arc or oceanic crust accreted onto the New Ireland region, perhaps when the Pacific plate was subducted at the Kilinailau Trench.

The slope down into Manus and Kilinailau Trenches is generally uneven, and seismic-reflection profiles show numerous diffractions and few coherent beds beneath the slope. Southward-dipping reflectors beneath the slope, which might be expected from a melange, are absent, suggesting that the slope is underlain by volcanic basement. Gulf profile SI-1B (Figure 5) shows the Kilinailau Trench slope to be almost free of young sediment. *S.P. Lee* profile 409 shows the irregular nature of the Manus Trench slope and the abundance of diffractions (Figure 6). Ryan and Marlow (this volume) note that the lower slope of the Manus Trench is poorly reflective, and that the upper slope is underlain by high-amplitude reflectors with interval velocities of 5-6 km/s. These velocities suggest that the slope contains igneous bodies.

Basinal Areas

We define "basinal areas" as the areas of thick

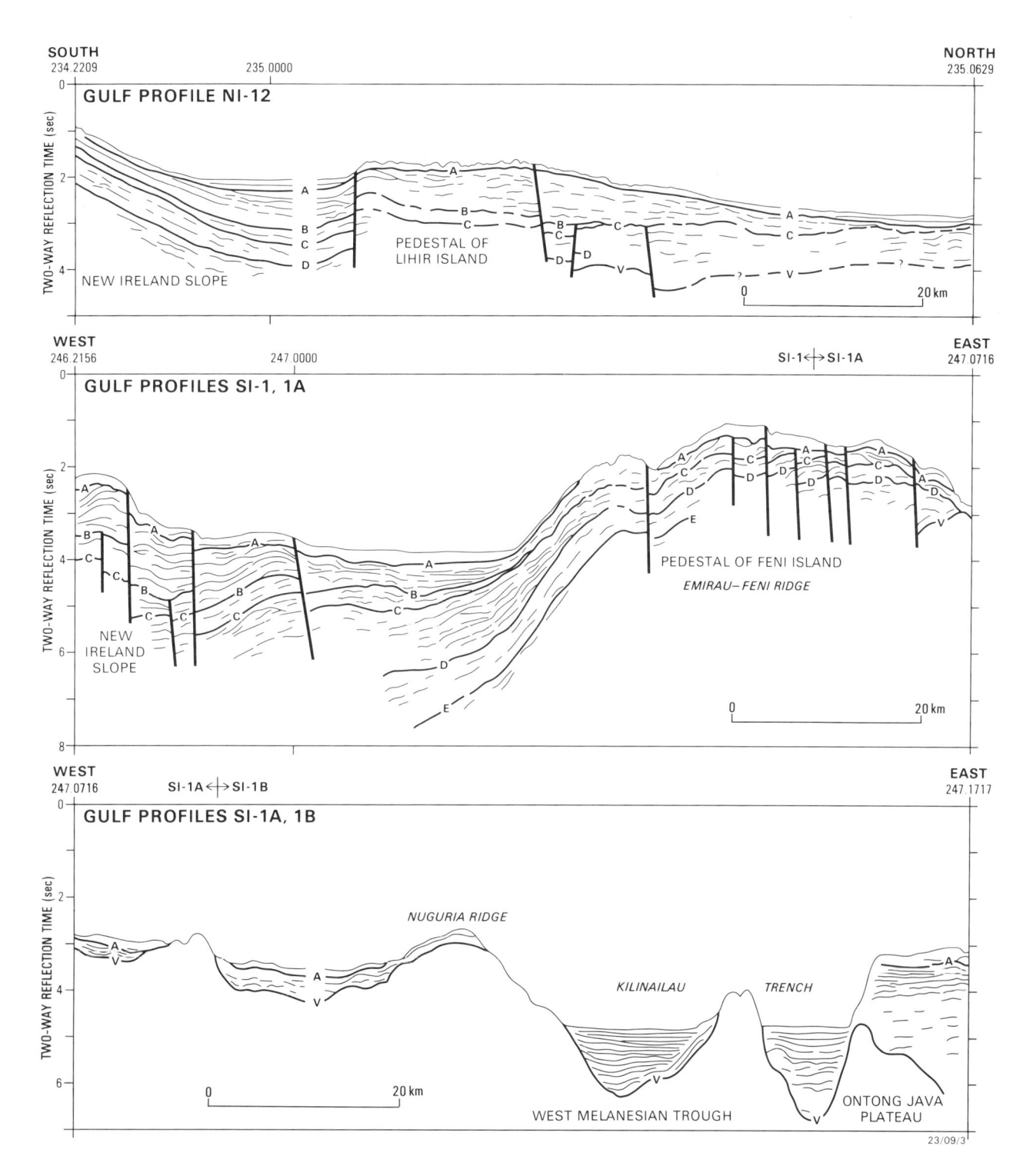

Figure 5. Line drawings of Gulf seismic-reflection profiles NI12 and SI-1, 1A and 1B in the New Ireland Basin. Locations shown on Figures and 1 and 3. After Exon and Tiffin, 1984.

sedimentary strata bounded to the north by outer-arc highs or the trench slope, and truncated to the south by young marginal basins—the Manus and New Guinea Basins—whose average depth is 2000 m (Figure 1). The "basinal areas" extend from Green Island in the east (southeast of Feni Island) almost to the Papua New Guinea mainland, and include the New Ireland Basin. In the south, the basinal areas

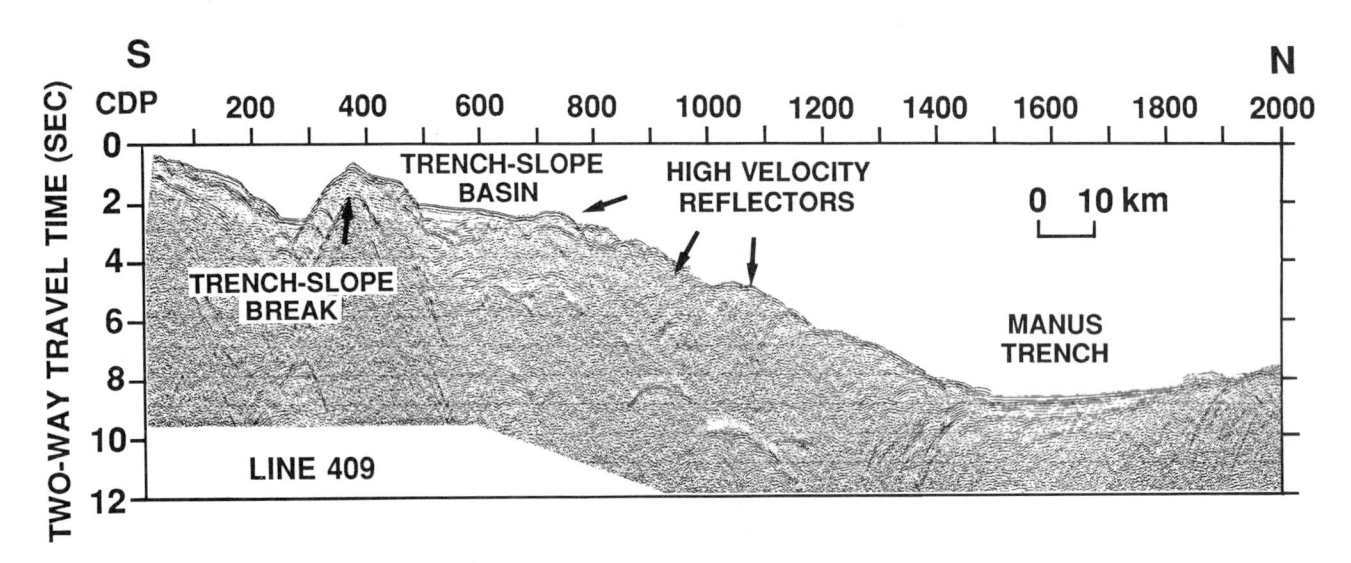

Figure 6. *S.P. Lee* multichannel seismic-reflection profile 409 across Manus Trench and the slope north of Mussau. Location shown on Figure 1. Profile shows an uneven slope with high-velocity basement(?) reflections and basin of younger sediments. There is some evidence of compression at the foot of the slope.

either emerge, or are covered by less than 1000 m of water. They have an general downward slope to the Manus-Kilinailau Trench in the north, but maximum water depth seldom exceeds 2200 m.

The New Ireland Basin—the area east of Mussau—forms a distinctive structural unit with thick sediment and simple structures. Its western third, between Mussau and Tabar Islands, lies between the rise of New Hanover and New Ireland to the south, and the Emirau-Feni Ridge to the north (Figure 3); water depth does not exceed 2000 m. In the western New Ireland Basin, fault systems are strongly developed on and near New Hanover and New Ireland (Stewart and Sandy, this volume), and also along the Emirau-Feni Ridge, but elsewhere faulting is insignificant (Figure 4). Both New Hanover and New Ireland are dissected into northeast-tilted blocks, by northwest-trending, high-angle normal faults. This pattern is complicated by northeast, north-northwest, and east-trending tensional and conjugate fractures. Brown (1982) concluded that the main fracture patterns first formed in the middle Tertiary, during development of the volcanic arc, and were reactivated during the Pliocene to Holocene uplift that formed the present-day islands. Faulting on the Emirau-Feni Ridge is less well defined, but appears to be dominated by trench-parallel normal faults that have grown over time and are still active.

Seismic-reflection profiles across the western New Ireland Basin (Figure 4) indicate that the Emirau-Feni Ridge generally has remained high, because most stratigraphic units in the basin thin toward the ridge. In contrast, the southern side of the presently preserved basin, which has been truncated by faulting, does not appear to have been high at any time during the basin history, because some sedimentary units maintain their thickness to its edge (Marlow, Exon, and Tiffin, this volume). Profile NI-31 (Figure 4) illustrates the large faults that throw the southern side of the basin down to the Manus Basin. These faults are believed to be left-lateral strike-slip faults with some dip-slip, which developed as Manus Basin started to open in the Pliocene (Taylor, 1979).

The eastern two-thirds of the New Ireland Basin, extending from Tabar to Green Island southeast of Feni Island (Figure 3), appears to have had a history similar to that of the western basin until the volcanic Tabar to Feni group of islands first formed in the Pliocene. The eastern basin deepens not only to the north, but also to the east, maximum axial water depths being about 2200 m in the west and 2800 m in the east. The basin is cut by horsts of the Tabar-Feni group of islands, which trend north or northeast (Exon and Tiffin, 1984). Canyons are cut in the slopes to the trench and to the deep south of Feni, and head between the volcanic islands.

Seismic-reflection profiles, such as Gulf NI-12 and Gulf SI-1, 1A and 1B (Figure 5), show the extensive faulting associated with the Tabar to Feni islands. Thinning of the younger sequences, across the horsts and toward the outer-arc highs, is also clearly apparent in these profiles. Structure contour and isopach maps of the eastern region suggest that a major northwest-trending fault zone bounds the volcanic islands to the south. This zone—the Tabar fault system—is downthrown to the southwest. We suggest that it may have a left lateral component of strike-slip, caused by the west-northwestward movement of the Pacific plate.

West of Mussau, in the depositional area of the Manus forearc, the density of bathymetric and geophysical profiles is much less than in the New Ireland Basin, the tectonic setting is more complex, and seismic reflectors are much harder to map for any distance (Marlow et al., this volume). We suggest that a basin like the New Ireland Basin, developed early, but a different tectonic regime has led to major differences since the Miocene.

Overall, the western area dips northward toward the Manus Trench (Figure 1). Just south of the trench are many volcanic edifices and other highs, but no continuous outer-arc high is apparent. Large canyons cut the trench slope in the area south of the Mussau Trench, but farther west none are known. The western area is bounded to the south by the Manus Basin, the Manus-Willaumez Rise, and the New Guinea Basin. The boundary with the Manus Basin is steep and normally faulted; it trends east-northeast parallel to the Manus Basin spreading axis (Figure 1).

The Manus-Willaumez Rise is a volcanic ridge trending northwest. It is parallel to the transform fault direction in the Manus Basin shown in Figure 1 (Johnson, Mutter, and Arculus, 1979). A major line of crustal weakness, marked by volcanoes, apparently extends west-northwest beyond the rise, through and southwest of Manus Island, and including Hermit Island and the Ninigo Group above the Manus Trench. The boundary between the depositional area near Manus Island and the Manus-Willaumez Rise is not clear, tectonically or even geomorphologically. The west-trending boundary between the depositional area west of Manus and the New Guinea Basin is cut by canyons, and is not well understood.

Seismic-reflection profiles across the western area illustrate a number of features. CCOP/SOPAC profile PN81-8-10 shows many normal faults on the southern margin of the depositional area (Figure 4). *S.P. Lee* profile 421, northeast of Manus Island, shows considerable deformation of forearc sediment just north of the island, with thrust and normal faults reaching the surface. Flat areas farther north are filled with what are assumed to be near surface lava flows (Figure 7). *S.P. Lee* profile 424, west of Manus Island, also shows the deformation zone and the flat areas assumed to be lava flows (Figure 8). The profile also shows young normal faults associated with the pedestal of Manus, including a fault scarp from which we dredged ocean-island basalts (Johnson et al., this volume).

Focal mechanisms of three earthquakes in the Manus region show significant thrusting with stresses in a west-northwest direction (McCue, this volume). These earthquakes, northwest of Manus Island and southwest of Mussau Island, confirm the evidence of recent compression noted in the seismic-reflection profiles. This compression could be related to the Pacific-Australian plate collision, renewed subduction at the Manus Trench, or spreading in the Manus Basin.

BASEMENT

Outcrops

On New Ireland, basement consists of 2000 m of Eocene(?) to Oligocene island-arc Jaulu Volcanics, and the Oligocene to middle Miocene Lemau Intrusive Complex (Stewart and Sandy, this volume). The Jaulu Volcanics and equivalents also are exposed on Mussau and the Tabar Islands (D'Addario, Dow, and Swoboda, 1976). Although Brown (1982) mapped "Jaulu Volcanics" on New Hanover, G. Francis (pers. comm.) considers them to be younger. The Jaulu Volcanics are mainly andesitic agglomerate, lapilli and crystal lithic tuffs, andesite lava, and ignimbrite; a variety of interbedded sediment occurs in small quantities.

Rocks of the Lemau Intrusive Complex are found in limited areas of New Ireland and New Hanover, as stocks and dikes of gabbro, diorite, quartz diorite, and various leucocratic rocks (Stewart and Sandy, this volume). The island arc apparently was affected by two periods of intrusion: (1) early to middle Oligocene and (2) early to middle Miocene.

Basement on Manus Island consists of the middle Eocene to lowermost Miocene island-arc Tinniwi Volcanics (Francis, this volume). The volcanic rocks consist largely of andesite, basaltic andesite, basalt, lava breccias, and pyroclastic rocks, and there are some interbedded sedimentary rocks. The sequence is intruded by batholiths and stocks of the early to middle Miocene Yirri Intrusive Complex, consisting largely of monzonite, diorite, and andesite.

Acoustic Basement

On offshore seismic-reflection profiles across the

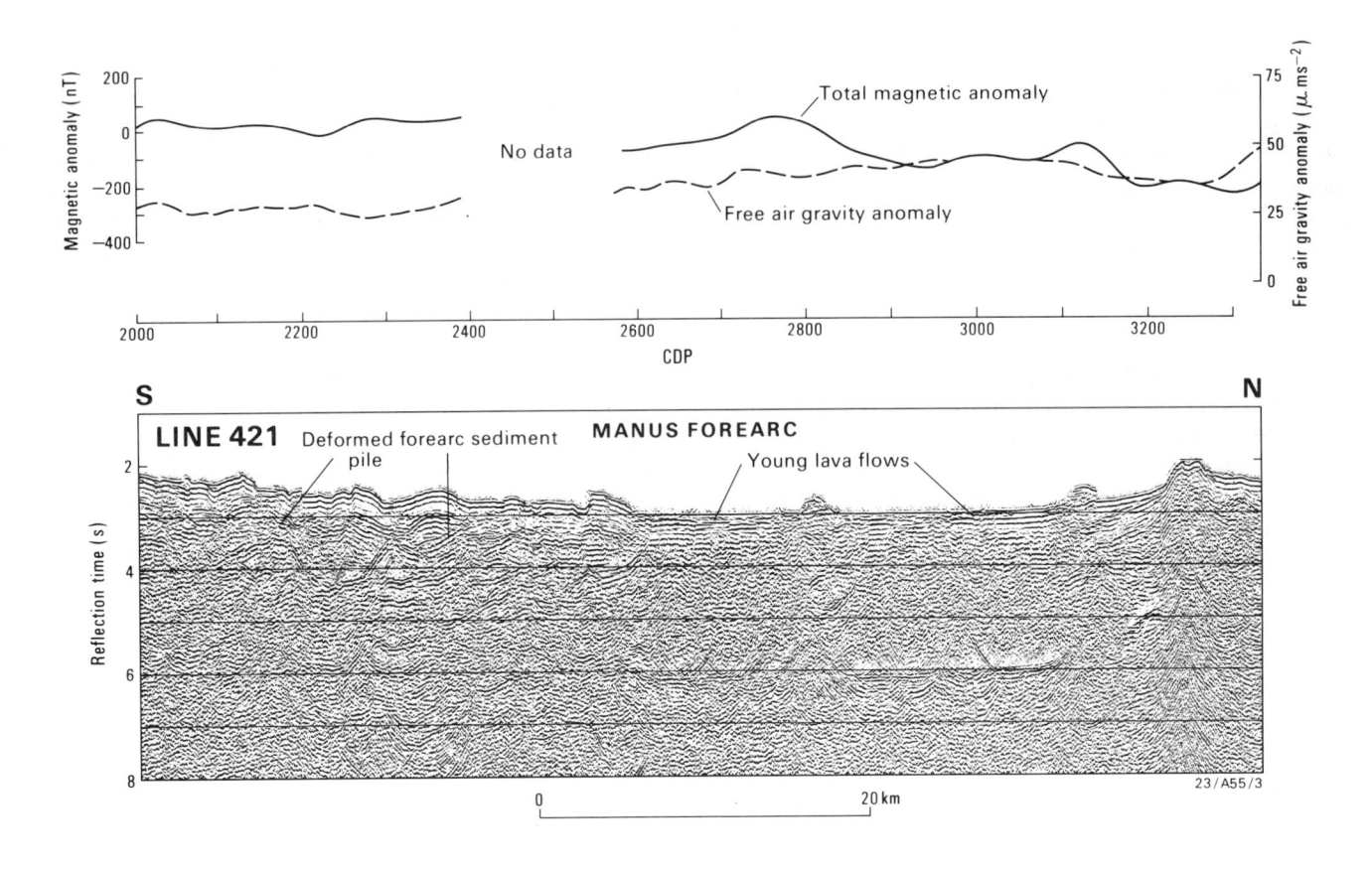

Figure 7. *S.P. Lee* multichannel seismic-reflection, magnetic, and gravity profile 421 across region northeast of Manus Island, showing deformation of sequence by thrusting and perhaps strike-slip faulting. Highly reflective strata near the sea bed are assumed to be lava flows. Location shown on Figure 1.

New Ireland Basin, the top of the Jaulu Volcanics is believed to be horizon E (Exon and Tiffin, 1984; Marlow et al., this volume). The E-V sequence is variably reflecting, poorly defined, and generally at the limit of acoustic resolution. The sequence is at least 1000 m thick in places. On some seismic-reflection profiles it appears to pinch out against the Emirau-Feni Ridge (e.g. Gulf Profile NI-20 in Figure 4). The section has not been sampled, but it is assumed to consist of volcaniclastic sedimentary and andesitic volcanic rocks, like those of the Jaulu Volcanics. As the offshore E-V sequence was laid down farther from the volcanic arc than the outcropping rocks, it is likely to contain more sedimentary rocks, especially more fine-grained ones.

Horizon V, beneath the E-V sequence, is a strong diffracting reflector at the top of acoustic basement. Horizon V crops out, or nearly so, along the basin margins (Figures 4 and 5). It may represent the top of oceanic basement on the Emirau-Feni and Nuguria Ridges and beneath the slopes down to the Manus and Kilinailau Trenches, and the top of the Jaulu Volcanics near New Ireland.

In the Manus Basin, the horizon clearly represents the Pliocene oceanic crust associated with back-arc spreading.

SEDIMENTARY SEQUENCES

Both outcrop and offshore seismic-reflection information indicate that thick sections of Miocene and younger sedimentary rocks rest on volcanic basement. The stratigraphy of the region is illustrated in Figure 2. The known maximum sedimentary thickness exceeds 6000 m in the New Ireland Basin (Figure 9), where acoustic penetration is good (Exon and Tiffin, 1984); farther west acoustic penetration is poor, and the known maximum thickness is less (Marlow, Exon, and Tiffin, this volume).

The outcrop geology of New Ireland is discussed in detail by Stewart and Sandy (1986; this volume). The composite maximum thickness of the sedimentary sequence on New Ireland cannot be directly measured, but appears to be 2000-2500 m. New Hanover has not been mapped in detail, but the

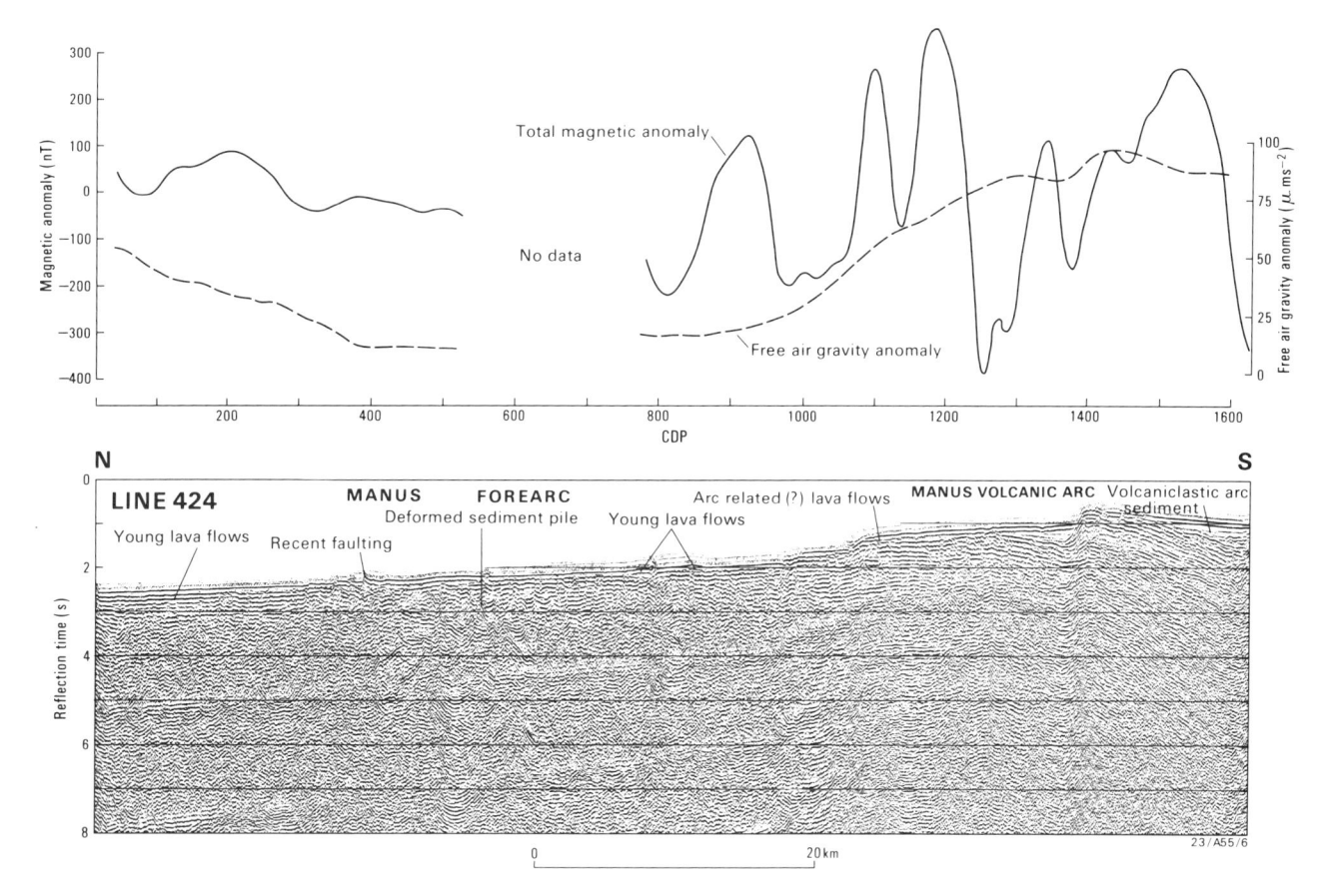

Figure 8. *S.P. Lee* multichannel seismic-reflection, magnetic, and gravity profile 424 across region northwest of Manus Island, showing deformation of sequence, assumed lava flows, and recent faulting. Dredging from the scarp of Manus volcanic arc (DR2) yielded ocean-island basalt. Note high-amplitude magnetic profile across the volcanic arc, in marked contrast to that on nearby profile 421 (Figure 6). Location shown on Figure 1.

maximum sedimentary thickness on the island probably does not exceed 1000 m. The stratigraphy of Manus Island is detailed by Francis (this volume); the maximum sedimentary thickness is more than 1000 m.

The ages of sedimentary sequences on the islands depend largely on studies of planktonic foraminifers; the zonation used is from Blow (1969). However, age-diagnostic planktonic foraminifers do not occur in inner-neritic samples, which must be dated by using longer ranging benthic foraminifers. Haig and Coleman (this volume) contribute significantly to these studies by analyzing outcrop material from New Ireland in order to better define age and paleobathymetry. The age range is early Oligocene (interbeds in the Jaulu Volcanics) to Pleistocene. Overall, water depths over New Ireland appear to have been shallow in the Oligocene and early Miocene, to have deepened to mid-bathyal in the Pliocene, and to have shallowed again in the Pleistocene. G. Francis (written comm.) has noted that water depths generally increase with distance from the volcanic arc.

Complementary studies of foraminiferal assemblages, from material dredged and cored from offshore sequences, were carried out by Belford (this volume). The ages of the samples from seismic sequences B-C, A-B and SB-A range from late Miocene (earliest Pliocene in the scheme of Haig and Coleman, in this volume) to Holocene. The faunas from the samples indicate that there have been no major changes in water depth since the enclosing sediments were deposited.

Sequences Exposed on New Ireland

On New Ireland, basement rocks are unconformably overlain by the Miocene Lelet Limestone, and its partial age-equivalents—the Lossuk River beds and Lumis River volcanics—in the northwest, and the Tamiu siltstone in the southeast. Above

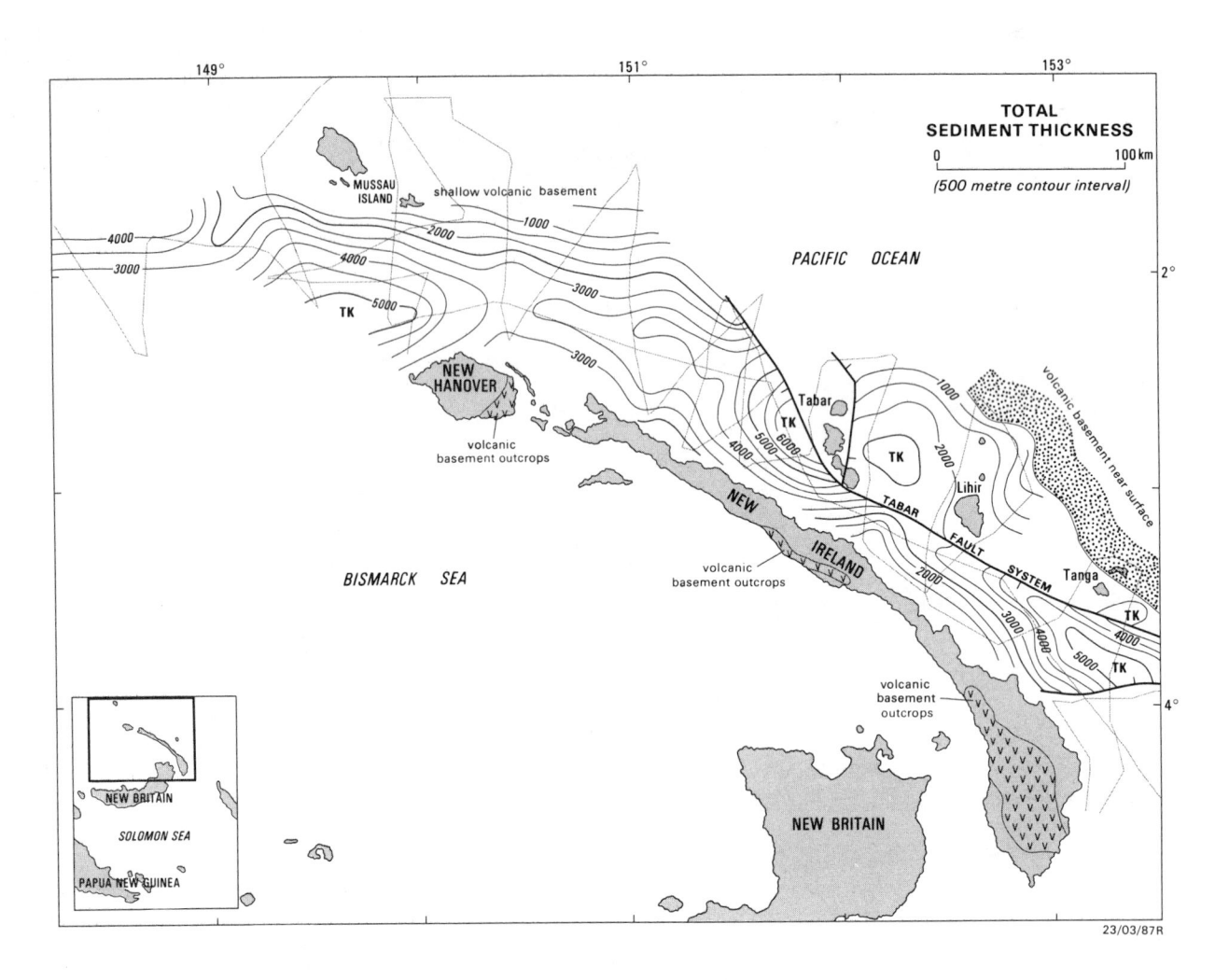

Figure 9. Total sediment thickness map of the New Ireland Basin, showing depocenters northwest of New Hanover and northeast of New Ireland. Drawn using Gulf and *S.P. Lee* seismic-reflection data. TK represents areas of maximum thickness.

these sequences are the Miocene to Pliocene Punam Limestone, the Pliocene to Pleistocene Rataman Formation, and a variety of younger units including the Pleistocene Maton Conglomerate (Stewart and Sandy, this volume).

The Lelet Limestone consists largely of lagoonal and shelfal biomicrites, that are generally lacking in reef-building massive corals. Local subunits include the tuffaceous and neritic Bagatere unit, the thin sandy lagoonal Kimidan unit, and the thin muddy lagoonal Matakan unit. The Lelet Limestone is thickest (ca. 1000 m) in the south, where it was probably deposited during early Miocene (N5) to late(?) Miocene (N17?) time.

The Lossuk River beds of northwestern New Ireland are lateral equivalents of the lower Lelet Limestone, and are early Miocene in age. They are predominantly calcareous carbonaceous mudstone, siltstone, and tuffaceous sandy siltstone, and contain abundant planktonic foraminifers and a variety of benthonic organisms. They were deposited in outermost neritic to upper bathyal water depths and are about 150 m thick.

The Tamiu siltstone of southeastern New Ireland is also a lateral equivalent of the lower Lelet Limestone, and is unconformably overlain by the upper Lelet Limestone along the Tamiu River. The Tamiu siltstone is of early to middle Miocene age, and is at least 300 m thick. The predominant rock types are calcareous siltstone, muddy sandstone, and tuffaceous biomicrite. Benthic organisms predominate and planktonic foraminifers are rare. The unit was deposited in water depths that decreased from upper bathyal to inner neritic over time.

The Lumis River volcanics of northwestern New Ireland are laterally equivalent to the upper

Lelet Limestone and are middle to late Miocene in age. They conformably overlie both the lower Lelet Limestone and the Lossuk River beds, and are predominantly andesitic tuff, agglomerate, breccia, sandstone, and siltstone. Foraminiferal assemblages indicate deposition in middle neritic to lower bathyal water depths. The maximum thickness of the unit is estimated to be 800 m.

The Punam Limestone of central New Ireland is latest Miocene to earliest Pliocene in age. It is a pelagic foraminiferal biomicrite, as much as 200 m thick, and contains some tuffaceous material, which was deposited in bathyal water depths. The Rataman Formation conformably overlies the Punam Limestone and is Pliocene to Pleistocene in age. It was laid down during a major volcanic episode and consists of three subunits characterized, going up the sequence, by andesitic to dacitic tuff, foraminiferal biomicrite, and epiclastic sandstone and siltstone. The formation probably has a maximum thickness of more than 1000 m, and was deposited in bathyal water depths.

Sequences Exposed Elsewhere

Although the outcrops on New Ireland are the keys to understanding the sequences offshore, the sequences on New Hanover and Manus are also important (Figure 2). On New Hanover no limestone equivalent to the Lelet Limestone has been found. Volcaniclastic Miocene sequences similar to the Lumis River volcanics — the pyroclastic Lavongai Volcanics and the epiclastic Matanalaua Formation — were laid down in bathyal water depths. The Pliocene Elsong Formation on New Hanover consists of calcareous tuffaceous clastic materials deposited in upper bathyal water depths.

On Manus Island (Francis, this volume; Figure 2), the oldest sedimentary sequence is the thin and laterally limited lower Miocene Louwa unit that consists of bathyal calcareous siltstone and biomicrite. The Mundrau Limestone, of late early Miocene to earliest middle Miocene age, is confined to the central part of the island and is about 200 m thick. It formed a carbonate platform of algal-foraminiferal biomicrite and biomicrudite, and so is similar to the Lelet Limestone of New Ireland.

The Tasikim Agglomerate is mostly middle Miocene in age and is widespread, it is as much as 400 m thick. It consists largely of andesitic agglomerate, lapilli tuff, and tuff possibly subaerially deposited, with minor amounts of neritic volcaniclastic sedimentary rocks. The widespread upper Miocene Lauis Formation consists predominantly of bathyal siltstone, conglomerate, and tuff, and is as much as 500 m thick. It includes the Lorengau basalt of eastern Manus Island. The lower Pliocene Naringel Limestone consists of raised reefs or carbonate banks. The Pliocene to Pleistocene Likum Volcanics form a caldera in the southwest, and are largely olivine basalt and mafic andesite of ocean-island affinities (Jaques, 1980).

Offshore Sequences

The seismic sequences within the New Ireland Basin, and their interpretation, are summarized in Figure 2 and Table 1. They are believed to consist largely of carbonate and volcaniclastic sedimentary rocks, except for the lower sequence (E-V), which is mainly volcanic rocks. The following brief summary of depositional sequences is drawn largely from Marlow et al. (this volume).

The upper 2-3 m of sequence SB-A have been cored and were found to consist of calcareous ooze (Carlson, Rearic, and Quinterno, this volume). The A-B and B-C sequences were dredged at Station 10 just west of the Tabar Islands (Figure 1) and were found to consist of upper Miocene (or lowermost Pliocene), Pliocene, and lower Pleistocene sedimentary rocks, including marls, indurated calcilutite and interbedded calcareous mudstone, and volcanogenic sandstone. The A-B and B-C sequences prograde and thin northeastward from New Ireland and New Hanover (Figures 4 and 5). The seismic-reflection characteristics described in Table 1, and the dredging results, suggest that the A-B sequence is predominantly volcaniclastic rock equivalent to the Rataman Formation on New Ireland, and that the B-C sequence is mainly marls and chalks equivalent to the Punam Limestone of New Ireland.

The C-D sequence has mixed seismic-reflection characteristics, with weak to strong parallel reflectors, dipping reflectors, and buildups or areas of complex and curved reflection events (Figure 4, Table 1). The sequence maintains its thickness over wide areas, although thinning toward the Emirau-Feni Ridge (Figures 4 and 5; Marlow et al., this volume, Fig. 11). It was interpreted by Exon and Tiffin (1984) and Exon et al. (1986) as being a carbonate-platform sequence containing reefs, back-reef deposits, and fore-reef rubble. They suggested that the carbonate deposits were equivalent to the Miocene Lelet Limestone and could be possible source and reservoir beds for hydrocarbons (see Exon and Marlow, petroleum potential, this volume).

Table 1. Seismic Sequences in New Ireland Basin (after Marlow et al., this volume).

Sequence	Thickness (m)	Character	Interpretation
SB-A	200-300	Semi-transparent, well-bedded sequence immediately beneath sea bed (SB).	Pleistocene and Holocene hemipelagic oozes.
A-B	500-1000	Well-bedded sequence with interbedding of strong reflectors and semi-transparent intervals.	Pliocene volcaniclastic turbidites interbedded with marls and chalks. Equivalent to Rataman Formation.
B-C	200-500	Well-bedded sequence like A-B.	Upper Miocene and lower Pliocene chalks and marls, interbedded with volcaniclastic turbidites. Equivalent to Punam Limestone.
C-D	1000-2500	Mixed seismic character. Parallel reflectors, dipping reflectors, and curved reflectors suggestive of carbonate buildups.	Probably largely lower and upper Miocene platform and upper slope limestones equivalent to Lelet Limestone. Basinward, may contain volcaniclastic turbidites.
D-E	1000	Reflecting sequence with variable amplitude, locally containing dipping and curved reflectors.	Probably lower Miocene outer shelf and slope volcaniclastic sediment, with some channels and possible carbonate banks. Equivalent to Lossuk River beds and Tamiu siltstone.
E-V	1000	Variably reflecting, poorly defined sequence at limit of acoustic penetration.	Probably largely Eocene to lower Miocene volcanics and volcaniclastic rocks equivalent to Jaulu Volcanics on nearby islands; younger Miocene rocks locally.

Refraction velocities within the C-D sequence average 3500 m/s over a wide range for 10 refraction stations (Childs and Marlow, this volume). The average refraction velocity, and the 3600 m/s interval velocity derived from multichannel stacking velocities, are within the range measured for compressional velocities in limestone (Clark, 1966), although there is such a wide range of possible velocities for limestone that velocities are not really diagnostic.

The deepest resolvable sequence, D-E, is variably reflecting, moderately well bedded, and locally contains dipping reflectors, channels, and buildups (Table 1). In general, we correlate the sequence with the lower Miocene marine Lossuk River beds of New Ireland, which stratigraphically underlie the Lelet Limestone; off southern New Ireland we correlate the sequence with the Tamiu siltstone (Stewart and Sandy, this volume). Its thickness is highly variable (Marlow et al., this volume, Fig. 8).

The section below reflector E (E-V in Table 1) is poorly defined and at the limit of acoustic resolution, V not being discernible in most places. Reflector E is a low-frequency event exhibiting many diffractions and overlain unconformably by the D-E sequence. In general, we correlate the section below reflector E with the Eocene to lower Miocene volcanic and volcaniclastic rocks of the Jaulu Volcanics exposed on nearby islands. However, in some cases reflector E is clearly younger.

The seismic or depositional sequences described above are not mappable west of Mussau or around Manus, but various unconsolidated Holocene sediments, and older sedimentary and igneous rocks, have been dredged and cored in those regions (see Tables 1 and 2 in Introduction of this volume). The cored Holocene sediments are foraminiferal ooze (Carlson, Rearic, and Quinterno, this volume; Belford, this volume). The sedimentary rocks dredged around Manus and Rambutyo, and west of New Hanover, are similar to the Punam Limestone and distal equivalents of the Rataman Formation; Belford (this volume) dates these rocks as late Miocene (N18) to late Pliocene (N21) in age. They are dominantly pelagic foraminiferal chalk, marl, and calcareous mudstone, but also include interbedded volcanic sandstone, ashy siltstone, and volcanic marl.

Small quantities of volcanic rocks were recovered from fault scarps west-northwest of Manus Island, at dredge stations 2 and 3 (Figures 1 and 8). Johnson et al. (this volume) describe the single pebble from dredge 2 as an ocean-island ferrobasalt, much like those in the Pliocene to Pleistocene Likum Volcanics of Manus Island. Dredge 3 recovered highly weathered and nondescript volcanic rocks. An important unanswered question is whether either of these rock types represents the widespread, flat-lying, seismically highly reflective material present just below the sea floor north of Manus Island, which Marlow, Exon, and Tiffin (this volume) believe to be lava flows.

Volcanic rocks were recovered at dredge station 6A, along with upper Miocene (N18) and younger sedimentary rocks, from the scarp leading down to the Manus Basin (Figure 1). *S.P. Lee* profile 421 suggests that sedimentary rocks overlie a volcanic section, and hence that the volcanic rocks are late Miocene or older in age. G. Francis (pers. comm.) correlates these volcanic rocks with the Tasikim Agglomerate and, using Manus Island geological relationships, argues that they are probably middle Miocene in age. Johnson et al. (this volume) describe one volcanic sample as an "orogenic andesite", which is markedly different from the ferrobasalt from dredge 2, but similar to andesites from eastern Manus Island.

GENERAL BASIN GEOMETRY AND STRUCTURE

Seismic-reflection profiles outline the geometry of the basinal section in the New Ireland Basin (Figure 10), but west of Mussau the geometry is not clear. New Ireland Basin contains as much as 6000 m of sedimentary section with three major depocenters: (1) between New Hanover and Mussau, (2) west of Tabar Island, and (3) south of Tanga Island. The basin fill thins markedly toward New Hanover and New Ireland, on both of which basement crops out, as well as toward the Emirau-Feni Ridge to the northeast, where volcanic basement either crops out or lies near the surface.

The best-defined seismic reflector is the C horizon, believed to be generally equivalent to the top of the Miocene Lelet Limestone on New Ireland; its structure reflects the deep structure where that is known. The structure contour map of the C horizon (Figure 11) is reflected in the bathymetry (Figure 3), and shows highs around islands formed largely of Miocene rocks, such as Mussau, New Hanover, and New Ireland, as well as highs in the Tabar-Feni group formed of younger intrusive igneous rocks punched through the older basinal sedimentary rocks. The map indicates the structural importance of the New Ireland-New Hanover high; the top of the equivalent Lelet Limestone is more than 1000 m high in parts of central and southern New Ireland (Exon et al., 1986, Fig. 2b). The Emirau-Feni Ridge incorporates the islands of Mussau, Emirau, and Feni, but runs north of Tabar, Lihir, and Tanga (Figures 3 and 11). Along the ridge, the C horizon generally lies less than 2500 m below sea level, deepening southeastward. Along the basin axis between the New Ireland-New Hanover high and the Emirau-Feni Ridge, the depth of the horizon always exceeds 2500 m, and deepens southeastward overall. However, the general basin structure is complicated by the highs associated with the horsts of the Tabar-Feni group.

The Tabar fault system separates the Tabar-Feni island groups from the main basinal areas to the southwest (Figure 11). We know little about the Tabar fault system, but it does appear to represent a major structural line. The age of faulting is uncertain, and the relationship of the fault system to the north- and northeast-trending horsts, upon which sit the Tabar-Feni group, is unclear. The system may possibly represent strike-slip movement caused by the movement of the Pacific plate west-northwestward past New Ireland. A marked struc-

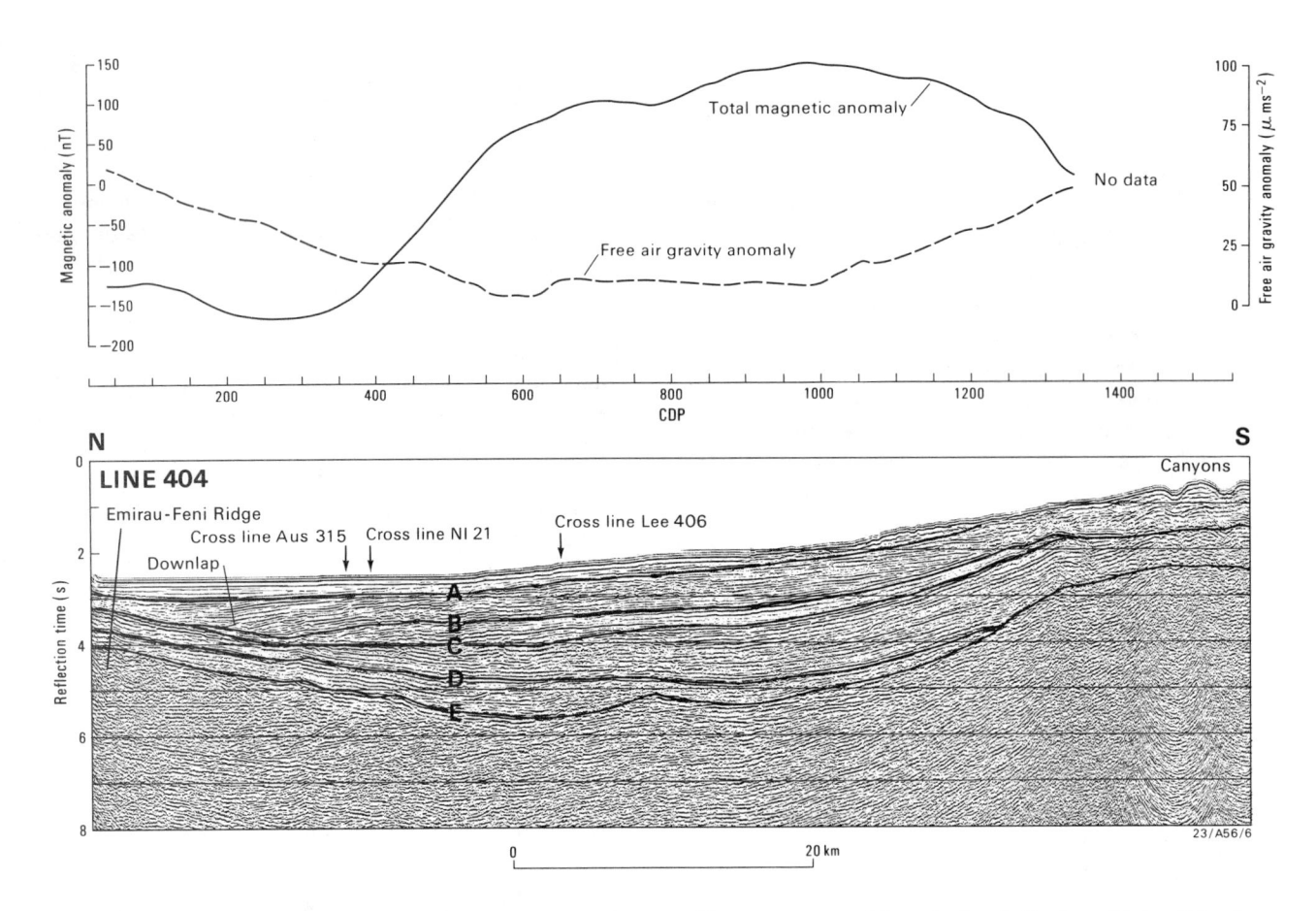

Figure 10. *S.P. Lee* multichannel seismic-reflection profile 404 across New Ireland Basin. Note thinning or pinching out of most sequences toward Emirau-Feni Ridge at northern end of line, and also toward New Ireland at southern end of line. Location shown on Figure 1.

tural low separates Tabar and Lihir Islands, reaching 3500 m at its deepest.

As pointed out by Exon and Tiffin (1984), many offshore faults formed recently and are probably still active, although some are long-lived growth faults on which movement directions have changed. A major fault system trends northwestward and forms the southern flanks of New Hanover and New Ireland (Stewart and Sandy, this volume). Onshore, this fault system is extensively developed on both islands (Brown, 1982; Hohnen, 1978), especially along the Weitin and Sapom faults (Figure 11), but also in high-angle faults, downthrown to the south, on New Hanover (Brown, 1982). Movement on the system has varied from fault to fault, both transcurrent and vertical movements being important. This fault system separates the arc crust of the New Ireland Basin from the oceanic crust of the Manus Basin (Taylor, 1979), and is interpreted as being a transform fault system (Curtis, 1973; Connelly, 1976). A lesser, parallel set of normal faults occurs farther north, both on land and offshore. These seem to have formed as adjustments to differential loading in the New Ireland Basin and to post-Pliocene tilting of the basin.

In the area of the Manus forearc west of Mussau, multichannel seismic penetration of as much as 3 s suggests that, in places, the sedimentary section is about 4 km thick. The overall lack of data, and the lack of continuity of reflectors, clearly illustrated on the seismic-reflection profiles (Figures 7 and 8), prevent us from commenting in detail on geometry and structure. However, a southern wedge of sedimentary rocks appear to thin northward for some distance. The wedge is no longer recognizable in the north because of considerable deformation, or because it is overlain by flat-lying highly reflective strata (probably lava flows) that prevent further seismic penetration (Marlow, Exon, and Tiffin, this volume).

OUTLINE OF GEOLOGICAL HISTORY

The New Ireland-Manus region probably first developed in the Eocene, as the northern part of the

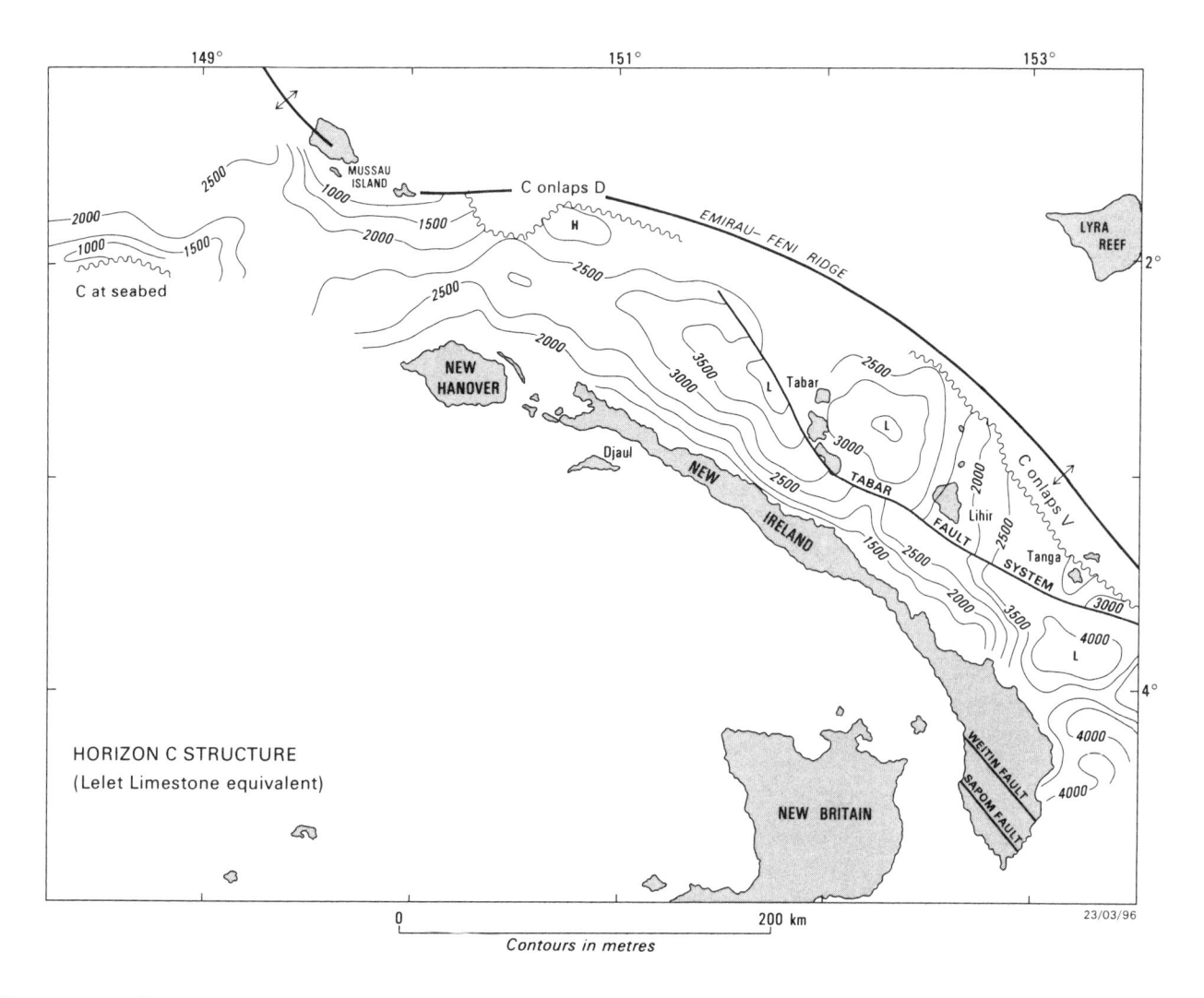

Figure 11. Structure contour map of C horizon in the New Ireland Basin, which is believed to represent the top of an offshore sequence generally equivalent to the Miocene Lelet Limestone of New Ireland. This map illustrates the importance of the outer-arc Emirau-Feni Ridge (Figure 3), the New Hanover-New Ireland high in the southwest, and the faulting associated with the Pliocene to Holocene volcanism that formed the Tabar-Feni group of islands. The Tabar fault system is poorly mapped but is important.

West Melanesian volcanic arc of which New Britain formed the southern part. The arc faced a subduction zone to the north—the Manus-Kilinailau Trench—and beyond that the oceanic crust of the Caroline Basin. The following geologic history is drawn largely from the papers in this volume and is illustrated by cross sections showing the structural evolution in the western New Ireland Basin (Figure 12).

Eocene and Oligocene Arc Volcanism

During the Eocene, Oligocene, and early Miocene, a large andesitic volcanic pile built up on oceanic crust especially near volcanic centers in the south. The pile was dominantly volcanic in the south and volcaniclastic in the north, and as much as 2000 m thick. It includes the Jaulu and Tinniwi Volcanics, and their presumed offshore equivalent, the E-V sequence.

Miocene Decline in Volcanic Activity

Subduction, and most of the related volcanism, ceased in the early Miocene. There followed a period of relative quiescence during which Miocene clastic and carbonate sediments were laid down in the New Ireland Basin in the east, while episodic volcanism

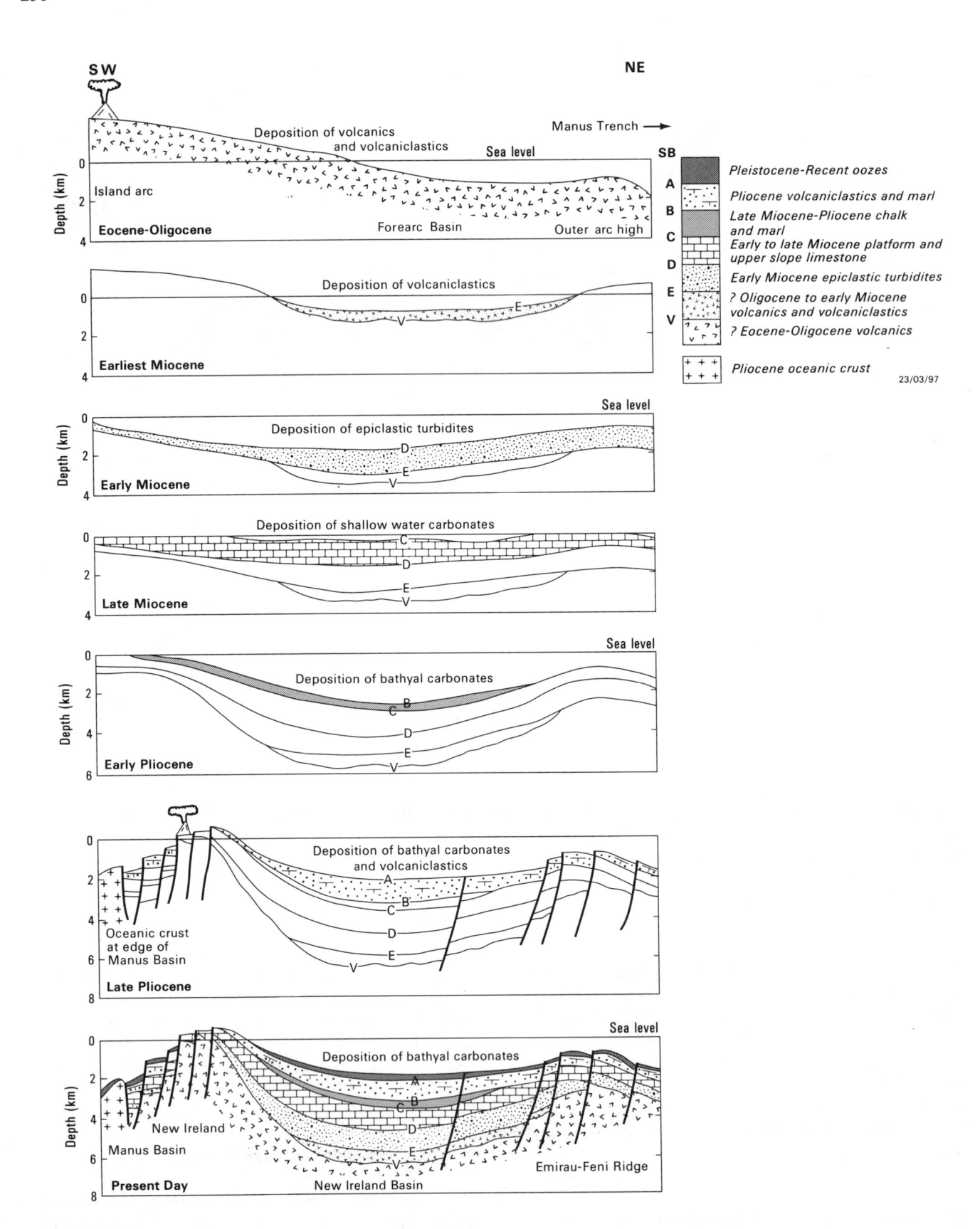

Figure 12. Schematic cross sections derived from multichannel seismic-reflection profiles in the western New Ireland Basin. These sections show the postulated development of the basin from the Eocene to the present day.

persisted in the west. The maximum thicknesses of the Miocene sequences were about 1000 m on New Ireland and New Hanover, less on Manus Island, and up to 3500 m in the offshore New Ireland Basin. A resurgence of volcanism in the middle or late Miocene in some areas was probably related to subduction at the New Britain Trench in the south. The Miocene also was a time of widespread intrusion of basic and intermediate rocks on Manus Island, New Hanover, and New Ireland (e.g. Lemau and Yirri Intrusive Complexes).

The most active volcanic region appears to have been New Hanover, where as much as 1000 m of the Lavongai Volcanics were laid down. These rocks consist of andesite, basaltic andesite, agglomerate, breccia, and tuff (Brown, 1982). On New Hanover, the distal equivalents of this sequence are the volcaniclastic sedimentary rocks and minor tuffs and agglomerates of the Matanalaua Formation, which were laid down in environments ranging from subaerial to bathyal.

On Manus Island, there were depositional hiatuses in the Miocene, but the detrital Louwa unit, the Mundrau Limestone, the Tasikim Agglomerate, and the detrital Lauis Formation were all deposited (Francis, this volume). Volcanism in the early middle Miocene produced the Tasikim Agglomerate. Environments were bathyal for Louwa deposition, shelfal for Mundrau deposition, inner shelfal and possibly terrestrial for Tasikim deposition, and upper bathyal for Lauis deposition.

On New Ireland, detrital deposition tended to predominate in the early Miocene, and shelf carbonate deposition in the middle and late Miocene, but both occurred synchronously in different areas (Stewart and Sandy, this volume). The detrital Lossuk River beds in the northwest, and the Tamiu siltstone in the southeast, are probably the oldest sequences. Both are fine-grained units and were laid down on the upper slope and shelf. Where water was shallow and detrital input small, the shelfal carbonate strata of the Lelet Limestone were laid down. Deposition continued through most of the Miocene in southeastern and central New Ireland, but only in the early Miocene in northwest New Ireland. The early disappearance of the Lelet Limestone in northwest New Ireland can be attributed to andesitic volcanism; the middle and upper Miocene Lumis River volcanics, consisting of agglomerate, tuff, and volcaniclastic sedimentary rocks, were laid down on shelf and slope. These volcanic rocks are clearly related to the pyroclastic Lavongai Volcanics and epiclastic Matanalaua Formation of New Hanover.

In the offshore New Ireland Basin, two seismic sequences are believed to represent Miocene deposition. The older, D-E, probably consists of volcaniclastic sedimentary deposits laid down on the slope. The younger, C-D, is apparently a mixture of platform carbonate rocks equivalent to the Lelet Limestone, and bathyal volcaniclastic sediments. Miocene shelf carbonate deposits were laid down near Mussau, and Miocene fore reef carbonate materials near the Tabar Islands.

Pliocene to Pleistocene Events

Little volcanism occurred in the earliest Pliocene, as evidenced by the widespread deposition of shelf and bathyal carbonate units (Punam Limestone, Elsong Formation, Naringel Limestone, Kulep Limestone). There was a resurgence of regional volcanism during the early Pliocene—probably related to subduction along the New Britain Trench. Lava flows and pyroclastic materials were deposited during this time on Djaul Island, and tuff and reworked tuff on New Ireland (Rataman Formation).

In middle Pliocene time, Manus Basin started to open (Taylor, 1979), and New Britain commenced its movement southeastward away from Manus and past New Ireland. Volcanism ceased for a time, and pelagic carbonate deposits were laid down, along with epiclastic material shed in large quantities from the southern islands (Manus, New Hanover, New Ireland) as they were uplifted.

In the Pleistocene, extensive alkaline volcanic rocks erupted in the Tabar-Feni islands, which were uplifted as a series of horsts. Contemporaneous volcanism on Manus Island formed the ocean-island Likum Volcanics. Rapid uplift brought northern and central New Ireland and Manus Islands to shelf depths, allowing the formation of fringing reefs and associated shelf sediments.

Three offshore seismic sequences (B-C, A-B and SB-A) were laid down in the New Ireland Basin during latest Miocene to Holocene time. The oldest apparently consists of bathyal carbonate rocks equivalent to the Punam Limestone, the middle of pelagic carbonate and volcaniclastic sedimentary rocks equivalent to the Rataman Formation, and the youngest of hemipelagic oozes.

RESOURCE POTENTIAL

The New Ireland-Manus region includes a major gold deposit on Lihir Island, and perhaps some other metals, but our studies have dealt only with hydrocarbons (Exon and Marlow, petroleum potential, this volume) and offshore hydrothermal metal deposits (Bolton and Exon, this volume).

Hydrocarbons

The onshore geology indicates that the basinal region is underlain by volcanic rocks. In the New Ireland Basin, which has the best petroleum prospects, these are overlain in turn by Miocene shallow-marine volcaniclastic sedimentary rocks, Miocene shelf and bathyal limestone, and uppermost Miocene to Holocene pelagic carbonate and volcaniclastic sedimentary deposits. There has been no drilling in the region, but 2600 km of multichannel seismic-reflection data, supplemented by dredge samples, allow correlation of the thick offshore New Ireland Basin sequence (as much as 6000 m thick) with the relatively thin onshore sequences. The poorly known Manus area has a thick offshore section, extensive faulting, and some roll-overs into faults. The New Ireland Basin's hydrocarbon potential probably depends on whether the rocks offshore are similar to the onshore Miocene sequences. The four onshore sequences are (1) the Lossuk River beds, (2) the Tamiu siltstone, and (3) the Lumis River volcanics, which all contain potential sources of hydrocarbons, and (4) the Lelet Limestone, which is a potential reservoir. Outcrop samples from New Ireland have been studied for potential hydrocarbon source rocks by Glikson (this volume) and Rigby (1986). The samples generally have moderately high total organic carbon values of 1-5%, dominated by woody material. The studies suggest that many outcropping rocks are capable of generating gas and waxy crude oils.

Rocks that crop out on New Ireland are immature to almost mature as petroleum sources, but north and west of New Ireland the depths of burial of equivalent rocks are much greater, suggesting that they should be mature. Geothermal gradients have not been measured in the New Ireland Basin, but given the almost mature nature of some outcrop rocks, and information from similar basins in the region, Exon and Marlow (petroleum potential, this volume) conclude that the onset of hydrocarbon production (65°C) might occur at a depth of 2500 m, and the onset of mature oil production (90°C) at a depth of 3500 m. These figures suggest that the offshore D-E sequence—our assumed equivalent of the Lossuk River beds and Tamiu siltstone—is within the present oil-generation zone over wide areas offshore the New Ireland Basin. If similar thermal gradients have prevailed in the past, possible source rocks would not have entered the hydrocarbon generation zone before the Pliocene.

Potential reservoir rocks are likely to be either epiclastic sandstones that have been reworked in the wave zone or in turbidites or, more probably, shelf limestones. Exon and Marlow (petroleum potential, this volume) follow Exon and Tiffin's (1984) earlier interpretation that the offshore C-D seismic sequence contains thick layers of shelf carbonate rocks like the onshore Lelet Limestone. The offshore C-D sequence is widespread, retains its reflection characteristics, and varies little in thickness, suggesting that it is not volcanic. Buildups in the C-D section appear to be separated by basin strata and chaotic zones, as is common on carbonate platforms. However, the seismic character of the C-D sequence is similar to that of Miocene units in the Cape Vogel Basin, which are dominantly fine tuffaceous clastic rocks, with subordinate amounts of carbonate turbidites and volcaniclastic sandstones (G. Francis, pers. comm.). Thus, it is possible that shelfal carbonate deposits are restricted in their distribution. The suspected reefal buildups, and associated fore-reef carbonate rocks, are the main targets for exploration in the region.

With the limited amount of seismic data, and the complete lack of well data, only generalized exploration plays can be recognized. We assume that fine volcaniclastic and pelagic carbonate rocks would form the seals in all cases. Stratigraphic traps may include carbonate banks, reefs, and fore-reef limestones of Miocene age. Potential structural traps, such as anticlines and fault traps, are generally limited to the basin margins. Potential structural/stratigraphic traps are confined to pinch-outs against highs—again largely along the basin margins.

Six cores of offshore surface sediment, taken by the *S.P. Lee*, were sampled for headspace gas, in the hope of detecting hydrothermal gas seeps. Gas concentrations were small, and no evidence was found for petroleum-related hydrocarbons (Kvenvolden, this volume). A "bright spot" occurs on *S.P. Lee* profile 401, near New Ireland (Exon and Marlow, petroleum potential, this volume, Fig. 9), and it may well represent a gas/oil or gas/water contact. No sediment cores were collected near the bright spot.

In summary, positive features of the New Ireland Basin for petroleum exploration include a thick sedimentary section, adequate structures, probable source and reservoir rocks, and a probable hydrocarbon indication (the bright spot) on a seismic section. Problems include the complete lack of well data, and the limited extent of shallow-water areas.

Hydrothermal Metals

The Manus Island area, and the Tabar to Feni islands region, are potential sites of hydrothermal deposits related to recent volcanism, hot springs, and faulting. Hydrothermal manganese was found at dredge station DR3, encrusting a weathered volcanic rock recovered from a fault scarp 1100-1250 m deep, west-northwest of Manus Island (Bolton and Exon, this volume; location in Figure 1).

Dredge sample 3-2 has a slightly gritty surface layer less than 1 mm thick, which rests on 15-20 mm of hard, shiny, bluish-black, finely laminated crust, which rests, in turn, on a 5-15 mm thick, basal grayish-black unlaminated crust. The crust as a whole is 49% Mn, less than 1% Fe, and less than 1% of all other metals combined. All the evidence indicates a hydrothermal origin. If normal hydrothermal systems are present in this region, polymetallic sulphides may exist, at or beneath the sea bed (Bolton and Exon, this volume).

DISCUSSION

The large and complex New Ireland-Manus region has not been sufficiently studied to allow a full understanding of its tectonic history, volcanic history, or basin fill, especially west of the New Ireland Basin. The New Ireland Basin appears to have moderate hydrocarbon potential, although it is untested. The resource potential of the western area around Manus Island cannot be assessed with the present data.

The Manus forearc, near Manus Island, is different from the New Ireland Basin in the extensive occurrence of near surface, highly reflective layers, which may be lava flows (Marlow, Exon, and Tiffin, this volume). Fault scarps have been dredged, apparently at the southern limit of the reflective strata, and the rocks recovered were ocean-island basalt (DR2), or highly weathered volcanic rocks (DR3). The thin sediment cover suggests that the suspected flows are too young to be forearc lavas related to Oligocene to early Miocene subduction at the Manus Trench. If they were related to later subduction at the New Britain Trench to the south, they are in a most unusual, back-arc position. Drilling or submersible operations are needed to sample the highly reflective layers.

Multichannel seismic-reflection data and earthquake focal mechanisms from north of Manus Island, indicate that the area is in compression now, with active thrust-faulting; however, some faulting may be strike-slip. A more detailed multichannel seismic-reflection survey of the area, and the deployment of ocean-bottom seismometers, would help determine fault trends and the direction of motion.

The plate-tectonic setting of the region is only moderately understood, and some drilling and more marine surveys are needed. We do not know whether the Manus or Kilinailau Trenches are active, what effect the newly commenced subduction at the Mussau Trench is having south of the Manus Trench, or whether the Tabar fault system is dominated by strike-slip movement related to the westward motion of the Pacific plate.

Other unknowns of considerable scientific interest include the origin of the Nuguria Ridge beyond the outer-arc Emirau-Feni Ridge, the origin of the Pleistocene alkaline volcanic rocks in the Tabar-Feni island group, the bathymetric deep south of Feni between New Ireland and Bougainville, and the hydrothermal vents and their deposits near Manus Island.

ACKNOWLEDGMENTS

This synthesis relies heavily on the work of all the authors who have contributed to this volume. Special mention must be made of the geologists of the Geological Survey of Papua New Guinea—G. Francis, M.J. Sandy and W.D. Stewart—whose help and advice were invaluable. The work was carried out as part of a Tripartite (Australia, New Zealand, United States of America) program in co-operation with Papua New Guinea and CCOP/SOPAC. The structure contour and isopach maps shown in this paper are simplified from maps developed by J. Pinchin of Flower, Doery, Buchan Pty. Ltd. in conjunction with the present authors. The figures were drawn by A. Murray and the text was typed by J. Brushett both of the Bureau of Mineral Resources. The paper has been reviewed by G. Francis of the Geological Survey of Papua New Guinea and H. Ryan of the U.S. Geological Survey. Exon publishes with the permission of the Director, Bureau of Mineral Resources.

REFERENCES

Blow, W.H., 1969, Late middle Eocene to Recent planktonic

foraminiferal biostratigraphy, *in* Bronnimann, P. and Rene, H.H., eds., Proceedings First International Conference on Planktonic Microfossils: Leiden, E.J. Brill, p. 199-422.

Brown, C.M., 1982, Kavieng, Papua New Guinea - 1:250,000 Geological Series: Geological Survey of Papua New Guinea, Explanatory Notes SA/56-9, 26 p.

Bruns, T.R., Vedder, J.G. and Culotta, R.C., in press, Structure and tectonics along the Kilinailau Trench, Bougainville-Buka Region, Papua New Guinea: Circum-Pacific Council for Energy and Mineral Resources Earth Science Series.

Clark, S.P., Jr., 1966, Handbook of physical constants: Geological Society of American Memoir 97, 583 p.

Connelly, J.B., 1976, Tectonic development of the Bismarck Sea, based on gravity and magnetic modelling: Geophysical Journal of the Royal Astronomical Society, v. 46, p. 23-40.

Curtis, J.W., 1973, Plate tectonics of the Papua New Guinea - Solomon Islands region: Geological Society of Australia Journal, v. 20, p. 21-36.

D'Addario, G.W., Dow, D.B., and Swoboda, R., 1976, Geology of Papua New Guinea, 1:2,500,000: Bureau of Mineral Resources, Canberra.

de Broin, C.E., Aubertin, F., and Ravenne, C., 1977, Structure and history of the Solomon-New Ireland region: International symposium on geodynamics in south-west Pacific: Paris, Edition Technip, p. 37-50.

Erlandson, D.L., Orwig, T.L., Kilsgaard, G., Mussells, J.H., and Kroenke, L.W., 1976, Tectonic interpretations of the East Caroline and Lyra Basins from reflection-profiling investigations: Geological Society America Bulletin, v. 87, p. 453-462.

Exon, N.F., and Tiffin, D.L., 1984, Geology and petroleum prospects of offshore New Ireland Basin in northern Papua New Guinea, *in* Watson, S.T., ed., Transactions of the Third Circum-Pacific Energy and Mineral Resources Conference, Honolulu, 1982, p. 623-630.

Exon, N.F., Stewart, W.D., Sandy, M.J., and Tiffin, D.L., 1986, Geology and offshore petroleum prospects of the eastern New Ireland Basin, northeastern Papua New Guinea: Bureau of Mineral Resources Journal of Australian Geology and Geophysics, v. 10, p. 39-51.

Falvey, D.A., and Pritchard, T., 1984, Preliminary paleomagnetic results from northern Papua New Guinea: evidence for large microplate rotations, *in* Watson, S.T., ed., Transactions of the Third Circum-Pacific Energy and Mineral Resources Conference, Honolulu, 1982, p. 593-599.

Hamilton, W., 1979, Tectonics of the Indonesian region: U.S. Geological Survey Professional Paper 1078, 345 p.

Hegarty, K.A., Weissel, J.K. and Hayes, D.E., 1983, Convergence of the Caroline and Pacific plates - collision and subduction, *in* Hayes, D.E., ed., The tectonics and geologic evolution of Southeast Asian seas and islands: part 2: American Geophysical Union Monograph Series, v. 27, p. 326-348.

Hohnen, P.D., 1978, Geology of New Ireland, Papua New Guinea: Bureau of Mineral Resources Australia Bulletin v. 194, 39 p.

Jaques, A.L., 1980, Admiralty Islands, Papua New Guinea - 1:250,000 Geological Series: Geological Survey of Papua New Guinea Exploratory Notes SA/55-10 and SA/55-11, 25 p.

Johnson, R.W., 1979, Geotectonics and volcanism in Papua New Guinea: a review of the late Cenozoic: Bureau of Mineral Resources Journal of Australian Geology and Geophysics, v. 4, p. 181-207.

Johnson, R.W., Mutter, J.C. and Arculus, R.J., 1979, Origin of the Willaumez-Manus Rise, Papua New Guinea: Earth and Planetary Science Letters, v. 44, p. 247-260.

Karig, D.E. and Sharman, G.F., 1975, Subduction and accretion in trenches. Geological Society of America Bulletin, v. 86, p. 377-389.

Kroenke, L.W., 1972, Geology of the Ontong Java Plateau: Hawaii Institute Geophysics Report HIG 72-5, 119 p.

Kroenke, L.W., 1984, Cenozoic tectonic development of the southwest Pacific: Committee for Coordination of Joint Prospecting for Mineral Resources in South Pacific Offshore Areas (CCOP/SOPAC) Technical Bulletin, v. 6, 122 p.

Kroenke, L.W., Jouannic, C., and Woodward, P., 1983, Bathymetry of the Southwest Pacific: Geophysical Atlas of the Southwest Pacific, Chart 1, Suva, CCOP/SOPAC.

Rigby, D., 1986, The source and isotopic composition of organic and carbonate carbon in rock outcrop samples from New Ireland: CSIRO (Australian Commonwealth Scientific and Industrial Research Organisation) Institute of Energy and Earth Resources Restricted Investigation Report 1648 R, 32 p.

Stewart, W.D., and Sandy, M.J., 1986, Cenozoic stratigraphy and structure of New Ireland and Djaul Ireland, New Ireland Province, Papua New Guinea: Geological Survey of Papua New Guinea Report 86/12, 96 p.

Taylor, B., 1975, The tectonics of the Bismarck Sea region: University of Sydney, Bachelor of Science Thesis, unpublished.

Taylor, B., 1979, Bismarck Sea : evolution of a back-arc basin: Geology, v. 7, p. 171-174.

Vedder, J.G., Bruns, T.R. and Cooper, A.K., in press, Geologic framework of Queen Emma Basin, eastern Papua New Guinea. Circum-Pacific Council for Energy and Mineral Resources Earth Science Series.

Weissel, J.K. and Anderson, R.N., 1978, Is there a Caroline plate?: Earth and Planetary Science Letters v. 41, p. 143-158.

Part 4 – Appendix

Marlow, M.S., Dadisman, S.V., and Exon, N.F., editors, 1988, Geology and offshore resources of Pacific island arcs—New Ireland and Manus region, Papua New Guinea, Circum-Pacific Council for Energy and Mineral Resources Earth Science Series, v. 9: Houston, Texas, Circum-Pacific Council for Energy and Mineral Resources.

APPENDIX 1.

SUBMARINE TOPOGRAPHY OF NORTHEASTERN PAPUA NEW GUINEA

T.E. Chase, B.A. Seekins, J.D. Young, S.V. Dadisman
U.S. Geological Survey, Menlo Park, California 94025, U.S.A.

INTRODUCTION

A series of submarine topography maps have been prepared for northeastern Papua New Guinea. The maps are shown in Figures 1-5 and are all Mercator projections reduced to double-page size. Figure 1 is a regional map showing all the major bathymetric features in the area. Figures 2-5 are more detailed and show the area covered in 1984 by the R/V *S.P. Lee* cruise (cruise L7-84-SP). Additional information regarding procedures used in preparing the maps can be obtained by contacting the authors.

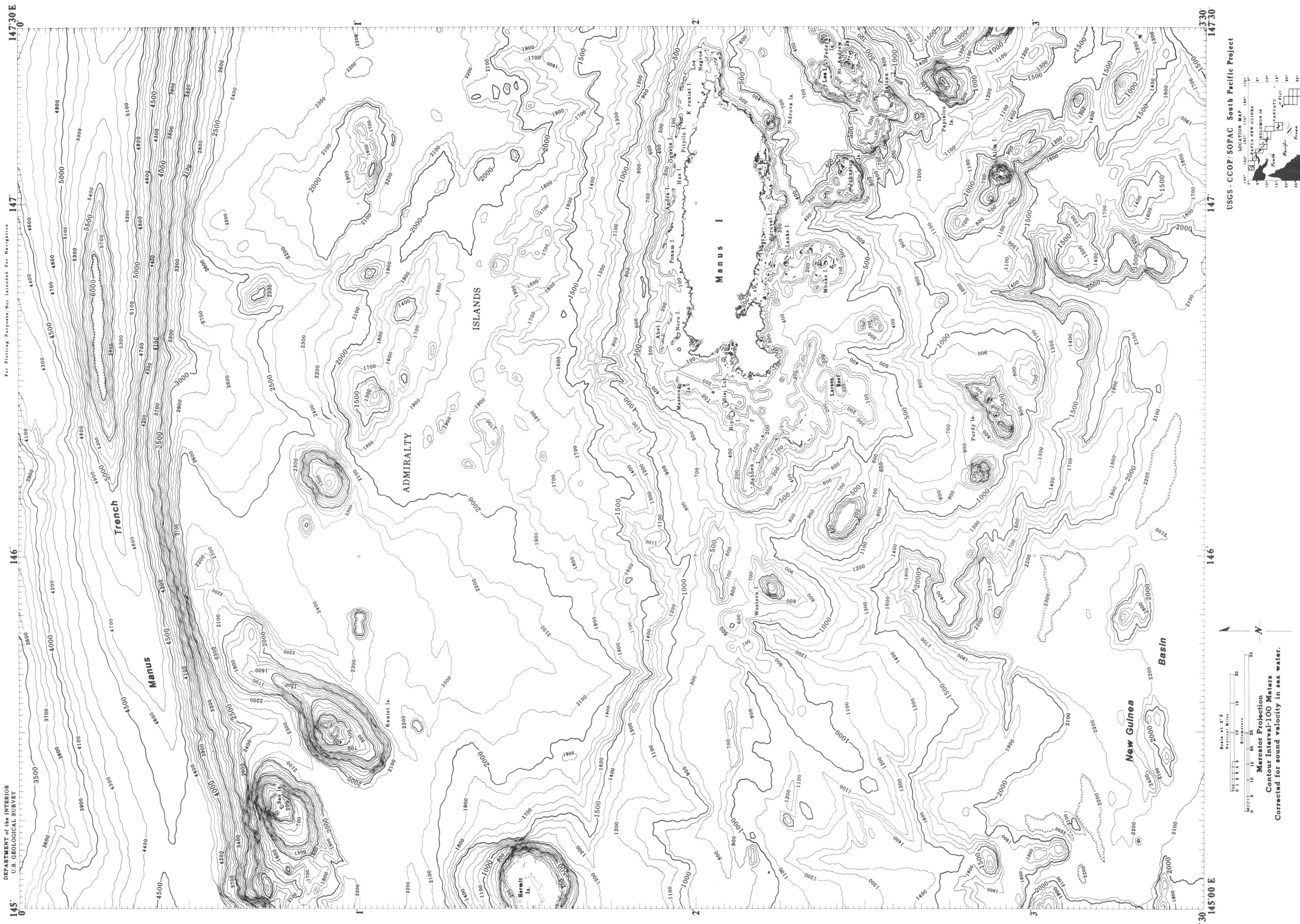

Figure 2. Detailed submarine topography of part of the northern region of Papua New Guinea. Contour interval is 100 m (based on depth-corrected acoustic soundings). Prepared by T.E. Chase, B.A. Seekins, J.D. Young, and S.V. Dadisman, 1986.

CHASE ET AL APPENDIX 1. SUBMARINE TOPOGRAPHY

Figure 1. Submarine topography of northeastern Papua New Guinea. Contour interval is 500 m (based on depth-corrected acoustic soundings). Detailed topography of the northern and eastern regions of Papua New Guinea is shown in Figures 2-4. Prepared by T.E. Chase, B.A. Seekins, J.D. Young, and S.V. Dadisman, 1987.

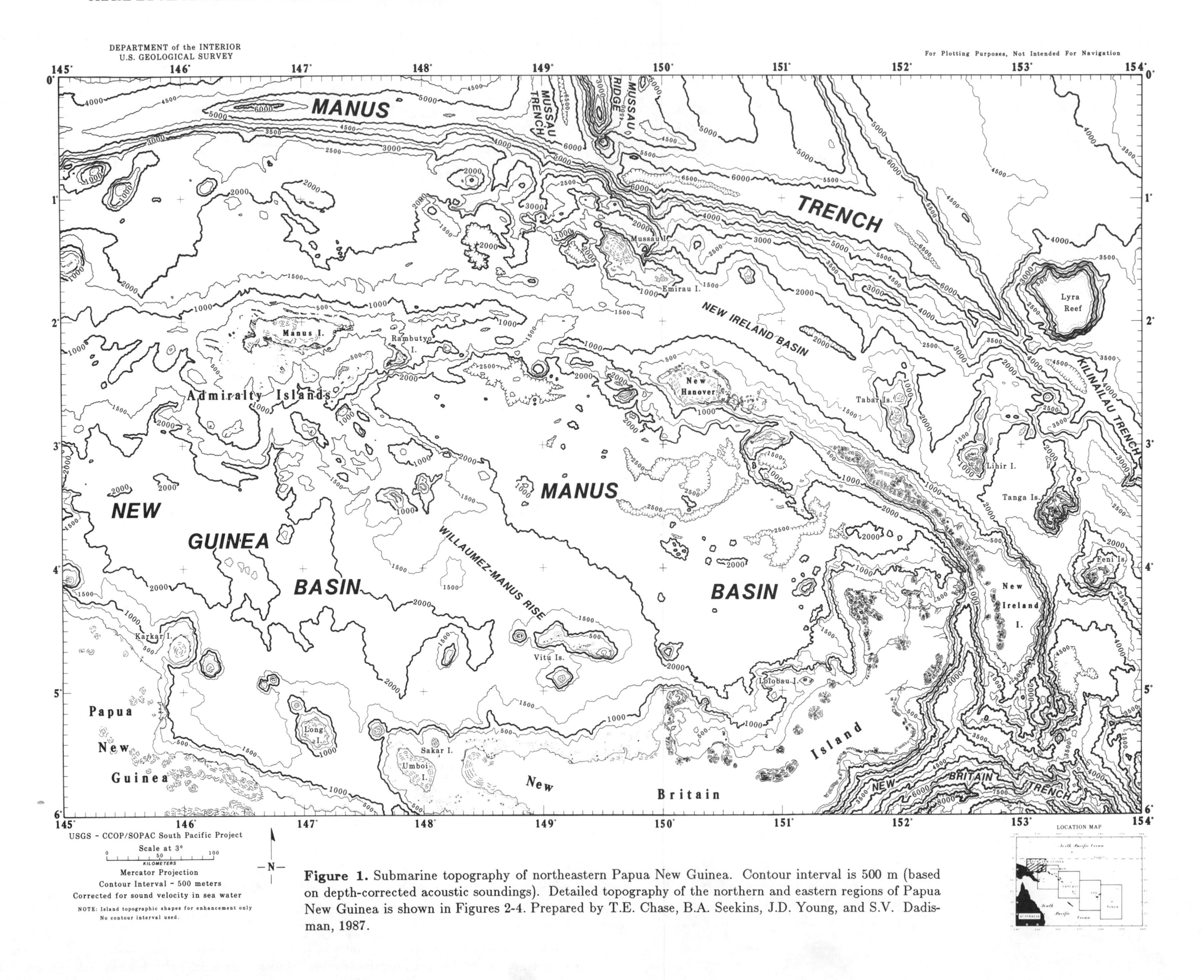

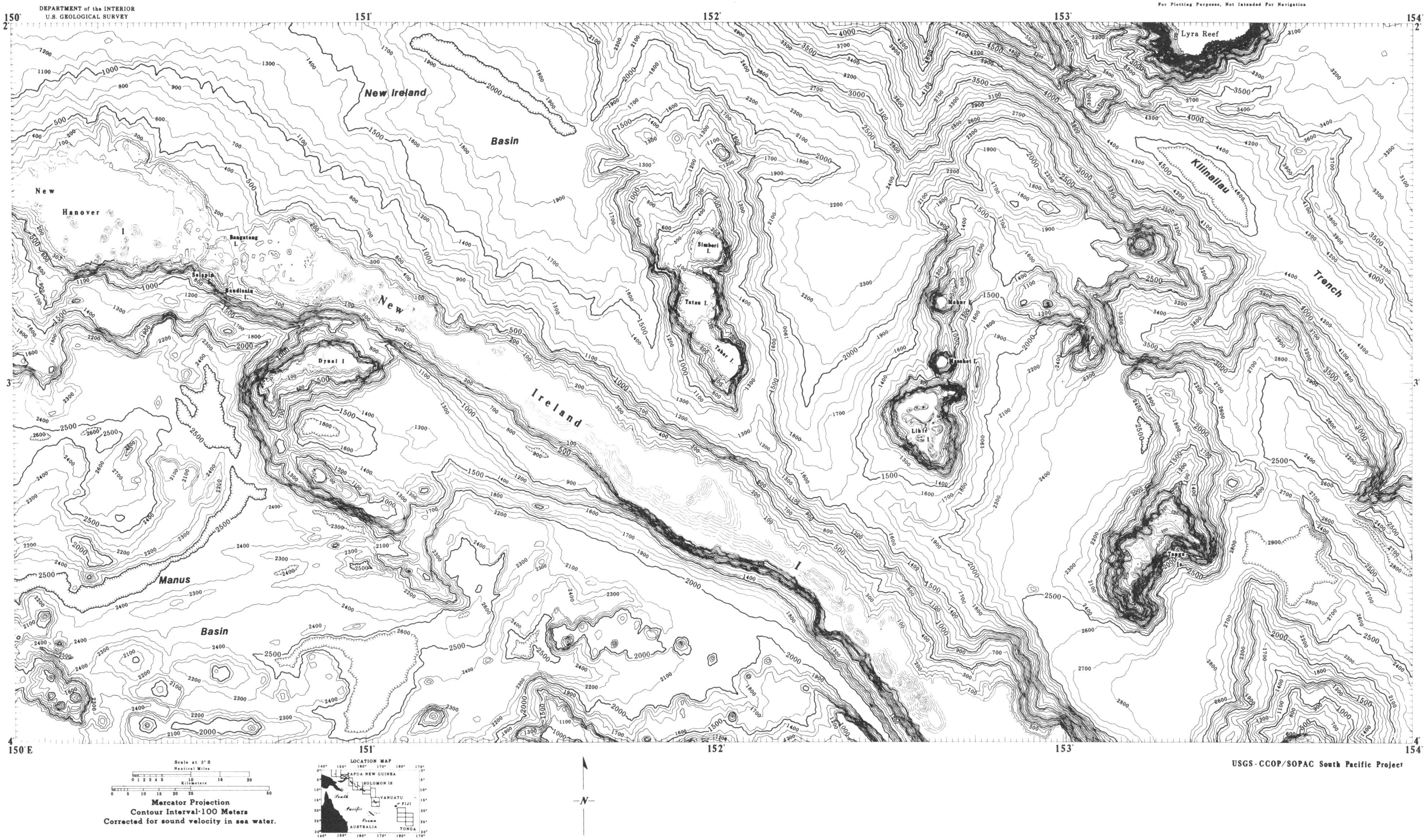

Figure 4. Detailed submarine topography of part of the eastern region of Papua New Guinea. Contour interval is 100 m (based on depth-corrected acoustic soundings). Prepared by T.E. Chase, B.A. Seekins, J.D. Young, and S.V. Dadisman, 1986.

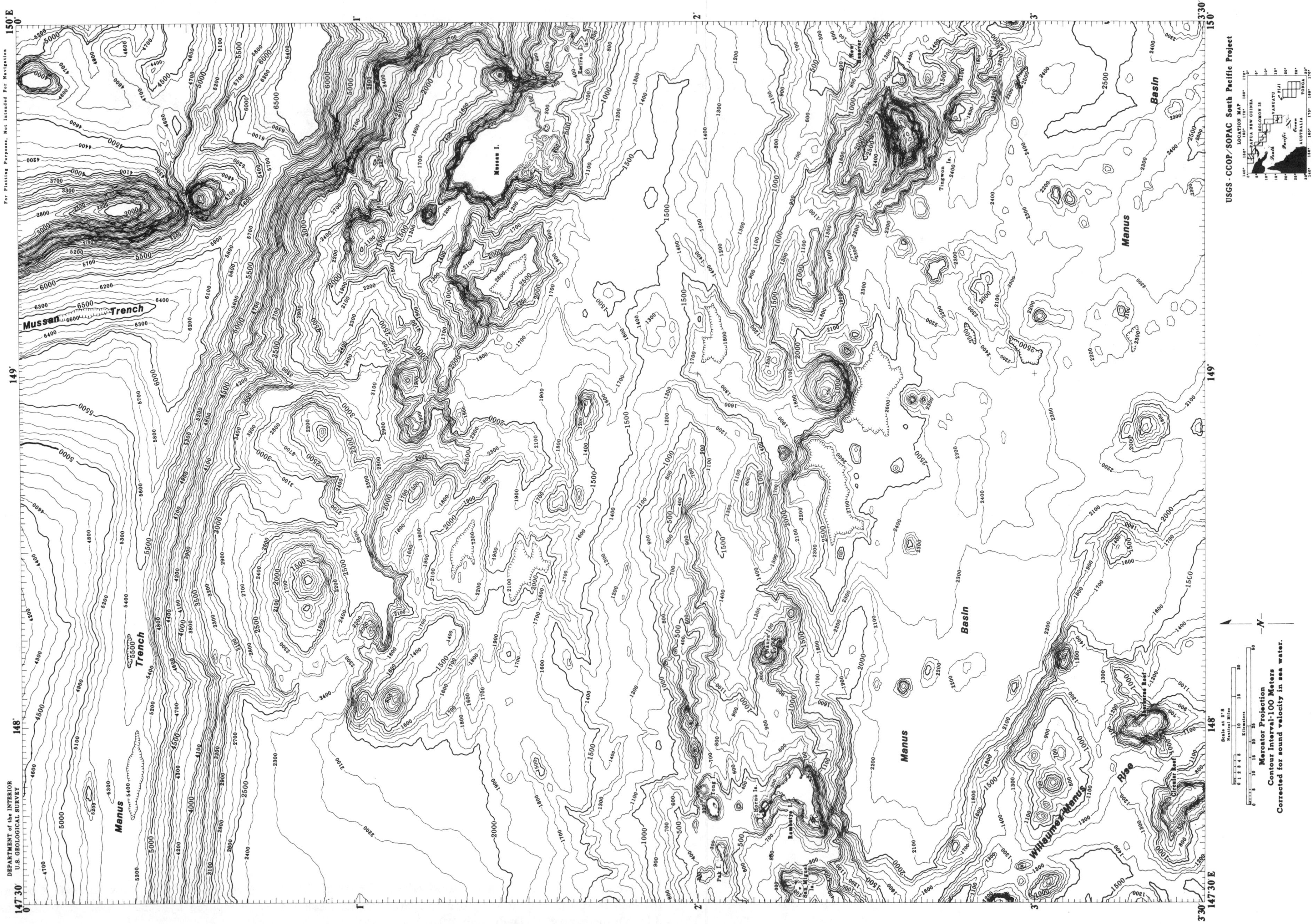

Figure 3. Detailed submarine topography of part of the northern region of Papua New Guinea. Contour interval is 100 m (based on depth-corrected acoustic soundings). Prepared by T.E. Chase, B.A. Seekins, J.D. Young, and S.V. Dadisman, 1986.

CHASE ET AL APPENDIX 1. SUBMARINE TOPOGRAPHY

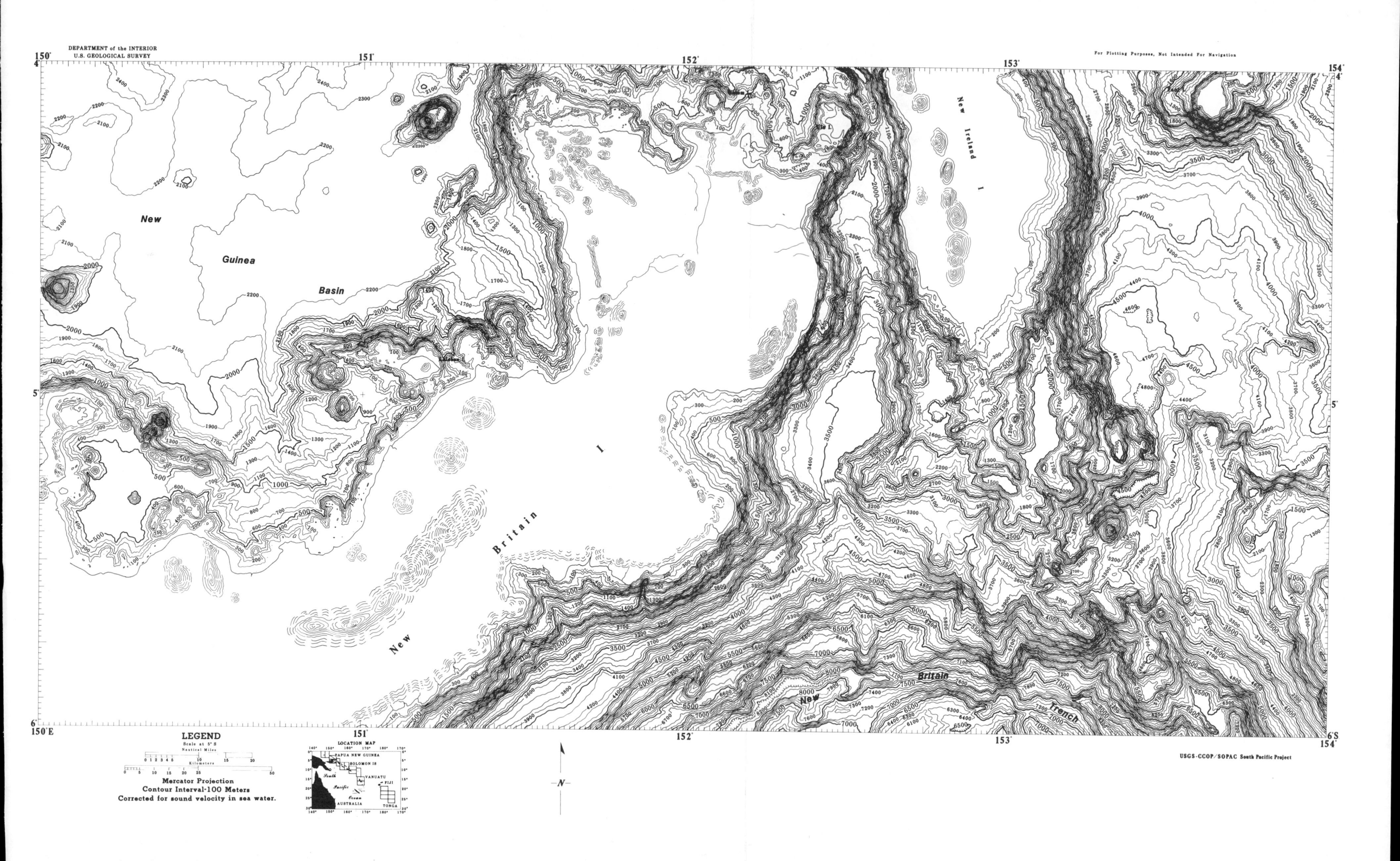

Figure 5. Detailed submarine topography of part of the eastern region of Papua New Guinea. Contour interval is 100 m (based on depth-corrected acoustic soundings). Prepared by T.E. Chase, B.A. Seekins, J.D. Young, and S.V. Dadisman, 1986.

Marlow, M.S., Dadisman, S.V., and Exon, N.F., editors, 1988, Geology and offshore resources of Pacific island arcs—New Ireland and Manus region, Papua New Guinea, Circum-Pacific Council for Energy and Mineral Resources Earth Science Series, v. 9: Houston, Texas, Circum-Pacific Council for Energy and Mineral Resources.

APPENDIX 2.

GEOPHYSICAL DATA NEAR MANUS AND NEW IRELAND ISLANDS, PAPUA NEW GUINEA

J.R. Childs, M.S. Marlow
U.S. Geological Survey, Menlo Park, California 94025, U.S.A.

INTRODUCTION

During the 1984 geophysical survey conducted by the U.S. Geological Survey in cooperation with CCOP/SOPAC near Manus and New Ireland Islands of Papua New Guinea, seismic-refraction (sonobuoy), magnetic and gravity profile data were collected in addition to multichannel seismic-reflection data. These data have been processed and edited, and are presented here to assist in interpretation of the multichannel seismic-reflection profiles.

The data include:

(1) Seismic-refraction data: a total of 12 sonobuoy stations were shot, extending in range between 15 and 40 km. The acoustic source for these profiles consisted of the 1300 cubic inch 5-airgun, tuned array; and receivers were U.S. Navy SSQ-41 sonobuoys. Solutions for sediment and crustal horizons are presented in Tables 1 and 2.

(2) Magnetic data were recorded continuously with a Geometrics proton-precession marine magnetometer, and reduced to residual anomalies.

(3) Gravity data were recorded continuously with a LaCoste and Romberg sea gravimeter on a two-axis stabilized platform, and reduced to free-air anomalies.

SONOBUOY DATA

Of the 15 refraction stations attempted during the survey, 12 were successful in recording refractions from sedimentary or crustal layers. The locations of these stations are shown in Figure 1. Recording and reduction of the data were done according to methods outlined by Childs and Cooper (1978). The inversion of the refraction data was by conventional ray-tracing calculations (Dobrin, 1976; Chapter 5) using travel times picked from the analog records. Because the sonobuoys were all collected in shallow water (usually less than 2 km), the wide-angle reflection data were of limited value, and were not used. The refraction solutions were corrected for dipping interfaces, calculated from time differences measured on the seismic-reflection records, which were converted to depth using the approximate velocity solutions calculated before dip correction. The refraction solutions are shown in Tables 1 and 2, and in Plates 1-5, and 7-9.

MAGNETICS

The magnetic field was measured with a proton-precession magnetometer towed 300 m aft of the ship, at a depth of about 20 m. Data were collected at a sampling interval of 4 seconds, with a sensitivity of 1 nanotesla (nT). Total-field measurements have been reduced to residual or anomalous

Table 1. Sonobuoy refraction velocity solutions. V and H represent velocity and thickness of successive layers, in km/s and km. Asterisks indicate an assumed the sea-floor refraction velocity.

SONO BUOY	WATER DEPTH	SEA FLOOR VSF	(H1)	V2	(H2)	V3	(H3)	V4	(H4)	V5	(H5)	V6	(H6)	CRUST V7
1	2.443	1.6*	(0.400)	2.99	(0.960)	3.44								
2	1.801	1.6*	(0.610)	2.25	(0.813)	3.09	(1.602)	4.65	(2.753)					5.38
4	1.724	1.6*	(0.562)	2.26	(0.748)	2.66	(2.167)	4.36						
5	1.876	1.6*	(0.392)	1.99	(0.411)	2.65	(1.669)	4.16	(2.728)					5.77
6	1.326	1.6*	(0.362)	1.89	(0.503)	2.65	(1.975)	4.53	(3.689)					6.06
8	0.997	1.6*	(0.803)	2.00	(0.305)	2.75	(1.369)	3.88	(1.109)	4.85	(1.421)			5.96
9	1.314	1.6*	(0.732)	2.41	(0.467)	2.93	(0.843)	3.75	(2.001)					5.86
11	1.537	1.6*	(0.338)	2.18	(0.752)	2.92	(0.642)	4.06	(0.802)	4.90	(2.205)			7.00
12	1.162	1.6*	(0.497)	1.82	(0.266)	2.45	(0.732)	2.68	(0.746)	3.53	(1.164)	4.29		
13	1.387	1.6*	(0.254)	2.41	(0.982)	3.32	(1.459)	4.57	(3.275)					5.64
14	1.012	1.6*	(0.534)	2.07	(0.515)	3.95	(1.485)							6.28
15	1.065	1.6*	(0.355)	2.18	(0.200)	2.77	(1.428)	4.70						

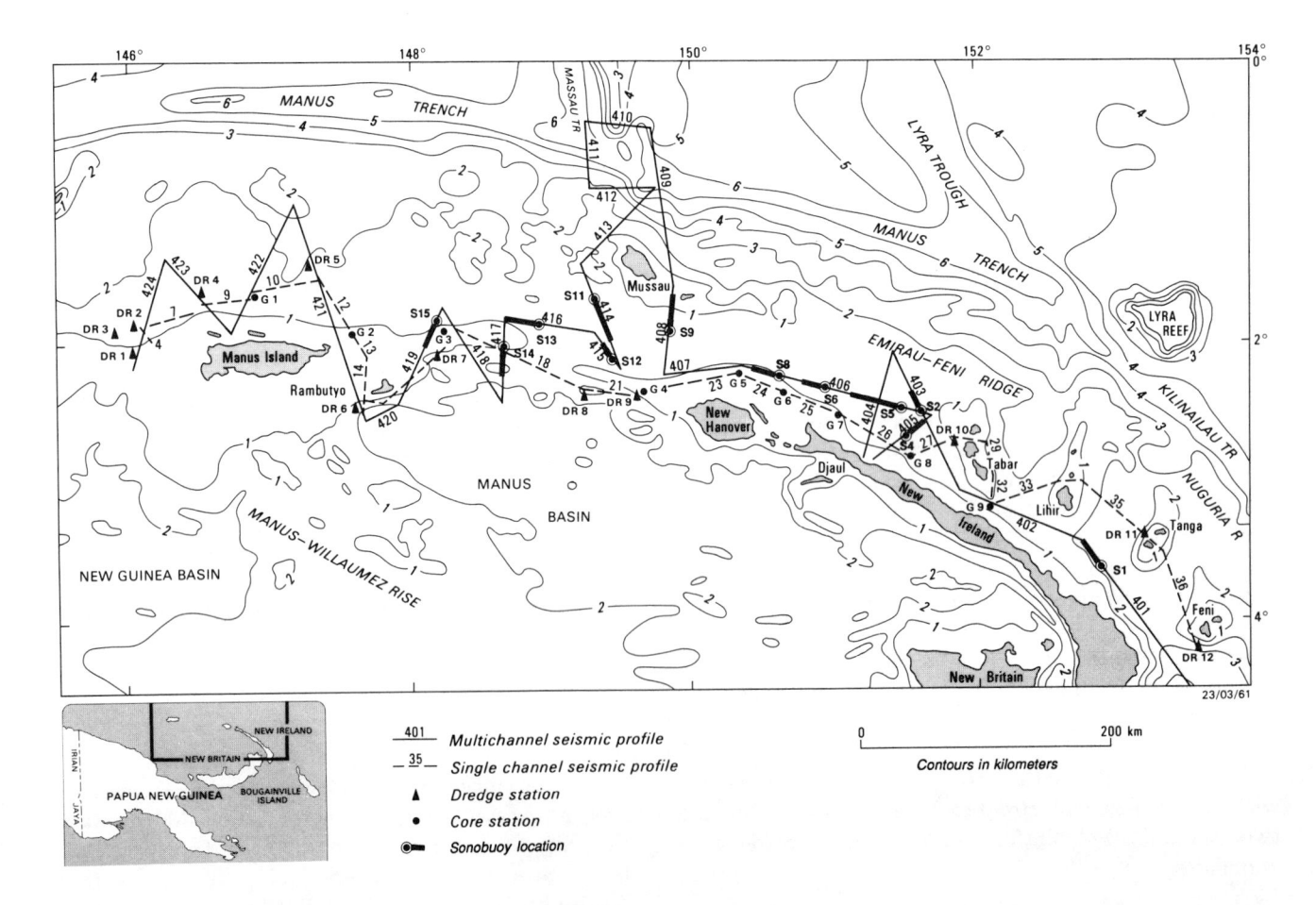

VSF = velocity of the seafloor; H1 = thickness of horizon 1 (km); V5 = velocity of horizon 5 (km/s).

Figure 1. Map of the study area showing geophysical tracklines and locations of sonobuoy refraction profiles.

APPENDIX 2. GEOPHYSICAL DATA

Table 2. Refraction results.

Sono # / Line # / SP #	Layer	Intercept (s)	Depth (km)	Thickness (km)	Velocity (km/s)	Travel Time Inc (s)	Sum Travel Time (s)	Slope (Degrees)
1 / 401-402 / 3310-0010	1	1.134	2.443	2.443	1.600	3.257	0.000	0.000
	2	3.239	2.843	0.400	2.989	0.500	0.500	0.917
	3	3.693	3.803	0.960	3.443	0.643	1.142	0.000
2 / 403 / 1226-1588	1	0.835	1.801	1.801	1.600	2.401	0.000	-0.215
	2	2.323	2.411	0.610	2.247	0.762	0.762	0.137
	3	3.247	3.224	0.813	3.086	0.724	1.486	0.128
	4	4.398	4.826	1.602	4.646	1.038	2.524	0.000
	5	5.142	7.578	2.752	5.383	1.185	3.709	4.431
4 / 405 / 757-1082	1	0.801	1.724	1.724	1.600	2.298	0.000	-0.516
	2	2.216	2.285	0.562	2.262	0.702	0.702	0.229
	3	2.803	3.033	0.748	2.655	0.662	1.363	-0.451
	4	4.670	5.200	2.167	4.355	1.632	2.996	0.300
5 / 406 / 485-1255	1	0.870	1.876	1.876	1.600	2.502	0.000	-0.300
	2	1.931	2.268	0.392	1.987	0.490	0.490	-0.321
	3	2.728	2.679	0.411	2.652	0.414	0.904	0.000
	4	4.119	4.349	1.669	4.158	1.259	2.163	0.000
	5	5.302	7.077	2.728	5.768	1.312	3.475	0.000
6 / 406 / 1723-2155	1	0.616	1.326	1.326	1.600	1.768	0.000	0.000
	2	1.322	1.688	0.362	1.894	0.452	0.452	0.000
	3	2.188	2.191	0.503	2.647	0.531	0.983	0.000
	4	3.786	4.166	1.975	4.534	1.492	2.475	0.000
	5	5.076	7.855	3.689	6.061	1.627	4.102	0.000
8 / 406 / 2534-3075	1	0.463	0.997	0.997	1.600	1.330	0.000	0.258
	2	1.479	1.801	0.803	1.997	1.004	1.004	0.000
	3	2.139	2.106	0.305	2.745	0.305	1.309	0.000
	4	3.108	3.474	1.369	3.882	0.997	2.306	0.000
	5	3.655	4.583	1.109	4.850	0.571	2.878	0.000
	6	4.201	6.004	1.421	5.960	0.586	3.464	0.000
9 / 408 / 568-1040	1	0.608	1.314	1.314	1.599	1.752	0.000	1.289
	2	2.058	2.046	0.732	2.413	0.916	0.916	1.373
	3	2.489	2.513	0.467	2.925	0.387	1.302	0.000
	4	3.088	3.355	0.843	3.746	0.576	1.879	2.139
	5	4.247	5.357	2.001	5.863	1.068	2.947	0.000
11 / 414 / 640-1220	1	0.714	1.537	1.537	1.600	2.049	0.000	-0.430
	2	1.772	1.875	0.338	2.178	0.423	0.423	-0.458
	3	2.572	2.627	0.752	2.921	0.691	1.113	0.369
	4	3.180	3.269	0.642	4.057	0.440	1.553	-1.315
	5	3.543	4.071	0.802	4.897	0.395	1.949	-1.549
	6	4.434	6.277	2.205	6.995	0.901	2.849	1.887
12 / 415 / 160-460	1	0.540	1.162	1.162	1.600	1.550	0.000	-0.430
	2	1.169	1.659	0.497	1.817	0.621	0.621	1.008
	3	1.890	1.925	0.266	2.447	0.293	0.914	0.419
	4	2.243	2.658	0.732	2.682	0.599	1.513	-0.428
	5	3.000	3.404	0.746	3.530	0.556	2.069	0.313
	6	3.593	4.568	1.164	4.288	0.660	2.728	0.416
13 / 416 / 588-1052	1	0.644	1.387	1.387	1.600	1.849	0.000	0.086
	2	1.680	1.640	0.254	2.411	0.317	0.317	-5.393
	3	2.487	2.623	0.982	3.320	0.815	1.132	4.878
	4	3.339	4.082	1.459	4.565	0.879	2.011	2.454
	5	4.378	7.357	3.275	5.642	1.435	3.446	6.077
14 / 417 / 493-870	1	0.471	1.012	1.012	1.600	1.349	0.000	-2.147
	2	1.353	1.545	0.534	2.072	0.667	0.667	2.107
	3	2.280	2.061	0.515	3.948	0.498	1.164	0.236
	4	3.008	3.546	1.485	6.278	0.752	1.917	2.256
15 / 419 / 55-530	1	0.493	1.065	1.065	1.600	1.421	0.000	-1.030
	2	1.330	1.421	0.355	2.175	0.444	0.444	-0.733
	3	1.670	1.620	0.200	2.767	0.184	0.628	0.483
	4	2.761	3.048	1.428	4.698	1.032	1.660	1.829

magnetic values by removal of the 1980 International Gravity Reference Field (IGRF) (IAGA, 1976). Crossing differences at 12 line intersections averaged 36 nT, with a standard deviation of 39 nT. Absolute differences were between 1 and 125 nT. The greatest crossing errors were much too large to be explained by navigational errors. Significant temporal variations (magnetic storms and diurnal changes) in the local magnetic field must have occurred during the survey. In the absence of gradiometer data or a nearby station magnetometer, the magnetic data must be interpreted with caution.

GRAVITY

Gravity data were originally collected at a sampling interval of 20 seconds, and desampled to two minutes or greater for plotting and modelling. The data have been Eötvös corrected, and reduced to free-air anomalies. Cross-coupling analysis and correction were considered unnecessary because of calm sea conditions during the survey. Absolute gravity values were obtained by tying to the International Gravity Standardization Net of 1971 (IGSN-71) (International Association of Geodesy, 1974). Free-air anomalies were calculated by removal of the 1967 reference field (International Association of Geodesy, 1971). The data were edited to remove spurious values at turns, and to remove erroneous Eötvös corrections introduced by navigational errors. Crossing differences at 18 line intersections averaged 4.9 milligals (mgal), with a standard deviation of 5.5 mgal. Absolute differences were between 0 and 20 mgal. Crossing errors are assumed to result from navigational errors, although cross-coupling errors may have also contributed to the differences.

PROFILES

Magnetic and gravity profiles for each survey line are shown in Plates 1-12. Refraction solutions are also displayed with interpretated line drawings of selected multichannel seismic-reflection lines.

REFERENCES

Childs, J. C. and A. K. Cooper, 1978, Collection, reduction, and interpretation of marine seismic sonobuoy data: U.S. Geological Survey Open File Report 78-442, 219 p.

Dobrin, M. B., 1976, Introduction to Geophysical Prospecting: New York, McGraw-Hill Book Company, 630 p.

International Association of Geodesy, 1971, Geodetic reference system 1967: Special Publication 3, 116 p.

International Association of Geodesy, 1974, The International Gravity Standardization Net 1971 (IGSN-71): Special Publication 4, 194 p.

IAGA Division 1 Study Group, 1976, International geomagnetic reference field: Transactions: American Geophysical Union, v. 57, p. 120-121.

APPENDIX 2. GEOPHYSICAL DATA

Plate 1. Refraction results for sonobuoy 1 and magnetic and gravity profiles along with an interpreted line drawing of multichannel seismic-reflection line 401. Top profile shows refraction data plotted over a line drawing of the sea-floor and subbottom reflection events. The refraction solutions from thin layers have been deleted for legibility; refer to Table 1 for complete solutions. Middle profile shows free-air gravity anomaly data. Bottom profile shows residual magnetic anomaly data.

Plate 2. Refraction results for sonobuoy 2 and magnetic and gravity profiles along with interpreted line drawings of multichannel seismic-reflection lines 402 and 403. Top profiles show refraction data plotted over line drawings of the sea-floor and subbottom reflection events. Middle profiles show free-air gravity anomaly data. Bottom profiles show residual magnetic anomaly data.

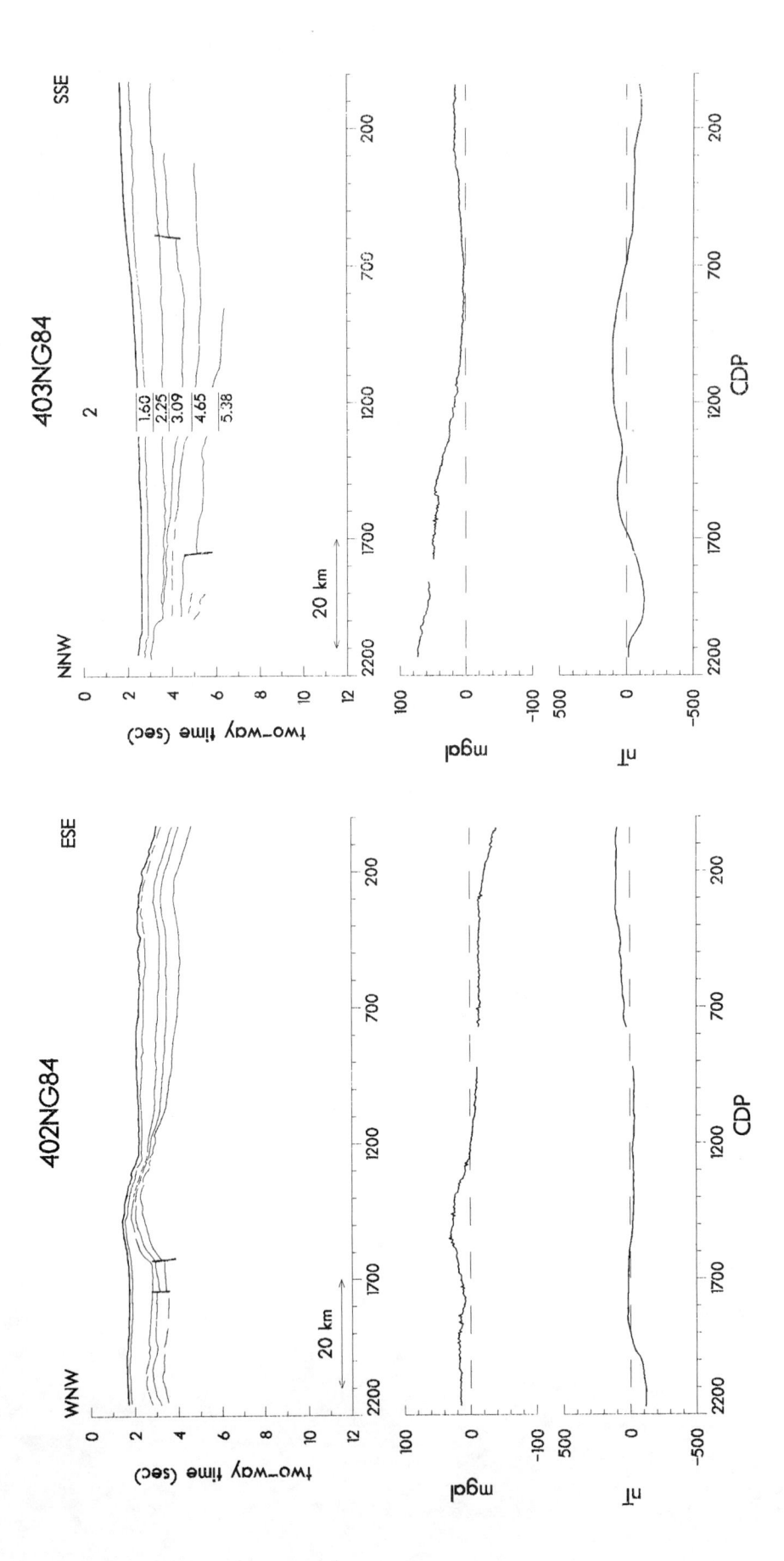

APPENDIX 2. GEOPHYSICAL DATA

Plate 3. Refraction results for sonobuoy 4 and magnetic and gravity profiles of lines 404 and 405 along with interpreted line drawing of multichannel seismic-reflection line 404. Top profiles show refraction data plotted or line drawings of the sea-floor and subbottom reflection events. Middle profiles show free-air gravity anomaly data. Bottom profiles show residual magnetic anomaly data.

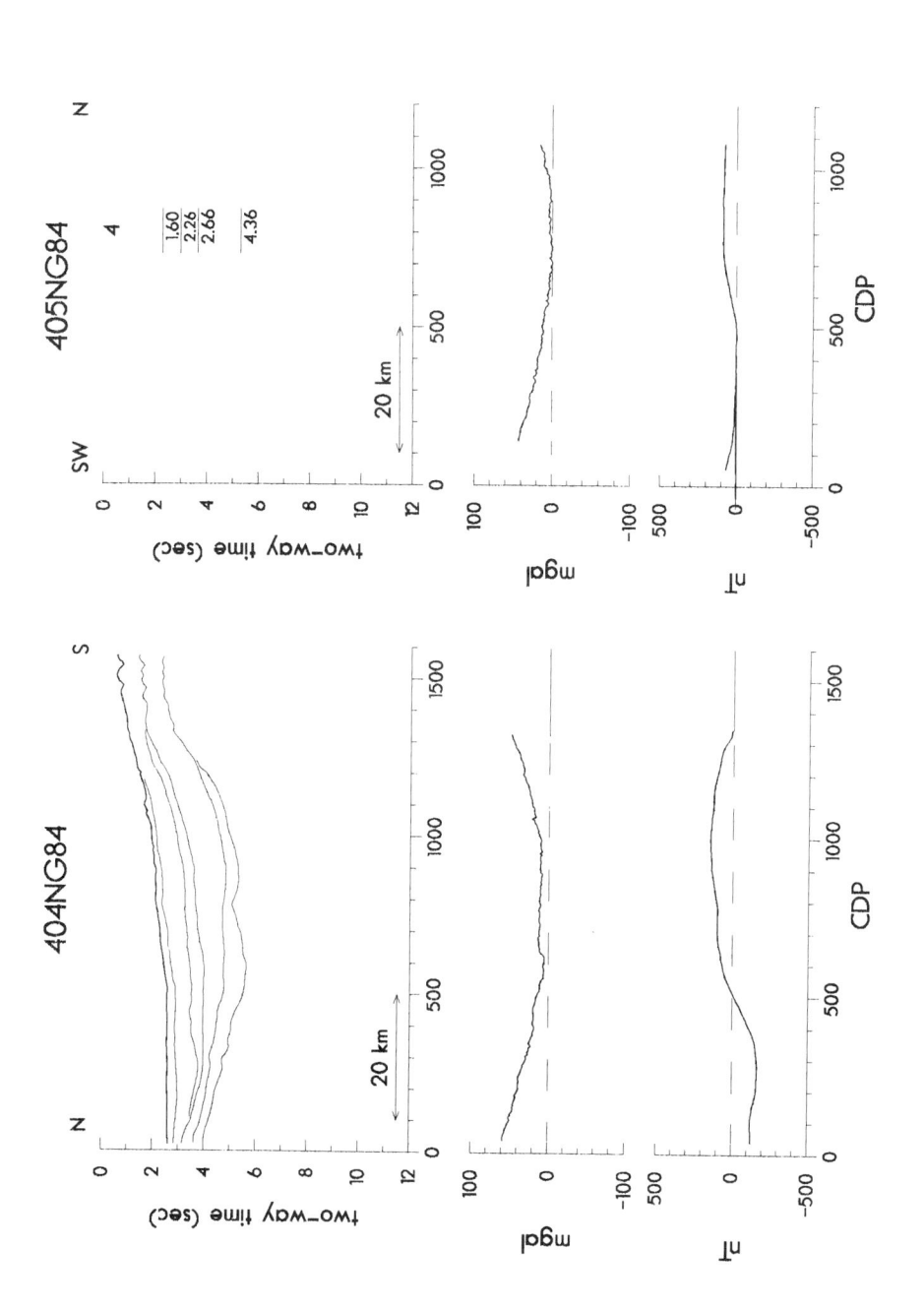

Plate 4. Refraction results for sonobuoys 5, 6, and 8 and magnetic and gravity profiles along with interpreted line drawings of multichannel seismic-reflection lines 406 and 407. Top profiles show refraction data plotted over line drawings of the sea-floor and subbottom reflection events. The refraction solutions from thin layers have been deleted for legibility; refer to Table 1 for complete solutions. Middle profiles show free-air gravity anomaly data. Bottom profiles show residual magnetic anomaly data.

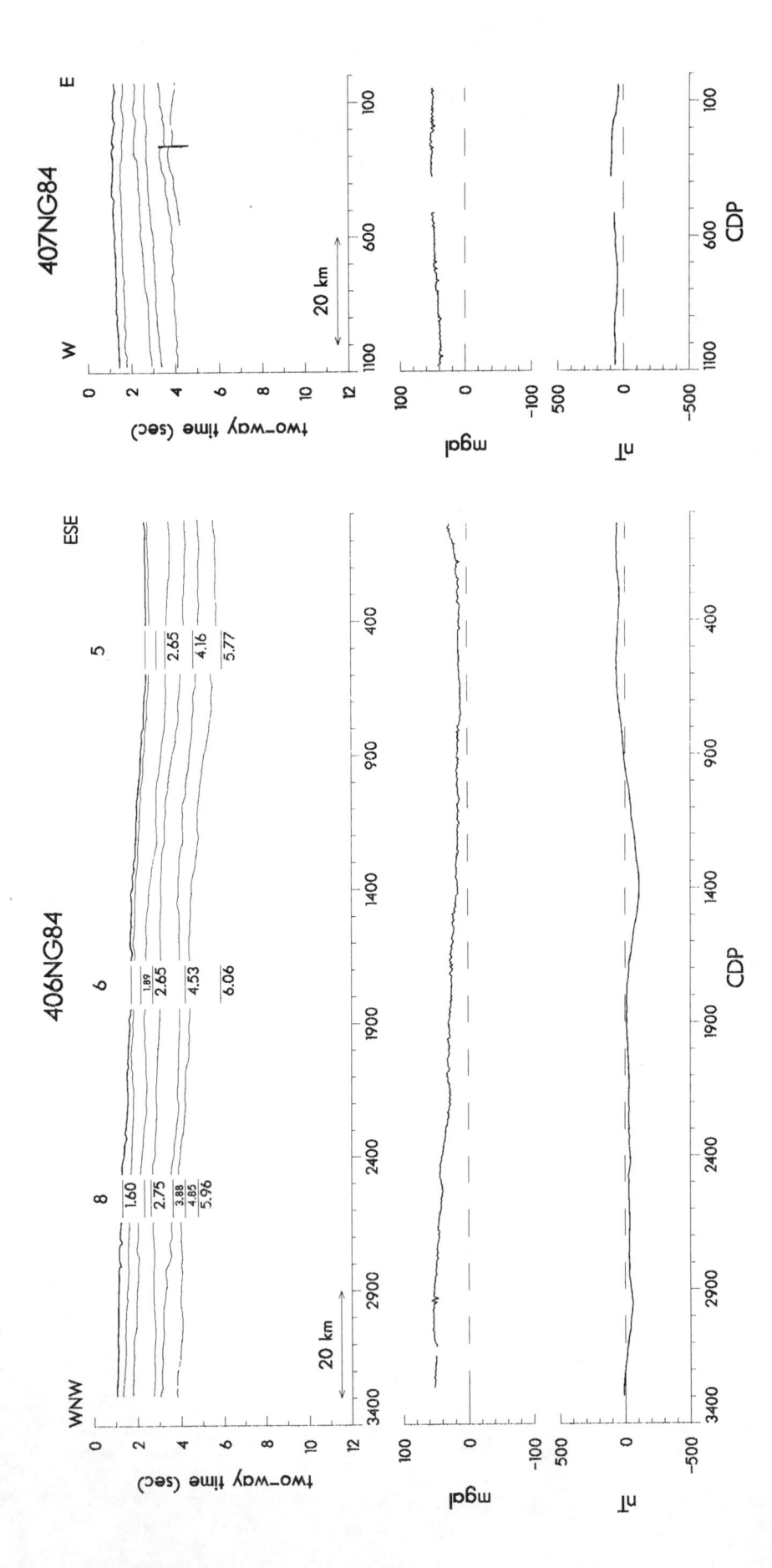

APPENDIX 2. GEOPHYSICAL DATA 281

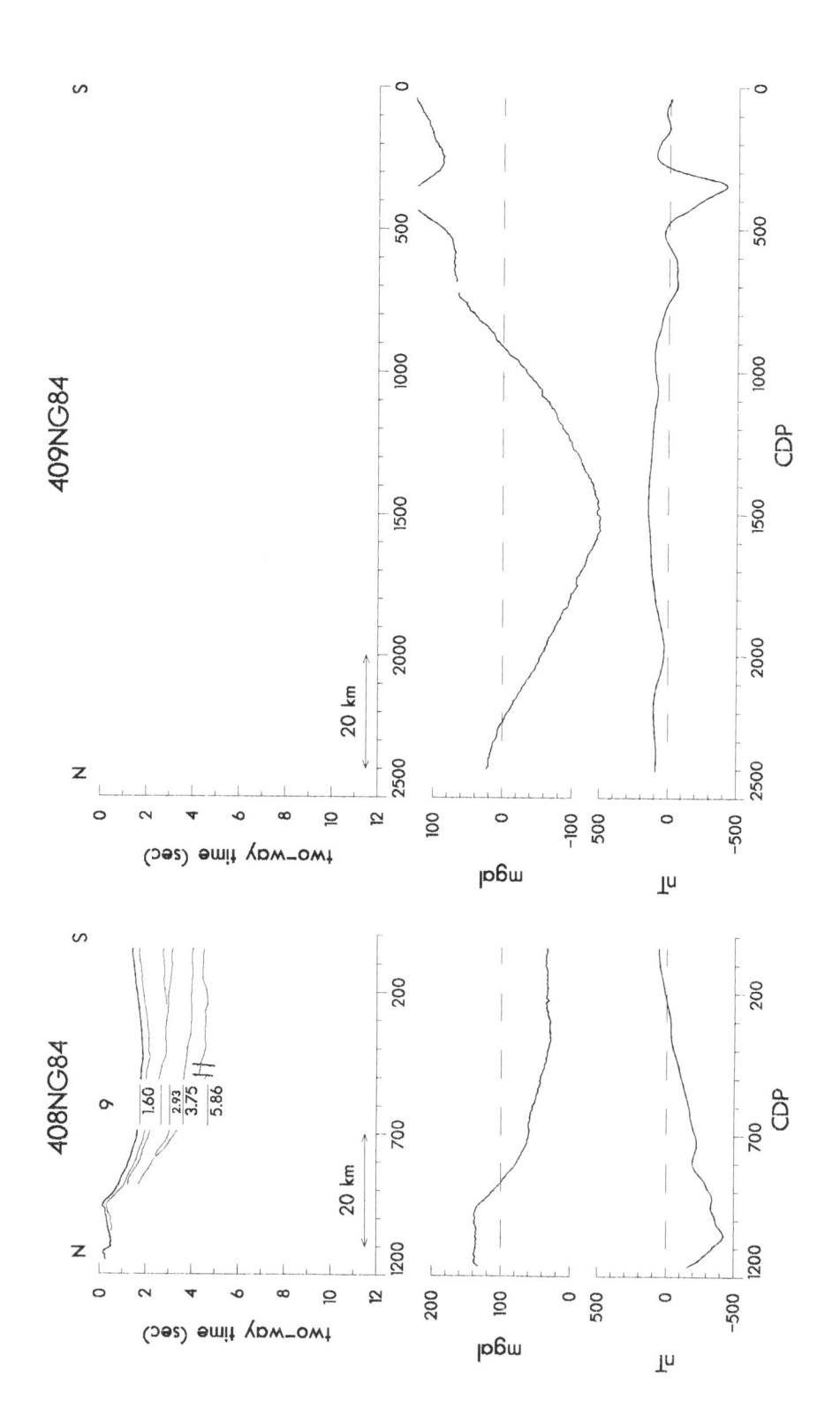

Plate 5. Refraction results for sonobuoy 9 and magnetic and gravity profiles of lines 408 and 409 along with an interpreted line drawing of multichannel seismic-reflection line 408. Top profile shows refraction data plotted over a line drawing of the sea-floor and subbottom reflection events on line 408. The refraction solutions from thin layers have been deleted for legibility; refer to Table 1 for complete solutions. Middle profiles show free-air gravity anomaly data. Bottom profiles show residual magnetic anomaly data.

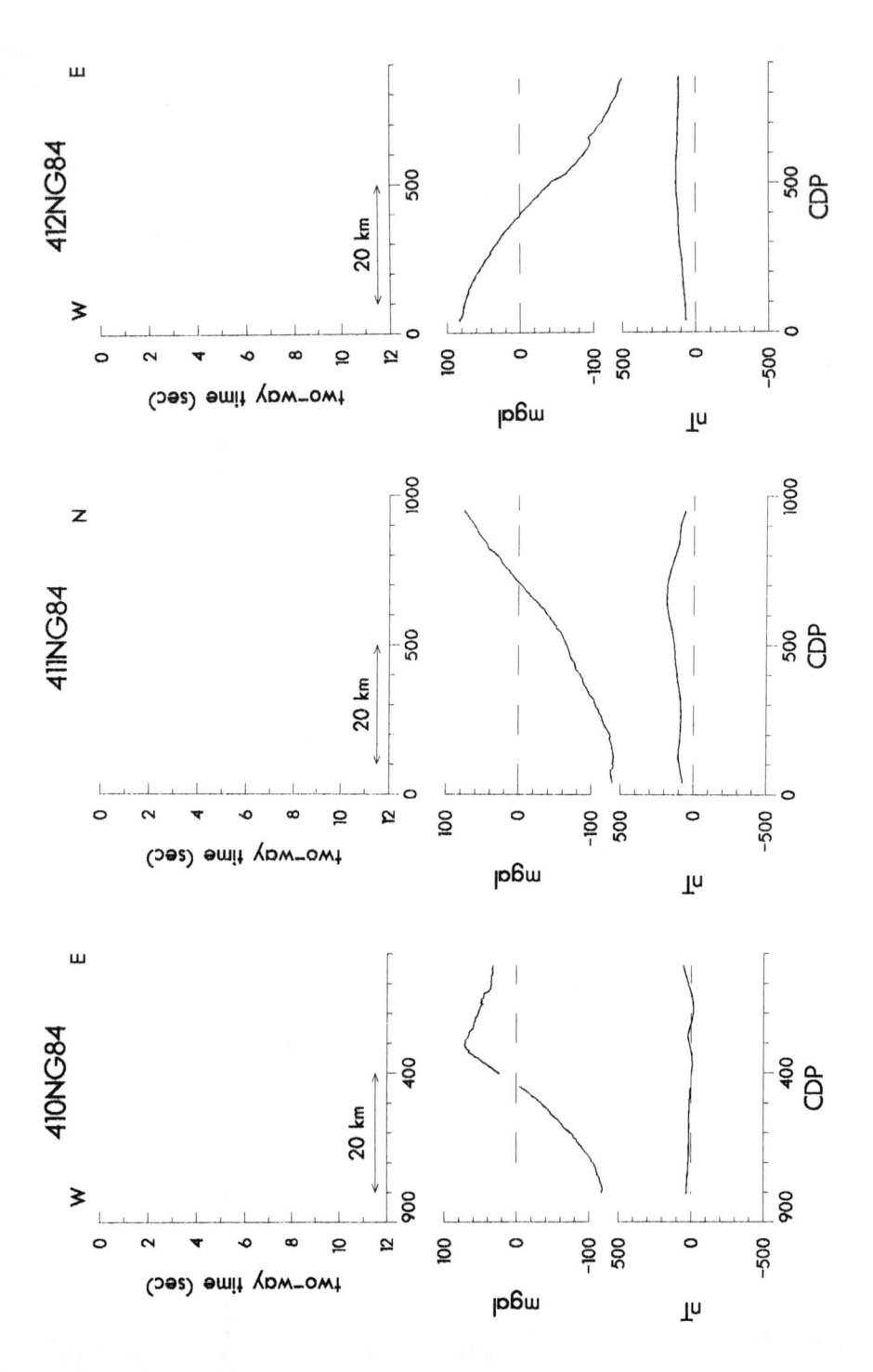

Plate 6. Magnetic and gravity profiles of lines 410, 411 and 412. Middle profiles show free-air gravity anomaly data. Bottom profiles show residual magnetic anomaly data.

APPENDIX 2. GEOPHYSICAL DATA

Plate 7. Refraction results for sonobuoys 11 and 12 and magnetic and gravity profiles of lines 413, 414, and 415 along with interpreted line drawings of multichannel seismic-reflection lines 414 and 415. Top profiles show refraction data plotted over line drawings of the sea-floor and subbottom reflection events on lines 414 and 415. The refraction solutions from thin layers have been deleted for legibility; refer to Table 1 for complete solutions. Middle profiles show free-air gravity anomaly data. Bottom profiles show residual magnetic anomaly data.

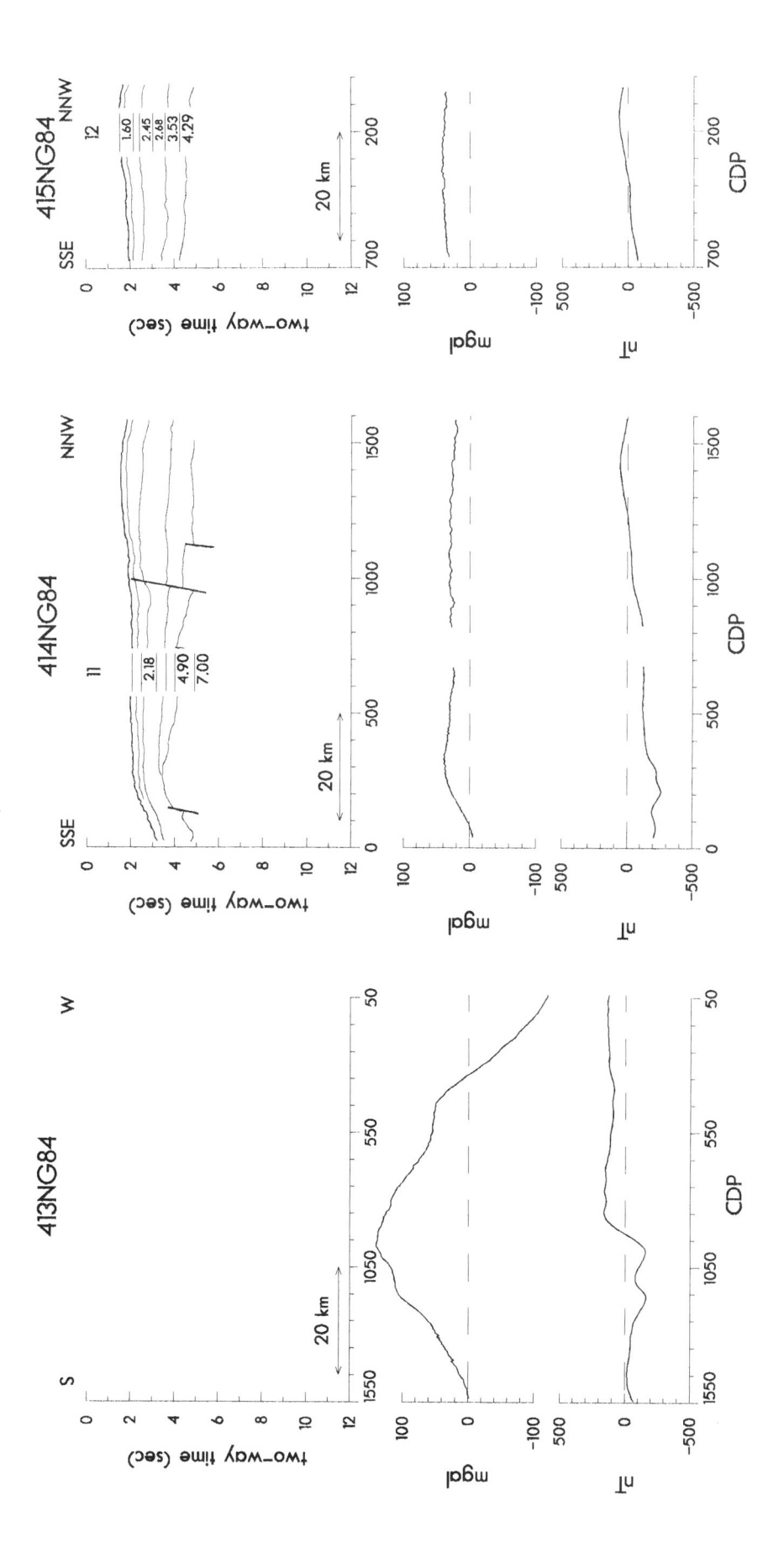

Plate 8. Refraction results for sonobuoys 13 and 14 and magnetic and gravity profiles along with interpreted line drawings of multichannel seismic-reflection lines 416, 417 and 418. Top profiles show refraction data plotted over line drawings of the sea-floor and subbottom reflection events. The refraction solutions from thin layers have been deleted for legibility; refer to Table 1 for complete solutions. Middle profiles show free-air gravity anomaly data. Bottom profiles show residual magnetic anomaly data.

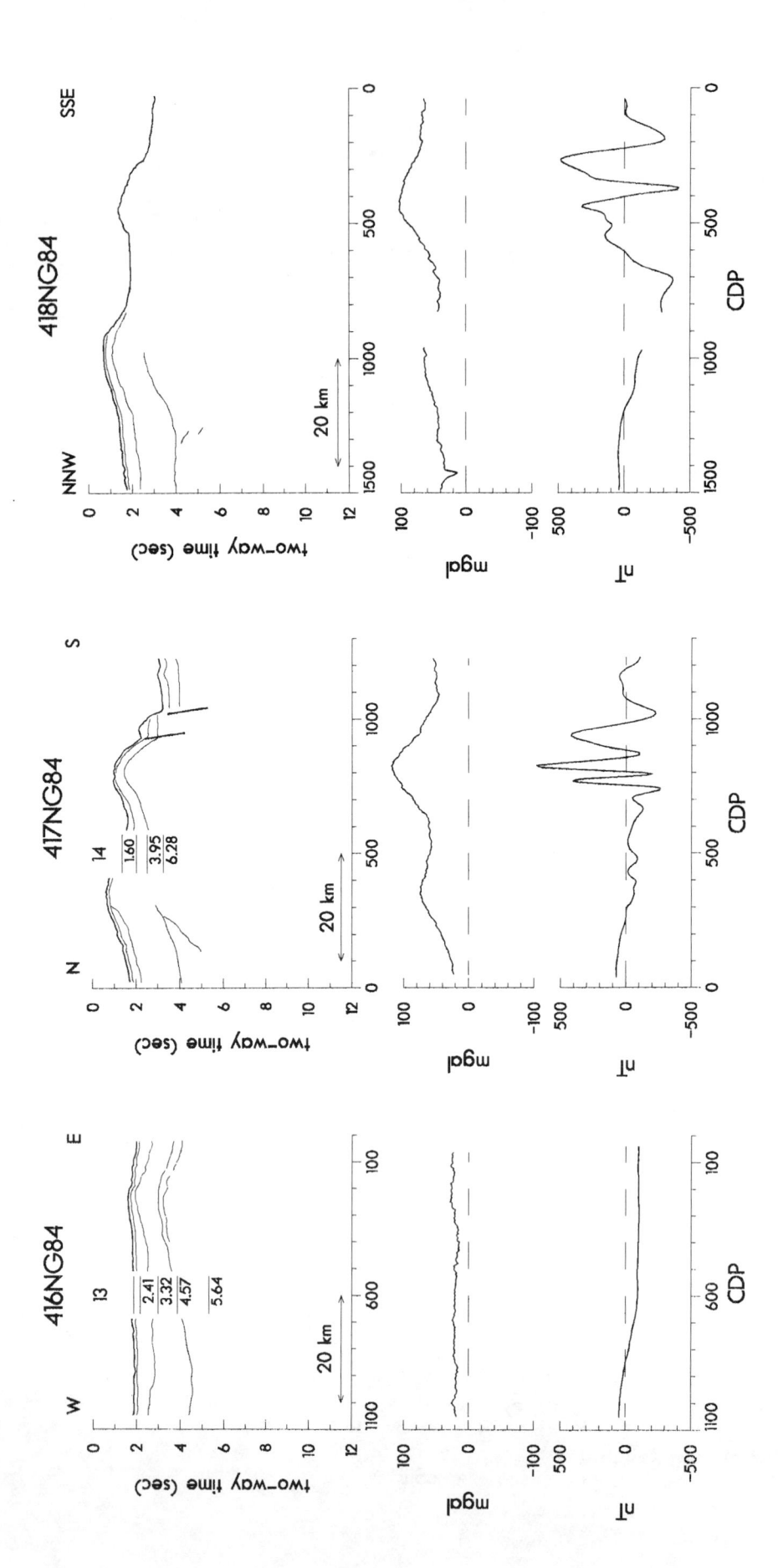

APPENDIX 2. GEOPHYSICAL DATA

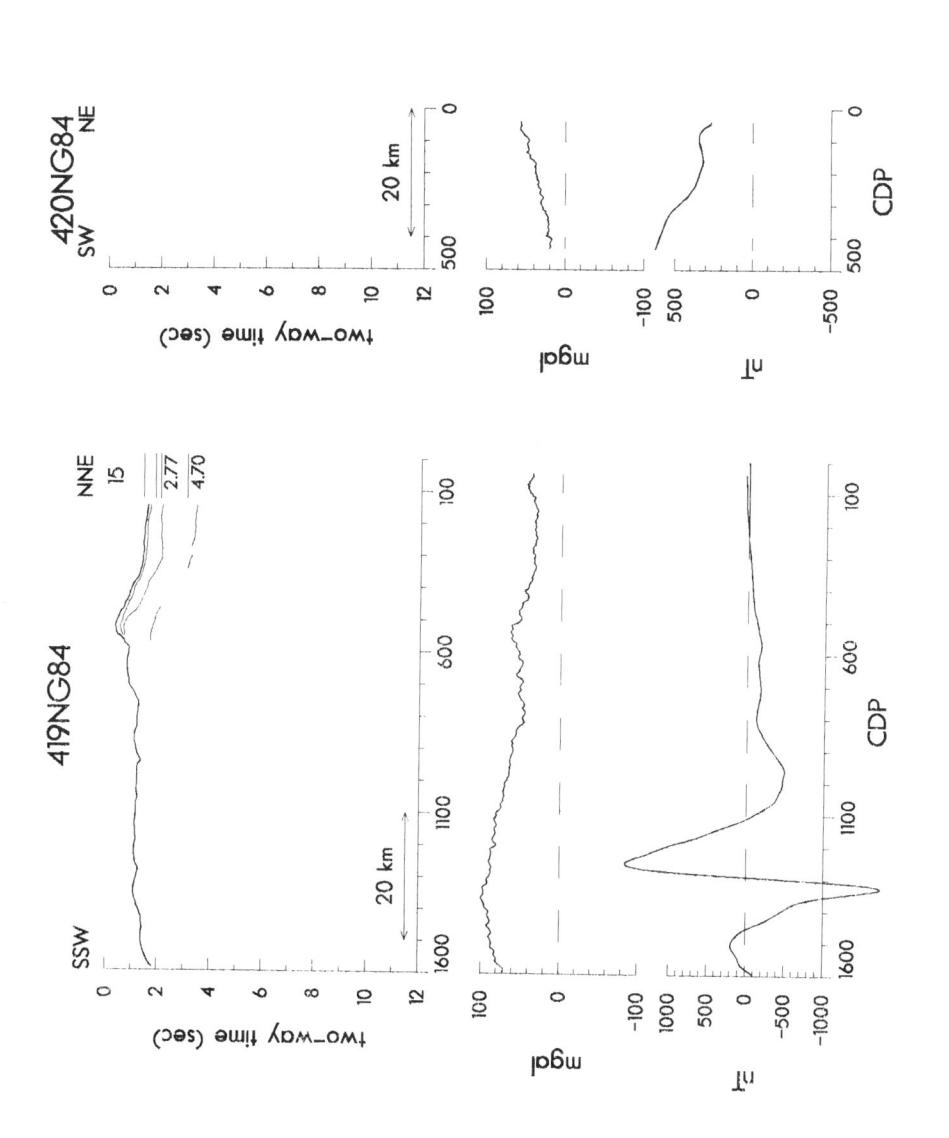

Plate 9. Refraction results for sonobuoy 15 and magnetic and gravity profiles of lines 419 and 420 along with an interpreted line drawing of multichannel seismic-reflection line 419. Top profile shows refraction data plotted over a line drawing of the sea floor and subbottom reflection events. The refraction solutions from thin layers have been deleted for legibility; refer to Table 1 for complete solutions. Middle profiles show free-air gravity anomaly data. Bottom profiles show residual magnetic anomaly data.

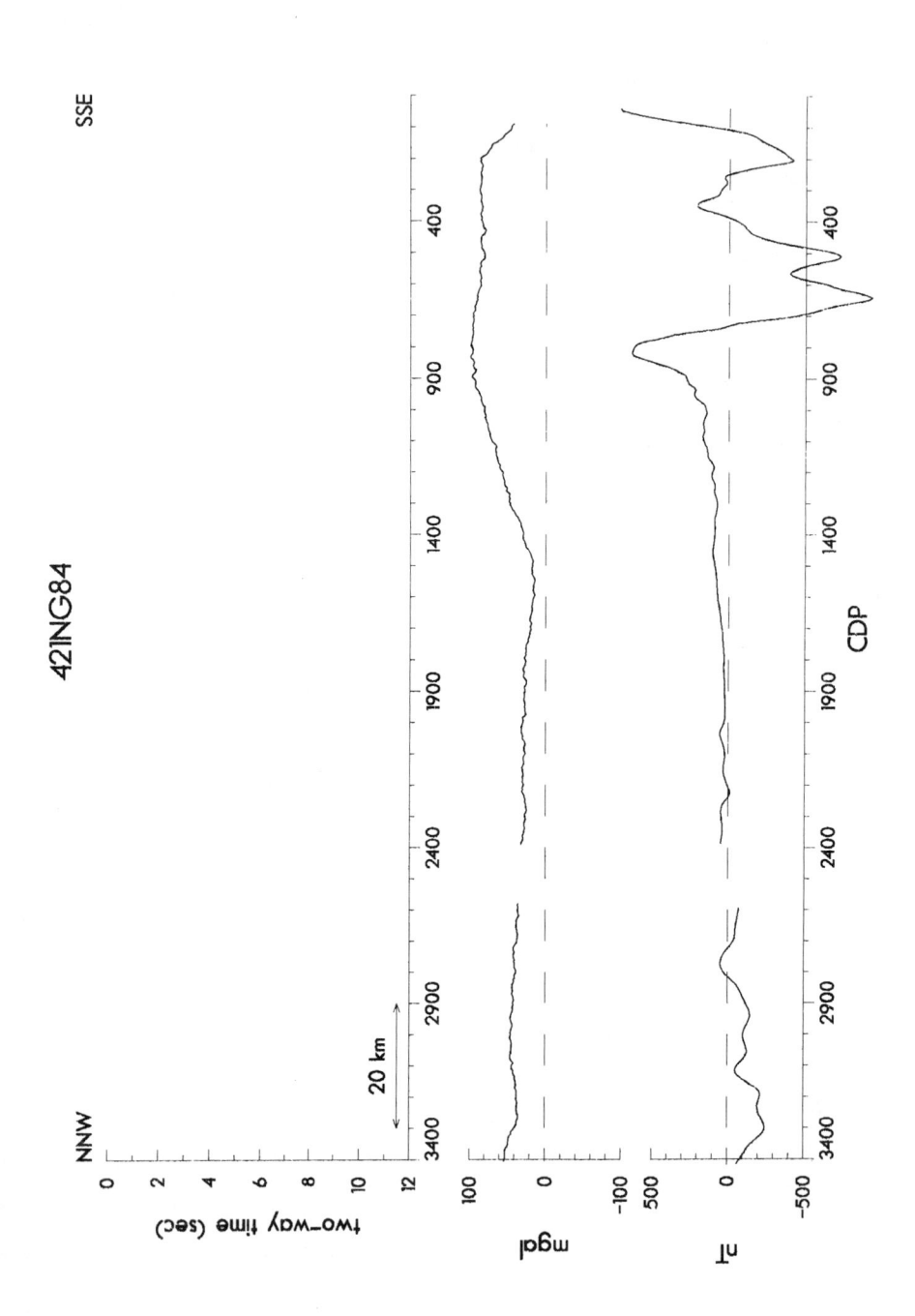

Plate 10. Magnetic and gravity profiles of line 421. Middle profile shows free-air gravity anomaly data. Bottom profile shows residual magnetic anomaly data.

APPENDIX 2. GEOPHYSICAL DATA

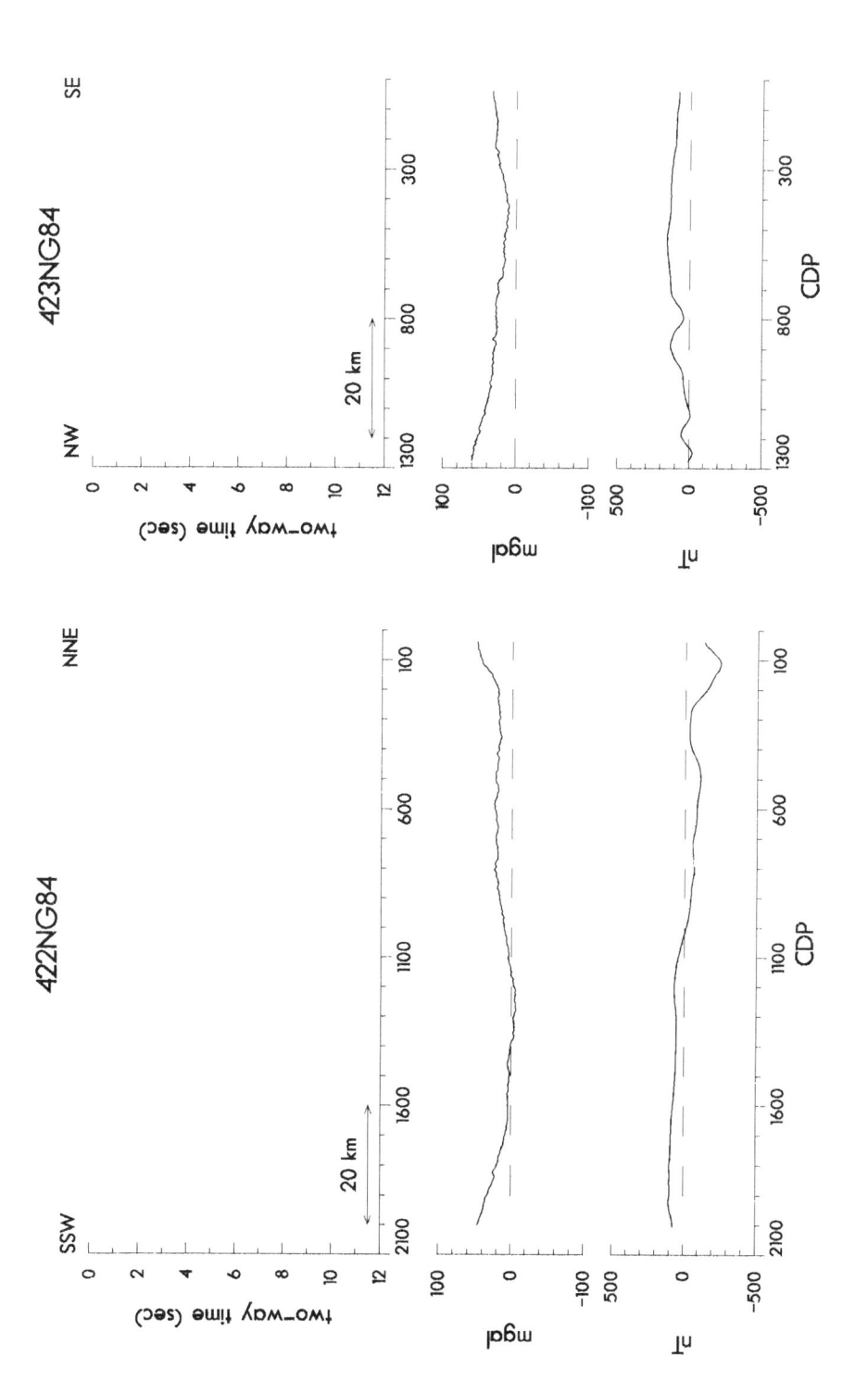

Plate 11. Magnetic and gravity profiles of lines 422 and 423. Middle profiles show free-air gravity anomaly data. Bottom profiles show residual magnetic anomaly data.

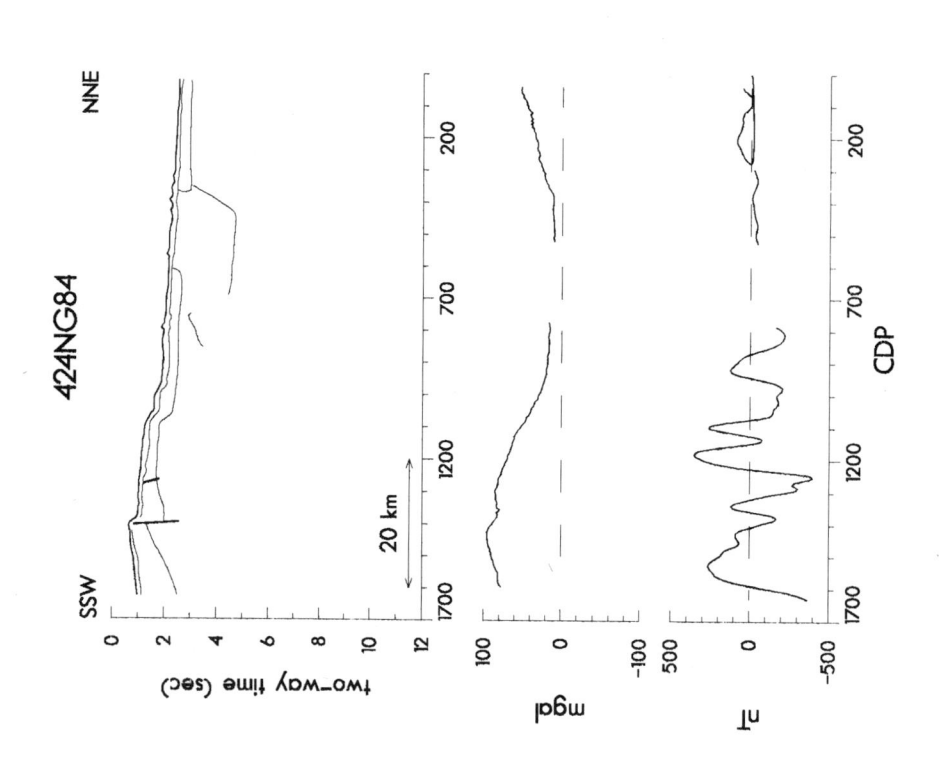

Plate 12. Magnetic and gravity profiles with an interpreted line drawing of multichannel seismic-reflection line 424. Top profile shows a line drawing of the sea-floor and subbottom reflection events. Middle profile show free-air gravity anomaly data. Bottom profile show residual magnetic anomaly data.

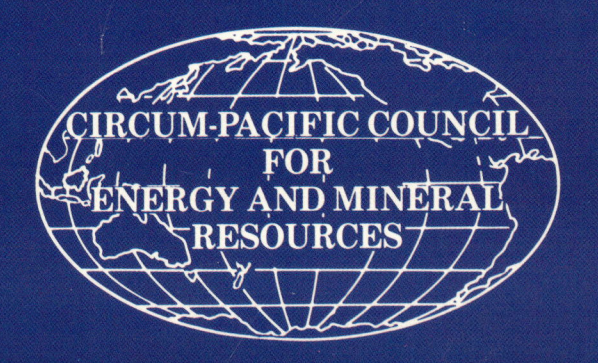

CIRCUM-PACIFIC COUNCIL
FOR
ENERGY AND MINERAL
RESOURCES

ISBN: 0-933687-10-9

DUE DATE

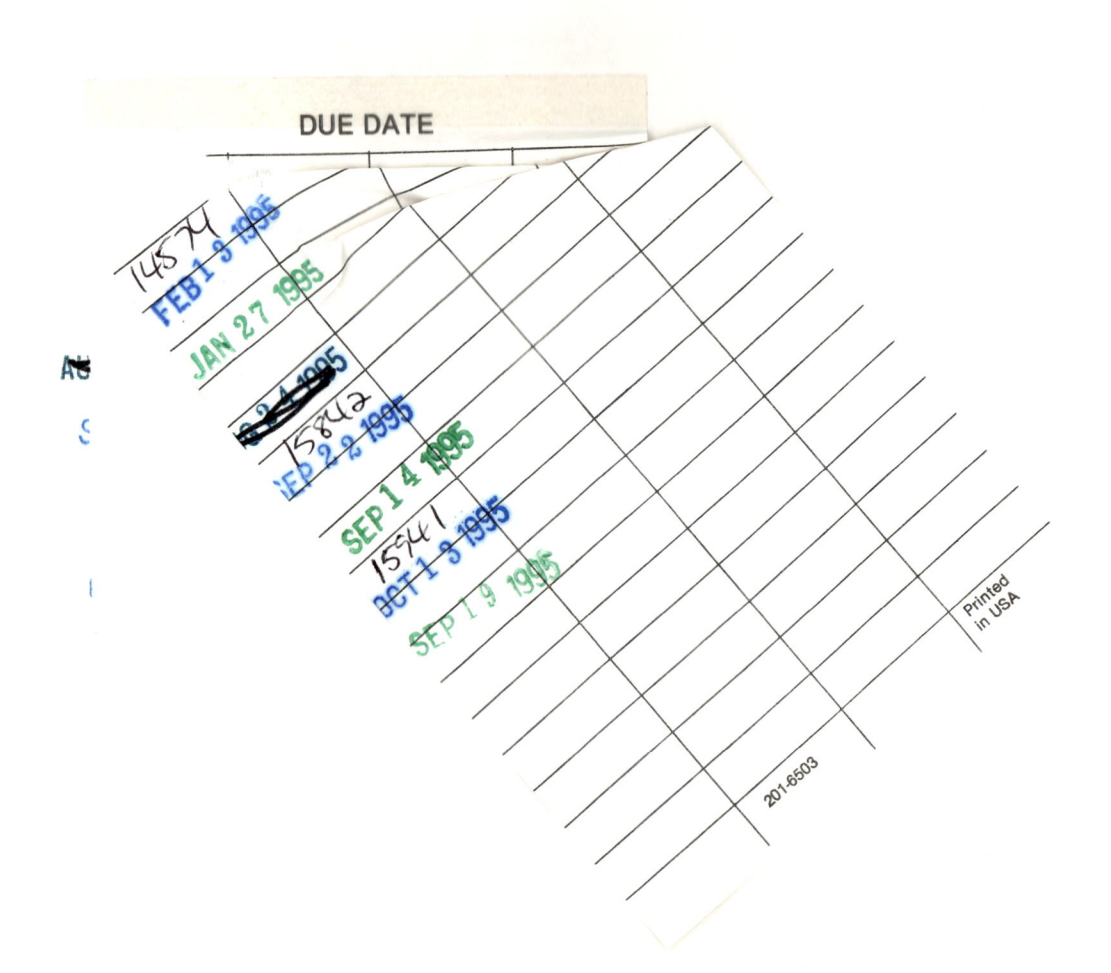

14574
FEB 13 1995
JAN 27 1995
15842
SEP 14 1995
15741
OCT 13 1995
SEP 19 1995

201-6503

Printed in USA